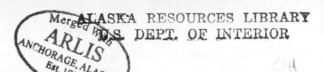
EVOLUTION AND THE GENETICS OF POPULATIONS

A Treatise in Four Volumes

VOLUME 3
EXPERIMENTAL RESULTS AND EVOLUTIONARY DEDUCTIONS

Sewall Wright

THE UNIVERSITY OF CHICAGO PRESS
CHICAGO AND LONDON

Sewall Wright is professor emeritus of genetics at the University of Wisconsin and professor emeritus of zoology at the University of Chicago. Among his many honors are nine honorary degrees and memberships in the National Academy of Sciences and the American Academy of Arts and Sciences. In 1966, he was awarded the National Medal of Sciences. He is a foreign member of the Royal Society, London, and an honorary fellow of the Royal Society, Edinburgh.

THE UNIVERSITY OF CHICAGO PRESS, CHICAGO 60637

The University of Chicago Press, Ltd., London

© 1977 by The University of Chicago

All rights reserved. Published 1977

Printed in the United States of America

80   79   78   77   987654321

Library of Congress Cataloging in Publication Data
Wright, Sewall, 1889–
    Evolution and the genetics of populations; a
treatise.
    Bibliography: v. 1,   p. [432]–456.
    Includes indexes.
    CONTENTS.—v. 1.   Genetic and biometric
foundations.—v. 2.   The theory of gene frequencies.
v. 3.   Experimental results and evolutionary
deductions.
    1. Population genetics.   2.   Evolution.
I.   Title.
QH431.W79   575.1   67-25533
ISBN 0-226-91051-2

To the memory of my wife Louise

CONTENTS

CHAPTER 1

Introduction

Volume 1 began with a brief account of ideas on evolution before the rediscovery of Mendelian heredity in 1900. In the main, it was concerned with various aspects of biological variability as the foundation on which population genetics must be built. The genetic basis was reviewed briefly with special attention to the complex relations between primary gene actions and the characteristics of organisms. Several chapters were devoted to the mathematical description of variation and the theory of compound variables. Path analysis was gone into at considerable length as a method to be used extensively in later volumes for dealing quantitatively with the relations among multiple variables from any desired point of view. The last substantive chapter made use of the preceding chapters in discussion of the genetic theory of quantitative variability.

Volume 2 was concerned primarily with the theory of gene frequencies, but two chapters were added on the division of observed variability into genetic and nongenetic components. The point of view was that formulas should be designed to take account of the simultaneous actions of all of the evolutionary factors as far as available data permit.

I have recognized from the first that a high degree of precision is out of the question in dealing with evolutionary phenomena in nature, or even ordinarily in laboratory experiments, because of the complexity of the interactions among the factors and the consequent excess of the number of parameters required for full description over that determinable from observation or even from well-controlled experiments.

I have also recognized from the first that what are considered single genes in natural populations must generally be supposed to consist of groups of alleles with more or less similar effects. Consequently, "constants" such as the rate per generation of the mutational origin of any given "gene," "may be expected to rise to a maximum as genes allied in structure become frequent and to fall off to zero as changes accumulate in the locus" and selection coefficients for a particular favorable gene, relative to its represented alternative, "should fall off and become negative as the group of allelomorphs comes to include still more useful genes" even under constant conditions (Wright 1931). It was recognized that this situation could be remedied only partially by taking account of as many alleles as could be distinguished by ordinary means.

Another practical difficulty has to do with the complexity of the selection process, which involves differences in prenatal viability (that often modify gene frequencies before phenotypic frequencies are available) as well as differences in postnatal viability, mating frequencies, tendencies to emigrate, productivity, and sometimes meiotic drive. The change of gene frequency is affected by the order of occurrence of the various factors: mutation, the various aspects of selection, immigration, and accidents of sampling. The selection pressure on any gene is practically always affected by nonadditive interactions with other genes. It is also probably always a function of its own and other gene frequencies.

In this situation, it seemed best to start from the case in which selection pressures on each gene are so weak that the order of events becomes sufficiently unimportant to be ignored and that the same is true of deviations from random combination due to interaction effects.

It also seemed advisable to start with the assumption of constant relative selective values of genotypes. The simplest model for change of a gene frequency, Δq, takes account of rates per generation of mutation to and from the gene in question (v and u, respectively), replacement to the extent m per generation by immigration at gene frequency Q, and selective change according to a coefficient s:

$$\Delta q = v(1 - q) - uq - m(q - Q) + sq(1 - q)/\overline{W}.$$

The mean selective value of the genotypes, $\overline{W}$, can be treated as 1 if s is small. The rate of change of gene frequency is subject to variation, $\sigma^2_{\Delta q}$, due to accidents of sampling in the population of effective size N, and to fluctuations in all parameters:

$$\sigma^2_{\Delta q} = q(1 - q)/2N + \sigma^2_m(q - Q)^2 + m^2\sigma^2_Q + \sigma^2_s q^2(1 - q)^2.$$

These lead to a steady state distribution that describes the so-called random drift about the equilibrium frequency, $\hat{q}$, the gene frequency at which $\Delta q = 0$.

$$\phi(q) = (C/\sigma^2_{\Delta q}) \exp [2 \int (\Delta q/\sigma^2_{\Delta q}) \, dq]; \int_0^1 \phi(q) \, dq = 1.$$

In the case with Δq as above but $\sigma^2_{\Delta q} = q(1 - q)/2N$,

$$\phi(q) = Ce^{4Nsq}q^{4N(mQ + v) - 1}(1 - q)^{4N[m(1 - Q) + u] - 1}$$

(if selection is so weak that $\overline{W}$ can be treated as 1).

These formulas do not take cognizance of multiple alleles. Multiple simultaneous equations are required to determine the equilibrium gene frequencies with multiple alleles with specified mutation rates. The mutation terms are, however, ordinarily negligible in comparison with the immigration terms.

In connection with the latter, it should be recognized that immigrants come, in general, from neighboring populations, which usually differ only slightly from the population in question, so that effective m is in general only a small fraction of the actual replacement by immigrants; effective N is also, for various reasons discussed in volume 2, chapter 8, only a fraction of the census number even of adults.

Selection relative to gene frequency q_x can be dealt with more realistically by replacing s by $(\partial \overline{W}/\partial q_x)/2$, assuming that $\partial q_i/\partial q_x = -q_i/(1 - q_x)$ in evaluation in the case of multiple alleles. This is under the assumption of constancy of relative selective values of the multifactorial genotypes but involves no restriction on gene interaction except that the ratio of interactive selection to the amount of recombination be so small than random combination is not appreciably disturbed.

$$\phi(q_x) = C\overline{W}^{2N} q_x^{4N(mQ+v)-1}(1 - q_x)^{4N[m(1-Q)+u]-1}.$$

This is the stochastic distribution along a cross section of the total multi-dimensional distribution under specified frequencies for all genes other than the one in question. The total multidimensional distribution, ignoring mutation rates, is

$$\phi(q_1, q_2, \ldots, q_n) = C\overline{W}^{2N} \prod q_i^{4NmQ_i - 1}.$$

In this equation, the q's include all alleles at all loci.

In general the multidimensional "surface" of genotypic selective values, $\overline{W}$, has multiple peaks that, as may be seen, are reflected in the stochastic distribution.

An absolute selective value, W_i, may be replaced here by w_i, the value relative to a specified standard. The absolute selective value, W_i, of a specified genotype under a given array of genes at other loci and under given environmental conditions, has been defined as such that the mean for all the genotypes in question is the ratio of the effective population number to that in the preceding generation. Thus $(\overline{W} - 1)$ is the growth rate of the population under the given genetic and environmental conditions. Letting N_i be the effective number of representatives of the ith genotype and using primes for the preceding generation,

$$\overline{W} = \sum N_i W_i \Big/ \sum N_i' = N/N'.$$

A statement near the beginning of chapter 5, volume 2, that implied that

$W_i = N_i/N'_i$, is correct only for haploids, but may be disregarded since no subsequent use was made of it.

Contrary to the assumption made above for simplicity, there is probably always more or less frequency dependence in actual populations, if only because the total amount of selection permitted by the reproductive excess is limited and is thus apportioned among the loci. If all individuals of a species successfully occupying a certain niche are handicapped, the least handicapped are favored by selection. The selection term in Δq, still assuming no appreciable deviation from random combination, is now $(1/2)q(1 - q) \sum [(w_i/\bar{w}) \, \partial f_i/\partial q]$, in which $f_i(=N_i/N)$ is the relative frequency of the genotype of which w_i is the selective value. In some cases, this can be written in the form $(1/2)q(1 - q) \, \partial F(w/\bar{w})/\partial q$, implying a multidimensional "surface" of population "fitnesses," $F(w/\bar{w})$, typically with many peaks that, however, are generally wholly different from those in the surface of mean selective values, $\bar{w}$. Again ignoring mutation,

$$\phi(q_i, q_2, \ldots, q_n) \approx Ce^{2NF(w/\bar{w})} \prod q_i^{4NmQ_i - 1}.$$

In general, no fitness function exists where multiple alleles or multiple loci are under consideration. There are still, however, usually multiple distinct "selective goals" toward a particular one of which mass selection tends to move the set of gene frequencies according to the initial array. An evolutionary step beyond that implied by mere change of gene frequency depends on a shift from control by one fitness peak or selective goal, as the case may be, to control by a superior one. Such a shift depends on the occasional crossing of a "saddle" under random drift, followed by firm establishment of the new peak (or goal) by mass selection.

These were the central concepts throughout volume 2. The formulas cited here are those for diploid heredity. The analogous ones for sex-linked and polysomic heredities were considered, but with special emphasis on the nearly universal case in nature of an intermediate optimum for multifactorial quantitative variability. Various sorts of frequency dependence were also considered.

Special attention was devoted in volume 2 to population structure, including at one extreme inbreeding within completely isolated small populations and, at the other, partial isolation merely by distance in a large continuous population in which diffusion is severely restricted. The intermediate case in which there is subdivision of a large population into numerous incompletely isolated clusters is the most general.

There have been many attempts to deal with problems of population genetics by other mathematical models. Some of these were discussed briefly in volume 2, but it has not been practicable to discuss all of them. No attempt will be made here to discuss the more sophisticated mathematical attacks on problems developed in recent years. These, while highly desirable, theoretically, are difficult to apply to actual populations in the laboratory or in nature, the subjects, respectively, of volumes 3 and 4. (These were originally intended to be parts of a single volume.)

Many of the recent developments have been concerned with the enormous increase in complexity where selection is sufficiently strong in relation to the recombination index to overbalance significantly the return toward random combination. These complications were illustrated in relatively simple cases in volume 2, but much more complicated cases have been investigated recently, some by use of computers. Loci in different chromosomes lose at once 50% of whatever deviation there may be from random combination. For such loci there is no appreciable error in ignoring the so-called linkage disequilibrium if selection coefficients are of the order 0.01 or less, necessarily characteristic of most loci if, as seems clear, most are usually strongly heterallelic. On the other hand, there is undoubtedly much that should be done to clarify the cases of strongly selected and closely linked loci. Thorough treatment of linkage is especially important in considering evolution in the species of *Drosophila* because of the small number of chromosomes and the absence of crossing-over in males.

The point of view and methods presented in volume 2, with some supplementation here, suffice for the review of the experimental results and discussion of their implications for evolution to which volume 3 is devoted and for a discussion of variation and evolution in nature, which is to be the subject of the final volume.

CHAPTER 2

Inbreeding Depression and Heterosis: Plants

Pre-Mendelian Beliefs and Experiments

It is not possible in experimental studies to make a sharp separation between those concerned with systems of mating and those concerned with selection since both are always involved. This and the three following chapters will deal with experiments in which the center of interest is in systems of mating.

Opinions on the effects of inbreeding and crossbreeding far antedate scientific experiment. Most, though not all, peoples have had taboos on incest and the belief that physically and mentally defective children are the typical consequence has been widespread (Zirkle 1952). How far such customs and beliefs are traceable to observation of biological effects, and how far to social experience, is difficult to say. Closely related, however, has been the worldwide belief of livestock breeders that continued inbreeding within a herd or flock leads to deterioration and that an outcross restores vigor; here there has doubtless been frequent opportunity for checking belief with experience.

This traditional belief was challenged in the eighteenth century by certain livestock breeders in Great Britain, led by Robert Bakewell (1725–95 Wallace [1923]). Bakewell built up superior strains of Longhorn cattle and Leicester sheep by careful choice of breeding stocks, followed by inbreeding. Others followed his example, and some at least, such as the Colling brothers, founders of the Shorthorn breed of cattle, were notably successful.

Darwin (1868) brought together an impressive array of testimony from authorities on the breeding of all kinds of livestock. He found a consensus that inbreeding leads to reduction in constitutional vigor and fertility but not necessarily in conformation. He noted (quoting Youatt) that even Bakewell's Longhorn cattle "had acquired a delicacy of constitution inconsistent with common management." He found breeders practically unanimous on the immediately beneficial effects of an outcross.

That any evil directly follows from the closest interbreeding has been denied by many persons; but rarely by any practical breeder; and never, as far as I know, by one who has largely bred animals which propagate their kind quickly.... The evil results from close interbreeding are difficult to detect, for they accumulate slowly, and differ much in degree with different species; whilst the good effects which almost invariably follow a cross are from the first manifest. It should, however, be clearly understood that the advantage of close interbreeding, as far as the retention of character is concerned, is indisputable, and often outweighs the evil of a slight loss of constitutional vigour.

Among plants, the exceptional vigor of many hybrids between species and varieties and the loss of this vigor on subsequent inbreeding was demonstrated by Koelreuter (1766), Gaertner (1849), and others. Darwin (1876) himself made extensive experiments. These involved no less than 57 species, distributed among 30 families. The number of individuals, however, was often rather limited, something over 1,000 selfed plants altogether were compared with a similar number of crossbred ones. Darwin's method consisted, for the most part, in planting, on opposite sides of a pot, two seedlings that had germinated simultaneously from self-fertilized and cross-fertilized plants, respectively, and comparing growth and productivity. The species ranged from those that were habitually self-fertilizing in nature to those that normally cross. The latter suffered severely from self-fertilization, but much the greatest effect came in the first generation. In the species that had long been maintained by self-fertilization, crossbreds within strains did not differ appreciably from plants produced by self-fertilization, but again crossing between different strains gave a considerable advantage that persisted in part after one generation of selfing.

The result that puzzled Darwin most was the occasional appearance of selfed lines of relatively high vigor and fertility. A plant "Hero" appeared in the sixth self-fertilized generation of *Ipomoea purpurea* (morning glory) that was slightly taller than its crossbred control. Its selfed descendants inherited this advantage. There were other similar cases.

A result to which Darwin called special attention was the remarkable uniformity in tint and form of flower, soon reached in each selfed line.

Darwin held that the "advantages of cross-fertilization do not follow from some mysterious virtue in the mere union of two distinct individuals, but from such individuals having been subjected during previous generations to different conditions or to their having varied in a manner commonly called spontaneous, so that in either case, their sexual elements have been in some degree differentiated and secondly that the injury from self-fertilization follows from want of such differentiation in the sexual elements." He combatted the view that the evil of inbreeding "is the result of some morbid tendency or weakness of constitution common to the closely related parents or to the two sexes of a hermaphroditic plant."

On the whole, Darwin's results agree well with those of more recent investigators. His interpretation might have been greatly sharpened if he had been aware of Mendel's (1866) analysis of the effects of self-fertilization, implied by his principles of heredity. Mendel showed that the amount of heterozygosis is halved in each generation, but his principles did not become generally known until 1900.

The most important pre-Mendelian experiments on the effects of inbreeding in animals were those of Crampe (1883) and Ritzema-Bos (1894) with rats. Both, starting from crossbreds, encountered serious deterioration in fertility and vitality in a few generations of inbreeding. In Crampe's strain there was also considerable decline in weight and the appearance of many abnormalities, which was not the case in Ritzema-Bos' experiment.

Von Guaita (1898, 1900) and Weismann (1904) noted decline in fertility in mice inbred brother with sister. The latter did not relate this to Mendelian heredity.

Early Post-Mendelian Experiments and Theories

With the recognition of Mendel's results, it became obvious that the appearance of many abnormal traits, such as albinism and alkaptonuria in man (Garrod 1902, 1908), were merely the consequence of segregation of a rare recessive, an interpretation borne out by the excess prevalence of cousin marriages in the parents of affected individuals.

Castle and associates (1906) inbred *Drosophila melanogaster* for 59 generations of brother-sister mating. While much sterility and low fertility appeared in the early generations, it was found possible to maintain high fertility by selection among lines. The segregation of recessive factors for low fertility was indicated. Castle's results were confirmed in later experiments with the same fly by Moenkhaus (1911), Wentworth (1913a), Hyde (1914), and others.

Davenport (1908) called attention to the fact that for most then-known pairs of alleles, the recessive tended to be the less vigorous, and pointed out that this could explain the usual but not invariable deterioration from inbreeding. At that time, however, this interpretation was somewhat lacking in substance as a general explanation. The earlier work on Mendelian heredity had naturally been confined to conspicuous differences. Thus the light thrown on the appearance of conspicuous abnormalities after inbreeding was, as noted, quickly recognized. It was not clear at first that any light was thrown on the gradual decline in size, fertility, and constitutional vigor, which are more typical consequences. It was necessary to reach the viewpoint that observable hereditary differences may be due to the summation of individually inconspicuous effects of many Mendelian differences and that the Mendelian mechanism is the prevailing one under sexual reproduction.

Experiments with Maize

This viewpoint was provided especially by the experiments with maize conducted by G. H. Shull (1908, 1910, 1911) and East (1908–36), which led to the firm establishment of the multiple-factor theory of quantitative variability (vol. 1, chap. 15) in conjunction with Johannsen's (1903) demonstration of the ineffectiveness of selection in pure (long self-fertilized) lines.

Shull found that on self-fertilization, an ordinary, supposedly homogeneous variety of maize broke up into strains, each extraordinarily uniform, but different from all others in numerous minute characteristics that had not been recognized as having a genetic basis. There was also a marked decrease in size and productivity in all strains in the early generations of selfing, but an approach to stability was soon reached in those that survived.

East encountered essentially the same results in more extensive and longer-continued experiments in which several lines were inbred in each of some 30 varieties of maize. Three of these lines, 1-6, 1-7, and 1-9, started by East in 1905 (East and Jones [1919]), were continued by Jones (1918), who recorded the results to the 11th generation of selfing. These were relatively superior lines since many others deteriorated so rapidly that they could not be maintained. Line 1-7 was subdivided after two generations, 1-7-1-1 and 1-7-1-2, and other subdivisions were made later in all of them. Table 2.1 shows the means of several characters in the random bred foundation stock no. 1 and two branches of each of the inbred lines that had seven or eight generations of common ancestry, at which time heterozygosis should be reduced to $(1/2)^7$ or $(1/2)^8$ of its initial value. Yield in bushels per acre is given for two years, 1914 and 1916, in which conditions differed considerably. That for the original strain in 1908, a good year, is given instead of that for 1914.

The branches that separated after seven or eight generations of selfing differ little in any respect and give a basis for judging the departure from the foundation stock and the amounts of differentiation. There is clearly deterioration of all lines in height (average amount, 26.9%), in length of ears (28.0%), and in number of nodes (12.8%). The yields in 1916 under somewhat unfavorable conditions show a marked decline in all lines (61.4%). The yields in 1914 (under more favorable conditions) show little if any decline of line 1-6 and an average decline of only 34.7%. The number of kernel rows is not an obvious aspect of vigor and shows no consistent decline, since one line (1-7-1-1) clearly had more rows than the foundation stock. The average

TABLE 2.1. Effects of self-fertilization on characters of maize. Mean values in the foundation stock in 1917 (except yield), in the random bred foundation stock, and in two branches of each of four inbred strains after nine or ten generations of selfing. The branches separated two generations earlier. Yield of the foundation strain (1) is given for 1908 instead of 1914.

Line	Generations of Selfing	Common Generations	No.	Height (inches)	Length of Ear (inches)	No. of Nodes	No. of Rows	Yield (bushels/acre) 1914	1916
1	0		213	117.3	7.5	14.1	18.4	(88.0)	74.7
1-6	10	8	56	97.8	6.1	12.7	16.9	78.8	32.5
	10		61	96.7	6.2	11.5	15.7		
1-7-1-1	10	8	52	78.5	4.3	11.8	21.8	47.2	32.7
	10		58	82.2	4.2	11.6	22.0		
1-7-1-2	10	8	54	82.6	5.1	12.1	15.9	58.5	19.2
	10		51	91.2	5.3	13.0	15.9		
1-9	9	7	59	77.0	6.1	12.9	15.5	45.4	30.6
	9		60	80.3	5.9	13.1	15.4		
Average inbred				85.8	5.4	12.3	17.4	57.5	28.8
% Decline				26.9	28.0	12.8	(5.4)	34.7	61.4
% per 10% F (inbreeding coefficient)				2.7	2.8	1.3	(0.5)	3.5	6.1

Source: Data from Jones 1918.

decline is only 5.4% and it may be accidental that three lines came to have fewer rows and only one, more rows than originally.

There were clearly significant differences among lines, not only in number of rows, but also in height, length of ear, and yield, although the rank in the last depended on conditions. There were probably also differences in numbers of nodes, but differentiation was least in this respect.

On the average, crosses between inbred lines merely restored the vigor of the foundation stock. Table 2.2 gives the average of 17 crosses among the

TABLE 2.2. Means of 17 crosses and F_1 (maize). The crosses were in rows between the inbred parents. The four inbred strains of table 2.1 were used with various numbers of replications of reciprocal crosses so that A and B differ only slightly.

	Height (inches)	Length of Ear (inches)	No. of Nodes	No. of Rows	Yield (bushels/acre)
Inbred A	88.0	5.8	12.5	16.2	28.8
Cross ($A \times B$)	112.4	7.2	13.2	17.9	78.4
Inbred B	88.3	5.3	12.3	17.7	27.2
% Reduction of inbreds	21.6	22.9	6.1	5.8	64.3

SOURCE: Data from Jones 1918.

four inbred strains recorded by Jones (1918). The F_1's were grown in rows between ones of the inbred parents A (ovule parent) and B (pollen parent) (which as groups are essentially the same because there were nearly equal numbers of reciprocal crosses). As may be seen, the percentage inferiority of the inbred parents in the crosses is comparable to their average inferiority to the foundation stock.

The most striking result observed by Shull and East was, however, the extraordinary size and productivity of the first cross from particular inbred lines. They found that much of this superiority was lost in the next generation. Shull (1908) proposed the systematic utilization of the hybrid vigor of selected first crosses between inbred lines. This, however, suffered from the handicap that the inbred ovule parent produced relatively few seeds and these were small and gave a poor start. Jones (1918) made the following suggestion:

A way to overcome this handicap suggests itself which is to cross two vigorous first generation hybrids whose composition is such that the resulting cross will not be less heterozygous than either parent and, therefore, theoretically no less vigorous and productive. This is easily accomplished by taking four distinct inbred strains which are of such a composition that a cross between any two of them gives a vigorous product.... These doubly crossed plants, however, starting from large seeds produced on large, vigorous hybrid plants would be freed from the handicap which their parents had and although less uniform should be more productive. While it may be out of place to say anything about this method until it has been thoroughly tested it is a method which is more promising than the plan originally advocated because by this method crossed seed for general field planting, is produced much more abundantly than when non-vigorous inbred strains are crossed.

Jones' plan was tested. It proved so successful that it completely revolutionized American agriculture in the course of two or three decades.

Shull (1914) coined the term *heterosis* "as a descriptive term for hybrid vigor, irrespective of mechanism." He had a theory of the mechanism, however.

The essential features of the hypothesis may be stated in more general terms, as follows: The physiological vigor of an organism, as manifested in its rapidity of growth, its height and general robustness, is positively correlated with the degree of dissimilarity in the gametes by whose union the organism has been formed. In other words, the resultant heterogeneity and lack of balance produced by such differences in the reacting and interacting elements of the germ cells act as a stimulus to increased cell division, growth, etc. The more numerous the differences between the uniting gametes—at least within certain limits—the greater on the whole, is the amount of stimulation. These differences need not be Mendelian in their inheritance, although in most organisms they probably are Mendelian to a prevailing extent.

This view is essentially that of Darwin, except in relating the differentiation of the gametes largely to Mendelian genes.

The principal alternative hypothesis at this time was the extension of Davenport's suggestion: that the deleterious effects of inbreeding, and conversely the vigor of crossbreds, are due to a prevailing tendency for recessiveness to be associated with reduction in vigor in specific ways. This view was restated by Bruce (1910). Keeble and Pellew (1910) described experiments in which a cross between two pure strains of pea produced hybrids taller than either, while segregation in F_2 proved that each strain furnished a dominant factor, lacking in the other. In this case, the effects of the two dominant genes happened to be visibly different in that one increased the number of internodes, the other increased their length.

Under this hypothesis, injurious recessive genes are largely carried out of sight in a random bred stock but become manifest with increased frequency on inbreeding. With gene frequency array $(pa + qA)$, in which p and q are frequencies $(p + q = 1)$, the proportion of recessive zygotes (aa) rises from p^2 toward p as fixation proceeds, in the absence of selection. More generally the mean, M_F, of the character under consideration in an array of lines with inbreeding coefficient F is $M_0 - (M_0 - M_1)F$ where M_0 is the mean of the random bred stock and M_1 is that under complete fixation (vol. 2, chap. 15), assuming that the measurements are on a scale on which the genes have additive effects. This would apply to a single line if the variability depends on many minor factors.

Under both the stimulation hypothesis and the dominance hypothesis, the change in character is proportional (on the appropriate scale) to the inbreeding coefficient. The difference is the lack of specificity of effect among loci in the former, and specificity in the latter. Under the dominance hypothesis, moreover, it should be possible to isolate inbred strains, homozygous in all factors conducive to vigor, but not under the stimulation hypothesis. Again, under the dominance hypothesis, F_2 of a cross should produce a distribution with negative skewness of the favorable character, if the latter is one exhibiting inbreeding depression and heterosis, while symmetry is expected under the stimulation hypothesis.

The apparently invariable deterioration following self-fertilization that both Shull and East observed and the failure to find F_2 distributions with significant negative skewness led East as well as Shull to favor the stimulation hypothesis.

The objections to the dominance hypothesis were met when Jones (1917) pointed out that, owing to linkage, the consequences of the dominance hypothesis would be closer to those of the stimulation hypothesis than had been recognized. With many factors affecting yield and other aspects of vigor, there would generally be both favorable and unfavorable genes in each chromosome of a random breeding population, and it would be unusual to find all of the favorable genes in one line and much more difficult to sort them out by selection into one line. The segregation of chromosome regions carrying both favorable and unfavorable genes would also increase the symmetry of F_2 distributions.

In this suggestion Jones did not add a new hypothesis, he merely pointed out the consequences of an established fact, linkage. The dominance hypothesis was thus strengthened. It should be added, however, that, as pointed out by Collins (1921), even among unlinked genes the objections to the dominance hypothesis on the basis of the absence of any good inbred lines and of F_2 skewness largely disappear if there are a very large number of more or less equivalent factors or, in the case of skewness, if there is much nongenetic variability.

This by no means ended discussion of the causes of inbreeding decline and heterosis, however. The assumption that there is a correlation between dominance and vigor required more support than the empirical observation that it seemed to hold as a rule for major Mendelian differences. There were various views on the possible theoretical basis. The phenomenon of over-dominance (heterozygotes superior to both homozygotes), not known at the time, introduced a third hypothesis. Finally the recognition of the relativity of grades of dominance—the recognition that a gene that is dominant in some combinations may be semidominant or even recessive in others, and ones that are dominant in some combinations may be overdominant in others (vol. 1, chap. 5)—indicates that the interpretation should be in terms of interaction systems rather than of separate genes. We will come back to these questions later.

Quantitative Studies with Maize

It will be well to go into an experiment with maize on the quantitative relation between yield and amount of inbreeding, modified slightly by selection. Richey and Sprague (1931) made crosses among a number of long-established, stable inbred lines. The first crosses ($F = 0$) were back-pollinated from one to six generations to one of the parents (R = recurrent parent). On the simple additive theory of gene effects, $F = 1 - (1/2)^n$, where n is the number of these generations. Each generation was subjected to selection both within and between progenies for vigor and yield.

Two sets of experiments were made in the final test conducted in one season with suitable replication to obviate effects of soil heterogeneity and to minimize competition effects. In one test, the seed came from crosses in which the nonrecurrent parent (N), F_1, and plants of all backcross generations were used as pollen parents on the recurrent strain (R). The latter was also selfed. The use of the inbred strain as ovule parent prevented systematic differences

TABLE 2.3. Comparisons of yields of maize (in pounds per replication) from seed produced by pollinating the recurrent inbred strain (R) (set 1) and the non-recurrent inbred strain (N) (set 2) with sources of pollen indicated in the first column, and also F_2.

Pollen Parent	Ovule Parent: R Set 1	Expected Values	Ovule Parent: N Set 2	$N \times R$	Expected Values
N	19.7 ± 0.66	(19.7)			(3.3)
$N \times R$	11.7 ± 0.54	11.65		9.4 ± 0.55	10.55
$(N \times R)R$	8.2 ± 0.37	7.62	13.5 ± 0.29		14.17
$((N \times R)R)R$	7.2 ± 0.30	5.61	15.7 ± 0.36		15.99
$(((N \times R)R)R)R$	5.8 ± 0.31	4.61	17.5 ± 0.35		16.89
$((((N \times R)R)R)R)R$	4.5 ± 0.19	4.10	18.3 ± 0.38		17.35
$(((((N \times R)R)R)R)R)R$	4.6 ± 0.20	3.85	17.4 ± 0.22		17.57
R	3.6 ± 0.16	(3.6)	17.8 ± 0.48		(17.8)

SOURCE: Data from Richey and Sprague 1931.

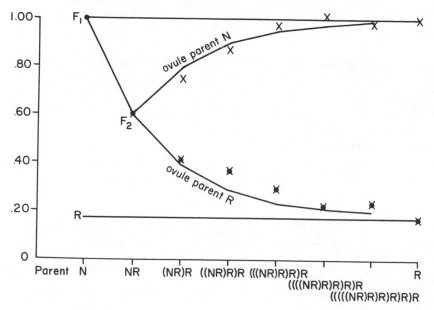

FIG. 2.1. Average yields of maize as percentages of F_1, from seed produced by pollinating the recurrent inbred strain R (set 1), or the nonrecurrent inbred strain N (set 2), with pollen from the sources indicated at the bottom, and also F_2. From data of Richey and Sprague (1931).

in size of seed, an important factor at the beginning of growth. In the other set the nonrecurrent parent (N) in each case had been pollinated by the lines that had two to six generations of back-pollination and F_2 was also included in this set. The average yields (pounds per replication) for the two experiments are shown in table 2.3 and, as percentages of F_1, in figure 2.1. The two sets of yields are not directly comparable since they were raised in different fields.

Assuming here that the effects of factors on yield are additive and that there was no selection, the difference between F_1 and the recurrent strain R should be halved in each generation of backcrossing to the latter. Yield in pounds per replication equals $3.6 + (1/2)^n \times (16.1)$. These expected values are given in a separate column. The first backcross (11.7) is almost exactly as expected, but the later generations are all greater, and significantly so in most cases, indicating some retention of favorable dominant genes by selection.

In the second set with ovule parent (N), the difference from F_1 should be halved in each generation. As the yields of the inbreds, N, were not obtained in this set, it was assumed to be the same percentage of F_1 as in the first set, giving 3.3, for the purpose of estimation. Yield in pounds per replication equals $17.8 - (1/2)^n \times 14.5$. The observed values are somewhat lower than expected in F_2 and $N \times (NR)R$ but rise above expectation in two other cases, suggesting that some of the favorable factors retained by selection in the pollen line are even more favorable as homozygotes than as heterozygotes. This is not confirmed in the last case, but at least there is no evidence for inferiority of the homozygotes of such genes (no support for overdominance). On the whole, the data support the simple dominance theory on making reasonable allowance for selection.

TABLE 2.4. Comparison (bushels per acre) of F_1 maize hybrids, three-way hybrids, and double (four-way) hybrids with their inbred ancestors.

	Hybrids	Random Bred Offspring	Inbred Ancestors	Loss (%) of Heterosis	F (%)
10 single hybrids (AB, etc.)	62.8	44.2	23.7	47.6	50
4 three-way hybrids ($AB \cdot C$, etc.)	64.2	49.3	23.8	36.8	37.5
10 double hybrids ($AB \cdot CD$, etc.)	64.1	54.0	25.0	25.8	25.0

SOURCE: Data from Neal 1935.

Some data (table 2.4) on yields (bushels per acre) given by Neal (1935) illustrate the expected similarity of single, three-way, and double hybrids and the fairly close parallelism, on the scale of untransformed yield, between the loss of hybrid vigor and the inbreeding coefficient of the random bred offspring in each case (vol. 2, pp. 336–39).

Lindstrom (1941) found that 677 useful inbred lines had been produced at experiment stations in the United States up to that time. This was about 2.5% of all started. These on the average yielded only about 30% as much as their

TABLE 2.5. Data on yield for all extensively used inbred lines of maize in American experiment stations to 1941 and their crosses.

	RELATIVE YIELD (BUSHELS/ACRE)	
	Observed	Expected
Inbreds (A, B, C, etc.)	36.2	(36.2)
F_1 (AB, etc.)	100.0	(100.0)
Three-way ($AB \cdot C$, etc.)	102.3	100.0
Backcrosses ($AB \cdot A$, etc.)	71.7	68.1
F_2 (($AB)^2$, etc.)	65.8	68.1

SOURCE: Data from Lindstrom 1941.

F_1 hybrids. The average relative yields of lines used in the more extensive tests, of F_1, three-way crosses, backcrosses, and F_2 were found to be as given in table 2.5. Here again there is moderately close agreement with theory, although F_2 was somewhat low, excess over I 46.4% of $(F_1 - I)$ where I is inbred, and the backcrosses were somewhat high, 55.6%, where 50% is expected in both.

There were similar deviations from expectation in experiments reported by Stringfield (1950). He made all possible crosses among four good inbred lines and compared groups into which all entered equally. The results for yield in bushels per acre are shown in table 2.6.

TABLE 2.6. Yield of maize in bushels per acre of four inbreds, and of crossbred groups into which all entered equally.

Heterozygosis (%)	Type	No. of Entries	Yield Observed	Expected
0	Inbred (sib × sib)	4	41.5 ± 0.45	...
50	F_2, $(AB)^2$, etc.	6	62.2 ± 0.98	63.2
	B_1, $(AB \cdot A$, etc.)	12	67.9 ± 1.20	63.2
75	$(AB)(AC)$, etc.	12	75.2 ± 1.00	74.1
100	F_1, AB, etc.	6	84.9 ± 1.14	...
	3-way, $AB \cdot C$	12	84.9 ± 1.18	84.9
	4-way, $AB \cdot CD$	3	81.9 ± 0.74	84.9

SOURCE: Data from Stringfield 1950.

The earliness of silking and the height of ear node were also obtained throughout. In the former, both F_2 and B_1 were well above expectation, with respect to excess over the inbreds (to which the following figures all apply). F_2, as usual, was lower (F_2 59.6% and B_1 66.7% of $[F_1 - I]$). The crosses $(AB)(AC)$ at 77.6% $(F_1 - I)$ were slightly above the expected 75%. The three-way and four-way crosses were 105.3% and 100.0% of $(F_1 - I)$, respectively. In the case of ear node height, F_2 and B_1 differed little from each other or from expectation (51.8%, 49.6%) but $(AB)(AC)$ was again high (91.5%). The three-way and four-way crosses were high (106.4%, 109.0%).

An experiment that showed much less agreement was that of Sentz, Robinson, and Comstock (1954), also made for the explicit purpose of testing the relation between yield and other characters to the percentage of heterozygosis. They made two different crosses between long inbred lines, making ten tests at several locations over four years in one case and six tests over three years in the other case. In each case, simultaneous comparisons were made of the inbreds with no heterozygosis, F_1 and both backcrosses (B_1) with 50% heterozygosis, both second backcrosses to the same inbred (B_2) with 25% heterozygosis, and both cross-backcrosses to the inbred parents in succession (type $[(AB)A] B$) with 75% heterozygosis (table 2.7).

TABLE 2.7. Relative yields on a scale of 100 for F_1 of inbred lines of maize and crosses, according to percentage of heterozygosis.

Cross	HETEROZYGOSIS (%)				
	0	25	50	75	100
CI 21-N7	24.4	57.6	65.4	77.6	100
NC16-NC18	23.9	62.4	68.4	78.6	100
Average	24.2	60.0	66.9	78.1	100
Expected	(24.2)	43.15	62.1	81.15	(100)

SOURCE: Data from Sentz, Robinson, and Comstock 1954.

The yields of the first backcrosses were stated to have exceeded F_2 by 3.9% and 7.0% in the two crosses but not sufficiently consistently for significance at the 0.05 level.

The results deviate greatly from a linear relation to percentage of heterozygosis, the increment for 25% heterozygosis being very much too great and from 25% to 75% very much too small, and from 75% to 100% somewhat large. There was not, however, consistency in this respect since there was near-linearity in two years (1948, 1952), enormous departure in the other two (1950, 1951). Analysis of variance revealed much more genotype-environment interaction with respect to year than to location.

The nonlinearity was not peculiar to yield. The number of ears per plant rose by more than 30% in the first backcross, with 25% heterozygosis, but then rose little or not at all from 25% to 100% heterozygosis. This might suggest that a ceiling with respect to number of ears, reached at 25% heterozygosis, was the cause of the relatively small rise in yield beyond this; this does not, however, account for remarkably close parallelism throughout between yield and earliness of maturity. Moreover, the other characters studied (plant height, ear height, ear length, and ear diameter) all showed increments from 0 to 25% heterozygosis almost as great or greater than that from 25% to 100%, and within this range the increment from 75% to 100% was disproportionately great in all cases. The results suggest a peculiar genotype-environment interaction in certain years, affecting all characters alike, rather than specific gene interactions.

Heterosis in Self-fertilizing Species

From Darwin on, those who have made crosses between varieties or species that normally self-fertilize have found relatively little heterosis. Some data on various characters of tobacco (Matzinger 1963) from all crosses among eight diverse varieties and F_2 based on single F_1 plants illustrate this point (table 2.8).

TABLE 2.8. Average percentage excess of F_1 over the midparent, relative to F_1, and of F_1 over F_2, also relative to F_1, in crosses among eight varieties of tobacco.

Character	$100(F_1 - \overline{P})/F_1$	$100(F_1 - F_2)/F_1$
Yield	1.16*	3.49*
Plant height	2.17*	1.80*
% Nicotine	−1.50	−1.22
Leaf length	0.60	0.38
Leaf width	1.46†	1.64*
Number of leaves	−0.50	−0.17
Days to flowering	−1.20*	−1.59*
Leaf axil suckers	3.73*	1.94

SOURCE: Data from Matzinger 1963.
* Significant at 0.01 level.
† Significant at 0.05 level.

There are some significant cases of heterosis and of inbreeding depression, but the percentages are extremely small in comparison with similar characters in crossbreeding species.

Robinson, Mann, and Comstock (1954) reported slightly greater increments of F_1 over midparent (2% in leaf yield, 6% in plant height relative to F_1) in other crosses of tobacco (data of Patel in Robinson, Mann, and Comstock 1954).

There appears to be considerably more heterosis with respect to fruit yield in crosses between tomato varieties than in any of the characters of tobacco, but still much less than in yield in maize. Williams and Gilbert (1960) studied the behavior of all hybrids among 18 pure lines with respect to fruit number, fruit weight, earliness of flowering, and the resultant, early yield. The averages of the 153 F_1's were about the same as those of their parents in the component characters, but there was about 17% superiority of F_1 in early yield. The parent lines varied greatly in fruit number and fruit weight, with a strong negative correlation (−0.66) between these characters. The yield of hybrids as the product of predominantly intermediate averages for number and weight exceeded the yields of parents with large numbers of small fruits or with a small number of large fruits for the same reason, as the authors note, that 2×2 exceeds 3×1 or 1×3. The explanation is thus more mathematical than physiological. There was little heterosis where both parent lines were intermediate in both respects and thus high in yield.

Some crosses had, however, been sufficiently high yielding to be commercially successful. Williams (1959) tested whether this was an essentially unfixable heterosis by selecting for superior performance in F_2 and later generations, in two cases. He was able to produce inbreds equal to the F_1

ancestors in four generations in marked contrast with the failure of such attempts with maize. He concluded that the number of genetic differences among the superior varieties was relatively small, and that the favorable genes were readily combinable.

Probably the most thoroughly controlled study of heterosis that has been made with a self-fertilizing plant was that of Griffing and Langridge (1963) with *Arabidopsis thaliana*, a small plant that can be grown aseptically on an inorganic nutrient salt solution solidified with agar, in continuous fluorescent light, and with controlled humidity and temperature. They studied 38 races, originally collected over a wide range, Holland to Japan, Sweden to North Africa. The character studied was the logarithm of the fresh weight, 14 days after germination. Growth rates were determined at temperatures at intervals of 3°C from 16°C to 31°C. The races differed considerably in both their logarithmic means and variances at each temperature and in their optimal temperatures.

In crosses between one of the strains and four others, the increase in log weight (with equivalent percentage in parentheses) over the midparent was 0.094 (25%) in F_1, reduced in F_2 to 0.026 (6%) at temperatures from 16°C to 25°C. At 28°C the excess of F_1 over the midparent in log weight was 0.294 (97%) in F_1, reduced in F_2 to 0.210 (62%). At 31°C it was 0.671 (368%) in F_1, reduced in F_2 to 0.391 (146%). The average temperature optimum rose from 24.0°C to 24.6°C.

The authors interpreted their results in terms of multiple loci at which the alleles determine enzymes, differing with respect to thermolability and optimal temperature. Heterozygotes, with two products at each locus, have more versatility than homozygotes, manifested especially at high temperatures, at which each strain is likely to suffer from thermolability of one or more enzymes.

The consensus of these and other studies of self-fertilizing species is that there is usually some heterosis but that the amount is less than that in cross-fertilizing species and that superior combinations are more easily obtained following crosses.

Tetraploids

Experiments with tetraploids supply evidence on the nature of inbreeding depression and heterosis of a different sort from that provided by diploids.

There are complications from the disturbed nuclear-cytoplasmic ratio. Thus Randolph (1941, 1942) found that autotetraploids induced in relatively vigorous, long inbred lines of maize, were of much reduced vigor and fertility, although ones induced in crossbred maize were fully vigorous and fertile.

The most thorough comparison of the properties of tetraploids and the diploids from which they were derived seems to have been made by Lundqvist (1966, 1969). He obtained six tetraploid lines of Steel rye sufficiently vigorous to be maintained, out of 13 induced by colchicine in lines with 25 years of selfing. He attempted to compare the inbreds, the 15 possible F_1's, the 15 F_2's, the 45 possible ($F_1 \times F_1$) of type AB/CD, called (4), and the 60 possible ($F_1 \times F_1$) of type $AB \times AC$, called (3), all in a single year, in suitably replicated patterns of both diploids and tetraploids. He succeeded, except for one tetraploid F_1 and derivatives. The gaps were replaced by appropriate interpolation.

There was considerable differentiation among the six inbred strains. The smallest averaged about 70% of the largest in six morphological traits in both diploids and tetraploids. There was much more differentiation in the amount of tillering (smallest was 19% of the largest in diploids, 32% in tetraploids). In total yield, the smallest was only 11% of the largest in diploids, 29% in tetraploids.

The rank correlation between the diploid and tetraploid inbreds in the nine characters studied was only moderate (0.49). The average correlation between the two F_1's was 0.59, but the correlations in the various $F_1 \times F_1$ generations were considerably lower.

Table 2.9 shows the values in inbreds and F_1's, the amounts of heterosis as indicated by the ratio of F_1 to inbred, and the ratios of tetraploid to diploid in both generations. It may be seen that the tetraploids had lower averages than the diploids in all respects except straw diameter, and mean kernel

TABLE 2.9. Averages of characters (Steel rye) of inbreds (I), F_1, and the ratio F_1/I in diploids and tetraploids, and ratios of tetraploids to diploids. Characters 4–8 refer to the biggest ear.

	DIPLOID			TETRAPLOID			$4n/2n$	
	I	F_1	F_1/I	I	F_1	F_1/I	I	F_1
1. Plant height (cm)	123.7	168.8	1.36	102.8	153.1	1.49	0.831	0.907
2. Straw diameter (mm)	4.20	5.04	1.20	4.47	5.53	1.24	1.064	1.097
3. No. of spikes	3.09	5.67	1.83	2.37	3.36	1.42	0.767	0.592
4. Straw weight (gm)	8.9	26.7	3.00	7.9	19.3	2.44	0.888	0.723
5. No. of spikelets	30.9	43.8	1.42	27.4	35.1	1.28	0.887	0.801
6. Seed set (%)	66.7	86.5	1.30	27.7	53.3	1.92	0.415	0.616
7. Mean kernel weight (gm)	21.7	28.4	1.31	18.9	35.0	1.85	0.871	1.232
8. Kernel yield (gm)	0.96	2.23	2.32	0.33	1.40	4.24	0.343	0.628
9. Total kernel yield (gm)	2.86	11.08	3.87	0.86	4.34	5.05	0.301	0.392
Average, excluding 4, 8, and 9			1.40			1.53	0.804	0.874
Average for 4, 8, and 9			3.06			3.91	0.511	0.581
Total average			1.96			2.33	0.707	0.776

SOURCE: Data from Lundqvist 1966.

weight in F_1. The tetraploids suffered most in total yield (30% of the diploid inbreds, 39% in F_1). The grand average of the ratios was 71% in inbreds, 78% in F_1.

There was heterosis in all respects in both diploids and tetraploids. This was least in straw diameter (20% excess in F_1 diploids, 24% in F_1 tetraploids) and most in the composite characters: straw weight (200% in F_1 diploids, 144% in F_1 tetraploids), kernel yield on the largest ear (132% in F_1 diploids, 324% in F_1 tetraploids), and especially total yield (287% in F_1 diploids, 405% in F_1 tetraploids). The average amount was somewhat greater in tetraploids (53%) than in diploids (40%) on excluding the three composite characters. For the composite characters, the increase was 291% in the tetraploids, 206% in the diploids. The excess performances in crosses in diploids and tetraploids are compared in table 2.10 as percentages of the excess in F_1.

Among the diploids, the four-way crosses ($AB \cdot CD$) differ from the inbreds by about the same amount (101%) as F_1, as expected if interaction effects are unimportant, since all chromosome segments are equally heterozygous. The

TABLE 2.10. Percentage differences between crossbreds (Steel rye) (F_1, (4), (3), and (F_2)) and inbreds in diploids and tetraploids. Characters 4–8 refer to the biggest ear.

	DIPLOID				TETRAPLOID			
	$F_1 - I$	$(4) - I$	$(3) - I$	$F_2 - I$	$F_1 - I$	$(4) - I$	$(3) - I$	$F_2 - I$
1. Plant height (cm)	100.0	104.4	86.5	65.2	100.0	124.3	107.2	90.5
2. Straw diameter (mm)	100.0	101.2	79.8	51.2	100.0	122.6	103.8	89.6
3. No. of spikes	100.0	111.2	82.9	49.2	100.0	246.5	197.0	179.8
4. Straw weight (gm)	100.0	102.2	71.9	47.8	100.0	187.7	143.9	121.9
5. No. of spikelets	100.0	95.3	72.9	45.7	100.0	145.4	115.6	100.0
6. Seed set (%)	100.0	104.5	90.9	66.7	100.0	126.6	110.5	93.7
7. Mean kernel weight (gm)	100.0	92.5	64.2	35.8	100.0	131.0	108.1	86.3
8. Kernel yield (gm)	100.0	95.3	69.3	42.5	100.0	150.5	114.9	89.7
9. Total kernel yield (gm)	100.0	103.9	69.2	37.6	100.0	221.3	161.2	121.3
Average, excluding 3, 4, and 9	100.0	98.9	77.3	51.2	100.0	133.8	110.0	91.6
Average for 3, 4, and 9	100.0	105.8	74.7	44.9	100.0	218.8	167.4	141.0
Total average	100.0	101.2	76.4	49.1	100.0	161.8	129.1	108.1

SOURCE: Data from Lundqvist 1966.

three-way cross ($AB \cdot AC$) shows 76.4% of F_1 heterosis, close to the expected 75% with three-fourths of the chromosome segments heterozygous. The F_2 ($AB \cdot AB$) shows 49.1%, close to the expected 50%. The differences between the two groups of characters are trivial.

Among the tetraploids, the four-way crosses go far beyond F_1, with 134% of F_1 heterosis in the six characters that do not depend on tillering and 219% for the three that do. The three-way crosses also show excess heterosis, though only about half as much excess as the four-way crosses on the average, with 110% and 167%, respectively, of F_1. Even F_2 shows more heterosis on the average, with 108% of that of F_1, although the group that does not involve tillering shows less (92%). The three that involve tillering average 141%.

Lundqvist estimated the amounts of heterosis shown by the five classes of tetraploid genotypes from those shown by the progeny of the mating types. F_1 individuals ($AABB$) produce gametes in the ratio $1AA:4AB:1BB$ with respect to genes close enough to the centromere to undergo no double reduction. The ratio otherwise depends on various factors (Mather 1936). If the chromatids are distributed wholly at random it is $1AA:2.67AB:1BB$, but it is $1AA:2AB:1BB$ if there is always just one crossover. The differences in the conclusions were not great, and only the case of no double reduction will be considered here.

Taking the heterosis of $AAAA$ as 0, and that of $AABB$ as 100, that of $AAAB$ can be estimated from the genotypic frequencies in F_2 in table 2.11, and the heterosis of F_2 in table 2.10. This permits estimates of the heterosis of $AABC$ from the genotypic frequencies and heterosis of (3). Finally the heterosis of $ABCD$ can be estimated from the genotypic frequencies and heterosis of (4).

TABLE 2.11. Expected frequencies of offspring-genotypes from tetraploid matings.

MATING TYPES	OFFSPRING-GENOTYPES					
	$AAAA$	$AAAB$	$AABB$	$AABC$	$ABCD$	Total
I $AAAA \times AAAA$	36					36
F_1 $AAAA \times BBBB$			36			36
(4) $AABB \times CCDD$			4	16	16	36
(3) $AABB \times AACC$	1	8	3	24		36
F_2 $AABB \times AABB$	2	16	18			36

These estimates (table 2.12) fall into the expected order for five of the characters and are not seriously divergent in another (number of spikelets on the largest ear), but they give wholly unreasonable results for the number of spikes and its dependents, straw weight and total yield. Excluding these, there appears to be little difference between $AAAB$ (including $ABBB$, etc.) and $AABB$. Qualitatively these involve only interaction effects of the same genomes, so that little difference is expected. The genotypes represented by $AABC$ and $ABCD$ involve interactions with additional genomes.

In working out the effects of inbreeding on polyploids (vol. 2, chap. 7), it was convenient to deal with the average correlation between pairs of alleles, drawn at random from the set, as the inbreeding coefficient. The proportion of heterozygous pairs in the five classes of genotypes are 0 from $AAAA$, 3/6 from $AAAB$, 4/6 from $AABB$, 5/6 from $AABC$, and 6/6 from $ABCD$. If heterosis were proportional, its values should be 0, 75, 100, 125, and 150, respectively. There is rough agreement except for $AAAB$. This definition of F for polysomics was, however, merely a device to permit path analysis, and there is no such clear reason for it to be in accord with phenotypic values as in the case of diploids.

Lundqvist suggested that the aberrant results for number of spikes may be due to an abnormally low average for F_1 in this character. It is possible that the effect of $AABB$ in F_1 is not fully comparable with the genotypic values in the segregating $F_1 \times F_1$ progenies. Since little difference is observed between $AAAB$ and $AABB$ in most cases, it may be better to assign 100 to the heterosis of these genotypes collectively and calculate those for $AABC$ and $ABCD$

TABLE 2.12. Estimated grades $(AABB = 100)$ of characters of various tetraploid genotypes, first assuming $AAAB$ to differ from $AABB$, using the F_1 data, and then assuming that $AAAB = AABB$ and not using the F_1 data. Characters 4–8 refer to the biggest ear.

	F₁ Used					F₁ Not Used			
							$AAAB$		
	$AAAA$	$AAAB$	$AABB$	$AABC$	$ABCD$	$AAAA$	$AABB$	$AABC$	$ABCD$
1. Plant height (cm)	0	91	100	111	137	0	100	122	144
2. Straw diameter (mm)	0	89	100	114	137	0	100	118	146
3. No. of spikes	0	(292)	100	(185)	(344)	0	100	109	182
4. Straw weight (gm)	0	(162)	100	(145)	(248)	0	100	121	186
5. No. of spikelets	0	112	100	123	179	0	100	118	167
6. Seed set (%)	0	98	100	121	139	0	100	121	141
7. Kernel weight (gm)	0	82	100	122	148	0	100	132	164
8. Yield (gm)	0	89	100	130	183	0	100	136	195
9. Total yield (gm)	0	(160)	100	(176)	(297)	0	100	142	226
Average excluding 3, 4, and 9	0	93.5	100	121.3	153.0	0	100	124.5	159.5
Average	0					0	100	124.3	172.3

SOURCE: Data from Lundqvist 1966.

without any use of F_1. This is done in the last two columns of table 2.12. The estimates for $AABC$ become relatively uniform with an average of 124. Those for $ABCD$ are still highly variable, with an average of 172. While $ABCD$ has only about twice the excess heterosis of $AABC$ for four of the characters—plant height, straw diameter, seed set, and kernel weight—it shows a relatively enormous excess in the other five. The actual genotypes, including the number of alleles, are not known. The situation may differ greatly among the characters, and in any case, there is no clear theory for polyploids.

For our present purpose, the most important finding is the qualitative difference in the relation of heterosis to generation of crossing between tetraploids and diploids, which parallels the difference with respect to the genome.

Interpretations

While only pre-Mendelian and early post-Mendelian observations on the effects of inbreeding and crossbreeding on animals have yet been considered and effects on variability in plants have only been touched on, it seems desirable to discuss here interpretations of inbreeding depression and heterosis, based on the observation in self-fertilized lines and crosses between them in plant species, which have obvious advantages for this purpose over observations on animals.

Role of Cytoplasm

It will be well to consider briefly the possibility of a role of cytoplasmic heredity, which as noted, Shull left open to some extent in his pioneer attempt to apply the principles of Mendelian heredity to the problem. At a time at which many biologists maintained that there was a contrast between a general developmental effect of cytoplasmic differences and highly specific ones determined by genes, it seemed likely that the former had more to do with the differences in general vigor that seemed to be the primary consequences of inbreeding and crossbreeding.

Cytoplasmically transmitted differences affecting viability, growth, and fertility undoubtedly exist (vol. 1, chap. 4; vol. 2, chap. 6). These, however, are transmitted almost exclusively down the female line, while inbreeding effects typically depend equally on both parents, as shown by the usual similarity of reciprocal crosses between inbred lines. Moreover, the striking difference between diploids and tetraploids with respect to the generation after a cross in which heterosis is greatest is in accord with the nuclear phenomena but not at all with the cytoplasmic ones.

Nucleocytoplasmic interactions are no doubt involved in all gene action, and disturbances may be manifested in particular crosses; and thus may contribute to inbreeding depression in the same way as segregation of any other deleterious recessive. It is likely, however, that all of the cases studied should be considered as crosses at most at the subspecific level. There is much to be said for the view that disturbance of nucleocytoplasmic interaction is a major factor in the defective development and infertility that often occur in crosses between distinct species. This, however, is the opposite of heterosis. The extraordinarily vigorous growth that occurs in some species hybrids, for example the radish-cabbage hybrid, one plant of which can fill a greenhouse (Gravatt 1914), is probably a quite distinct phenomenon from heterosis in varietal crosses. It may have more to do with defective regulation of growth than with the summation of favorable growth effects of both parents. Dobzhansky (1952) has distinguished it as luxuriance.

The Dominance Theory

As noted earlier, the first attempt to account for inbreeding depression on a Mendelian basis was by attributing it to the segregation of conspicuous deleterious recessives, each rare in the general population, but so numerous as a group that one or more is likely to be carried in heterozygous form by the common ancestors of the founders of the line. Such segregation is a common event, but it was by no means obvious that it had anything to do with the gradual but almost universal deterioration in all aspects of vigor under inbreeding, or with the restoration of vigor that follows the crossing of two inbred lines, neither of which is suffering from any conspicuous recessive defect. We are concerned here with these latter phenomena and with the dominance theory in the form given it by Jones (1917).

The basic assumption is that in natural species, as well as in cultivated plants and domestic animals, there is a positive correlation between favorability and degree of dominance among the alleles at the loci that contribute to quantitative variability in viability, growth, and fecundity. As noted, Jones' special contribution was to point out the unlikelihood that a completely favorable set would ever be sorted out by either natural or artificial selection within chromosome segments containing many such loci. Inbreeding would inevitably fix many of the slightly deleterious recessives, while in crossbreeding most of those from each parent would be covered up by more or less completely dominant alleles from the other parent.

The postulated correlation must have an immediate physiological basis, whatever its evolutionary origin. This basis has been discussed in volume 1, chapter 5; and will be discussed in chapter 15 of this volume.

The important point here is that there is good reason to expect a positive correlation between degrees of dominance of the genes and the amounts of positive effect of their loci on any quantitatively varying characters.

There are, however, aspects of degree of dominance, if not complete, which are mathematical rather than physiological. One of these is the effect of proximity of the phenotype, as determined by the rest of the genome, to a mathematical limit (0 or 100% of the character) where a pair of alleles have intermediate heterozygotes away from this limit. Of more significance for quantitative variability, however, is the effect of transformation of scale. Many cases were discussed in volume 1, chapter 15, in which the prevailing degree of dominance was greatly affected by scale transformation. The desirability was emphasized of using the scale on which all factors, environmental as well as genetic, have effects as nearly additive as possible. Full additivity is indeed never attainable by any transformation because of interaction systems, but additivity of the effects of independent systems may be approached. This usually means a logarithmic scale, sometimes modified, for characters involving growth. An unmodified logarithmic scale is desirable where one character is scored as the product of two or more others, in order to reduce the product to a sum.

This raises a difficulty in connection with the interpretation of the data presented earlier in this chapter. Yield is a character of a sort to which the contribution of any factor might be expected, a priori, to have a multiplicative effect rather than an additive one, except, perhaps, for damping as an upper limit is approached (Rasmussen 1933). This implies the appropriateness of a logarithmic transformation, $V' = \log V$ where V is the observed value (or more generally $\log [(V - L_1)/(L_2 - V)]$ where L_1 and L_2 are lower and upper limits [vol. 1, eq. 10.48]). Actually there is much more parallelism of percentage of heterozygosis with actual yield than with any logarithmic function of yield.

The assumption of multiplicative action above a lower limit does not allow for increasing sensitivity of yield to adverse factors as when the totality of these brings about an approach to a breaking point below which there is no yield at all (compare table 2.1). There could be exact cancellation of the multiplicative effect so that the effects of factors become additive on the scale of observed yields, except for possible damping near an upper limit.

This merely shows, however, that it is not wholly unreasonable to use the untransformed scale. It is far from being a compelling reason to suppose it to be the best. The rough parallelism of percentage of heterozygosis and observed yields must be considered to be an essentially empirical observation.

There is one obvious consideration that shows that the genes supplied by the inbred parents cannot have the additive effects attributed to them in the crosses, in the development of the parents themselves. If the heterosis of F_1 depended merely on the total of additive dominant effects supplied by the parents, added to the base of common factors possessed by both parents, the F_1 average would have to be less than the sum of the inbred averages by this base. Actually F_1 tends to be much more than twice the parental average,

2.80 times in Jones' data, 5.47 times in that of Richey and Sprague, 2.60 times in Neal's data, 2.76 times in Lindstrom's survey, and also 2.76 times in the data of Kinman and Sprague (vol. 1, chap. 15), 2.05 times in Stringfield's data, and 4.13 times in the data of Sentz, Robinson, and Comstock.

This difficulty does not exist if the factors have multiplicative effects; in inbred parents BX_1 and BX_2, $F_1 = BX_1X_2$ where B is the base and X_1 and X_2 are the multiplicative factors, restricted to each parent. If these potential effects are reduced exponentially by increasing sensitivity with a decrease in the values, a considerable array of favorable factors must be present to permit any yield at all. There are complementary sets, different in different strains, up to this point, additivity only beyond. Deviation from expectation may reflect either systematic imperfections of the scale or specific interactions among genes. For the moment, we will assume that gene effects are additive on the scale of actual yield.

Difficulty in Accounting for Extreme Heterosis

So far the observations can be interpreted fairly well to a first order by the dominance theory of inbreeding depression and heterosis. An apparently serious difficulty was first brought out clearly by Crow (1948, 1952). The results of Shull and East, discussed earlier, indicated that crosses between inbred lines derived at random (except for inevitable loss of the weakest) from open-pollinated varieties of maize merely return to the vigor of the open-pollinated strains. Particular crosses, however, went far beyond this and have indeed made possible revolutionary progress. Sprague (1963) states: "A measure of the success achieved can be appreciated from the knowledge that currently in excess of 95 percent of the nation's corn acreage is planted to hybrid corn. Total production has increased by at least 25 percent and this increase has been achieved on 25 percent fewer acres." This amounts to an increase of more than 67% (which, however, does not allow for increase due to other than genetic causes).

If we now suppose that substantial equilibrium had been reached in the foundation populations with respect to mutation of deleterious recessive genes (rate per locus u) and adverse selection (rate per gene s), the frequencies of each mutant gene would be $\hat{q} = \sqrt{(u/s)}$. The mean selective value of the population is $\overline{W} = 1 - s\hat{q}^2 = 1 - u$ (vol. 2, eq. 3.45) and thus depends merely on the mutation rate, as first pointed out by Haldane (1937). Mutation rates in maize are typically of the order 10^{-5} or less. Even with as many as 5,000 pertinent loci, $\overline{W}$ would be depressed by only 5%.

The depression in yield due to these genes is not, indeed, the same thing as the depression in selective value, but since the foundation stock had long been subjected to selection for yield, they should be closely related. Crow considered that 5% was probably much greater than the true mutational load with respect to yield and thus greater than the improvement possible from assembling dominant alleles by any cross under the dominance hypothesis and the assumption of equilibrium.

He noted the occurrence of heterozygous superiority over both homozygotes in a number of described mutations. For such loci, the reduction in selection value, $\overline{W} = 1 - sq^2 - t(1 - q)^2$, leads to an equilibrium at $\hat{q} = t/(s + t)$ (vol. 2, eq. 3.42), which is enormously greater than that from balancing of mutation by selection if s and t are not very different. Thus hybrids between lines in which different alleles of this sort have been fixed may go far beyond the random bred foundation stock. A relatively small number of such loci at equilibrium in the latter may suffice to give an enormous amount of hybrid vigor from crosses between appropriate inbred lines.

It is to be noted that strong repulsion linkage of dominant favorable genes (Jones' hypothesis) is not equivalent since such linkage does not prevent ultimate combination at random in the foundation stock and equilibrium according to Haldane's principle. It may, however, greatly delay the attainment of equilibrium (vol. 2, chap. 2).

The great success of hybrid corn breeding implies either that overdominance has contributed greatly or that the foundation stock was actually so far from equilibrium that the possibility existed of producing open-pollinated strains or even inbred lines approaching the best hybrids by the assemblage of all favorable dominant genes.

Overdominance and Heterosis

East (1936) strongly rejected the concept that heterosis is due to an appreciable extent to elimination of defective recessive mutations and proposed that it depends largely on complementary effects of positive alleles that thereby give heterozygous advantage. Others have, however, found heterozygous advantage of occasional defective mutations. Karper (1930) found a probable case in an albino mutation in inbred Sorghum. Jones (1945) found four cases in long inbred lines of maize in which heterozygotes of defective mutations were significantly superior to the homozygous normals in plant height (3% to 9%) and weight of ears per plot (21% to 104%). Gustafsson (1946) reviewed earlier cases and described two defective mutants from a pure line of barley, albino 7, xantha 3, that increased the vigor of heterozygotes and he later (1947a) showed that this was especially the case in double heterozygotes.

It should be said that further study by Jones (1957) of one of his cases led him to doubt the importance of this sort of case in heterosis. The mutation in this case appeared to have pleiotropic effects (crooked stalk, pale tip), but these proved to be separable. An exhaustive study led to the conclusion that neither of these gave any heterozygous advantage and indicated that the original mutation involved multiple genes, including at least one that affected growth and yield favorably. East's concept seems more plausible as a basis for strong heterosis than does a heterozygous advantage of mutants causing overt defect when homozygous.

Hull (1945, 1952) became the strongest advocate of this phenomenon, for which he coined the term *overdominance*, as the basis of heterosis, and became the proponent of a method, "recurrent selection," for adapting a strain to give maximum heterosis in crosses with an inbred tester line. He attempted to demonstrate the importance of overdominance from regression analyses of sets of diallel crosses among inbred lines. He found the regression on the other parent in each set of F_1's with a common parent. These should all be equally positive with intermediate heterozygotes, and should range from positive, with low-grade common parent, to zero with the highest ones if there is complete dominance, but to significantly negative in the latter case with overdominance. He found indications of overdominance, but not very decisive ones in the diallel sets that he analyzed. Even when overdominance was indicated, the method was incapable of distinguishing between true overdominance at loci and overdominance of chromosome segments due to repulsion linkage of favorable dominant genes. It became clear that more positive methods were needed.

Degrees of Dominance

There are various ways, useful in different connections, for describing the degree of dominance. In connection with selective advantage, it is convenient (vol. 2, eq. 3.29) to take s as that of AA over aa and hs as that of Aa over aa (scale I). Here h measures dominance in the sense implied when a lethal mutation is referred to as showing some dominance if the heterozygotes have a slight deleterious effect, or when (as often in this treatise) the case of exact intermediacy of the heterozygote is referred to as semidominance. This measure is somewhat awkward in connection with pure overdominance ($Aa > AA = aa, h = \infty$).

Where it is desired to exhibit the effects of different grades from exact intermediacy to pure overdominance, it is convenient to take aa as the base (0), the contribution of a single A as α_1, and the increment from a second A as α_2, usually following the convention that α_1 and α_2 are assigned to alleles so that $\alpha_2 \lessgtr \alpha_1$ (vol. 2, fig. 4.8–4.13, table 15.7). Dominance here is most naturally measured by $(\alpha_1 - \alpha_2)/\alpha_1$ (scale II), which is 0 for exact intermediacy, 1 for complete dominance, and 2 for pure overdominance. If alleles are, instead, assigned according to direction of effect, α_2 may be greater than α_1 and the coefficient negative.

The scale (III) most frequently in mind, however, when the case of exact intermediacy of Aa is referred to as one of no dominance is one in which the excess of AA over aa is taken as primary ($2u$) and that of Aa over aa is represented by $u + au$, in the symbolism of Comstock and Robinson (1948). This is especially convenient where the frequencies of all unfixed genes are 0.5, as in an F_2 population, since in this case u is the additive effect and the dominance deviations are $\pm au$. This is essentially the same scale as that used by Mather (1949) except for the symbols ($D = u$, $H = au$).

These scales are summarized in table 2.13. Concrete examples are given in table 2.14.

TABLE 2.13. Three ways of representing phenotypic deviations from the grade of aa for alleles A and a.

Zygotes	Frequency	I	II	III
AA	q^2	s	$\alpha_1 + \alpha_2$	$2u$
Aa	$2pq$	hs	α_1	$u + au$
aa	p^2	0	0	0

Comstock and Robinson (1948) and Comstock, Robinson, and Harvey (1949) devised an index $\dot{a}$ for the average dominance in F_2 populations in design III that can be calculated from the additive genetic and dominance variances, obtained by analysis of variance. These components are as follows in the general case (vol. 2, eq. 15.33, 15.34, 15.35) and in that in which all q's are 0.5:

$$\textit{General} \qquad\qquad\qquad p = q = 0.5$$

$$\sigma_G^2 = 2\sum[pq(p\alpha_1 + q\alpha_2)^2] \qquad \sigma_G^2 = (1/8)\sum(\alpha_1 + \alpha_2)^2 = (1/2)\sum u^2$$

$$\sigma_D^2 = \sum[p^2q^2(\alpha_1 - \alpha_2)^2] \qquad \sigma_D^2 = (1/16)\sum(\alpha_1 - \alpha_2)^2 = (1/4)\sum a^2u^2$$

TABLE 2.14. Examples of the three modes of describing dominance.

	I $h = \alpha_1/(\alpha_1 + \alpha_2)$	II $(\alpha_1 - \alpha_2)/\alpha_1$	III $a = (\alpha_1 - \alpha_2)/(\alpha_1 + \alpha_2)$
1. $\alpha_1 = 0$	0	$-\infty$	-1
2. $\alpha_2 = 2\alpha_1$	0.33	-1	-0.33
3. $\alpha_2 = \alpha_1$	0.5	0	0
4. $\alpha_2 = 0.5\alpha_1$	0.67	0.5	0.33
5. $\alpha_2 = 0$	1	1	1
6. $\alpha_2 = -0.5\alpha_1$	2	1.5	3
7. $\alpha_2 = -\alpha_1$	∞	2	∞

Comstock and Robinson's index is

$$\dot{a} = \sqrt{[2\sigma_D^2/\sigma_G^2]} = \sqrt{\left[\sum (a^2 u^2) \Big/ \sum u^2\right]}.$$

This is a special sort of weighted average. It does not become infinite where there is pure overdominance at any locus unless this is true of all loci, because of the separate summations for numerator and denominator. Several hypothetical cases all having the same amount of heterosis, $\bar{p} = 10$, $\overline{F}_1 = 30$, but having mixtures of loci that differ in magnitudes of effect and dominance, are shown in table 2.15.

TABLE 2.15. Properties of six mixtures of loci. The ratio of number of loci with no dominance, I, (in the sense $a = 0$) or with complete dominance, D, $(a = 1)$ to the number of purely heterotic ones, H, $(a = \infty)$ are at the tops of the columns. The phenotypic effects of single locus zygotes $2u$ for BB, $u + au$ for Bb, and 0 for bb are given in the table. The weighted average degree of dominance, $\bar{a}$ $(= \sum au/\sum u)$, and the dominance index, $\dot{a}$ $(= \sqrt{[\sum (a^2 u^2)/\sum u^2]})$, are given for each mixture in the two lower rows.

Zygotic	Effect	$1I$:$10H$		$1D$:$10H$		$1D$:$1H$		$1I$:$1H$		$10D$:$1H$		$10I$:$1H$	
BB	$2u$	20	0	20	0	20	0	20	0	2	0	2	0
Bb	$u + au$	10	2	20	1	20	10	10	20	2	10	1	20
bb	0	0	0	0	0	0	0	0	0	0	0	0	0
	$\bar{a}$	2		2		2		2		2		2	
	$\dot{a}$	0.63		1.05		1.41		2.00		3.32		6.32	

Since the ratio of $\overline{F}_1$ excess to $\overline{P}$ excess in the absence of the overdominant genes is only 1.0 or 2.0, according to whether those present have intermediate heterozygotes or are dominant, the overdominant genes play an essential role in all of the cases shown in raising the ratio to 3.0. There is, however, a tenfold difference between the dominance index $\dot{a}$, in the first and last cases. In the former, a large number of overdominant genes with small effects are overbalanced as far as $\dot{a}$ is concerned by a small number of genes with intermediate heterozygotes but large effects. The situation is the opposite in the last case. The values of $\dot{a}$ for the separate loci are 0, 1, or ∞ according to the state of dominance. The actual average in all cases is thus ∞. It is evident that the weighting used in calculating $\dot{a}^2$ is such that $\dot{a}$ reflects the state of dominance of the genes with greatest effect and gives little indication of the nature of minor factors, even though these are so numerous that they play a major role in determining the degree of heterosis. A more appropriate weighting system could be devised if the actual genes and their effects could be known but, as Comstock and Robinson stated, "the experiments under consideration offer no choice of the measure to be estimated." In general, a high index implies the presence of major genes with overdominant effects (or the equivalent in terms of repulsion linkage of favorable dominants), while a low index implies the absence of major overdominant effects but is compatible with a multiplicity of minor ones. An index significantly less than 1 indicates that neither complete dominance nor overdominance can be characteristic of the genes with major effects and, in view of the probable positive correlation between magnitude of effect and degree of dominance, indicates that pairs of alleles with intermediate heterozygotes prevail throughout, with only sufficient incomplete dominance to permit such heterosis as exists.

Experimental Estimates of Dominance

In applying this index to data, the authors noted a number of assumptions: (1) regular diploid behavior at meiosis, (2) gene frequencies of test population all 0.5, (3) no multiple allelism, (4) random combination of loci, (5) no interaction among locus effects, (6) random choice of individuals for production of experimental populations, and (7) random distribution of genotypes relative to environmental variation. Assumptions 1–3, 6, and 7 are satisfied by properly designed tests of random F_2 individuals from crosses between diploid inbred lines, but both assumptions 4 and 5 are likely to be violated to extents that are difficult to determine.

The first experiments to which the methods were applied (Robinson, Comstock, and Harvey 1949) served mainly to show that very large numbers

TABLE 2.16. Mean grades of offspring with respect to a single locus of backcrosses of representative F_2 individuals to the parents.

Inbreds	AA	Aa	Aa
AA aa	$\alpha_1 + \alpha_2$ α_1	$\alpha_1 + 0.5\,\alpha_2$ $0.5\,\alpha_1$	$\alpha_1 + 0.5\,\alpha_2$ $0.5\,\alpha_1$
M_C $M_C - \bar{M}$	$\alpha_1 + 0.5\,\alpha_2$ $0.25\,(\alpha_1 + \alpha_2)$	$0.75\,\alpha_1 + 0.25\,\alpha_2$ 0	$0.75\,\alpha_1 + 0.25\,\alpha_2$ 0

	aa	M_R	$M_R - \bar{M}$
AA aa	α_1 0	$\alpha_1 + 0.5\,\alpha_2$ $0.5\,\alpha_1$	$0.25\,(\alpha_1 + \alpha_2)$ $-0.25\,(\alpha_1 + \alpha_2)$
M_C $M_C - \bar{M}$	$0.5\,\alpha_1$ $-0.25\,(\alpha_1 + \alpha_2)$	$0.75\,\alpha_1 + 0.25\,\alpha_2$ 0	0 0

SOURCE: Comstock and Robinson 1952 (design III).

NOTE: The table gives the row M_R and column M_C means, and the deviations of these from the mean of means $\bar{M}$.

are needed in the design of experiments used, namely, testing of random F_2 plants as pollen parents on four random F_2 ovule parents in a suitably randomized pattern.

A more efficient method was described later (Comstock and Robinson 1952). This consisted in using each F_2 plant to pollinate both parental inbred lines.

Tables 2.16 and 2.17 show the core of this pattern for determination of the variance components.

TABLE 2.17. Sums of squares of deviations from the grand average of all entries (table 2.16) and their meanings in terms of σ_G^2 and σ_D^2.

Inbreds (R)	$4 \sum\limits^{2} (M_C - \bar{M})^2$	$= 0.5\,(\alpha_1 + \alpha_2)^2$	$= 4\sigma_G^2$
F_2 (C)	$2 \sum\limits^{4} (M_C - \bar{M})^2$	$= 0.25\,(\alpha_1 + \alpha_2)^2$	$= 2\sigma_G^2$
Interactions ($R \times C$)	$\sum\limits^{8} (M - M_R - M_C + \bar{M})^2$	$= 0.25\,(\alpha_1 - \alpha_2)^2$	$= 4\sigma_D^2$
Total	$\sum\limits^{8} (M - \bar{M})^2$	$= \alpha_1^2 + \alpha_1\alpha_2 + \alpha_2^2$	$= 6\sigma_G^2 + 4\sigma_D^2$

With multiple independently distributed loci with additive effects, the sum of squares for inbreds, F_2, interactions, and total would be the sum of the contributions for each locus.

The experiments testing n random F_2 individuals consisted of s sets of n pairs of progenies with r replications of each progeny in a randomized block arrangement. The pattern for analysis of variance was as follows:

	DF	Mean Square
Sets	$s - 1$	
Replications in sets	$s(r - 1)$	
Inbred lines in sets	s	
F_2 parents in sets	$s(n - 1)$	$M_{31} = \sigma_E^2 + 2r\sigma_m^2 = \sigma_E^2 + 0.5r\sigma_G^2$
$F_2 \times$ inbred in sets	$s(n - 1)$	$M_{32} = \sigma_E^2 + r\sigma_m^2 = \sigma_E^2 + r\sigma_D^2$
Among plots	$s(2n - 1)(r - 1)$	$M_{33} = \sigma_E^2$
Total	$2nrs - 1$	

$$\sigma_G^2 = 2(M_{31} - M_{33})/r \qquad \sigma_D^2 = (M_{32} - M_{33})/r \qquad \dot{a} = \sqrt{[2\sigma_D^2/\sigma_G^2]}$$

Gardner et al. (1953) analyzed data for several characters from two crosses between inbred lines of maize; data from one sample of each cross were obtained in two years and from another in two locations. Only weighted averages of $\dot{a}$ are given in table 2.18.

TABLE 2.18. Average estimates of $\dot{a}$ from two crosses between inbred lines of maize.

Character	NC7 × CI21	NC33 × K64
	$\dot{a}$	$\dot{a}$
Yield	1.97	1.44
Number of ears	0.72	0.83
Ear length	0.74	0.78
Ear diameter	1.08	0.77
Kernel rows	0.51	0.53
Plant height	0.71	0.94
Ear height	0.69	0.89
Days to flowering	0.79	0.70

SOURCE: Data from Gardner et al. 1953.

Yield in both crosses, and ear diameter in one, gave indexes significantly greater than 1, thus indicating prevailing overdominance at the major loci. In a few of the other cases, complete dominance appears to be characteristic of major loci, while in most of them incomplete dominance seems to have been the rule, with many determinations in separate experiments significantly less than 1. There was least dominance (average 0.52) in the case of number of kernel rows. This agrees with the results of Emerson and Smith (vol. 1, chap. 15) and with Jones' data in tables 2.1 and 2.2.

Gardner (1963) reviewed all previous estimates of variance components and dominance in experiments with yield in maize, made by the above and related designs (Lindsey et al. and Gardner and Lonnquist in Gardner 1963) in addition to those already referred to (table 2.19).

TABLE 2.19. Average values of σ_G^2, σ_D^2, and $\dot{a}$ for yield from F_2 experiments with five crosses between inbred strains of maize.

Cross	No. of Determinations	σ_G^2	σ_D^2	$\dot{a}$
CI21 × NC67	7	0.0028	0.0034	1.56
NC33 × K64	2	0.0022	0.0024	1.48
NC34 × NC45	4	0.0031	0.0079	2.26
NC16 × NC18	1	0.0070	0.0036	1.01
M14 × 187-2	3	0.0034	0.0019	1.06

SOURCE: Gardner 1963.

All but the fourth of these (the statistically inadequate first case reported) are well within the overdominance range. As Comstock and Robinson (1948) emphasized from the first, however, a value in this range by no means necessarily means true overdominance. Apparent overdominance because of repulsion linkage is also possible.

They and others tested this question by carrying F_2 populations with high $\dot{a}$ values through several generations of random mating to give an opportunity for recombination: Comstock, Robinson, and Cockerham (1957), Gardner and Lonnquist (1959), Robinson, Cockerham, and Moll (1960), Lindsey, Lonnquist, and Gardner (1961), again summarized (table 2.20) by Gardner (1963).

TABLE 2.20. Estimates of $\dot{a}$ for grain yield from two crosses in F_2 and later generations of random mating.

	F_2	F_4	F_8	F_{12}	F_{16}
CI21 × NC67	1.68	...	1.24	1.09	...
M14 × 187-2	1.98	1.04	0.72	...	0.62

SOURCE: Gardner 1963. Reprinted with permission of the National Academy of Sciences.

There is clearly a downward trend of $\dot{a}$ from the overdominance range and thus evidence that the major gene differences in these crosses are not characterized by overdominance, or even complete dominance in the second case. The implication is that there was repulsion linkage of major more or less dominant genes favoring yield in the parental inbred lines to a much greater extent than in inbred lines drawn at random from an equilibrium population. This situation could be the result of the drastic selection that had been practiced among inbreds for those that give the highest yielding hybrids, and were thus complementary to the maximum extent with respect to more or less dominant favorable genes that they could supply. The results do not, however, exclude slight degrees of overdominance at numerous loci, a situation that could play a major role in the heterosis of F_1.

An additional line of evidence comes from determinations of the additive genetic variance, σ_G^2, and the variance of dominance deviations, σ_D^2, in

TABLE 2.21. Determinations of σ_G^2, σ_D^2, and σ_D^2/σ_G^2 for grain yield in three open-pollinated strains of maize.

Strain	Tests	σ_G^2	σ_D^2	σ_D^2/σ_G^2
Jarvis	6	0.0030	0.0005	0.17
Weekley	6	0.0037	0.0018	0.49
Indian Chief	2	0.0023	0.0010	0.35
Mean		0.0030	0.0011	0.37

SOURCE: Robinson, Comstock, and Harvey 1955.

open-pollinated strains. Robinson, Comstock, and Harvey (1955) studied three varieties (table 2.21).

The estimates for σ_G^2 compare well with those for F_2 from crosses (table 2.9), but those for σ_D^2 are relatively small. No calculation of $\dot{a}$ can be made because the distribution of gene frequencies is not known. It is possible, however, to calculate σ_D^2/σ_G^2 for a single locus for different degrees of dominance and different gene frequencies. This was done by Robinson, Comstock, and Harvey (1955). Table 2.22 covers a wider range of the variables than their table.

The gene frequency of the favorable allele was presumably much greater than 0.5 in the open-pollinated varieties if there was not overdominance. The observed values of σ_D^2/σ_G^2 in table 2.21 are consistent with incomplete dominance. They also are consistent with slight overdominance and gene frequencies well over 0.95, or pure overdominance and gene frequencies of about 0.90. With overdominance, however, the equilibrium frequency of q in a random bred population is $(a + 1)/2a$ and thus much less than 0.90. The major genes could thus hardly have been predominantly overdominant.

TABLE 2.22. Values of σ_D^2/σ_G^2 for a single pair of alleles corresponding to different degrees of dominance as measured by $(\alpha_1 - \alpha_2)/\alpha_1$ or by a (of table 2.14) at different frequencies, q, of the favorable allele.

	DEGREE OF DOMINANCE								
$(\alpha_1 - \alpha_2)/\alpha_1$	0	0.25	0.50	0.75	1.00	1.25	1.50	1.75	2.00
q　　　　a	0	0.141	0.333	0.600	1.000	1.666	3	7	∞
0.10	0	0.003	0.013	0.030	0.056	0.092	0.140	0.203	0.281
0.20	0	0.006	0.025	0.062	0.125	0.222	0.367	0.580	0.889
0.50	0	0.010	0.056	0.180	0.500	1.389	4.5	24.5	∞
0.80	0	0.008	0.056	0.281	2.0	∞	4.5	1.531	0.889
0.90	0	0.005	0.037	0.240	4.5	4.5	0.827	0.417	0.281
0.95	0	0.003	0.022	0.162	9.5	1.056	0.296	0.166	0.117
0.99	0	0.001	0.005	0.042	49.5	0.137	0.047	0.028	0.021
1.00	0	0	0	0	∞	0	0	0	0

The Stimulation Hypothesis

The hypothesis that heterozygosis per se tends to stimulate metabolic activity and hence all aspects of vigor implies pure overdominance for all pairs of alleles, the homozygotes of which do not differ with respect to vigor. The contributions of individual pairs to heterosis from this cause must thus be exceedingly small and easily overbalanced by small numbers of genes with major effect with respect to the index $\dot{a}$ even though their total contributions cause most of the heterosis. The stimulation hypothesis is thus not excluded by any evidence based on this index.

It is not, however, a very plausible hypothesis from the physiological standpoint. It is more plausible to suppose with East that overdominance arises from the versatility of heterozygotes with two positively acting but different alleles. The decline of $\dot{a}$ from the overdominance range tends to rule out pairs of alleles with a high degree of overdominance of this sort as a major factor in heterosis but does not rule out a multiplicity of pairs with specific but minor effects.

Interaction Systems

A priori, specific interactions are expected to be present and complicate interpretations since, as discussed in volume 1, chapter 5, the examination of the combination effect of known genes that affect the same character always reveals many such cases. Among such systems, reaction chains in which the effect of each gene depends on the presence of the genes that control all prior links, and joint or complementary effects that fail if either component fails, are perhaps the most common, but other systems—duplicate factors, intermediate optimum, and so on—are also common.

The consequences of various systems in F_2 and backcrosses, relative to the grades of parents, and F_1 were discussed in volume 1, chapter 15, in connection with the interpretation of the results of crosses between the extremes of quantitative variability.

At this point, it is advisable to examine the data for deviations from the linear relation of yield to percentage of heterozygosis that may suggest patterns of interaction rather than second-order modifications of scaling. The results of Kinman and Sprague (1945) on P_1, F_1, and $\overline{F}_2$ in 45 diallel crosses (vol. 1, chap. 15) are included.

TABLE 2.23. Excess yields over inbreds in maize as percentages of the excess of F_1 over the inbreds. The bottom row shows the ratio of actual yields of inbreds to F_1.

Heterozygosis (%)	Type of Mating 1	Richey and Sprague (1931)	Neal (1935)	Lindstrom (1941)	Kinman and Sprague (1945)	Stringfield (1950)	Sentz, Robinson, and Comstock (1954)
0	Inbred (base)	0	0	0	0	0	0
25	$\overline{B}_2$	28.6	...	...	...	...	47.2
50	$\overline{B}_1$	50.3	...	55.6	...	60.8	57.3
50	$\overline{F}_2$	42.1	52.4	46.4	43.3	47.7	54.5
75	$\overline{B}_{2x}\,(AB)(AC)$	70.3	...	...	...	77.6	71.1
100	$\overline{F}_1$	100.0	100.0	100.0	100.0	100.0	100.0
Inbred yield / F_1 yield		0.183	0.377	0.362	0.362	0.489	0.242

Table 2.23 shows the differences in yield as percentages of the difference between F_1 and the inbred parent lines, in relation to the percentage of heterozygosis. The entries for 75% heterozygosis consisted of matings of the type $(AB)(AC)$ in the data of Stringfield but $[(AB)A]A$ in the other cases.

The most pronounced deviations were in the results of Sentz, Robinson, and Comstock (1954), as noted earlier. The excessively high grade at 25% heterozygosis, not only in yield but in all other characters in certain years, may perhaps reflect an excessive depression of the inbred strains due to an approach to the breaking point of no growth or yield at all to a greater extent in these years than postulated in the transformation theory presented earlier. The average yield was only 24.2% of F_1, much less than in the other data except for that of Richey and Sprague (18.3%), which, however, do not show an excessively high value at 25% heterozygosis. These can be reconciled only by assuming different breaking points.

The difficulty of accounting for the 47% increment over the first 25% increase in heterozygosis in the data of Sentz, Robinson, and Comstock is moreover hardly greater than that of accounting for an increment of only 24% over the next 50% increase in heterozygosis, followed by a 29% increment over the last 25% increase in heterozygosis. The best explanation seems to be in terms of strong genotype-environment interactions that modified the scale in such a way as to increase response near both ends relative to the middle.

The most consistent deviation from expectation in table 2.23 is the low yield of F_2 in comparison with the first backcrosses in all cases where both were obtained. Since both of these have 50% heterozygosis, their differences cannot be accounted for by any distortion of the scale, and so must indicate interaction. It is probably not constant in view of the smallness of the difference in the data of Sentz, Robinson, and Comstock and the relatively high yield of F_2 and close agreement with the expected relation to heterozygosis in that of Neal. The fact that the low value for F_2 in Kinman and Sprague's data (discussed in vol. 1, chap. 15) was due almost wholly to six of their diallel crosses, all of which were significantly less than $0.5 \, (\overline{P} + \overline{F}_1)$ at the 0.05 level, again indicates that specific interactions, not present in all crosses, are responsible for the unexpectedly low F_2 yields.

However this may be, the highly significant differences between the two types of mating at 50% heterozygosis in several cases seem to require gene interaction of some sort, sufficiently common to stand out from the canceling effect of mixtures. Where important enough so that B_1 with only one of the parental interaction systems intact has a significantly higher yield than F_2 with both broken up, F_1 with both intact should have about twice the excess of B_1 over F_1 (on an additive basis). Since this excess is, on the average, 8.3% ($56.0 - 47.7$) in the four cases in table 2.23 in which both yields were obtained, it is implied that F_1 would have some 17% more heterosis than a population in which interaction systems were broken up as much as in F_2. F_1 has, however, merely the sum of the interaction systems supplied by each parent, absent in the other. The heterosis on this account is thus on the same basis as that supplied by favorable dominants with additive effects.

If there are multiple selective peaks, there may be equilibrium at a peak that is lower than others possible from genes already present. Crosses between particular inbred strains that differ in opposite ways from the current peak genotype may give F_1 near a peak considerably higher than this, and on interbreeding may lead to a new equilibrium at the new peak. This may play some part in accounting for extreme heterosis from crosses selected from a large number of trials.

Open-pollinated Maize

The preceding discussion has led to three possible explanations of the observed great superiority of the better crosses between inbred lines of maize over the ancestral open-pollinated strains. If these strains had attained equilibrium between mutation and selection, there must either be a great deal of overdominance, which must for the most part be slight per locus but at a great many loci, or the equilibrium must have been at a lower selective peak than one possible from the genes present. The third possibility is that the ancestral strains were far from equilibrium in any sense. All three may play important roles.

There is indeed little a priori likelihood that there was any sort of equilibrium. The origin of the cornbelt strains, from which most of the inbreds were drawn, is known to have involved extensive crossing during the last century between widely different varieties: Southern Dents, of largely Mexican origin, and Northern Flints, long established on the east coast of the United States (Anderson and Brown 1952). The sorting out by selection of the genes for the most favorable combination from a mixture of ones, favorable and unfavorable for this purpose, would have been a very slow process at best (vol. 2, chap. 2), and methods of selection for yield were not very effective at the time.

That the favorable possibilities present in different strains had not actually been assembled is shown by the considerable heterosis found on crossing strains. Pollak, Robinson, and Comstock (1957) demonstrated heterosis in F_1 yield of about 16% above strain averages, using the same three strains in which variance is analyzed in table 2.21. Half of this was lost in F_2 and backcrosses as expected. Smaller amounts of heterosis were found in other characters: 10% in ear height and 5% in ear diameter. Moll et al. (1965) made crosses between two strains from each of four regions and crosses between these regions. Those within regions gave yields 11% above the strain averages. Those between regions gave increases ranging from 9% between Mexican and Puerto Rican strains, about 20% in each cross between strains from northeastern United States, midwestern United States, and Mexico, and about 38% between either strain from the United States and Puerto Rico.

There is also still much residual potential for improvement within single strains, as shown by the additive variance in the three strains referred to in table 2.21. Robinson, Comstock, and Harvey (1955) estimated that the use of more effective methods of selection than hitherto should be capable of increasing yields by 12% to 15% in these strains. Attempts to accomplish this (Comstock, Robinson, and Cockerham [1957]; review by Gardner [1963]) soon resulted in about 10% increase. Taking into account both the heterosis between strains and the additive genetic variance within them, there would seem to be the potential for some 30% additional yield before the equilibrium state is reached.

The actual superiority of the best crosses seems, however, to be so much greater than this that a substantial contribution from slight tendencies toward overdominance at many loci, and perhaps from the attainment of higher selective peaks, seems probable.

Summary

A gradual decline in viability, growth, and fecundity as a result of close inbreeding, and an abrupt restoration of vigor from crossing inbred lines, were widely recognized long before the period of scientific investigation. These beliefs were confirmed by experiments with plants and animals in the eighteenth and nineteenth centuries, but, in the absence of any valid knowledge of the mechanism of heredity, there could be no theory beyond the vague hypothesis of a stimulus from a difference between the cells that unite at fertilization.

Experiments with both animals and plants after the rediscovery of Mendel's principles soon led to the enunciation of various theories that have been the basis for much later experiment and discussion, without leading to any single resolution of all difficulties. The present chapter has been largely devoted to experiments with self-fertilization and crossing of selfed lines of plants, which have obvious advantages over the concurrent experiments with animals discussed in the next chapters.

It was soon recognized that the frequent appearance of major defects after inbreeding could be accounted for by the segregation of Mendelian recessives, each rare in the general population but so numerous as a group that one or another was likely to be present in common ancestors of closely related parents, both of which would thus often be heterozygotes. A rather common view at first was that the gradual decline in aspects of vigor in inbred lines was a wholly different phenomenon that depended on cytoplasmic change rather than genic segregation. It was found, however, that cytoplasmic differences were almost wholly transmitted down the female line and could not be responsible for the typical effects of inbreeding and crossbreeding, which depended equally on both parents. The conclusion that the primary site of change must be nuclear in no way invalidated the importance of genotype-cytoplasm interaction in the physiology of inbreeding and cross-breeding effects, or the probability that genotype-cytoplasm incompatibility is a major cause of developmental defect and infertility in species crosses.

The development about 1910 of the multiple factor theory of the heredity of quantitative variability led to the view that inbreeding depression and the vigor of first crosses depend on a positive correlation between degree of dominance and favorable effect on aspects of vigor, and thus that there is no essential difference between inbreeding decline and the segregation of defective recessives except in the magnitudes of the individual gene effects. The occurrence of such a correlation is to be expected on physiological grounds. The recognition of the difficulty imposed by linkage on the assemblage of all favorable genes into a single inbred line removed the principal objection to the dominance theory.

Quantitative studies of inbreds and derivatives with various percentages of heterozygosis revealed a first-order agreement between the grades of various characters of maize, especially yield, and the percentage of heterozygosis. This supported the theory, subject to considerations of the appropriateness of the scale of measurement. There have been, however, some major discrepancies, notably a usual inferiority of F_2 to the first backcross average, in spite of identity (50%) in amount of heterozygosis. These indicate the prevalence of certain types of interaction systems.

Meanwhile, it came to be recognized that there was a serious difficulty with the dominance theory in accounting for the extraordinary superiority of certain crossbreds (notably in maize) over the ancestral random bred stock under the assumption that the latter was in equilibrium with respect to the pressures of recurrent mutation and selection. The superiority should probably always be less than 5%.

This shifted attention to the role of overdominance in the vigor of hybrids, since there is no theoretical limit to their superiority. Overdominance has been demonstrated in some heterozygotes of otherwise recessive defects, but this can have nothing to do with the heterosis from crosses between inbred lines that lack such defects. Overdominance at major loci, causing great differences in vigor, has, however, been ruled out as a major cause of heterosis in maize hybrids by experiments in which it was shown that an indicator of overdominance in F_2 disintegrated under random mating in such a way as to indicate repulsion-linkage rather than overdominance.

The hypothesis of numerous pairs of alleles with only slight overdominance was not, however, ruled out. The theory of a slight stimulus from hetero-zygosis at any locus, essentially a revival of the old stimulation hypothesis at a higher level of knowledge of heredity, is thus possible but purely speculative and rather improbable on physiological grounds. The less extreme hypothesis that there are numerous pairs of isoalleles with positive but qualitatively slightly different effects, because of the greater versatility of heterozygotes than of either homozygote, is more acceptable physiologically and has considerable support. It merges into the dominance hypothesis in the application of the latter to compound loci.

The objection to the dominance hypothesis on the basis of the high degree of superiority of crossbred maize from selected inbred lines over the open-pollinated ancestral populations has been removed to a considerable extent by recognition that the latter were far from equilibrium. Equilibrium could hardly have been reached in populations derived rather recently from crosses between very different varieties, because of linkage and the presence of interaction systems. It has indeed become clear that equilibrium has not been at all closely approached since rather strong heterosis has been demonstrated in crosses between open-pollinated strains, and large stores of additive genetic variance have been found within them. Considerable progress in yield has been achieved by improved methods of selection. The production of much more vigorous inbred lines than formerly has indicated the possibility of assembling favorable genes in homozygous form to a much greater extent than once seemed probable.

There is still a degree of superiority of the best hybrids over the best open-pollinated strains, and of these over the best inbreds, which indicates that the prevailing dominance of favorable genes is probably supplemented by slight overdominance at numerous loci.

While the precise weights to be given to dominance of favorable factors and overdominance remain uncertain, the data are in essential harmony with the Mendelian theory of inbreeding and crossbreeding in the broad sense. The same is true of experiments with self-fertilizing species. In these, heterosis from crossing varieties is much less and the simpler genetic differences make the assemblage of all favorable dominants in single lines less difficult than in crossbreeding species.

In experiments with tetraploids, maximum heterosis is not reached in F_1. Four-way and three-way crosses are much superior, and even F_2 shows little or no inbreeding decline. These results are expected if heterosis depends on the assemblage of dominant chromosomal genes.

CHAPTER 3

Inbreeding in Animals: Differentiation and Depression

The primary theoretical effect of inbreeding is the tendency toward *random* fixation in each line of some combination of the genes that are heterallelic in the foundation population. Differentiation among the lines is an immediate consequence. Inbreeding depression and heterosis in F_1 are secondary consequences of random fixation, in association with a prevailing positive correlation between dominance and favorable effect, and complementation of positive alleles. This chapter will continue the discussion of these phenomena, as manifested in studies of animals, based largely on sib-mated lines and crosses, but it will devote special attention to the differentiation among lines. I will begin with an account of inbreeding and crossbreeding experiments that I conducted many years ago with guinea pigs.

Inbreeding and Crossbreeding in Guinea Pigs

Experiments on the effects of inbreeding on guinea pigs were started in 1906 by G. M. Rommel of the U.S. Department of Agriculture. Thirty-five lines were started but only 23, maintained exclusively by sib mating, persisted long enough to give substantial data. One of these became extinct in 1911 and four more by 1915, when I took charge to analyze the results and continue the experiments (Wright 1922*b–d*). After 1917, new matings were made only in the five most vigorous strains in order to make room for more intensive study of these and for crossbreeding experiments. Lagging branches were disposed of in order to produce strains, each descended from a single mating with as many generations of sib mating back of it as practicable. At the end of 1915, when the average number of generations was 11.5, strain 2 had two large branches tracing to a common mating in the sixth generation. Strain 13 had branches tracing to one in the third, but soon all traced to one in the eighth. Strains 32, 35, and 39 traced to single matings in the eighth, apart from a few that were soon eliminated. The random bred foundation stock (B) was carried along with avoidance of even second cousin mating, but records began to be kept only in 1911. Through 1924, 29,310 inbreds and 5,105 control young had been recorded.

The vital characters studied consisted of two components of fecundity (young per litter and litters per year) and two components of perinatal viability (percentage born alive [BA] and percentage raised to 33 days of those born alive [RBA]). Studies were also made of birth weight, gain to 33 days, and, less extensively, the total growth curves. The interrelations of the perinatal characters were discussed in volume 1, chapter 14. The factors affecting the growth curve of an inbred strain (no. 2) were also discussed in that chapter. Data on the relations of weight at 33 days to size of litter and to year were given for the principal inbred strains for the years 1916–24 in volume 1, chapter 10. Data on other characters in these strains, especially white spotting, polydactyly, and otocephaly, were also given in volume 1.

Size of litter was found to have such important effects on the other perinatal characters that allowance seemed desirable. An index was based on the averages in litters of one, two, three, and four by weighting these in the ratio 1:3:4:2 in a study of decline and differentiation of the inbred strains in various aspects of vigor through 1920 (Wright 1922b,c). Figures 3.1–3.3 give the results (extended through 1924) for fecundity (average number of young per mating year), viability (percentage of all young born that reached 33 days), and weight at 33 days. Each of these is the product or sum of two components referred to above (Wright and Eaton 1929).

Unfortunately, these graphs are very irregular because of differences in the environmental conditions from year to year. The animals were maintained

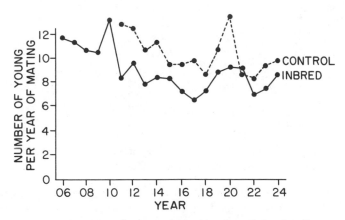

FIG. 3.1. Number of young per year of mating, from inbred and control guinea pigs from 1906 to 1924. From data of Wright and Eaton (1929).

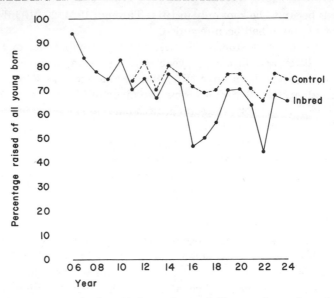

FIG. 3.2. Percentages raised to 33 days of age of all young born, from inbred and control guinea pigs from 1906 to 1924. From data of Wright and Eaton (1929).

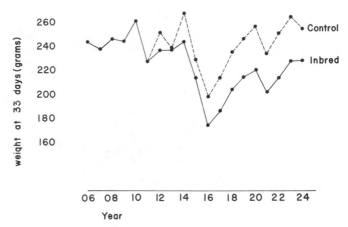

FIG. 3.3. Average weights at 33 days of inbred and control guinea pigs, from 1906 to 1924. From data of Wright and Eaton (1929).

in a barn that always reached unfavorably low temperatures in winter and often was disastrously hot for long periods in summer. The available green feed was inadequate in the winters of 1916–18. The records indicate that these years were the most unfavorable.

No certain conclusions can be drawn from the apparent downward trend of the inbreds. Valid comparisons can only be made between contemporaries. Unfortunately the control stock itself cannot be assumed to be wholly free from inbreeding depression. The number of pens devoted to it was presumably reduced during the expansion of the inbred strains, but how much is not clear until records began to be kept in 1911. From this time, the average number of breeding pairs was 35 with harmonic mean 33, indicating an effective population number of 66 and an increase in the inbreeding coefficient of 0.75% per generation. The average interval between generations of inbreds was about 10 months. The inbred strains reached an average of about 16 generations by the end of 1919 and 22 generations by the end of 1924. This leads to an average value of F of 84% for the period 1911–15, 95% for the period 1916–19, and 99% for the period 1920–24. The random bred matings were retained somewhat longer, giving an interval of about a year. The amounts of inbreeding in the above three periods would be about 5%, 10%, and 14%, respectively. Thus inbreeding of the inbred strains relative to the controls would have ceased in the second of those periods. Beyond 1924, F would rise more in the controls than in the inbreds (but probably with more effective balancing by selection for vigor).

Table 3.1 gives a comparison of the inbreds as a whole with the controls for the period 1911–15. There were 10,674 inbreds and 2,071 controls born in this period. The figures are the unweighted averages for the five years.

TABLE 3.1. Comparison of total inbred and control stocks of guinea pigs, 1911–15. Unweighted averages of annual averages.

1911–15	Size of Litter	Litters per Mating Year	% Born Alive (Index)	% Raised of Born Alive (Index)	Birth Weight (gm) (Index)	Gain (gm) (Index)	33-Day Weight (gm) (Index)
Control	3.03	3.67	87.0	85.6	87.2	150.2	237.4
Inbred (total)	2.54	3.36	85.7	84.2	84.7	147.5	232.2
Difference	0.49	0.31	1.3	1.4	2.5	2.7	5.2
Percent decline	16.2	8.5	1.5	1.6	2.9	1.8	2.2

SOURCE: Wright 1922b.

There is not much decline of the inbred strains relative to the controls in anything but the two components of fecundity, especially size of litter. The product of the first four entries, the effective fecundity, shows a very considerable decline, 25.6%.

Since there were some changes in the relative frequencies of the inbred strains, a better estimate of the decline per strain can be obtained by using the unweighted average of the annual *strain* averages (table 3.2).

TABLE 3.2. Comparison of unweighted averages of annual strain averages with control averages from the same data as those on which table 3.1 is based.

1911–15	Size of Litter	Litter per Mating Year	% Born Alive (Index)	% Raised of Born Alive (Index)	Young Raised per Mating Year	Birth Weight (gm) (Index)	Gain to 33 Days (gm) (Index)	33-Day Weight (gm) (Index)
Controls	3.03	3.67	87.0	85.6	8.28	87.2	150.2	237.4
Inbred (mean)	2.49	3.30	85.0	83.0	5.80	85.0	146.5	231.5
Difference	0.54	0.37	2.0	2.6	2.48	2.2	3.7	5.9
Percent decline	17.8	10.1	2.3	3.1	29.9	2.5	2.5	2.5

SOURCE: Wright 1922b.

The decline per inbred strain is slightly greater than that in the inbreds as a group in all respects but birth weight, and the decline in effective fecundity is nearly 30%.

In the period 1916–19—including the years 1916–18, which have the poorest records, due to a large extent to disastrous winters—it was desirable to make comparisons on the basis of seasons instead of whole years. The averages for the total inbred stock, now consisting 75% of the five best strains, were obtained for each of the 16 three-month periods (table 3.3). The frequencies of the controls born in each three-month period were used as weights for the inbred averages to estimate the averages for the contemporary inbreds. The viability percentages are here indexes for which litters of one, two, three, and four were weighted 1:3:3:1. The actual percentages raised are used in calculating young per mating year. There were 4,611 inbreds and 1,559 controls born in this period.

The birth weight, gain to weaning, and viability—both at birth and up to weaning—here show declines somewhat similar to those in fecundity, which are not very much changed. Since F was already 84% in 1911–15 (95% in

TABLE 3.3. Comparison of control guinea pigs (1916–19) with contemporary inbreds (three-month averages of the latter weighted by the numbers of controls).

1916–19	Size of Litter	Litters per Mating Year	% Born Alive (Index)	% Raised of Born Alive (Index)	Young Raised per Mating Year	Birth Weight (gm) (Index)	Gain to 33 Days (gm) (Index)	33-Day Weight (gm) (Index)
Controls	2.65	3.68	86.8	83.8	6.71	84.6	143.5	228.1
Inbred (contemporary)	2.32	3.22	78.4	73.3	4.19	77.3	120.7	198.0
Difference	0.33	0.46	8.4	10.5	2.52	7.3	22.8	30.1
Percent decline	12.5	12.5	9.7	12.5	37.6	8.6	15.9	13.2

SOURCE: Wright 1922b.

1916–19), the changes may be attributed to the excess sensitivity of the inbreds to more unfavorable conditions.

Conditions in the period 1920–24 (table 3.4) were somewhat better than in 1911–15, judging from the increase in weight of the controls, although not better with respect to fecundity and viability. There were 5,265 inbreds and 1,475 controls born in this period.

TABLE 3.4. Comparisons of unweighted annual strain averages of inbred guinea pigs with unweighted annual averages of controls (1920–24).

1920–24	Size of Litter	Litters per Mating Year	% Born Alive (Index)	% Raised of Born Alive (Index)	Young Raised per Mating Year	Birth Weight (gm) (Index)	Gain to 33 Days (gm) (Index)	33-Day Weight (gm) (Index)
Controls	2.83	3.52	87.0	84.0	6.62	91.2	160.0	251.2
Inbreds (means)	2.51	3.32	83.5	76.0	4.99	81.7	136.6	218.3
Difference	0.32	0.20	6.5	8.0	1.63	9.5	23.4	32.9
Percent decline	11.3	5.7	7.5	9.5	24.6	10.4	14.6	13.1

SOURCE: Wright and Eaton 1929.

The averages used for the controls in this case are the unweighted averages for each of the five years. Those for the inbreds are the unweighted averages for the five strains in each of the five years. The averages for the actual totals do not differ appreciably. The birth weight and gain show about the same percentage total decline from the controls as in 1916–19, the viabilities at birth and up to weaning decline a little less. Size of litter deviated from the control about as much as in 1916–19 (considerably less than in 1911–15), while frequency of litters deviated much less than in either preceding period. No further decline from the controls is expected where F rises only from 95% to 99% in the inbreds and is probably also rising by about 4% in the controls. The agreement between the latter two periods is perhaps as much as can be expected in view of environmental conditions, which were probably unfavorable in somewhat different ways. The important point is that inbreeding (approaching 100% in comparison with about 14% in the controls) causes a considerable decrease in early growth (some 13%) and decreases in both elements of fecundity and in both elements of viability, which combine to give a decrease in the number of young raised per year of about 25%.

During the period 1916–19, comparisons were made between various crossbreds and the contemporary inbreds in the way described for the random bred stock (Wright 1922c,d). Most of the possible first crosses were made among the five main inbred strains and averaged (C0). Three-way crosses were made between C0 males and females of a third inbred strain (CA). The reciprocal (AC) was also made. Four-way crosses were made between C0 males and females, involving four different inbred strains (CC). There was also renewed sib mating from the first cross (C1), and this was carried to the next generation (C2). During this period 4,611 inbreds were born, and 1,334 in C0, 410 in CA, 571 in AC, 617 in CC, 629 in C1, and 182 in C2.

Some of the results are shown in table 3.5 and figures 3.4 and 3.5 as percentage deviations from the contemporary inbreds. The figures include the percentage deviation of the five principal inbred strains, the miscellaneous inbred strains (M), the various crosses, and the control stock (B). It is to be noted that the percentage deviations shown here differ from those in the preceding tables in being from the inbreds instead of from the controls. The larger differences were highly significant (Wright 1922d).

TABLE 3.5. Percentage deviations of inbred strains of guinea pigs and crossbreds from contemporary inbreds (1916–19) by the method described in the text.

	Size of Litter	Litters per Mating Year	% Born Alive (Index)	% Raised of Born Alive (Index)	Young Raised per Year	Weight at Birth (gm) (Index)	Gain (gm) (Index)	Weight at 33 Days (gm) (Index)
Controls	+14.2	+14.3	+10.7	+14.3	+60.1	+10.8	+18.9	+15.2
2	−3.4	+13.2	−4.7	+3.9	+13.1	−9.4	−19.4	−14.4
13	+7.7	0	−1.6	−1.9	+2.1	+3.4	+13.2	+10.5
32	−6.4	+1.5	+2.8	−8.1	−7.8	+2.3	+0.3	+0.5
35	+4.3	+9.8	+0.9	+5.6	+16.1	+5.8	+11.8	+9.0
39	+0.9	−8.5	+5.7	−7.6	−12.0	−4.6	−1.1	−2.6
Other (M)	−1.8	−11.9	+1.5	−2.8	−15.2	+2.8	+0.6	+1.0
C0	−1.7	+0.9	+1.9	+14.4	+9.5	+1.7	+13.2	+9.0
CA	+3.7	+19.5	+2.6	+10.5	+47.4	+3.6	+11.7	+8.6
AC	+28.6	+9.5	+7.4	+9.9	+63.7	+8.8	+11.9	+10.5
CC	+14.1	+36.0	+8.0	+16.4	+82.5	+12.9	+21.0	+16.6
C1	+10.5	+33.5	+10.5	+12.2	+73.1	+10.7	+15.9	+12.5
C2	+4.3	+25.3	+4.0	+8.4	+43.7	+6.7	+12.7	+9.8

SOURCE: Wright 1922d.

Birth weight (here of all young born) clearly depends largely on the breeding of the dam (inbred in C0 and CA; first cross in AC, C1, and CC; and 50% inbred in C2). Crossbreeding of the dam essentially restores the level of the control stock B.

The gain from birth to weaning (33 days) depends, on the other hand, largely on the inbreeding of the individuals, since the averages for C0 and CA are significantly above that of the inbreds. There is probably some increase from crossbreeding of the dam (average of CC and AC above that of C0 and CA, and C1 much higher than expected if dependent wholly on its own breeding). The average for C2 is, however, also much higher than expected, but as noted it is the least reliable statistically of the crossbreds. Crossbreeding restores the level to that of the controls.

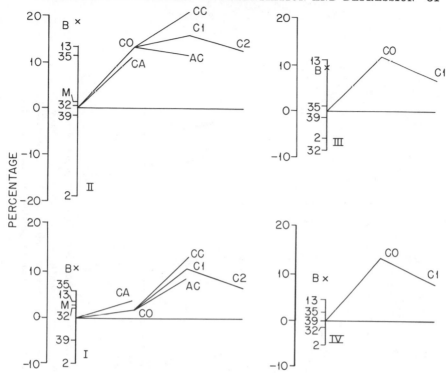

FIG. 3.4. The percentage deviations from the contemporary inbred average (1916–19) in birth weight of all young born (I), and in gain to 33 days (II), among five inbred strains, miscellaneous inbred strains (M), controls (B), and various crossbreds: CO(= F_1), CA(= CO ♂ × unrelated inbred ♀), AC(= inbred ♂ × unrelated CO ♀), CC(= CO ♂ × unrelated CO ♀), C1(= CO × CO littermates), C2(= C1 × C1 littermates); in weights of males (III) and of females (IV) at one year of age. Reprinted, by permission, from Wright (1960b, fig. 5).

The weights of males and females at one year of age show the first crosses considerably beyond the controls while C1 loses half of the F_1 gain as expected, assuming complete dependence on the breeding of the individual.

The frequency of litters, not surprisingly, shows no improvement in C0 in which both parents are inbred, but in three-strain crosses shows twice as much improvement from crossbreeding of the sire (CA) as of the dam (AC). If both parents are first crosses (CC, C1), the improvement over the inbreds is about the sum of those in AC and CA. In this case, there is improvement far beyond the level of the controls. There is some falling off in C2, although less than expected.

The absence of significant differences among the inbreds, C0, and CA in

52

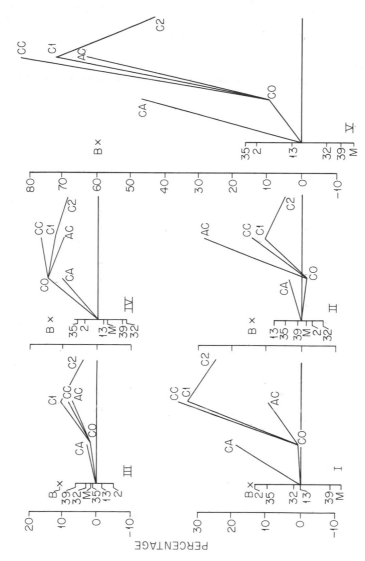

Fig. 3.5. The percentage deviations from the contemporary inbreds, in litters per year (I), size of litter (II), percentage born alive (III), percentage raised of those born alive (IV), both percentages corrected for size of litter, and the composite variable; young raised per mating year (V). Symbols as in figure 3.4. Reprinted, by permission, from Wright (1960b, fig. 6).

size of litter indicated dependence on the dam. This is in accord with an unpublished study by Haines (cf. Wright 1934b) of counts of corpora lutea and of fetuses in 338 females from the colony, largely inbreds. Only 8.7% of the corpora lutea were unaccounted for and there were only 7.9% dead fetuses. This indicates that size of litter depends largely on amount of ovulation. There is, however, a highly significant excess of AC over CC and C1, all with C0 dams. This is probably because of a negative relation between amount of ovulation and the interval since the preceding litter, which is less than a day where litters succeed each other as rapidly as possible. The much lower frequency of litters from AC as compared with CC and C1 causes them to be considerably larger when they do occur. The control level is restored in CC and C1 and is greatly exceeded in AC. Renewed inbreeding of the dam in C2 causes loss of half the improvement of C1 as expected.

The results for percentage born alive and percentage raised of those born alive are rather closely parallel to those for birth weight and gain to weaning, respectively (except in ranking of the inbred strains, not here under consideration). This is as expected if the percentage born alive is largely a maternal character, while percentage raised of those born alive depends some 75% on the breeding of the individuals.

Finally, the effective productivity (the number of young raised per mating year) is a resultant of the preceding four characters. These agree well enough to bring about very great percentage increases over the inbreds in most cases. There is not much improvement (10%) in C0 in which the only component that improves appreciably is percentage raised of those born alive. There is a 47% improvement in CA in which there is also much improvement in the frequency of litters. The improvement in CC, AC, and C1, with notable improvement in all four components, averages 73%, well above that of the controls (60%). Table 3.6 compares the four-way cross (CC) as a reconstructed random bred strain, with the contemporary inbreds.

TABLE 3.6. Comparison of four-way crosses (CC) (1916–19) with contemporary inbreds (three-month averages of latter weighted by the numbers of four-way crosses). Similar to table 3.3 except that CC appears instead of controls.

	Size of Litter	Litters per Mating Year	% Born Alive (Index)	% Raised of Born Alive (Index)	Young Raised per Mating Year	Birth Weight (gm) (Index)	Gain to 33 Days (gm) (Index)	33-Day Weight (gm) (Index)
CC	2.65	4.34	84.1	84.0	7.61	84.5	146.0	230.5
Inbreds	2.33	3.18	77.7	72.1	4.17	77.6	120.2	197.8
Difference	0.32	1.16	6.4	11.9	3.44	6.9	25.8	32.7
Percent	12.1	26.7	7.6	14.2	45.2	8.1	17.7	14.2

SOURCE: Wright 1922d.

The figures are not very different from those comparing the inbreds with the controls, except for the very much greater difference in frequency of litters (26.7%) and its derivative, the number of young raised per mating year (45.2%). The inbred strains to which CC traces were, however, somewhat superior to the array of random inbred strains derived from the controls because of natural selection among and within them. Their superiority seems to be manifested largely in a great improvement over the controls (at least over those of 1916–19, which are themselves slightly inbred) in the regularity in producing litters. On the whole, the figures support the view that heterosis is here merely the converse of the inbreeding decline.

An important question to consider is how far this decline was in separate aspects of vigor rather than in general vigor as an entity. Correlations were calculated between the means of the 22 strains present in the two early periods, corrected as described for direct effect of litter size (table 3.7).

TABLE 3.7. Correlations between components of vigor in 22 strains, in 1906–10 and 1911–15.

	Strain Averages Correlated	1906–10	1911–15
Components of fecundity	Size of litter–litter/year	+0.04	−0.03
Components of viability	% BA–% RBA	+0.03	+0.30
Components of growth	Birth weight–gain to 33 days	+0.75*	+0.59*
Fecundity-viability	Size of litter–% BA	−0.10	+0.12
	Size of litter–% RBA	+0.28	+0.17
	Litters/year–% BA	+0.04	+0.01
	Litters/year–% RBA	0.00	+0.23
Fecundity-growth	Size of litter–birth weight	+0.26	+0.62*
	Size of litter–gain to 33 days	+0.37	+0.62*
	Litters/year–birth weight	−0.05	−0.34
	Litters/year–gain to 33 days	+0.09	−0.22
Viability-growth	% BA–birth weight	−0.08	+0.01
	% BA–gain to 33 days	+0.03	−0.28
	% RBA–birth weight	+0.07	−0.21
	% RBA–gain to 33 days	+0.02	−0.23

SOURCE: Wright 1922c
NOTE: Born alive, BA; raised of born alive, RBA.
* Significant at 0.05 level.

There are only four correlations that are significant at the 0.05 level or less (denoted by asterisks). It is not surprising that birth weight and gain to 33 days are very significantly correlated in both periods. The only other significant correlations are between these measures of growth and size of litter (in 1911–15). This was positive and relatively high even though not at the 0.05 level in 1906–10. There is probably a physiological relation between growth capacity and amount of ovulation since similar correlations have been observed in rabbits (Gregory 1932) and mice (MacArthur 1944b; Falconer 1960a). The small values and inconsistencies among the remaining correlations, and the negative values of many of them, give no support for the concept of differentiation in general vigor. Evidence from later trends indicates, however, that differentiation in general vigor was not wholly absent.

Comparisons of the averages of the 22 inbred strains in the two early periods, 1906–10 and 1911–15, showed that there was much more variability among the strains than could be accounted for by accidents of sampling, and that this genetic differentiation was 50% greater in the second of these periods than in the first. While the strain characters were not fully established, even in the second period, the correlations between the strain averages are of interest (table 3.8).

TABLE 3.8. Correlations between strain averages in 22 strains, in 1906–10 and 1911–15.

Fecundity		Vitality		Growth	
Litters per mating year	+0.25	% BA	+0.51	Birth weight	+0.65
Size of litter	+0.65	% RBA	+0.32	Gain to 33 days	+0.64
Young per mating year	+0.41	% raised	+0.36	33-day weight	+0.60

SOURCE: Wright 1922c.
NOTE: Born alive, BA; raised of born alive, RBA; % raised refers to all young born.

All are positive and those above 0.38 are significant at the 0.05 level. This includes the growth variables, size of litter, young per mating year, and percentage born alive.

The differentiation of the major inbred strains in 1916–19 is shown in figures 3.4 and 3.5. Because of the independence in the main of ranking in the various respects, there is not much compounding of differences with respect to the compound character, number of young raised per mating year. The superiority of the controls in all of the four components leads, on the other hand, to very great superiority in the compound vairable.

More detailed evidence on the differentiation was obtained by applying Student's test to paired comparisons among the five main inbred strains and the controls for the nine years, 1916–24. An example of this test for weight at 33 days of the two heaviest inbred strains (no. 13 and 35) was given in volume 1, table 9.10. The difference was significant ($P = 0.011$), and in this case slightly more significant than the difference of the logarithms ($P = 0.014$).

In general, however, the latter was more significant and is used here (see vol. 1, tables 10.16 and 10.17).

The mean differences (Δ) in all pairs and the probabilities (P) from Student's formula are given for five characters in table 3.9. In the case of 33-day weight (logarithms), all differences are significant except that between strain 13 and the controls (B). The close parallelism of B and four of the inbred strains over these nine years was shown in volume 1, figure 14.3, and of the five inbred strains in volume 1, figure 10.7.

In size of litter, the controls were significantly superior (at the 0.05 level) to all of the inbred strains except no. 39. Seven of the ten comparisons between inbreds were significant. The moderately good parallelism in this case is also shown in volume 1, figure 14.3.

In percentage raised to weaning of young born alive, the controls were significantly superior to all but strain 2. This was significantly superior to all other inbreds. There were two other significant differences (strains 35 and 13 superior to 39).

In frequency of litters, the controls were significantly superior only to strains 32 and 39. There were six significant differences between inbred strains.

Finally, in percentage born alive, the controls were significantly superior to all but strain 32, but there were no differences among the inbreds significant at the 0.05 level in contrast with the significant correlation of +0.51 between the 22 strain averages of the earlier periods.

This does not mean that there were no genetic differences among the five inbred strains of table 3.9 in this respect. While there was no significant difference between the rising trends of controls and inbreds as a group during these nine years, there were some highly significant differences in trend among the latter. The differences in trend, calculated by Fisher's (1925) extension of Student's method, are shown in table 3.10 together with probabilities of origin by accidents of sampling.

There are no significant differences in trend in 33-day weight. In size of litter, on the other hand, there are six significant differences. On the whole, strains 39, 13, and 2 show rising trends in this respect, and 32 and especially 35 falling ones, relative to the average (essentially shown by B).

The only significant differences in trend with respect to the percentage raised to weaning of those born alive was the rise of 2 relative to both 35 and B. The averages were, however, rather erratic (especially 39).

In the case of frequency of litters, strain 13 showed a significant rising trend relative to all others including the controls.

Finally, in percentage born alive, there were four significant differences. These were due to significant rise of 2 and 32 relative to 39 and 35.

TABLE 3.9. Mean differences and their probabilities by Student's method.

1916–24	Size of Litter		Litters per Year		% Born Alive (Index)		% Raised of Born Alive (Index)		Weight at 33 Days (Log Index)	
	Δ	P	Δ	P	Δ	P	Δ	P	Δ	P
B-2	+0.38	0.000	−0.10	0.43	+7.1	0.026	+1.5	0.50	+0.132	0.000
B-13	+0.10	0.016	−0.01	0.85	+7.4	0.029	+7.6	0.003	+0.005	0.28
B-32	+0.59	0.000	+0.41	0.006	+3.2	0.17	+12.1	0.001	+0.055	0.000
B-35	+0.32	0.001	+0.05	0.73	+7.5	0.011	+7.4	0.006	+0.025	0.001
B-39	+0.10	0.065	+1.10	0.001	+6.1	0.050	+17.6	0.002	+0.068	0.000
2–13	−0.27	0.000	+0.08	0.62	+0.3	0.89	+6.1	0.001	−0.127	0.000
2–32	+0.21	0.007	+0.50	0.000	−3.9	0.10	+10.6	0.001	−0.077	0.000
2–35	−0.06	0.49	+0.15	0.052	+0.4	0.92	+5.9	0.043	−0.107	0.000
2–39	−0.19	0.017	+1.10	0.000	−1.0	0.74	+16.1	0.002	−0.064	0.000
13–32	+0.48	0.000	+0.42	0.058	−4.2	0.082	+4.5	0.087	+0.050	0.000
13–35	+0.22	0.033	+0.07	0.73	+0.1	0.99	−0.2	0.95	+0.020	0.014
13–39	+0.09	0.31	+1.11	0.018	−1.3	0.51	+10.0	0.020	+0.063	0.000
32–35	−0.27	0.001	−0.36	0.009	+4.3	0.11	−4.6	0.15	−0.030	0.000
32–39	−0.40	0.004	+0.69	0.003	+2.9	0.25	+5.5	0.15	+0.014	0.043
35–39	−0.13	0.29	+1.05	0.002	−1.4	0.61	+10.1	0.004	+0.043	0.000

SOURCE: Wright and Eaton 1929.

TABLE 3.10. Differences in trend and their probabilities by Student's method.

1916–24	Size of Litter		Litters per Year		% Born Alive (Index)		% Raised of Born Alive (Index)		Weight at 33 Days (Log Index)	
	Δ	P	Δ	P	Δ	P	Δ	P	Δ	P
B-2	−0.02	0.29	−0.04	0.36	−1.4	0.18	−2.0	0.003	+0.002	0.17
B-13	−0.02	0.11	−0.21	0.011	−0.2	0.89	−1.3	0.072	+0.001	0.61
B-32	+0.03	0.22	+0.02	0.74	−0.5	0.58	−1.5	0.12	+0.003	0.064
B-35	+0.05	0.034	−0.02	0.75	+1.6	0.060	+0.2	0.79	+0.003	0.062
B-39	−0.05	0.13	+0.06	0.49	+1.3	0.24	−0.1	0.95	+0.001	0.61
2–13	−0.003	0.86	−0.17	0.000	+1.2	0.082	+0.7	0.10	−0.002	0.56
2–32	+0.05	0.02	+0.03	0.34	+0.9	0.29	+0.5	0.60	+0.001	0.65
2–35	+0.07	0.007	+0.03	0.35	+3.0	0.007	+2.2	0.008	+0.001	0.45
2–39	−0.04	0.16	+0.10	0.21	+2.7	0.006	+1.9	0.20	−0.001	0.61
13–32	+0.05	0.07	+0.19	0.000	−0.3	0.71	−0.2	0.98	+0.002	0.43
13–35	+0.07	0.013	+0.19	0.000	+1.8	0.17	+1.5	0.084	+0.003	0.33
13–39	−0.03	0.35	+0.27	0.012	+1.5	0.051	+1.2	0.27	0.000	0.94
32–35	+0.02	0.33	−0.003	0.95	+2.1	0.007	+1.7	0.13	0.000	0.95
32–39	−0.08	0.019	+0.08	0.26	+1.8	0.039	+1.4	0.34	−0.002	0.31
35–39	−0.10	0.005	+0.08	0.40	−0.3	0.76	−0.3	0.76	−0.002	0.28

SOURCE: Wright and Eaton 1929.

The changes in trend are hardly distributed among the strains at random. The number of cases of significantly rising trends ($+$), nonsignificant differences ($\pm$, excluding weight), and falling trends ($-$) are shown in table 3.11.

TABLE 3.11. Numbers of significant differences in trend in table 3.10 (excluding weight).

Strain	$+$	$\pm$	$-$
13	6	14	0
2	6	13	1
B	1	17	2
32	2	15	3
39	2	15	3
35	0	12	8

These figures, relating to the two components of fecundity and the two components of perinatal viability, look very much as if changes were occurring in a general response to environmental conditions. Strain 35 did relatively well under the conditions of the earlier years, under which strains 13 and 2 did relatively poorly, and conversely under the conditions of the later years. Strains 32 and 39 showed no such systematic tendencies. It should be said, however, that strain 39 did so very poorly in some of the later years in frequency of litters and both viability percentages that the very magnitude of the fluctuations prevented the demonstration of significance in differences in trend.

An overall evaluation of the relative importance of additive genetic differences, differences in annual conditions in the period considered, and residual factors (which must consist largely of genotype-environmental interactions) may be obtained by analysis of variance of the annual averages. Correction of the formulas as printed in volume 1, set 12.53 may be noted here. The first expression given there for the sums of squares for rows (R) and columns (C) should obviously be multiplied by C and R, respectively. In the analysis of the mean squares there is a confounding of the components for rows, $\sigma^2_{M(i\cdot)}$, and for columns, $\sigma^2_{M(\cdot j)}$.

Weight at 33 days has been adjusted by regression to a litter size of 2.50. The viability percentages are again averages for litters of one, two, three, and four, weighted by $1:3:3:1$. In this case, 80 gm has been subtracted from the 33-day weight before taking logarithms in order to maximize the parallelism of the strains (see vol. 1, table 10.17 and fig. 10.7). The overall productivity (number of young raised per year), which is the product of litters per year, litter size, and the two unadjusted viability percentages, is included among the characters (table 3.12).

TABLE 3.12. Mean squares: annual averages (1916–24) of five inbred strains.

	DF	Size of Litter	Litters per Year	% Born Alive (Index)	% Raised of Born Alive (Index)	Young Raised per Year	Weight at 33 Days – 80 gm (Log Index)
Inbred strains	4	0.3114‡	2.168‡	16.7	152.7‡	9.81‡	0.06011‡
Years	8	0.0987‡	0.368*	56.1†	165.8‡	5.26‡	0.01861‡
Residual	32	0.0269	0.151	14.7	13.0	0.71	0.00039

SOURCE: Data from Wright and Eaton 1929.
* $0.01 < P \lesssim 0.05$.
† $0.001 < P \lesssim 0.01$.
‡ $P \lesssim 0.001$.

The differentiation among the years was significant at the 0.001 level in all cases except percentage born alive (0.01 level) and litters per year (0.05 level).

The differentiation among the strains was significant at the 0.001 level in all cases except in percentage born alive, in which there was no significance. It should be recalled that these five inbred strains were the most vigorous of the original 23, so that the differentiation among all strains was undoubtedly much greater in most respects.

The variance components are given in table 3.13 as percentages in order to make comparisons between the characters.

TABLE 3.13. Variance components of the indicated characters, as percentages, based on table 3.12.

	Size of Litter	Litters per Year	% Born Alive (Index)	% Raised of Born Alive (Index)	Young Raised per Year	Weight at 33 Days – 80 gm (Log Index)
Inbred strains	43.4	53.6	0.9	26.3	38.5	62.2
Years	19.7	10.4	35.6	51.8	34.6	34.1
Residual	36.9	36.0	63.5	21.9	26.9	3.7
Total	100.0	100.0	100.0	100.0	100.0	100.0

This shows the remarkable contrast between weight at 33 days and percentage born alive. The former shows the greatest importance of additive heredity and the least of residual factors, while the situation is the opposite with percentage born alive. All components are important in the other cases, except for the rather small variance of yearly difference in the litters per year.

Resistance to Tuberculosis

A shifting relation of genotype to character was illustrated by experiments on length of life after inoculation with tuberculosis. Dr. Paul A. Lewis, who was experimenting along this line at the Phipps Institute, proposed a test of the inbred strains of the U.S. Department of Agriculture. Inbreds, crossbreds, and random breds were sent to him and he returned data from his earlier experiments (groups I and II) for statistical analysis. Those from the first five experiments (412 animals) gave highly significant statistical results (Wright and Lewis 1921). The analysis of the next three experiments, combined with the last of those referred to above, which was similar in mode of inoculation and relatively long average duration of life, yielded even more significant statistical results. These were presented at the meeting of the American Association for the Advancement of Science in 1921, but were published only in abstract form (Wright and Lewis 1922). This abstract gives the principal results of the statistical analysis up to this point.

The resistance to tuberculosis has been tested in over eleven hundred guinea pigs belonging to five closely inbred families. It has been found that sex and even threefold differences in age, rate of gain and weight have only a slight effect on length of life after inoculation. In striking contrast to these negative results, are the great differences among the inbred families, differences which are not correlated with the differences in fertility, weight and vitality. The progeny of crosses are in general at least equal to the better of the two parental families. Resistance is thus dominant over susceptibility. There is equal transmission by sires and dams to sons and daughters. In particular crosses, the average of the progeny is consistently superior to either parental line, indicating that the latter are susceptible for different reasons, each being able to supply the dominant resistance factor lacking in the other. In the whole crossbred stock, over 30 per cent of the observed variation is determined by the amount of blood of the best inbred family, as compared with less than 10 per cent due to age, weight and rate of gain combined, and leaving about 60 per cent due to conditions at or following inoculation.

I continued to send animals for Dr. Lewis to experiment with at the Phipps Institute and later at the Rockefeller Institute in Princeton, including breeding stock, but no further statistical analyses were made until many years later. After some time, however, Dr. Lewis informed me that inconsistencies in the ranking of the inbred strains had appeared and later that the original order had broken down completely in the experiments at the Rockefeller Institute. He suspected that this might be due to a drastic change in the ration. That at the Phipps Institute had been adequate for growth but not for reproduction. That at the Rockefeller Institute gave better growth and permitted reproduction. In his last series of experiments, he compared lots maintained on the two rations and obtained approximate return to the original ranking on the Phipps ration, both from animals sent from the laboratory of the U.S. Department of Agriculture at Beltsville, Maryland, and ones he bred himself.

After his untimely death in 1929, the whole body of data in the 25 usable experiments (over 4,000 animals) was sent to me for analysis. Because of the extreme heterogeneity in all factors, the analysis was necessarily long and complicated and the conclusions were still tentative in important respects. Not surprisingly, the resulting manuscript proved to be unpublishable. There was, however, no doubt of the existence of important genetic differences, even though differences in manifestation under different conditions had to be postulated. For this very reason, it seems desirable to present here the main statistical results on the strain and cross differences.

It will be convenient to label the experiments in order of time by the letters of the alphabet and put them into six groups, in spite of much residual heterogeneity. Experiments within groups often differed in the culture used (human or bovine) and in the mode of inoculation (experiment J intravenous in contrast with intraperitoneal in A–D in group I, both subcutaneous and intraperitoneal inoculation in group V, and these and intracutaneous in group VI). Table 3.14 gives data on the six groups. (Experiment V was not usable.)

TABLE 3.14. Conspectus of groups of experiments on length of life after inoculation with tuberculosis.

Group	Experiments	No. Treated	Ration	Mode of Inoculation	Mean Age (days)	Mean Weight (gm)	Mean Days Lived
I	A–D, J	267	Phipps	IP, IV	70	272	28
II	E–H	1,151	Phipps	SC	104	359	56
III	I, K	429	Phipps	SC	109	319	33
IV	L–O	934	Phipps	IC	106	354	76
V	P–T, Y	1,075	Rockefeller	SC, IP	126	428	58
VI	U, W–X, Z	552	Phipps	IP, IC, SC	146	405	92

It seemed best to use the logarithms of the days lived, excluding those animals that died in the first 15 days (of causes other than tuberculosis). Increments or decrements were calculated for each experiment within a group by the method of least squares in order to minimize differences for the same strain within a group. The adjusted logarithmic means are shown for each strain or cross in table 3.15 and in terms of geometric means in figures 3.6–3.9.

TABLE 3.15. Numbers and means of logarithms of days lived for each strain and cross in each group of experiments.

Strain	I No.	I Log GM	II No.	II Log GM	III No.	III Log GM	IV No.	IV Log GM	V No.	V Log GM	VI No.	VI Log GM
35	38	1.452	114	1.788	37	1.503	22	1.988	119	1.732	87	1.976
2	41	1.389	145	1.712	47	1.498	43	1.841	95	1.757	116	1.907
B	22	1.437	243	1.677	86	1.500	83	1.801	135	1.688	54	1.883
32	12	1.339	60	1.634	8	1.460	20	1.984	35	1.693	59	1.835
13	50	1.386	64	1.603	21	1.462	36	1.812	136	1.742	108	1.864
39	1	1.303	12	1.460	11	1.405	14	1.603	4	1.733	0	···
35 × 2			53	1.866	25	1.568	9	2.005	32	1.843		
× 32			31	1.843	30	1.544	17	1.980	6	1.828		
× 13			35	1.855	42	1.527	14	1.961	10	1.895		
× 39			13	1.817	6	1.532	8	1.952	15	1.915		
2 × 32			31	1.761	23	1.523	25	1.972	50	1.845		
× 13			86	1.732	23	1.510	7	1.931	52	1.795		
× 39			22	1.702	9	1.469	19	1.861	20	1.830		
32 × 13			43	1.738	28	1.516	9	1.917	26	1.850		
× 39			17	1.615	20	1.469	17	1.856	14	1.792		
13 × 39			14	1.598	5	1.460	9	1.814	13	1.850		

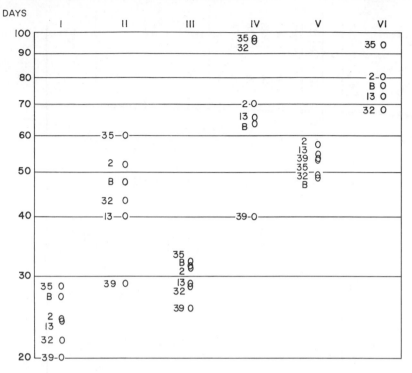

FIG. 3.6. Conspectus of lengths of life after infection with tuberculosis (log scale) of five inbred strains of guinea pigs and controls, B, in six sets of experiments. Deaths in first 15 days excluded. Means of experiments within groups adjusted. From unpublished data of Dr. Paul A. Lewis.

The average standard deviation of logarithms within the inbred strains of group I was 0.068, from which it may be seen that, in spite of much overlap, there were highly significant differences among the means. There was much more spread of the means in group II (that to which the above abstract refers) in spite of the logarithmic transformation. The standard deviation within strains was 0.1324, with about twice as much spread as in group I but with numbers so much larger that the significance of differences is much greater. The order of the means in the two groups agrees approximately: 35 > (2, B) > (32, 13) > 39.

There were only a few crossbreds (22) in group I, although enough to bring out the superiority of those involving strain 35. There were many in group II, including considerable numbers from all ten crosses (reciprocals not distinguished here). The superiority of those involving strain 35 over all others, and of those involving strain 2 over 32 × 39 and 13 × 39 is apparent.

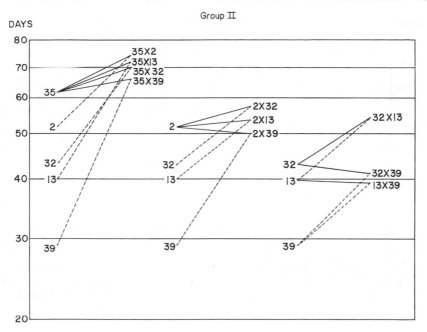

FIG. 3.7. Lengths of life (log scale) of inbred strains of guinea pigs and crosses in experiments of group II (means adjusted). From unpublished data of Dr. Paul A. Lewis.

The heredities of 32 and 13 are clearly complementary, and this seems also to be the case to a lesser degree where either 35 or 2 is crossed with any strain other than 39, which is the lowest, both in itself and in the crosses.

There was very little spread under the conditions of group III, but the order of the strains (inbreds and control) was the same as in group I and close to that in group II. The order of the crosses was surprisingly similar to that in group II, considering the smallness of the spread. The general conclusions are the same as above.

The spread was the greatest in group IV and the order was somewhat similar to those of the preceding groups with one notable exception, the relatively long mean life of strain 32, here practically equal to that of strain 35. The four crosses involving 35 are high as a class, but do not exhibit any heterosis as in all other cases. Those involving strain 2 but not 35 are lower as a class, with more heterosis in 2 × 13 than in the preceding cases. The cross 2 × 32 is high in rank but not as high as 32; the crosses of 32 with 13 and 39 show little more than semidominance. Thus the very high rank of 32 itself is only partially reflected in its crosses. The three crosses involving 39 but not 35 are,

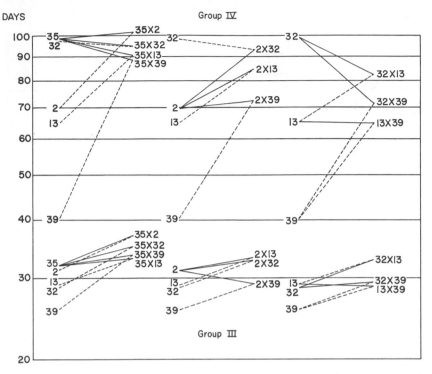

FIG. 3.8. Lengths of life (log scale) of inbred strains of guinea pigs and crosses in experiments of groups III and IV (means within each adjusted). From unpublished data of Dr. Paul A. Lewis.

as in the previous groups, the lowest. The only difference in the data that may be suggested as the cause of the anomalously high rank of 32, and to a lesser extent of its crosses, is the systematic difference between the mode of inoculation in this experiment (IV), intracutaneous instead of subcutaneous as in II and III or intraperitoneal or intravenous as in group I. Perhaps spread of infection in 32 was delayed by some genetically determined property of its skin. This hypothesis was not supported, however, in the only other experiment (W) with intracutaneous inoculation in which 32 was slightly inferior to B, 13, and 32 and considerably inferior to 35.

Group V was the one in which the ranking of the inbred strains in the earlier groups completely broke down. There is no difference between strains 35 and 39, which in other cases lived the longest and the shortest times, respectively. Here both are intermediate. Strain 32, which lived practically as long as 35 in group IV, is here the shortest lived. This is the group that

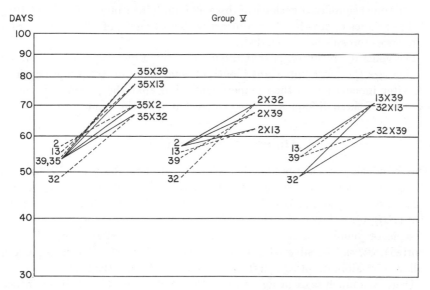

FIG. 3.9. Lengths of life (log scale) of inbred strains of guinea pigs and crosses in experiments of group V (means adjusted). From unpublished data of Dr. Paul A. Lewis.

was on a much better ration (Rockefeller) before inoculation, suggesting that the usual ranking had more to do with degree of success in coping with an inadequate ration while infected than with resistance to tuberculosis in a more specific sense.

This seems to be supported by the near-approach of group VI, on return to the Phipps ration, to the ranking in groups I–III (with the unfortunate qualification that the lowest strain [39] was not represented). The relatively long life of all strains in group VI, in comparison especially with groups I and III, indicates that this ranking was not associated with length of life.

The results from the crosses in group V do not differ as radically from the usual ranking as did those from the strains themselves. They all show an unusual amount of heterosis, which is rather surprising for the group on the best ration. The four involving 35 have the highest average of any set of the four involving a particular strain, as usual. The crosses involving 39, as well as 39 itself, did relatively well, in contrast with their unequivocal inferiority in all other cases. The reasons for the peculiar results in group V are obscure.

Analysis of all later groups with respect to the relation of days lived to age and weight at inoculation confirmed the conclusions from groups I and II that no function of these had much significance. The deviation of weight at inoculation from the typical growth curve of the strain to which it belonged was the most significant variable of this sort found; it accounted for about 10% of the variance in days lived, in contrast with an average of 20% due to strain differences (crosses here excluded).

The general conclusion is that differentiation of genotypes of the inbred strains was the most important identifiable factor in determining length of life after inoculation in these experiments, but that the phenotypic manifestation of these differences varied somewhat with the conditions.

Differentiation in Color

The foundation stock was homallelic in several of the known color factors: all animals were tortoiseshells, $e^p e^p$ (except for albinos), all dark-eyed, PP (again excepting albinos), and with black or sepia in eumelanic areas of the coat (BB), not grizzled and thus $GrGr$, and not showing exceptional dilution of sepia or yellow and thus $DmDm$. This stock was heterallelic in S, s (ss spotted); Si, si ($sisi$ silvered); A, a (A agouti); C, c^k, c^a (c^k dark sepia and yellow, $c^a c^a$ albino); and F, f (ff dilute yellow) (Wright 1922c).

There was much segregation in the early generations of most of the inbred strains, but lines soon became established that were homallelic, although there seems to have been no selection, except perhaps against albinism. With the gradual disposal of lagging branches in each strain, each became characterized by a single genotype. With one exception, no. 34 $SSe^p e^p$, all became tricolored, $sse^p e^p$. Most of the strains were agouti-red-white, $AACC$ (for example, 32, 39); black-red-white, $aaCC$ (for example, 2, 13); yellow agouti-yellow-white, $AAc^k c^k$ (for example, 35); or black-yellow-white, $aac^k c^k$ (for example, 18). Only two strains, 34 and 13, were certainly silvered, $sisi$, but it is possible that others were weakly silvered. Gene f was first demonstrated in the control stock, only after all but five of the strains had been disposed of. It may have been confused with c^k in ones in which tests were not made by crossing with albinos, $c^a c^a FF$ ($Cc^a Ff$ red spotted, $c^k c^a FF$ pale yellow spotted).

The automatic fixation of some combination of Mendelian genes is, of course, according to expectation. Of more interest was the fixation of some grade of intensity due to multifactorial systems that resisted Mendelian analysis. Thus strain 32 had uniquely dark mahogany red spots with twice as much phaeomelanin as strain 2, which had typical red spots; and much more than twice as much as in strain 38, which had light red spots, although all were $CCFF$ with respect to pertinent known genes. Similarly, there was a great difference at birth between the light red spots of strain 18 and the yellow spots of 35, although both were $c^k c^k FF$ by tests with albinos. In this case both faded to about the same light yellow in older animals.

The distributions of amounts of white in the piebald pattern in certain of the strains have been illustrated in volume 1, chapters 6 and 9. All varied widely but differed much in median. In their later histories, strains 38 and 39 varied from a trace of white to about 90%, with medians at less than 20%. Strains 2, 32, and 13 ranged from about 40% to 100% (black-eyed white) with medians over 90% white, that of strain 13 being 98%. Strain 35 was typical of those with intermediate medians (in this case 65%), but varied from a trace of white to black-eyed white. The distributions were consistent from year to year, and selection of extremes for mating had no effect (Wright 1920; Wright and Chase 1936).

Differentiation in Morphological Traits

There were differences in conformation in the strains after 1915 that enabled one to recognize them much as one recognizes persons (fig. 3.10–3.18) (Wright 1922b). Individuals of strain 2 had sharply pointed noses, much in contrast with the rounded noses of strain 13. Others were intermediate. Protruding eyes were characteristic of strain 35, sunken ones of strain 13. Individuals of strain 39 were swaybacked, in contrast with the usual straight backs. There were usually sharp folds in the ear margins of strain 13 that were not present in other strains.

Strandskov (1942) made anatomical studies of strains 2 and 13. The males of strain 13 were 30% heavier, the females 20%, than those of strain 2. As these figures imply, there was a much greater sex difference in strain 13 (11%) than in strain 2 (3%). The differences in body length were equally striking, but there was no significant difference in skull length. The small animals of strain 2 had absolutely longer hind legs and upper forelegs, so that they appeared to run much more off the ground. There were also striking differences in internal organs (Strandskov 1939). The loose, stringy thyroid of strain 2 could not be confused with the compact gland of strain 13. The heavy, smooth contoured adrenal gland of strain 2 contrasted with the flatter and deeply indented one of strain 13. There were striking differences in size and shape of spleen.

A character that was studied intensively was presence or absence of the little toe (Wright 1922c). In the records up to 1915, 12 strains with 8,404 young never had it, and 4 others showed it in less than 0.3% of their 4,626 young. Three strains showed it in 2% of their 3,591 young. Three other strains (31, 35, and 38) showed it in 13.5% of 2,913 young. There was segregation of widely different percentages in the early branches of these strains. A branch of strain 35 that started in generation 12 and displaced all others produced 31% with little toes up to 1925 but with variations from 9% to

10

12

11

13

Fɪɢ. 3.10. Male of strain 2, belonging to generation 12 of sib mating. From Wright 1922*b*, pl. 1.

Fɪɢ. 3.11. Female of strain 2, littermate of the male in figure 3.10. From Wright 1922*b*, pl. 1.

Fɪɢ. 3.12. Male of strain 13, belonging to generation 18 of sib mating. From Wright 1922*b*, pl. 2.

14

Fɪɢ. 3.13. Female of strain 13, littermate of the male in figure 3.12. From Wright 1922*b*, pl. 2.

15

Fɪɢ. 3.14. Male of strain 32, belonging to generation 17 of sib mating. From Wright 1922*b*, pl. 3.

Fɪɢ. 3.15. Female of strain 32, littermate of the male in figure 3.14. From Wright 1922*b*, pl. 3.

16

17

Fig. 3.16. Male of strain 39, belonging to generation 13 of sib mating. From Wright 1922b, pl. 5.

Fig. 3.17. Female of strain 39, littermate of the male in figure 3.16. From Wright 1922b, pl. 5.

Fig. 3.18. Four successive generations, 19–22 (right to left), of sib matings in strain 35. From Wright 1922b, pl. 6.

18

69% in subbranches, a situation that will be discussed in chapter 5. Crosses with a regularly four-toed inbred strain D (obtained from Prof. W. E. Castle) revealed an important subthreshold difference between equally three-toed strains 2 and 13. The former produced no polydactyls in F_1, the latter 33%. The genetics has been discussed in volume 1, chapter 15.

The situation was somewhat similar with a much more extreme type of abnormality, otocephaly, in which manifestation ranged from mere reduction of the mandible to almost complete headlessness (vol. 1, chap. 5). The frequency in the general colony was 61 in 110,000 or 0.06%, including 23 in seven inbred strains other than strain 13. None appeared in 15 other strains. In strain 13, 34 in 3,507 (1.0%) appeared in certain early branches and 203 appeared in 3,689 young (5.5%) among the descendants of a mating in generation 13 (excluding the descendants of a single mating six generations later). Abundant evidence indicated that this line was homallelic with respect to the pertinent genes. A dominant mutation, probably lethal when homozygous, appeared in the mating of generation 19 referred to above and was responsible for 212 in 1,168 young (vol. 1, chap. 6).

Differentiation in Temperament

There were marked differences in temperament. The pigs of strain 35 were nervous and struggled violently when picked up. Those of strain 13 were phlegmatic and could be handled like sacks of meal; the others were intermediate.

Differentiation in Histocompatibility

Differentiation of the inbred lines in histocompatibility factors was demonstrated in experiments by Leo Loeb with inbreds and crossbreds that were sent to him (Loeb and Wright 1927). Loeb graded the response of grafts of a great variety of tissues after intervals up to half a year, on a scale on which 6 was full acceptance (as in autotransplants). Transplants between littermates from the control stock (B) were rejected by reactions of average grade 3.3. Transplants between animals from different matings in this stock gave an average of 1.4, almost the same (1.3) as between the controls (B) and random bred animals from his St. Louis colony (SL).

The detailed results, involving five inbred strains and most of their first crosses, and different kinds of tissues (thyroid, cartilage, parathyroid, liver, spleen, pancreas, adrenal glands, bone, ovary, testis, salivary gland, striated muscle) are too complicated for presentation here. Fortunately the results fell into two sharply distinct groups according to whether the inbred strain I, or either of the parental inbred strains (I, J) of a cross was foreign to the hosts. Those in which the graft did not obviously involve anything foreign are on the left in table 3.16, those in which it did are on the right.

TABLE 3.16. Results of transplants in guinea pigs. I and J are different inbred strains; I × J is F_1 of a cross; B is the random bred stock to which inbreds trace; and SL is a random bred St. Louis strain.

FEW IF ANY FOREIGN GENES IN GRAFT					FOREIGN GENES IN GRAFT				
Graft	Host	Relationship	No.	Grade	Graft	Host	Relation	No.	Grade
I	I	Sibs	14	5.9	I	J		6	2.0
I	I	Not sibs	29	5.2	I × J	I		24	3.2
I	I × J		14	5.8	B	B	Sibs	15	3.3
I × J	I × J	Sibs	8	5.8	B	B	Not sibs	6	1.4
I × J	I × J	Not sibs	2	5.5	B	SL		4	1.3

SOURCE: Loeb and Wright 1927.

The average reaction where the graft does not involve a foreign inbred strain range from 5.2 to 5.9, close to that expected from autotransplants. Where the graft involved a foreign inbred strain or was from a random bred stock, none of the reaction grades is greater than 3.3 (the case of sibs from the control stock B).

The degree of approach to grade 6 should depend on the degree of fixation and hence on number of generations of common sib mating in the strain or strains involved in the graft and the corresponding strain or strains of the host. In the case of inbreds from the same mating, this varied between 17 and 23 (average reaction, 5.9). In the case of crossbreds (I × J) from the same mating, 13 to 21 (average reaction, 5.8). Animals from different matings of the same strain had as few as six generations in common in a few cases, usually nine or more. It is not surprising that the reaction grades were lower in these cases (5.2, 5.5, and 5.8).

The rule that foreign factors in the graft are required for a hostile reaction agrees with that which had been found by Little and Johnson (1922) with transplants of normal tissues and by Little and Strong (1924) with tumor transplants, both involving inbred strains of mice.

Inbreeding in Rats

An inbreeding experiment with rats was started by Helen D. King (1918a–c, 1919) a few years after that with the guinea pigs of the U.S. Department of Agriculture, but it proceeded much more rapidly because of the shorter interval between generations. The results were strikingly different from the usual ones, exemplified by the guinea pig experiments, and thus they present something of a problem.

Two lines (here called A and B) were started from a single litter of stock albinos and maintained for at least 106 and 92 generations of sib mating, respectively. The first six generations were produced on unfavorable rations and showed poor growth and much infertility. Their records were, however, no worse than those of the contemporary stock albinos. Thereafter the ration was radically improved, and growth and fertility became very good.

There was no deliberate selection during the first six generations, but after this all litters that seemed lacking in vigor were destroyed at once. There was also selection for excess of males in line A, of females in line B, with positive results described in chapter 7 of this volume. Some three times as many rats were saved to sexual maturity (three months) as needed and then rigorously selected for vigor. The experiment differed radically in this respect from the guinea pig experiment, in which nothing more than natural selection was practicable in trying to maintain many inbred strains from littermate matings with litters of average size about 2.5 or less instead of 7.5.

Average weights, at least of adults, were greatest in generations 7–9. King was inclined to attribute the slight decline in later generations to some deterioration of conditions rather than to inbreeding, since the inbreds continued to be superior to the stock albinos. The latter were less rigorously selected, however.

She found, indeed, no evidence (generation 25) of any impairment of the inbreds in fertility, longevity, or any other aspect of constitutional vigor. The average size of litter in both lines (7.5 in A, 7.4 in B) was superior to that of the stock albinos (6.7). The inbreds were, however, somewhat less active and more nervous. There was little difference between the lines (except in sex ratio) but the A's became slightly heavier, were more fertile, matured slightly earlier, and lived slightly longer on the average. King believed that these qualities were merely different aspects of constitutional vigor. This contrasts with the largely random combination of such qualities among the inbred strains of guinea pigs. Two lines, both strongly selected for vigor, do not, however, provide as much information on this matter as 23 largely unselected lines.

Considerable light was thrown on the apparent absence of inbreeding decline by transplantation experiments, conducted by Leo Loeb at the same time (in the first series) as his experiments with the inbred and crossbred guinea pigs referred to earlier. He made studies at widely separated intervals: generations 37–41, 46–47 (Loeb and King 1927), 60–67 (Loeb and King 1931), 91–92, and 104–106 (Loeb, King, and Blumenthal 1943).

In the first series, transplants between rats from the same line but different matings were rejected almost as vigorously (average grade, 3.0) as transplants between the lines (average grade, 2.6) or between noninbred rats (average grade, 2.7). Even transplants between littermates averaged only 4.0, well below the average (5.9) of such transplants in guinea pigs with only half as many generations of common sib mating, or of autotransplants in noninbred rats (or guinea pigs) with grade 6.0.

The results of all of the experiments with rats were summarized in the 1943 paper on a new scale in which autotransplants were graded 3.3. There was also some revision upward of the grades of the first series. These 1943 grades have been multiplied here by 1.8 in table 3.17 to make them comparable with the guinea pig averages. Some transplants from B to A of generations 91 and 92 were made, but B had been lost by generations 104–106, when transplants were made within A.

TABLE 3.17. Results of transplants of tissues in rats. A and B are two inbred strains studied in three periods. Strain B was extinct in the last period except for transplants B to A in generations 91 and 92.

GRAFT	HOST	RELATION	GENERATIONS 37–47		GENERATIONS 60–67		GENERATIONS 91–106	
			No.	Grade	No.	Grade	No.	Grade
A	A	Sibs	12	3.0	17	4.7	29	5.4
B	B	Sibs	24	4.6	19	5.1	...	...
A	A	Not sibs	24	3.3	16	2.1	17	5.0
B	B	Not sibs	27	3.5	33	3.0	...	...
A	B		18	3.0	32	2.5		
B	A		19	2.8			14	3.0

SOURCE: Loeb, King, and Blumenthal 1943.

The grades indicate more heterozygosis with respect to histocompatibility genes in A than in B, at least up to generations 60–67. Transplants between rats from different matings of the same strain showed no increase in grade by the second period, but in the third series the reaction to such grafts had reached 5.0, almost as high as in the guinea pigs but at a very much later generation. The grade reached 5.1 in littermates in strain B in the second series, but only exceeded 5 in strain A in the third series (grade 5.4). These compare with littermate reactions of 5.9 at only about 20 generations in the guinea pigs. It appears that fixation was proceeding exceedingly gradually in the rats.

The results suggest that the absence of any decisive inbreeding decline was due to persistent heterozygosis at least at some loci, resulting from the rigorous selection for vigor. It should be noted that the selection of individuals at birth and sexual maturity was inevitably supplemented by selection among the continually branching sublines. With only two instead of 23 inbred strains, there were many more such branches and much more rapid elimination of those that lagged in generation number because of low vigor in any respect. It is probable that any branches that happened to become unfavorably homallelic in any respect were rapidly eliminated, while those retained remained heterallelic or were ones that became favorably homallelic in whole chromosomes. Since it requires over three generations under sib mating to reduce heterozygosis by 50% in the absence of selection, it is possible to prevent unfavorable random fixation for long periods by very rigorous selection at all levels.

A test of whether it would be possible to produce crossbred rats, using the King inbreds as one parent yet superior to them, was made by Livesay (1930). Since his experiment began in 1927, at which time Loeb and King reported 47 generations of sib mating, his representatives (S_3) must have been at about this stage. As just indicated, they were far from homozygous, at least at an important histocompatibility locus. As the other parent in one cross, he used a strain of small pink-eyed pale (pp) rats (S_1) inbred for several generations, although not regularly by sib mating. There was sufficient parallelism throughout in the weights of the sexes at 90 days to use the unweighted average. $\bar{P}$ in table 3.18 is the average of the means. The standard deviations (SD) and coefficients of variation (CV) of the two strains are also given.

TABLE 3.18. Weight at 90 days of two inbred strains of rats and F_1 and F_2.

	No.	Mean (gm)	SD (gm)	CV
S_1 (pp)	198	155.4	35.5	22.3
S_3 (King)	142	180.1	39.6	21.6
$\bar{P}$		167.7		22.0
F_1	122	191.3	24.7	12.8
F_2	269	163.1	28.7	17.4
$F_1 - \bar{P}$		+23.6		9.2
$F_2 - \bar{P}$		-4.6		4.6

SOURCE: Livesay 1930.

F_1 (191.3 gm) was much above the average of the two parental strains (167.7 gm) and significantly heavier than the King inbred parent. F_2 was slightly less than the midparent, although expected to relapse only halfway. There seems to have been considerable interaction in this case. The greater variability of the inbred strains than of F_1 will be discussed in the next chapter.

A cross was also made between strain S_1 (pp) and a red-eyed pale (rr) strain S_2 that had been closely inbred. These rats were much smaller even than S_1 and difficult to maintain because of low fertility. The data at 90 days of age are given in table 3.19.

TABLE 3.19. Weight at 90 days of two inbred strains of rats and of F_1 and F_2.

	No.	Mean (gm)	SD (gm)	CV
S_1 (pp	198	155.4	35.5	22.3
S_2 (rr)	111	118.1	29.2	24.6
$\bar{P}$		136.7		23.5
F_1	85	184.2	24.3	13.3
F_2	115	166.0	25.1	15.1
$F_1 - \bar{P}$		+ 47.5		10.2
$F_2 - \bar{P}$		+ 29.3		8.4

SOURCE: Livesay 1930.

There is twice as much heterosis as in the preceding cross. In this case, F_2 relapses toward the midparent more nearly the expected 50%.

Inbreeding of Mice

The most extensive use of close inbreeding in mammals has been with mice. The Jackson Memorial Laboratory has developed a considerable number of strains, maintained exclusively by sib mating, that are sufficiently vigorous in all respects to be multiplied indefinitely. Each of these strains has its unique combination of traits with respect to color, morphology, and physiology. The imposing structure of mouse genetics has been, to a major extent, based on them and their mutants, and they have been used widely in experimental studies outside of genetics in which genetic homogeneity has been an important consideration. Excessive phenotypic variability in some respects, relative to crossbreds, and the degree of persistence of strain characters are matters that will be discussed in the next chapters.

A study of differences between inbred strains in length of life after inoculation with tuberculosis by Lynch, Pierce-Chase, and Dubos (1965) gave results that are interesting to compare with the experiments with guinea pigs described earlier. Two of the Jackson strains, C57BL and Swiss, were tested. Under one set of conditions (intravenous inoculation of a certain dose of the culture at four to six weeks of age), C57BL survived an average of 28.1 $\pm$ 0.6 days and Swiss 55.3 $\pm$ 0.6 days. This twofold difference was essentially confirmed in repetitions under the same conditions but disappeared in experiments that differed only in that inoculation was at four to ten months of age: C57BL, 51.2 $\pm$ 1.8 days; Swiss, 51.3 $\pm$ 2.9 days. At still greater ages, C57BL mice had somewhat greater survival times. The original order was also reversed by using one-tenth the original dosage at four to six weeks: C57BL, 125.1 $\pm$ 4.8 days; Swiss, 89.9 $\pm$ 2.6 days. Even more than with the guinea pigs, the relative degrees of resistance exhibited by different genotypes are thus functions of conditions.

In this case, an attempt was made to analyze the genetic difference between the two strains by repeated backcrosses of F_1 to each parent strain. Under the conditions of the first experiment cited above, the F_1 mice lived considerably longer than the Swiss mice (70.3 $\pm$ 1.0 days). The backcross to the short-lived C57BL strain gave a strongly bimodal distribution readily interpretable as one-half close to C57BL and one-half close to F_1. Later backcrosses also indicated segregation of a major unit and the expected decrease in frequency of the heterozygotes. The other backcross gave no indication of unitary segregation. The results can be interpreted on the basis of a single gene difference with overdominance under the conditions, although the supplying of minor favorable genes by the shorter-lived strain cannot be ruled out.

Differentiation in Histocompatibility in Mice

The genetics of the reaction to transplanted tissues was first worked out in crosses between inbred strains of mice. The theory had two principal roots. Leo Loeb (1901), who began studies in the 1890s with random bred mice and guinea pigs, found that degree of success was correlated with closeness of relationship: 100% with autotransplants, a usually hostile reaction but rare success between sibs or parent and offspring (syngenesiotransplants), a more violent reaction between unrelated members of a species (homoiotransplants), and a still more violent one between species (heterotransplants). He held that he was dealing with the heredity of protoplasmic specificity.

Little and Tyzzer (1916) obtained results that they interpreted as Mendelian from crosses between a strain of Japanese waltzing mice, susceptible to a tumor from an animal of this strain, and a wholly resistant strain of house mice. The results, F_1 all susceptible, F_2 nearly all resistant, could be explained on the hypothesis that susceptibility depends on the simultaneous presence of numerous dominant genes.

Later, Little and Strong (1924) combined the Mendelian interpretation with Loeb's concept of hereditary specificity. Under this theory, each pertinent gene determines the presence of an antigen in tissues that on transplantation induces a hostile reaction from a host that lacks it, while a similarly determined host antigen causes no such reaction in a graft that lacks it. These principles applied to normal tissues as well as to tumors (Little and Johnson 1922). Loeb's results with various normal tissues of different inbred strains of guinea pigs and their crosses were fully in conformity with this theory. His results with Dr. King's rats were in conformity on the additional hypothesis that strong selection greatly delayed the achievement of homozygosis.

The proportion of acceptances in F_2 of a graft from one of the parent strains could be interpreted as $(3/4)^x$ and in the first backcross as $(1/2)^x$, where x is the number of dominant "histocompatibility" genes by which the strain supplying the graft differed from the other. The modifications in relation to sex to be expected if one of the genes is sex linked were observed (Strong 1929; Bittner 1931). Complications in the case of tumor grafts will be discussed in the next chapter. It may be added that Gorer (1937) showed that the most important locus (H-2) concerned in reactions to tumor grafts in mice, was the same as one that determined a red blood cell antigen.

Skeletal Variations in Mice

Another class of characters in which the differences among inbred strains have been much studied consists of minor skeletal variants of the threshold type. Green (1941) made a survey of the inbred strains of the Jackson Laboratory that revealed wide variations in percentage incidence of types of vertebral columns. Grüneberg and his associates have described 30 skeletal variants (Deol et al. 1957), all within the range of normal variation.

Size of Litter in Mice

A clear illustration of the process by which vigorous inbred strains can emerge from an experiment that initially shows inbreeding depression is provided by an experiment of Bowman and Falconer (1960). The character used was the number of live young in first litters. Ten lines were started from noninbred double first cousins in order to produce offspring with only 12.5% inbreeding. In the next generation, each of these lines was divided into two: JU always from a litter of intermediate size and JS always selected from a large litter. In each line one male was mated with up to five littermates. If more full sisters were available as well as two males, two such groups were made but only one litter was used to continue the line, which became extinct if it produced no litter containing at least one survivor of each sex.

Figure 3.19 shows the average size of litter of the whole array for 21 generations. There was rapid decline (7.77 to 4.58) for five generations. There was then a rise followed by wild fluctuations from generation 5 to 13 about on an average of 6.3. From this point to generation 21 there was little variation about an average substantially the same as in the control stock (nearly 8.0).

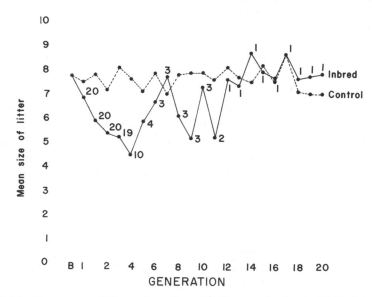

FIG. 3.19. Mean sizes of litter along lines of sib mated mice (*solid line*) in comparisons with controls (*dotted line*). Twenty inbred lines were started but only one persisted to generation 20 (with mean practically unchanged). The numbers of lines are indicated. From data of Bowman and Falconer (1960).

Obviously this is no simple process. The number of surviving lines must be noted. All persisted until one JU was lost in generation 4. The unweighted means of the lines are shown up to this point in relation to the inbreeding of the litters. There was some suggestion of an effect of the selection by litter size in the slightly higher averages of JS as compared with JU but the difference was not significant. The main result was a nearly linear decline at the rate of 0.6 of a mouse per 10% increase in F (or 7.7% in terms of litter size per 10% in F). This downward trend continued to generation 5 with only ten lines extant, but there was a sharp rise in generation 6 with only four left. There were only three lines from generations 7–11. There were two in generation 12. It is not surprising that these were lines that were well above the average when many lines were on hand. The one line that was on hand after the generation 12 had remained from the first substantially at the control level. This illustrates the enormously greater effectiveness of selection among many lines in which random fixation is occurring than of selection within lines (JS vs. JU).

The three surviving lines of generation 9 were crossed. Litter size rose from about 6.3 (generations 6–12) to 7.53 (mothers, $F = 80.6\%$; litters, $F = 0$) and in the next generation with minimal inbreeding (mothers, $F = 0$; litters, $F = 20.0\%$) it rose to 9.58 ± 0.36. Presumably these three surviving lines had become fixed in largely favorable genes but not wholly the same ones, so that each could supply one or more favorable largely dominant factors lacking in one or both of the others. Decline began in later generations (8.33, 7.28, 6.90, 6.89) as inbreeding of both mother and litter rose. Attempts at improvement of litter size by two more cycles of crossing and inbreeding failed to carry the average beyond the control level.

The inbreeding depression of the first generation (mother, $F = 0$; litter, $F = 12.5$) was about the same in relation to the inbreeding of the litter as in later generations. This implies that inbreeding of the litter is more important than that of the dam (in contrast with the situation in the guinea pig experiment). However, the comparison of the first crosses (mother, $F = 80.6$; litter, $F = 0$), with mean 7.53 ± 0.55, with that in the next generation (mother, $F = 0$; litter, $F = 20.0\%$), mean 9.58 ± 0.36 cited above, implies that inbreeding of the mother is also important.

Other data in which the effects of inbreeding of mother and offspring on litter size can be compared were obtained in an experiment by Roberts (1960) in the same laboratory. Ten lines each were derived from high, low, and control stocks from an experiment on selection for litter size. These 30 lines were sib mated for three generations (I_1, I_2, and I_3) to produce litters with 25%, 37.5%, and 50% inbreeding, respectively. The last were crossed at random for two generations (X_1, X_2) within the three categories. The principal results, combining all 30 lines, are shown in table 3.20.

TABLE 3.20. Relation of litter size to inbreeding coefficients of parents and offspring.

GENERATION	INBREEDING COEFFICIENT		LITTER SIZE
	Parents (%)	Offspring (%)	
0	0	0	8.12
I_1	0	25	6.73
I_2	25	37.5	5.82
I_3	37.5	50	5.69
X_1	50	0	6.20
X_2	0	0	8.47

SOURCE: Data from Roberts 1960.

The decline in I_1 must be due wholly to inbreeding of the litter (0.55 mouse [or 6.8%] per 10% decline in litter F). The total decline to I_3, at the rate 0.49 mouse per 10% decline in litter F, is not significantly lower despite supplementation by inbreeding of the mothers. On the other hand, inbreeding of the mother must be involved to account for the slightness of the improvement in the first cross with no inbreeding of the litter but 50% of the mother and the great improvement in the second cross when the mother also is not inbred. Taking X_2 as the base, the inbreeding decline implied by this gain for X_1 is 0.45 mouse [or 5.4%] per 10% change in maternal F. The effects of maternal and litter inbreeding seem to be roughly equal when each is alone, but no more if both are present. The difference between generations 0 and X_2 is not significant at the 0.05 level. Tabulation of the three groups gave results that were irregular, except that derivation from high selection gave the highest X_1 and X_2.

Falconer and Roberts (1960) studied the mode of action of inbreeding on litter size directly. Incomplete fertilization by inbred sires was shown to be negligible as a cause. Dissection of 16-day pregnant females with $F = 50\%$ to 60% or $F = 0$, both with noninbred litters, showed no difference in ovulation rate (corpora lutea) but sufficient difference in implantation rate to account for the difference in litter size. There were no demonstrable differences in postimplantation mortality of the embryos.

Generalization to all strains of mice from these experiments is not warranted, however. Eaton (1953), starting from nine highly inbred strains, found a highly significant increase in litter size at birth (1.15 ± 0.21) in the average of the 72 possible crosses on correcting for effects of dam's age. This indicates a prevailing tendency for heterozygosis of the offspring to increase litter size, but there were enormous differences among the crosses. On combining into nine groups according to the maternal line, the difference between the corrected averages of F_1 and the inbred parents varied from 0.23 to 2. Moreover, in a later experiment with four inbred strains (only one the same as before) and four first crosses (each strain used twice) there was no significant increase on crossing (0.19 ± 0.24). On backcrossing to a parent strain, there was an increase of 1.50 ± 0.35 attributable to maternal crossbreeding.

There have been many other experiments (reviewed by Liljedahl 1968). Most of these have given increased size of litter in F_1, but where further analysis has been attempted the reasons have varied widely. In some cases the differences were largely in preimplantation losses, in others in postimplantation mortality. There has also been much diversity in the importance of embryonic and maternal effects. The latter include not only effect on ovulation rate but also play a role in preimplantation and postimplantation losses.

In at least one case, a decrease in F_1 has been attributed to maternal-fetal incompatibility. In another case, small litter size has been attributed to low fertility of the inbred males.

These variations among experiments are, of course, quite as expected if the inbreeding depression and its converse, heterosis, depend on the specific genotypes that are fixed, rather than on general physiological effects of inbreeding and crossbreeding. The apparent restriction of the effects on litter size to differences in ovulation rate in the guinea pigs may be due to the much smaller litters.

Birth Weight in Mice

Several of the same experiments have shown favorable effects of crossbreeding on weight at birth and later (Liljedahl 1968). The gain from birth to weaning is, however, largely an effect of lactation of the dam, which is favorably affected by crossbreeding.

The difference between crossbreds and inbreds in number of live young in litters increases from birth to weaning (Eaton 1953; Liljedahl 1968). The crossbreeding of the latter clearly plays a role and that of the dam probably does also (Eaton 1953). There are increasing differences between the mortality curves later in life (Liljedahl 1968).

Tolerance of Extreme Temperatures by Mice

Barnett and Coleman (1960) compared two inbred lines of mice and F_1 at 21°C and −3°C (with cotton wool nests, in both cases). Table 3.21 gives some of the comparisons.

The standard errors are not given here, but they were sufficiently small to leave no doubt with respect to the deleterious effect of low temperature on both inbreds and F_1 and to the greatly increased heterosis (on a percentage basis) at low temperature.

TABLE 3.21. Comparison of two inbred strains and F_1 at 21°C and −3°C.

	21°C			−3°C		
	A2G	C57BL	F_1	A2G	C57BL	F_1
No. of pairs	15	15	11	19	10	12
Litters per pair	5.1	5.1	4.6	2.8	2.4	3.1
Litters weaned per pair	4.5	3.7	4.5	1.3	1.1	3.0
Young born per pair	27.1	33.5	43.8	10.9	12.3	22.7
Young weaned per pair	20.7	20.4	39.1	4.3	5.3	22.7
% loss, birth to weaning	23.6	39.1	10.7	60.6	56.9	18.1

SOURCE: Barnett and Coleman 1960.

Inbreeding in Rabbits

Attempts to inbreed rabbits have been notably less successful than with guinea pigs, rats, or mice. Chai (1969) has, however, reported on two lines in generation 20, two in 12, and one that became extinct in generation 14. These were maintained largely by sib mating, but occasional matings had to be made of parent with offspring or between half sibs or other close relatives. There was some decline in mating performance, a decline of average litter size in all lines from about eight to less than four (due more to increased zygote mortality than to depressed ovulation), an increase in mortality at birth and later, and a reduction in weaning weight.

Inbreeding in Livestock

The experience of livestock breeders as reported by Darwin has been referred to in chapter 2. The depression in all aspects of vigor from close inbreeding has been confirmed by many recent studies.

McPhee, Russel, and Zeller (1931) inbred Poland China swine through two generations of sib mating but were unable to continue because of the small litters and small percentage raised (table 3.22).

TABLE 3.22. Comparison of vital statistics of a herd of Poland China swine and the progeny of two generations of sib mating.

| | F | | | Size of Litter | % Born Alive | % Raised to 70 Days | Weight (lb) | | Sex Ratio |
	No.	Dam	Litter				At Birth	70 Days	
General herd	694	0+	0+	7.15	97.0	58.1	2.75	34.7	109.7
F_1 inbred	189	0.09	0.33	6.75	93.7	41.2	2.31	33.8	126.1
F_2 inbred	64	0.33	0.42	4.26	90.6	26.6	2.29	28.9	156.0

SOURCE: McPhee, Russel, and Zeller 1931.

The decline in the size of litters in the first generation, when the dams were probably no more inbred than average in the herd, was not significant, but there was a highly significant decline in the F_2 litters amounting to 19% (of the excess over 1) per 10% increase in inbreeding (ΔF) of the dams. The percent born alive declined somewhat in both generations, indicating effects of inbreeding of both dam and pigs. There were more serious losses from birth to 70 days, 12% in F_1, 16% in F_2 per 10% ΔF of the pigs. Birth weight declined significantly in F_1 (7% per 10% ΔF) but not appreciably more in F_2, probably because of balancing of inbreeding decline by a favorable effect of decline in litter size. Weight at 70 days decreased only 1% and 5% per 10% ΔF of the pigs in F_1 and F_2, respectively, the latter probably augmented by inbreeding of the dams. The average inbreeding of the herd was taken as 0.09 in all of these cases.

A peculiar color, sepia, segregated out as a recessive and a considerable number of morphological abnormalities appeared in F_2. The ratio of males to females showed a remarkable increase with inbreeding.

Such results, while typical, do not preclude the possibility that lines useful in crossbreeding may be obtained by mild inbreeding associated with selection. An extensive cooperative project involving 38 lines of various breeds of swine was conducted by the experiment stations of the North Central states of the United States. Dickerson et al. (1954) analyzed the data from 1932 to 1948. The inbreeding was such as to cause increases of 2% to 4% per year. The overall average for dams was 21%, for litters 24% (between 30% and 60% during the last years). The initial size of litter averaged 7.50 at birth, reduced to 4.48 at weaning (56 days). The average weight of pigs was 29.0 lb at weaning, 131.7 lb at 154 days.

Analysis indicated that there was no more than the expected automatic natural selection for litter size, but there was considerable deliberate selection for growth rate, which, it was shown, implied rather strong selection for heterozygosis. The declines for each 10% ΔF of the pigs were estimated from comparisons of 525 inbred litters with 325 line-crossed ones. Those for each 10% ΔF of the dams were estimated from comparisons of line-crossed litters from 50 inbred dams and 63 crossbred ones, representing the same lines.

Litter size showed considerable decline per 10% ΔF of pigs (-0.20 at birth, -0.38 at 56 days, -0.44 at 154 days), somewhat less per 10% ΔF of dams (-0.17 at birth, -0.28 at 56 days, -0.28 at 154 days). (There was little or no change in weight from inbreeding of both pig and dam at birth (0.03 lb) or 56 days (0.08 lb), but some at 154 days (-3.44 lb per 10% ΔF of pig). The authors note that selection might be more effective if directed toward the improvement of crosses between complementary strains, with selection of individuals or families based on performance of test-cross progeny.

The effects of mild inbreeding on various characters of range sheep were investigated by Hazel and Terrell (1946). There were considerable decreases in weaning weight per 10% ΔF in Rambouillets (5.4%) and in a flock of mixed breeding (4.1%). That for yearling weight of Rambouillets was 3.4%. There was less effect on fleece weight (2.7%) and little if any on staple length, face covering, or neck folds. Morley (1954), in a study of Australian Merinos, found a similar decrease in body weight (3.7% per 10% ΔF), more in fleece weight (5.9%) and neck score (14.6%), but again little in staple length (1.3%).

Tyler, Chapman, and Dickerson (1949), in a comparison of 42 inbred and 47 outbred Holstein-Friesian cows, found little effect of inbreeding on bodily dimensions, the greatest being in heart girth (0.6% per 10% ΔF). Milk production and butterfat production were, however, depressed (6.2% and 5.8% per 10% ΔF).

In poultry, Schoffner (1948) found a depression of 6.2% in egg production, 6.4% in hatchability, but only 0.8% in body weight, all in terms of 10% ΔF. There have been other similar results.

Sittmann, Abplanalp, and Fraser (1966) found more depression in Japanese quail (*Coturnix coturnix*). Fertility decreased 11% and hatchability 7% per 10% ΔF. The probability for a zygote's reaching maturity and leaving offspring was only 0.3 in the controls but was reduced to 0.1 by three generations of inbreeding.

Human Inbreeding

Much has been written about inbreeding in man. The highest degree for which large bodies of data are available is that for offspring of first cousins ($F = 0.0625$). The great difficulty in estimating inbreeding depression has been in obtaining reliable controls because of possible socioeconomic differences.

One effect was, however, obvious from the first studies of recessive traits such as albinism or alkaptonuria. The percentage of cousin marriage among the parents is far above that in the general population as, of course, expected. A high percentage of consanguinity is indeed usually the best evidence that a rare trait depends on a recessive gene rather than on a dominant gene of low penetrance or environmental accident.

The most exhaustive study of inbreeding depression in man seems to have been that of Schull and Neel (1965) of the populations of Hiroshima and Nagasaki, made in connection with an investigation of the genetic effects of the atomic bombing of these cities. The inbreeding study was restricted to children whose parents had not been in these cities at the time (60%), or who were estimated to have been exposed to less than 10 rep (40%).

Consanguineous marriages are relatively frequent in Japan (6% in Hiroshima, 8% in Nagasaki) in contrast with less than 1% in the United States. Records were obtained of 3,314 children of consanguineous marriages (58% first cousin) and of 3,570 controls, chosen by a random process from the same populations. The results were sufficiently similar from the two cities that they may be combined for most purposes. Among children of first cousins, there was a 60% excess above the control mortality of 0.035 in the first year of life. The percentage of stillbirths and of childhood mortality after the first year showed smaller and more inconsistent excesses. There were 37% and 24% excesses above the control frequencies of 0.085 for major defects and 0.079 minor defects, respectively.

An attempt was made to correct anthropometric and mental measurements for inflation by associated socioeconomic differences. It was estimated that about 20% on the average of the depression of the inbred children had this cause. With this correction, it was found that at five years of age, weight was reduced 0.9%, height 0.4%, and eight other linear measurements about the same. There were slight average delays in walking and talking in the inbreds. Six neuromuscular tests showed 1.85% decrease and IQ (Wexler) an average 4.2% decrease (4.75% in the verbal component, 3.65% in the performance components). Depression in school performance was about 3.6%.

Inbreeding in *Drosophila*

Early experiments with *Drosophila* by Castle and others, referred to in chapter 2, indicated the occurrence of differentiation of sib-mated lines, inbreeding depression, and crossbreeding heterosis, all interpretable on a Mendelian basis. There have been many more recent studies, a few of which will be considered.

Complications from linkage are especially serious in this case because of the small number of chromosomes and absence of crossing-over in males. Associations of recessive deleterious genes, due to mutation, with favorable ones are disentangled only very slowly by crossing-over and selection. In a species in which the same number of deleterious genes is distributed over a larger number of chromosomes, favorable genes are more likely to be free from deleterious linkage and to be capable of ready fixation and the unfavorable ones are more readily eliminated. Even under close inbreeding, the strong favoring of chromosome heterozygotes in species of *Drosophila* tends to slow down the fixation process.

Gowen (1952) emphasized the discrepancy between the theoretical loss of heterozygosis (at neutral loci) under continued sib mating and that which he actually found in *Drosophila* lines, as indicated by the continued segregation of lethals.

Nevertheless, he obtained abundant heterosis on crossing inbred lines, indicating that there had been much fixation in these, even if not as much as expected by the theory for neutral loci. He found 113% heterosis for length of the laying period, 154% for egg production in the three highest days, and 203% for lifetime egg yield. These corresponded to decreases of 5.3%, 6.1%, and 6.7%, respectively, per 10% increase in theoretical F, which is fairly typical of inbreeding depression.

Clayton et al. (1957) and Clayton, Morris, and Robertson (1957) started ten lines with four pair matings each in each generation to minimize losses of lines, but three had been lost by the generation 10 and two more by generation

22. A second series of 100 lines, started without reserves, had been reduced to only 14 at the eighth generation, after which reserve matings were made. There was clearly much depression of vital characters in spite of natural selection of individuals.

In these experiments, the numbers of chaetae in the fourth and fifth abdominal segments were counted in the tenth generation ($F = 87\%$) as well as before inbreeding (average number, 35.3). The counts in the first series from seven lines averaged 32.2 and in the second from 12 lines 36.4 (grand average, 34.9). There was differentiation among the lines but no evidence of inbreeding depression in this character, in agreement with much other evidence on chaetal counts. A great many crosses have been made between inbred lines as a basis for selection experiments on chaetal number (this will be discussed later), but no appreciable heterosis has been observed. This is not a character that has any obvious relation to fitness.

Wallace and Madden (1965) obtained similar results. They started from 60 pairs and made five replicates in each generation. Half of these lines were lost by the 19th generation, again indicating much fixation of deleterious genes. Two lines were continued from each of those left to restore the full number, but only 11 of these 60 were still extant at the 50th generation. Tests did indeed demonstrate the presence of many heterozygous lethals (largely nonallelic) in the persisting lines, although not so many that any of the lines arrived at a balanced lethal condition.

A great many experiments with *Drosophila* have required lines that are isogenic, at least with respect to the major chromosomes. It has often seemed best to produce these lines by segregation from crosses involving dominant markers and crossover inhibitors for each of these chromosomes as intermediary steps in making single chromosomes from the desired source homozygous. Many lines must usually be started, however, to make it reasonably certain that one sufficiently vigorous one can be obtained. Thus Fox (1949) made 20 attempts to produce an isogenic wild-type strain by this method. No viable lines were obtained and it became necessary to use a stock that was balanced in one of the chromosomes for the purpose of the experiment. In other cases, however, excellent isogenic lines have been obtained.

Straus (in Gowen 1952) used this method to produce lines isogenic in the major chromosomes for the purpose of analyzing the effects of all combinations of the pairs of major chromosomes of two inbred strains.

The eight possible types of homozygotes were obtained and all of the possible heterozygotes were produced. He studied the average daily egg yield (table 3.23).

Starting from the homozygotes, heterozygosis of one chromosome adds

TABLE 3.23. Average daily egg yield of *D. melanogaster* (according to the number of heterozygous chromosomes).

Type	No.	Average	Range
Wholly homozygous ($A_1A_1B_1B_1C_1C_1$, etc.)	8	38.9	27.8–51.3
Single heterozygote ($A_1A_1B_1B_1C_1C_2$, etc.)	12	51.5	35.3–60.6
Double heterozygote ($A_1A_1B_1B_2C_1C_2$, etc.)	6	62.6	51.5–66.5
Triple heterozygote ($A_1A_2B_1B_2C_1C_2$)	1	76.9	76.9

SOURCE: Straus (in Gowen 1952).

12.6 eggs per day, of a second chromosome adds 11.1 eggs per day, and of all three chromosomes adds 14.3 eggs per day, with an average of 12.7 eggs per 33% decrease in F or 4.9 per 10%. While the differences are not significant, it is perhaps worth noting that the smallest increment is between single and double heterozygotes.

F. W. Robertson (1954) made a similar chromosome replacement study using a strain N, derived from a wild stock by more than 100 generations of sib mating and a strain D derived from the same stock by long-continued selection for small size followed by more than 40 generations of sib mating. We will consider here only the replacements of chromosomes of the two original inbred lines. The characters studied were wing and thorax length. These are strongly correlated as indicators of general size (table 3.24).

The averages for wing and thorax are shown on different scales in figure 3.20. In the case of replacements in NNN, there is heterosis only in the sense

TABLE 3.24. Average deviations of wing and thorax lengths (in terms of 0.01 mm) from those of the Nettlebed line, N, *D. melanogaster*, on replacing one chromosome of one or more pairs by a chromosome of the D line and on replacing one chromosome of one or more pairs of the latter by a chromosome of the N line. The major chromosomes of the two lines are here represented in order (NNN, DDD), heterozygosis of one of them by X. The data here refer to females.

	Wing	Thorax		Wing	Thorax		Wing	Thorax
NNN	0.0	0.0	NNX	0.6	−0.1	NXX	0.0	−0.3
XXX	−5.1	−1.6	NXN	3.2	2.0	XNX	−4.1	−1.6
			XNN	−3.7	−0.9	XXN	−3.7	+0.6
			Average	+0.0	+0.3	Average	−2.6	−0.4
DDD	−46.3	−10.1	DDX	−27.3	−1.9	DXX	−24.8	−2.6
XXX	−5.1	−1.6	DXD	−26.9	−4.4	XDX	−10.1	−1.3
			XDD	−22.3	−7.3	XXD	−15.1	−6.2
			Average	−25.5	−4.5	Average	−16.7	−3.4

SOURCE: Robertson 1954.

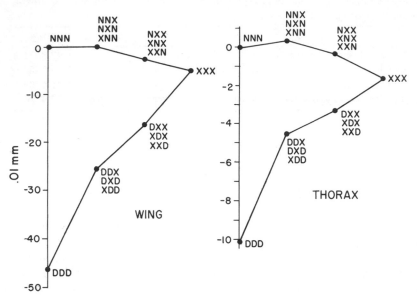

FIG. 3.20. Deviations of wing and thorax sizes of females (*Drosophila melanogaster*) of various sorts from those of a strain (NNN) with more than 100 generations of sib mating. DDD is a strain selected for small size and subsequently inbred for more than 40 generations. In the others, one or more chromosomes have been made heterozygous ($X = D/N$). From data of F. W. Robertson (1954).

that the triple heterozygotes are far above the average of the two inbred strains. One replacement (NNX, NXN, XNN) however, produces a slight excess in thorax length in one case and, in two of three cases, in wing length, although not in the average. It is interesting that one replacement in DDD produces considerably more effect (51% of XXX − DDD with respect to wing length, 66% with respect to thorax length) than additional single replacements, and that the step from two to three replacements adds more (28% and 21%, respectively) than the step from one to two (21% and 13%, respectively). Data from this experiment will be considered further in connection with selection.

Robertson and Reeve (1955a) made similar replacement experiments in seven crosses between inbred lines with respect to wing and thorax lengths and four with respect to egg production. The averages for thorax length and egg production in the parent lines and F_1 are shown in table 3.25, together with averages from several wild strains (thorax) and a particularly vigorous one with respect to egg production.

TABLE 3.25. Comparison of parental inbred lines and F_1 with respect to thorax length and egg production (for four days).

	Thorax Length (0.01 mm)			Egg Production over Four Days		
	P_1	P_2	F_1	P_1	P_2	F_1
$R_1 \times C_6$	107.3	106.3	111.3	108	114	292
$R_2 \times C_{10}$	106.1	102.8	107.3	148	124	342
$N_1 \times N_2$	96.0	103.9	105.0	190	122	305
$N_2 \times R_2$	105.5	103.6	107.2	174	181	284
$N_3 \times O_1$	104.2	98.8	107.6			
$O_2 \times E_4$	102.2	101.9	107.3			
$N_3 \times E_4$	104.5	100.9	107.0			
Average P_1 and P_2, F_1		103.1	107.5		145	306
Wild			107.6			320

SOURCE: Reprinted, with permission, from Robertson and Reeve 1955a.

The inbreds ($F = 100\%$) show an average depression of thorax length in relation to F_1 (or the wild strains) of 4.8%, with only moderate variation (4.1% to 6.1%). In wing length, the depression was 4.3% with wider variation (-0.8% to 19.0%). In egg production, the depression was enormous, 61% with variation from 47% to 73%. The amount of heterosis averaged 130%.

By the use of dominant markers and crossover inhibitors they introduced chromosomes from one parental strain into the other so as to produce single or double heterozygotes of each possible sort. They estimated the additive effect of each chromosome on the scale of actual lengths in the cases of wing and thorax lengths, of ten times the natural logarithms in the case of egg production, using the method of least squares. Analysis of variance indicated significant interaction in 22 of the 36 cases (19 at the 0.01 level).

There were highly significant differences in the patterns of interaction. These may conveniently be put into five groups for thorax and wing lengths and two for egg production. The averages for 0X, 1X, 2X, and 3X are shown in table 3.26.

The similarity of the patterns for thorax length and wing length are apparent. Group 5 deviates relatively little from linearity although the increments fall off slightly as heterozygosis rises. Group 2 has relatively small increments from 0X to 1X. Group 3 rises little or declines slightly from 0X to 1X. Group 1 and especially group 4, on the contrary, show the greatest increment from 0X to 1X, a decrement from 1X to 2X, and a small increment for 2X to 3X.

TABLE 3.26. Averages for groups in which the genomes of parental lines (OX) are replaced by one (IX), two (2X), or three (3X) heterozygotes of the major chromosomes of *D. melanogaster*. The donor line is indicated in parentheses.

Recipient and Donor Lines	Thorax Length (0.01 mm)				Wing Length (0.01 mm)			
	0X	1X	2X	3X	0X	1X	2X	3X
1. $N_2(R_2)$, $N_3(E_4)$, $C_6(R_1)$	105.5	107.6	107.5	108.5	208.7	212.0	211.3	212.7
2. $R_2(N_2)$, $R_1(C_6)$	105.5	106.3	107.2	109.3	204.1	206.2	210.0	212.8
3. $N_2(N_1)$, $N_3(O_1)$, $C_{10}(R_2)$	103.6	103.8	105.0	106.6	205.6	204.2	206.0	207.5
4. $E_4(O_2)$, $E_4(N_3)$	101.4	107.3	105.4	107.2	196.2	210.2	205.4	209.3
5. $O_2(E_4)$, $O_1(N_3)$, $N_1(N_2)$	99.0	102.6	104.8	106.6	190.2	197.8	203.3	207.3
Average	102.9	105.5	106.2	107.8	201.1	205.8	207.1	209.8
Average gain as percentage of (3X-0X)	0	53.1	67.3	100	0	56.3	68.9	100

Recipient and Donor Lines	Egg Production ($10 \log_e$ (eggs/day))			
	0X	1X	2X	3X
6. $N_2(R_2)$, $R_2(C_{10})$, $N_1(N_2)$	38.2	40.7	41.1	42.8
7. $C_6(R_1)$, $R_1(C_6)$, $N_2(N_1)$, $C_{10}(R_2)$	33.2	39.7	41.5	43.5
Average	35.3	40.2	41.3	43.2
Average gain as percentage of (3X-0X)	0	62.0	75.9	100

SOURCE: Data from Robertson and Reeve 1955a.

Overall, the increment from inbred 0X to heterozygosis in one chromosome (1X) accounts for more than half of the total increment in 3X, and that from 2X to 3X is greater than from 1X to 2X. This is the same pattern as in the cross analyzed by Robertson (1954) and is similar to the pattern found by Sentz, Robinson, and Comstock (1954) in maize with varying degrees of heterozygosis, discussed in the preceding chapter. This suggests a general stimulatory effect of any heterozygosis put into a completely fixed line rather than a specific pattern of gene interaction. On the other hand, the overall pattern in the present data depends so much on a group of two (group 4) with the same recipient line, E_4, as to suggest that, after all, it may be due to a specific peculiarity of this line, easily remedied by replacement of one or other of its chromosomes.

A paper by Tantawy and Reeve (1956) is interesting for its comparison of inbreeding decline under four systems of inbreeding: mating of sibs ($\lambda = 0.809$), of half sibs ($\lambda = 0.870$), of double first cousins ($\lambda = 0.920$) and three cyclic pairs ($\lambda = 0.951$). The inbreeding coefficient reaches 25% at 1, 2, 4, and 6 generations, respectively; 50% at 3, 5, 9, and 14 generations, respectively; and 75% at 7, 10, 17, and 25 generations, respectively. The authors studied the course of change for 18 generations of percentage survival from egg to imago, the total output of pairs over ten days (in three successive vials), and wing length. Wing length, unfortunately, changed very little in this series, only about 1% at $F = 80\%$ in the three systems carried this far.

The clearest evidence was provided by the data on total output of live adults from eggs laid over a period of ten days. This averaged 33 per day in the controls but fluctuated considerably. The inbred lines showed correlated irregularities, but the trend under all systems was clearly downward and the rates were roughly according to the rates of inbreeding. The percentage of the productivity lost per 10% ΔF were 2.0, 2.0, 1.7, and 1.5 under the four systems in order of slowness. The smaller decline under the slower systems is most easily explained by lower effectiveness in relation to opposed natural selection.

Summary

The differentiation expected among inbred lines because of the randomness of fixation was only touched on in the preceding chapter. It has been illustrated more extensively in this chapter, especially that observed in an experiment with guinea pigs, conducted in the early generations with a minimum of selection.

Differentiation was most conspicuous in color, as each subline of each line became automatically fixed in some combination of the major color genes that had been segregating in the foundation stock. Ultimately there was fixation of each line as a whole, after sublines, which were lagging in the generation of inbreeding, began to be systematically eliminated. Of more significance, however, was the differentiation in intensity of color among lines with the same genotype with respect to known genes. These differences depended on multiple factors with individual effects too slight for Mendelian analysis. Most of the lines were spotted. While enormous phenotypic variability persisted in all of these because of nongenetic factors, each came to be characterized by a different median percentage of white.

Marked differentiation occurred in weight, with differences of more than 30% whether at weaning or as adults. There were also marked differences in conformation (length of legs relative to body, straightness of back, and so on), in minor superficial morphology (shape of nose, ears, and so on), in forms of internal organs (thyroid gland, adrenal, spleen), in percentage with little toes, and in percentage incidence of various types of monsters (especially otocephaly).

There came to be highly significant differences in both size and frequency of litters and in percentage born alive and in percentage reared of those born alive. There was no significant correlation between the rankings of 22 inbred strains in these four respects, indicating the random fixation of genes affecting them independently. There was, however, strong correlation between rank in size of litter (due largely to ovulation rate) and in growth rate, indicating an important common factor in this case. While the ranking with respect to weight remained essentially the same under all conditions over the years, there were highly significant shifts in rank in the other aspect of vigor, indicating important genotype-environment interactions for these. Moreover, strains tended to rise or fall in rank in all of these respects simultaneously to some extent, indicating that there were, after all, common factors of considerable importance affecting the general vigor of strains under particular conditions.

There were highly significant differences in resistance to tuberculosis, transmissible in crosses among the five most studied lines, but again there were genotype-environment interactions that shifted the ranking under some conditions.

These same lines differed in the histocompatibility genes that became fixed in them. Animals accepted transplants from others of their own line almost as if antotransplants, but rejected those of all other lines. F_1 hybrids accepted transplants from either parent line, but the latter rejected transplants from their hybrids, as is usual with histocompatibility genes.

Finally, there were marked differences among the lines in temperament, ranging from extremely phlegmatic to very active or very nervous.

It appeared that these inbred lines of guinea pigs would be found to differ in any trait in which they were examined sufficiently carefully. A similar conclusion is indicated for inbred lines in other species of animals as well as of plants, although not discussed here in as much detail.

There is general agreement among experiments on sib mating with minimum selection that the rate of fixation is much slower than under self-fertilization. It is noted that fixation of a number of multifactorial series (histocompatibility genes, modifiers of white spotting, percentage occurrence of little toe) show almost but not quite complete fixation after some dozen generations of sib mating ($F = 0.923$). Comparison of different systems— mating of sibs ($\lambda = 0.809$), of half sibs ($\lambda = 0.870$), of double first cousins ($\lambda = 0.920$), and of three cyclic pairs ($\lambda = 0.951$)—showed roughly similar inbreeding declines at the theoretically equivalent numbers of generations in *Drosophila*.

That inbreeding brings about a decline in characters concerned with fitness (viability, fecundity, growth) has been confirmed by all sufficiently extensive experiments with animals, including even Dr. King's rats, the most notable for long persistence of high vigor. The typical results were illustrated by the guinea pig experiments, with marked decline in size and frequency of litters and slight declines in viability percentages and weight during a period in which F averaged about 84%. The later catching up of the declines in viability and weight with those in fecundity was probably due to the greater susceptibility of inbreds than controls to less favorable conditions and to extinction of sublines and lines that fell below a critical level in fecundity. There was full recovery of all aspects of vigor in F_1 crosses with respect to offspring characters and in offspring from F_1 mothers in characters with maternal heredity (size of litter, percentage born alive). There was dependence on both parents in frequency of litters.

More drastic declines in other cases, for example, rabbits, livestock, man, Japanese quail, may be traceable to lesser exposure to previous bottlenecks of inbreeding during which deleterious genes could have been eliminated. This applies to *Drosophila* experiments, supplemented by the effect of much more linkage as a result of the fewness of chromosomes and the absence of crossing-over in males. With strong linkage, all fixation tends to involve seriously deleterious genes, and the consequent elimination of most lines causes most of those that persist to be ones that happen to be much more heterozygous than expected by the theory for neutral genes. The differences in the decline in productivity among the lines in which different systems of mating were compared by Tantawy and Reeve (percentage decline of 2.0, 2.0, 1.7, and 1.5 per 10% increase in F in the order of citation above) may reflect the decreasing efficiency of the systems in the face of natural selection.

There seems to have been no more difficulty, and perhaps less, in establishing good homozygous inbred lines of mice than of guinea pigs. In rats, King's long-continued experiment seemed for many generations to exhibit a complete absence of inbreeding decline. Tests of transplantability by Loeb indicated that this was due to persistence of a high degree of heterozygosis up to generations 37–47 and of a degree at the 100th generation as great as in the guinea pigs after only a dozen generations (also tested by Loeb). The explanation of this persistence of heterozygosis, and perhaps also of fixation of exceptional favorable combinations of genes, seemed to be the very much more vigorous selection practiced by King within the large litters and among the many sublines in her two large lines of rats, in contrast with the absence of selection within the small litters or among the few sublines of the many inbred strains of guinea pigs in the first dozen generations. The efficiency of selection among lines was evident in connection with litter size in mice in the experiments of Falconer and Roberts. That the inbred rats had declined appreciably in weight by generation 47 is indicated by the strong heterosis found in outcrosses by Livesay.

The theoretical proportionality of inbreeding decline to the increase in the inbreeding coefficient is on the assumption of additivity of locus effects. The

experiments of Robertson and Reeve on the effects of chromosome replacements in strains of *Drosophila*, made isogenic by breeding methods involving
marker genes and crossover inhibitors, has revealed important nonadditive
interactions that cannot be overcome by any transformation of scale. The
most general interaction effect in this and other cases, including experiments
with maize, has been excess depression as 100% fixation is approached, but
this has not been invariable and other types of interaction were demonstrated.

CHAPTER 4

Variability under Inbreeding and Crossbreeding

This chapter will be concerned with the analysis of phenotypic variability into genetic and nongenetic components in random bred stocks, and with the effects of inbreeding and the crossing of inbred lines on such variability. Illustrations of patterns of components will be drawn largely from data from the guinea pig colony, to which frequent reference has been made, but comparisons of variability in inbred strains and their crosses will be made more broadly.

The simplest pattern is, of course, that of complete determination of phenotype by genotype, which is often manifested by "good" Mendelian genes. Of more interest here is the high degree of uniformity at birth of each of the inbred strains of guinea pigs, or their first crosses, in the intensity of whatever color is determined by major genes, where the differences among these is due to genetic factors with effects too slight for Mendelian analysis. It should be noted that the uniformity of the environment before birth plays an important role, since quality and intensity are much affected later by aging, temperature, and nutritional factors (vol. 1, chap. 5).

Among other characters, certain conformational differences referred to in chapter 3, while not measured, gave an impression of great uniformity within strains in contrast with the conspicuous differences among them.

The next category is that of characters that vary widely within isogenic lines and that do so largely independently of all tangible environmental factors. In mammals, great variation in such lines but the absence of any appreciable correlation between littermates at birth indicates a character that is greatly affected by intangible accidents of implantation or development.

White Spotting

The most intensively studied case of continuous variability of this sort was piebald spotting. Aspects have been discussed in volume 1, chapters 6, 7, 11, 14, and 15. The strains to be discussed here were all homozygous *ss* but varied from about 20% to 89% in median amount of white, while each varied over half or more of the range from near-self color to self-white. As shown in volume 1, figure 6.5c, the distributions were highly asymmetrical if the median deviated much from 50%, because of damping of variability toward each extreme. No completely self-colored occurred, however, in genotype ss, and self-white was not common except in one strain (no. 13), where there were about 9% in the period (1916–24) principally dealt with here.

Variabilities could not be adequately compared except by a transformation of scale. The best transformation arrived at was on the assumption that the potentialities for pigmentation in skin areas are distributed in the form of an isosceles triangle (vol. 1, fig. 11.6). This permits 100% color (in Ss or SS) or 100% white if the upper threshold falls wholly below or the lower one wholly above. Outside these thresholds, no transformation is possible. The assumption of a normal distribution of potentialities among areas of the coat and transformation of the percentage of white according to the inverse probability integral, $pri^{-1}p$, where p is the observed percentage of white, is more convenient and satisfactory here, except where there is a considerable percentage of self-white (Wright 1920). The probit transformation of Bliss (1935) is $5 + pri^{-1}p$. The amounts of white were recorded in units of 5% (or in some cases 10%), and transformation was made of these values. The terminal ranges, 0–2.5%, 97.5%–100%, where 5% units were used, were treated as if centered at 1.25% and 98.75%, respectively.

A branch of strain 35, tracing to a single mating in generation 12 of sib mating, was the most suitable for detailed study because of its intermediate median, 65%, and its expected high degree of genetic uniformity. It ranged from a trace of white to rare self-white. Near-fixation of all modifying factors was supported by the absence of significant differences among 21 subdivisions averaging about 100 individuals each (Wright and Eaton 1929) (fig. 4.1), as well as by the insignificance of the parent-offspring correlation, 0.024 ± 0.017. A large branch, 35D, descended from a single mating in generation 22, was maintained during a later period (1926–34) by random mating. This was about 6% whiter than the group studied earlier, indicating that fixation had not been quite complete in the earlier period. There was no significant parent-offspring correlation within it.

The females were consistently whiter than the males in strain 35 and in all other strains studied (vol. 1, tables 9.11 and 9.12). The pattern is determined so early that it is unlikely that this is traceable to any effect from the gonads. More probably it depends on the chromosomal difference, XX vs. XY, directly.

An unexpected increase in the amount of white with increasing age of the dam was the only tangible environmental effect found in a study of the Belts-

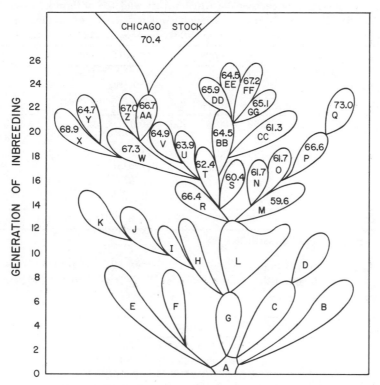

FIG. 4.1. Branches of strain 35 of inbred guinea pigs, showing percentages of white in the coat in the later generations. The Chicago stock was maintained by random mating. Largely from data of Wright and Eaton (1929).

ville branch of strain 35. Other possible indexes of environmental effect, month of birth, size of litter, birth weight, and perinatal mortality (Wright 1926), showed negligibly small effects. These results were confirmed in the Chicago branch (Wright and Chase 1936). The relation to sex and age of dam are shown in table 4.1 for the two periods.

The females were, on the average, 6.7% whiter than the males in the earlier period and 5.3% in the later. In the earlier period, the percentage of white of offspring rose from 58.4% from dams less than six months old to 70.1% from ones over 21 months old, or 11.7%. In the more recent period, with much smaller numbers, there was more irregularity but a total increase of 17.4%.

A comparison of the variabilities in strain 35 (1916–24) and the contemporary random bred controls (stock B) is instructive. The statistics here were

TABLE 4.1. Average percentages of white by sex and age of dam in strain 35 in two periods. The last column gives averages based on equal weighting of the sexes but weighting according to the total frequencies in the two periods.

AGE OF MOTHER (MO.)	STRAIN 35 (BELTSVILLE)				STRAIN 35D (CHICAGO)				AVERAGE
	Males		Females		Males		Females		
	No.	% White	No.	% White	No.	% White	No.	% White	% White
3–5	182	56.3	153	60.5	25	56.0	31	65.0	58.7
6–8	195	59.5	187	67.6	45	67.8	41	70.2	64.4
9–11	152	60.6	160	66.5	40	71.7	21	78.1	65.4
12–14	150	61.3	124	70.6	33	69.8	30	76.3	67.1
15–20	174	63.2	149	69.6	27	61.9	34	74.7	66.7
21–46	138	66.9	144	73.3	23	78.2	25	77.6	71.4
Total	991	61.1	917	67.8	193	67.8	182	73.1	65.4

SOURCE: Wright and Chase 1936.

based on the transformation ($pri^{-1}p$) of classes at 10% intervals, used in preparing correlation tables, and differ slightly from those based on 5% intervals discussed later.

Strain	No.	Mean	σ	σ^2
35	1,426	0.404	0.583	0.340
B	1,929	0.636	0.757	0.573

On the transformed scale, the variance of strain 35 is 59% of that of the random bred strain. Under the hypothesis that the variance of strain 35 is wholly nongenetic and that the random breds have the same nongenetic component but an additional genetic one, the latter determines 41% of its variability.

A wholly independent estimate of heritability can be made from the parent-offspring correlations in the random bred stock under the hypothesis of additivity of all factors, including semidominance of the genetic ones. These are shown by sex in table 4.2, together with the correlations between litter-mates, also by sex. The same correlations are given for the inbred strain 35, averaging those from the earlier and later periods. Correlations between sibs from different litters were obtained only for the later period.

There is no significant correlation between the parents in either case. The highly significant parent-offspring correlation in the random bred stock ($+0.191 \pm 0.016$), taken as $0.5h^2$ under the assumption of semidominance of modifiers, indicates a heritability of $h^2 = 0.38$. The absence of significant parent-offspring correlation in strain 35 ($+0.024 \pm 0.017$) supports the

TABLE 4.2. Correlations between parents, parents and offspring, and between littermates by sex in the random bred stock B (1910–24) and in the inbred strain 35 (1910–24), 1926–34 averaged. Correlations between sibs from different litters in strain 35D are shown for the later period.

| | | RANDOM BRED (B) | | INBRED (35) | |
		No.	r	No.	r
Sire-dam		143	+0.060	212	+0.026
Parent-offspring	♂ — ♂	973	+0.244	938	−0.003
	♂ — ♀	929	+0.187	878	+0.090
	♀ — ♂	1,014	+0.217	934	+0.021
	♀ — ♀	965	+0.116	873	−0.010
	Total	3,881	+0.191	3,623	+0.024
Littermates	♂ — ♂	537	+0.355	431	+0.106
	♂ — ♀	1,050	+0.288	902	+0.093
	♀ — ♀	493	+0.190	394	+0.132
	Total	2,080	+0.282	1,727	+0.106
Sibs, different litters	♂ — ♂			335	−0.061
	♂ — ♀			509	−0.049
	♀ — ♀			202	−0.021
	Total			1,046	−0.048

hypothesis of near-fixation in this strain. The estimate of 38% heritability in the random breds agrees fairly well with the independent estimate of loss of 41% variance by fixation in strain 35.

There is a small but significant correlation between littermates in the inbred strain (+0.106 ± 0.024) that indicates that influences from common environment were not wholly absent. The relation to the age of the mother is one such factor. The correlation ratio of 0.190 ± 0.015 indicates that it should be responsible for a correlation of +0.036 (=0.190²) between sibs. Apparently there are also other small common influences of the intrauterine environment. The absence of significant correlation between sibs from different litters (−0.048 ± 0.031) indicates that permanent differences among dams play no role.

A small influence from common environment is confirmed by the higher correlation between sibs (0.282 ± 0.022) than between parent and offspring (0.191 ± 0.016) in the controls. The excess 0.091 pertains to a larger variance (0.573) than in the case of the inbreds (0.340), and thus implies a larger variance component (0.052) than that (0.036) implied by the excess in the inbred strain, even if the parent-offspring correlation in the latter is treated as zero.

The excess in the random breds might, in itself, be interpreted as due to correlation of dominance deviations, but this does not account for the excess in the inbreds or for the fairly close agreement between the two estimates of the portion of the random bred variance lost on inbreeding. The relations of these quantities may be seen from the formulas for parent-offspring (r_{op}) and offspring-offspring ($r_{oo'}$) in the presence of some degree of dominance (vol. 2, eq. 15.39 and 15.40). Here g^2h^2 is the additive heritability and d^2h^2 is the component of total heritability h^2, due to dominance; e^2 measures the influence of common environment of littermates; and a^2 measures that of accidents of development;

$$g^2 + d^2 = h^2, \qquad h^2 + e^2 + a^2 = 1$$
$$r_{op} = 0.5\, g^2h^2$$
$$r_{oo'} = 0.5\, g^2h^2 + 0.25\, d^2h^2 + e^2$$

The loss of variance on inbreeding is h^2 ($=0.407$). Since the additive heritability is estimated to be $g^2h^2 = 0.382$, $d^2h^2 = 0.025$, and the contribution of dominance to $r_{oo'}$ is only 0.006, leaving $e^2 = 0.085$, the corresponding variance component (0.049) is still larger than the estimate from the inbreds (0.036).

Another possible complication is that homozygotes may vary more than heterozygotes (as in the case of some characters to be discussed later). If this were true, the inbreds should have more than 62% of the variance of the controls on the basis of the parent-offspring correlation of 0.38, assuming $d^2 = 0$, instead of less. There may, of course, be balancing of excess non-genetic variability of inbreds and a greater contribution from dominance to the loss of random bred variance on inbreeding than 0.025. As discussed later, however, there is no indication of dominance in crosses between inbred strains.

In view of the standard errors, there is no necessity to postulate any deviations from additivity of either genes or environment, but a slight contribution due to deviations of heterozygotes from exact intermediacy is quite probable.

There are significant differences among the correlations in the control stock in relation to sex. These are obviously not at all of the sorts that could be explained by maternal influence, cytoplasmic heredity, or special sex-linked genes. Like the sex difference in grade, they are probably due to the chromosomal difference, XX vs. XY. They clearly indicate a greater heritability in males than in females. The parent-offspring correlations indicate that g^2h^2 is about 0.49 in males but only about 0.23 in females. The littermate correlations indicate rather similar figures on subtracting the excess 0.09 (with no allowance for differences in d^2h^2) (0.52 and 0.20, respectively). These are based on like-sexed pairs, but the cross-sex correlations are intermediate in both the parent-offspring and littermate correlations. It appears that the interaction between sex and spotting is of a more complicated sort than the mere difference in median.

Rough estimates of the degrees of determination by factors and of the variance components are given in table 4.3. Those in the inbreds are as indicated above. In the controls, d^2h^2 is assumed to be such as to account exactly for the reduction of the random bred variance (0.573) to that observed in the inbreds (0.340).

TABLE 4.3. Estimated degrees of determination and variance components with respect to white spotting in the random bred stock B (by sex) and in the inbred strain 35. Somewhat less than half of the variance common to littermates is due to the age of the dam.

	RANDOM BRED (B)				INBRED (35)	
	♂(%)	♀(%)	%	σ^2	%	σ^2
Genetic						
Additive (g^2h^2)	50	25	38	0.219	0	0
Dominance (d^2h^2)	4	2	3	0.014	0	0
Sex	...	...	2	0.011	3	0.010
Nongenetic						
Common to littermates (e^2)	9	9	9	0.049	10	0.036
Accidental (a^2)	37	64	48	0.280	87	0.294
Total	100	100	100	0.537	100	0.340

This analysis is in terms of the animal as a whole. Actually, variation in strain 35 has been shown to behave as if due to largely independent chance events on each side of the head, in three dorsolateral areas on each side, and in a median rump area (vol. 1, table 14.13).

Next we will compare four inbred strains (39, 35, 32, and 13) with widely different median amounts of white, and their crosses, during the period 1916–24. Strains 39, 32, and 13 each traced to single matings in generation 8 and 35, as before, to one in generation 12. Substrain analysis (Wright and Eaton 1929) indicated little if any genetic heterogeneity in amounts of white in strains 35 and 13, only a slight amount in strain 32, but a larger amount in strain 39. The fifth large strain, no. 2, tracing, as tabulated for spotting, to a single mating in generation 8, turned out to be extremely heterogeneous, consisting of two roughly equal branches that centered at 63% and 85% white, respectively, but with considerable variability within these. The branch retained after 1924 traced to a single mating in generation 15. With the median at 94% white, it approached the contemporary branch of 13 (median 98%) in its whiteness. Crosses between 13 and this branch of 2 were intermediate and showed substantially the same variability in F_1 and F_2 (Wright and Chase 1936).

The mean and standard deviation of each sex was calculated for strains 39, 35, 32, and 13, for each of the reciprocal crosses, and for two F_2 populations, using the transformation $X = pri^{-1}p$ for ranges at 5% intervals. There was a slight matroclinous tendency in the reciprocal crosses (average difference of means, 0.086) but it was not sufficiently consistent for significance. Reciprocals are combined in table 4.4.

TABLE 4.4. Statistics of four inbred strains, their first crosses, and two F_2 populations with respect to white spotting with transformation of the observed percentages of white; $X = pri^{-1}p$.

Strain (1916–24)	Males			Females			Differences		
	No.	M_m	σ_m	No.	M_f	σ_f	$M_m - M_f$	$\sigma_m - \sigma_f$	
39	364	−1.001	0.685	295	−0.623	0.709	−0.378	−0.024	
35	751	0.299	0.597	709	0.490	0.587	−0.191	+0.010	
32	405	0.980	0.744	467	1.209	0.584	−0.229	+0.160	
13	636	1.615	0.604	642	1.720	0.548	−0.105	+0.056	
39 × 35	27	−0.400	0.693	17	−0.457	0.700	+0.057	−0.007	
39 × 32	72	−0.008	0.901	77	0.274	0.806	−0.282	+0.095	
39 × 13	67	0.405	0.806	90	0.623	0.690	−0.218	+0.116	
35 × 32	64	0.617	0.627	60	0.796	0.571	−0.179	+0.056	
35 × 13	76	1.036	0.629	80	1.105	0.483	−0.069	+0.146	
32 × 13	81	1.040	0.743	69	1.306	0.624	−0.266	+0.119	
$(39 \times 13)^2$	34	0.307	0.765	48	0.412	0.890	−0.105	−0.125	
$(35 \times 13)^2$	29	0.971	0.642	26	1.094	0.654	−0.123	−0.012	

In all but one of the entries, the females are whiter on the average than the males (average difference 0.202 in the four inbred strains, 0.188 in the six first crosses, and 0.112 in the two small F_2 populations). Here, as in all other adequate comparisons, the difference is highly significant (see vol. 1, tables 9.11 and 9.12).

The males were more variable than the females in most cases. There is significance at the 0.01 level of the difference 0.057, in the ten nonsegregating populations by Student's test. There is, however, a serious amount of inconsistency among the four large inbred strains, which seems to imply that there are specific differences among them in this respect. The F_2 populations do not agree at all, but this may be discounted because of segregation in small numbers. It appears that greater male variability is part of the complex effect of the X and Y chromosomes on spotting (more color, higher heritability, and greater variability in males).

Only two F_2 populations were produced that approached adequacy for estimating segregation indexes. That from 39×13 is of most interest as being from the cross between the extremes. F_2 variance (average of sexes) was 0.689. The nongenetic variance, estimated from the parental strains and F_1 in 1:1:2 ratio, was 0.486, giving 0.203 as the average genetic variance. This is significantly less than 0.773, the expectation under one-factor segregation. The difference, R, between the parental means was 2.480, giving an index of $S = R^2/8\sigma_G^2 = 3.8$ (vol. 1, eq. 15.9).

In the case of 35×13, the average F_2 variance was 0.420 and the average nongenetic variance was 0.328, giving $\sigma_G^2 = 0.092$ compared with 0.204 for one-factor segregation. The range, assuming no complementarity, was 1.274, giving an index of 2.2. Thus it is reasonably certain that the parents differed by more than a single locus and the estimated numbers are four similar loci in the extreme cross, two in the other, or equivalents in terms of unequal gene effects (vol. 1, table 15.6).

In table 4.5 the standard deviations of the inbred strains are compared in each case with the unweighted average from the three F_1's to which the strain in question contributed. The sexes are averaged.

TABLE 4.5. Comparison of the standard deviations of white spotting (transformed scale) of four inbred strains and the averages of the crosses into which they entered.

Strain	Mean	σ (Inbred)	σ (3F$_1$'s)	Difference
39	-0.812	0.697	0.766	-0.069
35	$+0.345$	0.592	0.617	-0.025
32	$+1.094$	0.664	0.712	-0.048
13	$+1.667$	0.574	0.663	-0.087
Average		0.632	0.689	-0.057

In each case the inbred strain is less variable than the F_1's of the crosses into which it entered, and the average difference is 0.057. The standard deviations of strain 13 and its crosses are, however, rather unreliable because of the high medians and the breakdown of the scale at self-white. The standard deviation of strain 39 and its crosses are unsatisfactory because of its considerable heterogeneity. The biggest differences must thus be discounted. The most reliable single inbred standard deviation undoubtedly is that of strain 35 (0.592) and the most reliable crosses are those to which it contributes (0.617), which exceed it by only 0.025. There is thus no clear evidence that the inbreds are less variable, much less that they are more variable than F_1.

Otocephaly

White spotting is a continuous character, except near limits that are rarely approached in most strains. A character that is similar in the absence of appreciable correlation between littermates, but that differs in always being dependent on the crossing of a threshold is otocephaly (vol. 1, figs. 5.16 and 5.17). This occurred in varying frequencies in ten of the inbred strains, was wholly absent in thirteen, while one case occurred in the control stock. It was most common in the early history of strain 32 (9/2,718) and the later history of strain 13, in which the main line, tracing to generation 13, produced an average of 5.5% without significant variation among seven large sublines over a period of 30 years, except for one, starting from a mating in generation 19, in which a partially dominant mutation raised the incidence to 18.2%. The frequencies in sublines are shown in figure 4.2.

The constancy of the percentage incidence and the absence of serious prenatal losses demonstrated that the main line was essentially isogenic, except, of course, for segregation of the sex chromosomes. The frequency in females was about twice that in males in all branches of strain 13 and also among the sporadics from all other stocks (vol. 1, table 6.3). This is even more certainly a direct effect of the chromosome difference, XX vs. XY, than is the case of spotting, in view of its determination by inhibition of the prechordal mesoderm not later than the medullary plate stage. Apart from this sex difference, occurrence within the main line of strain 13 could not have been the result of segregation of any genotype, but must have been the consequence of nongenetic causes (such as accidents of implantation) acting on a susceptible genotype almost wholly independently of other eggs in the same uterus. The evidence for this independence has been given in volume 1, tables 8.4 and 8.6. Reciprocal crosses with other strains showed that the trait was not determined by maternal influence.

As with white spotting, however, tangible external factors acting on the condition of the dam played a slight role. In the early Beltsville data, there was a significantly higher incidence (13.5%) from January to April than in the rest of the year (7.1%), in which nutritional conditions were more favorable. There was also a significantly greater incidence in litters of one or two than in larger litters, but no significant relation to age of dam, birth weight, or mortality of littermates (Wright and Eaton 1924). It should be noted that a small correlation with an environmental factor implies only the much smaller square of this correlation between littermates. The absence of any significant correlations in the later data under better controlled conditions indicates that seasonal or other tangible environmental factors ceased to play any appreciable role.

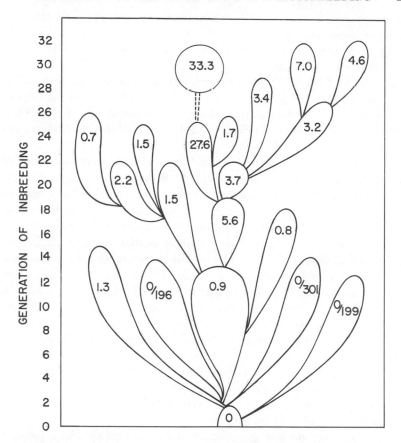

FIG. 4.2. Branches of strain 13 of inbred guinea pigs, showing the percentages of otocephaly. The branch at the top with 33.3% was maintained by random mating. Redrawn from Wright (1949*d*, fig. 11); used with permission.

The striking variability within litters containing both normals and monsters cannot be interpreted as a general consequence of inbreeding. It is a consequence of inbreeding only in the sense that the random fixation of genotypes, due to inbreeding, is likely to give rise to extremes that are inadequate for normal development under one or other stress.

Digit Numbers

The number of digits on the hind feet of guinea pigs is a character with two thresholds, a lower one for any development and an upper one for perfect development of an atavistic little toe and its associated plantar pad. Big toes and thumbs were restored only by a mutation, Px, which appeared in a single animal in an unrelated random bred strain I. The distribution of the little toe among the strains was rather similar to that of otocephaly, except that the particular ones in which it was present were, in general, different. Thus it was present in 11 of the 23 inbred strains in frequencies ranging from 0.3% to 19% (up to 1915) and it occurred, but only very rarely (four cases), in the random bred stock. It differed from otocephaly in not being lethal, and Castle (1906) produced an inbred strain, D, by selection in which the little toes were perfect in all individuals.

The mode of inheritance was studied principally in crosses between the four-toed strain, D, and three strains (2, 13, and 32) that were wholly three-toed, or one (strain 35) that had come to have an average of 31% incidence in the large branch developed from one mating in generation 12, referred to in connection with white spotting. Unlike the situation in the latter character, however, there was a large amount of substrain differentiation, 9% to 69% (fig. 4.3), with parent-offspring correlation 0.18 (tetrachoric) in the strain as a whole, 0 within sublines, during the period 1916–24 (Beltsville laboratory). There was no significant parent-offspring correlation in the large branch 35D from one mating in generation 22, maintained later in another laboratory (Chicago). As stated previously (vol. 1, pp. 93–95, 415–16), the results of crosses indicated continuous variability on an underlying physiological scale, interrupted by the two thresholds, referred to above. Normal distributions were assumed and means and standard deviations were determined by use of the inverse probability integral, taking the distance between the thresholds as the scale unit. The nongenetic standard deviation was taken as that within sublines of strain 35 (0.80 on this scale) (Wright 1934b).

There was a second striking contrast with the situation with respect to otocephaly of strain 13 in the strong correlation between littermates in strain 35 (vol. 1, tables 8.5 and 8.6). The tetrachoric correlation was 0.62 in the Beltsville data, reduced to 0.54 on removing the substrain differentiation. It was 0.54 without any reduction in the Chicago data, in which genetic differences were absent.

It was shown in the latter case that there was a correlation of 0.25 between individuals from consecutive litters of the same mating and essentially the same, 0.29, between nonconsecutive ones, indicating permanent nongenetic differences among dams. This leaves about 0.27 as the correlation from nongenetic causes peculiar to litters. Most of this was due to the age of the dam, with high incidence from immature females (vol. 1, table 6.5). The squared correlation ratio was 0.25 in the Chicago data, 0.14 in the Beltsville data. There were highly significant seasonal effects (squared correlation ratio, 0.034) in the Beltsville data (vol. 1, tables 9.6 and 12.2, fig. 12.2), wholly

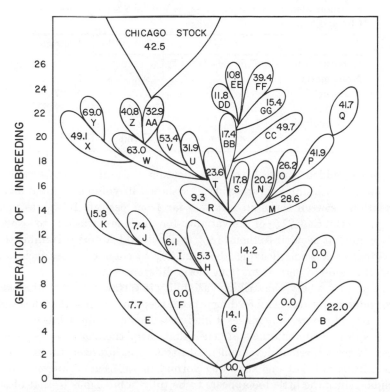

FIG. 4.3. Branches of strain 35 of inbred guinea pigs showing percentages of polydactyly. The Chicago stock was maintained by random mating. Redrawn from Wright (1949d, fig. 9); used with permission.

absent under the more uniform nutritional conditions at the Chicago laboratory.

Analysis of variability into components in both bodies of data were given earlier (vol. 1, table 12.5). It seems desirable to give a somewhat condensed version here (table 4.6) for comparison with the analysis of white spotting. The contribution from permanent differentiation of the dams is given for the Chicago data in which it was determined.

It may be noted that only 46% of the variability of an isogenic strain is due here to accidents of development, in comparison with 87% in white spotting and some 90% to 100% in otocephaly.

TABLE 4.6. Analysis of variance of digit numbers in earlier (Beltsville) and later (Chicago) branches of strain 35. In the former the component common to littermates includes that common also to nonlittermates, which are separated in the Chicago figures.

	Beltsville	Chicago
Genetic (substrain differentiation)	18	0
Nongenetic		
Dams	...	27
Litters	44	27
Individuals	38	46
	100	100

It may be added that, as with white spotting, the individual as a whole was not the ultimate unit of variability. As shown in volume 1, table 6.4, the correlation between the hind feet was far from perfect. It was estimated (Wright 1934c) that 27% of the variance for single feet was due to accidents of development peculiar to them, 14% to accidental factors pertaining to the individual as a whole, leaving about 29.5% to common litter factors and about the same to permanent nongenetic differences in dams.

As with otocephaly, the striking variability in certain inbred strains cannot be considered an effect of inbreeding per se, except in the sense that the random fixation due to inbreeding makes it likely that some strains will become fixed with respect to combinations of genes that carry the character beyond the threshold for abnormal development. In this case, however, the trait is not a monstrous one but one that was normal in the remote ancestry of the Caviidae, since the little toe appears to be fully homologous to that of other families of South American rodents (vol. 1, chap. 5). That it is not a matter of association of variability with inbreeding as such and of uniformity with crossbreeding, was illustrated by the cross 13 × D, two completely uniform inbred strains (three-toed and four-toed, respectively), and F_1, in which 31% were three-toed, 69% four-toed, with no genetic differences as shown by F_2 and backcross tests. In this case the inbreds were uniform because they were beyond lower and upper different thresholds, respectively, while the crossbreds fell between.

It may be well to clear up an apparent inconsistency with respect to figures given for the correlation between littermates in volume 1. In table 10.4 in that volume, the distribution of three-toed and four-toed individuals in litters was used as an illustration for fitting a binomial distribution, modified by correlation in occurrence. The estimates of the correlation that emerged were 0.45 in litters of three, 0.33 in litters of four, both from the

Beltsville representatives of strain 35, and 0.41 in litters of three from the Chicago branch. These are, as would be expected, identical with the easily determined point correlations. This is appropriate enough for the purpose of fitting, but for analysis of variance components under the assumption of normal distributions on an underlying scale, cut by thresholds, the appropriate correlation coefficient is Pearson's tetrachoric (vol. 1, p. 287), which, unfortunately, is very complicated. This comes out considerably larger, 0.67, 0.53, and 0.60 in the three cases above, with average 0.60 instead of 0.40. The average for all litter sizes was 0.62 in the Beltsville data, 0.54 in the Chicago data (vol. 1, pp. 293–94).

The standard error, also unfortunately complicated, was not given in volume 1. Pearson (1913) gave as a good approximation:

$$SE_r = f(r)\sqrt{(pp'qq')}/zz'\sqrt{N},$$

where

$$f(r) = \left[1 - \left(\frac{\sin^{-1} r}{90°}\right)^2\right](1 - r^2)$$

and where p and q are the proportions for one of the dichotomies of the 2×2 table, p' and q' are those for the other, z and z' are the corresponding ordinates of the unit normal curves at the cleavage points, and N is the number of entries. In the case of a symmetrical table (as for littermates) $p' = p$, $q' = q$, $z' = z$. In the case of littermates, $N = 0.5\sum n(n - 1)$ where n is the size of litter. A factor,

$$\sqrt{\{[1 + (n - 1)r]/(1 + r)\}},$$

must be used in the formula for SE_r to allow for repetition of individuals, if greater than 2 (vol. 1, 312–13). Convenient tables are given by Kelley (1923), except that his table for $f(r)$ introduces the factor 0.6745 to convert standard errors into probable errors, which were then customary.

Digits Restored by the Mutation Px

An atavistic little toe, identical with that in the strains discussed above, was usually present in heterozygotes, $Pxpx$, of a mutation, Px, associated with atavistic thumbs, less frequently with atavistic big toes, and very rarely with a nonatavistic sixth digit, usually beyond the fifth digit on the forefeet (Wright 1935a). The homozygotes $PxPx$ were monsters, only about 8% of which reached birth (and then died). Their broad paddle-shaped feet had up to 12 digits each (vol. 1, table 6.8), and there were many other grossly abnormal characters (vol. 1, p. 96).

This mutation appeared in an otherwise normal random bred three-toed strain (I). Crosses were made with the wholly four-toed strain D that showed that the heredity of the latter greatly increased the occurrence of thumb and big toe, as well as little toe. It proved to be impossible, however, to use the concept of a single underlying physiological scale with constant lower and upper thresholds for the occurrence of the little toe in $Pxpx$, much less for the other toes and the abnormalities of the homozygotes. With respect to the little toe, normals of strain I ($pxpx$) could be interpreted as having a mean at about the same scale value as three-toed strain 13 (-4 on the scale on which the lower threshold is 0, upper threshold 1, with strain D at $+3$) but much more variability, as expected from its mixed origin. The means and variabilities of F_1 (I × D) and first and second backcrosses to I selected for the presence of the little toe, but all $pxpx$ (as indicated by absence of thumbs) although from matings $Pxpx$ × $pxpx$, were approximately as expected under the simple multifactorial theory with constant thresholds. The distribution with respect to little toes (none, poor, good) in $Pxpx$ of strain I had a mean of about $+0.2$, but variability less even than in an isogenic strain.

This can be interpreted as implying either that the presence of Px reduces variability or that the thresholds are about 2.6 times as far apart as with $pxpx$ of the same strain. On this view there was 2.6 times as much canalization in the terminology proposed by Waddington (1942, 1955). These relations are shown in figure 4.4.

On taking the pollex and hallux into account, no single scale with diverse thresholds was at all adequate. The ranges of the three genotypes $pxpx$, $Pxpx$, and $PxPx$ in strains D, I, and crosses ID, (ID)D, AD, and (ID)A are shown in a two-dimensional pattern in figure 4.4 (Wright 1935a). Strain A ($pxpx$) was one that showed very few little toes in $pxpx$ from the cross with D, and on introduction of Px in cross (ID)A showed a very much stronger tendency to exhibit thumbs than little toes, and never had big toes (vol. 1, table 6.6). It was suggested that in figure 4.4 the abscissas represent time in development while the ordinates at a particular abscissa correspond to the scale with the constant thresholds of the simple one-dimensional theory. There may well be more than two dimensions.

Other Abnormalities

A considerable number of other types of morphological abnormality were observed among some 120,000 guinea pigs (Wright 1960b). Dependence on a single mutation with 100% penetrance was rare. There were only two or three probable recessive mutations of this sort (ones that cause micromelia [in

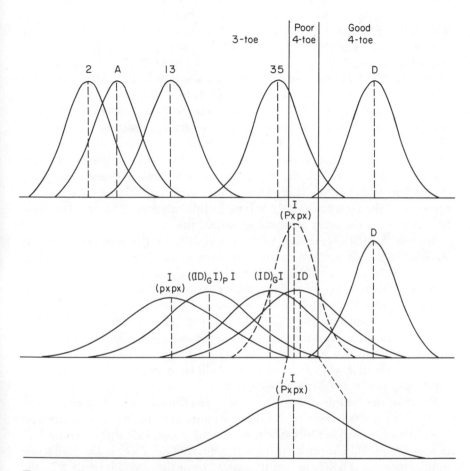

Fig. 4.4. *Top,* estimated frequency distributions of factor complexes for inbred strains of guinea pigs, deduced from frequencies of three-toed, and poor and good four-toed offspring in strain 35, and in crosses between strain D and the others. *Middle,* estimated frequency distributions of factor complexes for genotype *pxpx* in strain I, based on $F_1(ID)$, and of two successive backcrosses of the latter to I. The distribution for genotype *pxpx* from I is shown in broken lines. Its small scatter is inconsistent with that of *pxpx* from I unless the thresholds are further apart at the time of gene action in this case (as illustrated at the bottom). Redrawn from Wright (1935a, fig. 4).

strain 24] and perhaps dropsy and dwarfism from particular crossbred matings). The mutation, Px, causing polydactyl monsters when homozygous, had 82% penetrance in $Pxpx$ in the strain of origin, which was reduced to less than 20% by outcrosses to certain strains but raised to 100% by crossing with four-toed strain D. The combination of two color factors, silvering $sisi$ and diminished $dmdm$, resulted in whites with reduced eye pigmentation, males with diminutive testes and no spermatogenesis, 50% of females sterile, anemia (75% of normal hemoglobin), and a significantly higher mortality after birth than in normal littermates (vol. 1, table 9.5).

Lerner (1958) suggested that "phenodeviants," such as crooked toes in fowls, are due to combination of polygenes with cumulative physiological effects, causing abnormality beyond a threshold. In large inbred strains, however, it must be nongenetic factors that decide which embryos fall below the threshold of normal development, characteristic of their genotypes, except for sex differences and effects of rare mutations. Otocephaly and the presence of the little toe clearly belong in this category. This was certainly true of most of the other observed abnormalities.

Microphthalmia appeared in 1.6% of strain 38, 0.1% in seven other inbred strains, 0 in the other inbred strains, but four individuals (0.2%) in the random bred stock B, prior to 1916. Contorted legs appeared in this period in 20 individuals distributed among the inbred strains and in three random breds (Wright 1922c). Much later, seven monsters of a very peculiar type (anotia), superficially somewhat like extreme otocephaly but with defects that indicated a slightly later critical period in development, occurred among 160 animals descended from a single mating in generation 19 of strain 2 with such a distribution as to indicate a threshold character based on a mutation giving 4% incidence (Wright 1960b).

Study of the records of 76,000 births in the Chicago branch of the colony in which only 8,000 were from inbred strains, gave further information on the factors involved in phenodeviants. Among those (145) with microphthalmia or anophthalmia constituting 0.2% (excluding $PxPx$), the frequency in littermates was 6.8%, in nonlittermates from the same mating 1.2%, the excess of the former indicating the influence of tangible environmental factors, of the latter either a genetic tendency in the parents or persistent conditions of the dam. Other defects (contorted legs, abnormal feet or jaws, otocephaly) were associated with microphthalmia much more frequently (12%) than their overall frequency (0.2%), indicating a common developmental factor. This applied to a lesser extent (1.0%) to nonmicrophthalmics from the same matings. The sex ratio among the microphthalmics, 52 male: 90 female:3 unsexed, indicated an influence of XX-XY segregation as in otocephaly and white spotting (but in no other class of abnormalities).

Generally similar situations were found with the other types of abnormality. These included a chunky type (two-thirds normal length but normal weight) associated with leg and foot abnormalities, independent contortion of legs or flexure of the feet, hydrocephaly and the more extreme exencephaly, in sufficient frequencies to indicate that they were all threshold characters in which crossing of the threshold depended on both genetic and nongenetic factors. There were also cruciate double monsters; micromelia; dropsy; distorted, fused, and doubled digits; assorted facial defects; diminutive ears; umbilical hernia; and spina bifida. These were too infrequent for definite conclusions, but were probably all of the same sort.

There was a general tendency for association of different types in the same individual or in littermates and, to a lesser extent, in different litters from the same mating. It appeared that gene products and environmental disturbances acted similarly and largely nonspecifically, except for time of action, to inhibit one or more developmental processes that were in a critical state at the time (Wright 1960b).

Viability

The traits of being born alive, and of reaching weaning if born alive, are ones in which no parent-offspring correlations exist, but it is possible to find the correlation between littermates in order to estimate the importance of accidental factors not common to them. These are not true point variables but, like otocephaly and the presence of atavistic digits, they are dependent on thresholds and underlying continuous distributions. The appropriate correlation coefficient is again Pearson's tetrachoric.

TABLE 4.7. Tetrachoric correlations and their standard errors for pairs of littermates with respect to whether alive or dead at birth in the Chicago representative of three inbred strains.

LITTER SIZE	STRAIN 2		STRAIN 13		STRAIN 35		TOTAL
	No. of Pairs	r	No. of Pairs	r	No. of Pairs	r	r
2	151	0.55 ± 0.17	340	0.82 ± 0.08	62	0.67 ± 0.15	0.757 ± 0.063
3	363	0.79 ± 0.06	876	0.83 ± 0.05	177	0.63 ± 0.11	0.796 ± 0.035
4	276	0.64 ± 0.11	756	0.80 ± 0.05	150	0.34 ± 0.16	0.735 ± 0.044
5, 6	95	0.04 ± 0.16	130	0.32 ± 0.26	40	0.70 ± 0.27	0.594 ± 0.114
Average		0.719 ± 0.050		0.809 ± 0.031		0.590 ± 0.072	0.762 ± 0.024

Table 4.7 shows estimates of the correlations in the case of birth, dead or alive, based on concurrence, or otherwise in 2×2 tables from all possible pairs in litters of two or more in the Chicago representative of inbred strains 2, 13, and 35, with weighted marginal averages and the grand average. The

reciprocals of the squared standard errors are used as weights. Otocephalic monsters, nearly all dead at birth, were excluded from consideration in strain 13.

There is probably some falling off in the correlation, and thus an increasing role of accidental factors, in the larger litters. Accident seems to have played a greater role in strain 35 than in the others, especially 13. There is a general consistency, however, and the grand average indicates that 76% of the variability depended on environmental factors common to littermates, leaving 24% accidental.

Table 4.8 deals similarly with survival to weaning (in these data, 30 days of age) as opposed to death between birth and weaning. Stillbirths are excluded from litter size for this purpose. Here the accidental causes include external factors that act differentially during the rearing process. There is a more conspicuous increase in the role of accidental causes with the increase in the size of the litter surviving birth. There are also significant differences among the strains, but in the opposite direction from the situation at birth. The grand average indicates that 66% of the variability is due to factors acting differentially on litters and 34% to factors acting differentially within litters.

TABLE 4.8. Tetrachoric correlations and their standard errors for pairs of live-born littermates with respect to whether they reached weaning at 30 days of age.

LITTER SIZE	STRAIN 2		STRAIN 13		STRAIN 35		TOTAL
	No. of Pairs	r	No. of Pairs	r	No. of Pairs	r	r
2	149	0.81 ± 0.09	347	0.60 ± 0.09	69	0.83 ± 0.07	0.756 ± 0.048
3	285	0.68 ± 0.09	729	0.69 ± 0.05	96	0.78 ± 0.19	0.692 ± 0.045
4	102	0.27 ± 0.23	504	0.38 ± 0.10	36	0.48 ± 0.29	0.304 ± 0.086
5, 6	20	0.80 ± 0.27	75	−0.05 ± 0.27	...		0.377 ± 0.189
Average		0.685 ± 0.060		0.590 ± 0.041		0.803 ± 0.067	0.661 ± 0.030

Weight

The correlations between littermates within litters of each size with respect to birth weight and weight at weaning were obtained only in the Chicago branch of strain 35. The correlation for birth weight of all young was 0.623 ± 0.065, but that for young that reached weaning was only 0.306 ± 0.080. Factors acting through the condition of the dam thus determined twice as much of the variability if all young (stillborn, dying before weaning, and reaching weaning) are considered, as in the last group alone, which is of course as expected. The great role played by factors that act differentially on which littermates reach weaning (69%) is noteworthy.

In the case of weight at weaning (30 days) the role of factors common to the litter as a whole was rather large, this correlation being 0.778 ± 0.044; leaving 22% for ones that act differentially on littermates.

The correlations between birth weight and weight at weaning (here 33 days), and various other factors were determined in the large Beltsville branch of strain 35, descended from one mating in generation 12 during 1916–24. Logarithms of weights were used as giving somewhat more determination. There were 2,008 entries for birth weight and 1,240 for weaning weight. The squares of the correlations give the variance components.

The most important single factor in both cases was the negatively correlated size of litter. This determined 33.1% of the variance of individual birth weights and 29.7% of those of individual weaning weights. The mean logarithmic weights of classes of the other variables were adjusted to a mean litter size of 2.5. The variance components are given in table 4.9. The components due to accidental factors were not determined in these data, but a rough apportionment has been made by using the estimates from the Chicago data from the same strain: birth weight, 37.7% accidental within litters, and weaning weight, 22.2% accidental within litters, which reduce to 25.2% and 15.6%, respectively, on including litter size.

TABLE 4.9. Components of the variances of birth weight (all young born) and 33-day weight of strain 35 (Beltsville) due to litter size and various factors within litters, supplemented by the estimates of the total components due to factors, common to littermates, and to accidental causes within litters of each size in the Chicago branch of the strain.

	Birth Weight (%)	33-Day Weight (%)
Litter size	33.1	29.7
Year, within litter	4.1	9.5
Month, within litter	0.8	1.8
Age of dam, within litter	1.0	1.7
Others common to littermates	35.8	41.7
Individual	25.2	15.6
	100.0	100.0

Much the most important factor in determining these early weights is size of litter, as concluded previously (vol. 1, chap. 14). As to the factors that affected weights independently of litter size, the differences in conditions in the nine years 1916–24 played small but appreciable parts, the effects of average monthly differences and the age of the dam were decidedly minor, leaving fairly large residuals, common to littermates both at birth and weaning. A small part of this (vol. 1, chap. 14) was due to factors other than litter size that determined gestation period. Most of it was undoubtedly due to the condition of the dam. The average correlation ratio between weight of dam and birth weight within litters of a given size in the inbreds was 0.577 according to Eaton (1932) (0.556, 0.570, 0.576, and 0.623 in litters of one, two, three, and four, respectively, correcting an erroneous statement in vol. 1, p. 341). This implies 33% determination, as compared with 41.7% in data considered here from all factors common to littermates acting within litters of each size. These percentages do not refer to the same stock and do not refer to quite the same components, but the magnitude of the correlation in the data studied by Eaton tends to confirm the conclusion, which is logically almost inevitable in any case, that the category "others common to littermates" depends on variation in the conditions of the dam, independent of her age or of year or of season, except for a very small component due to variations in gestation period, independent of all of these factors.

Heredity played little or no role within this inbred strain. It played little role, indeed, within the control stock (1911–15). As noted in volume 1, table 14.14, the correlation between different nonconsecutive litters in mean birth weight of litters in the control stock was +0.060 ± 0.021, and between gains, 0 to 33 days, +0.063 ± 0.025. Consecutive litters with correlation of −0.052 ± 0.034 and +0.224 ± 0.030, respectively, are more subject to common unfavorable or favorable external factors. These correlations indicate at most 6% determination by heredity in the control stock, leaving no room for large persisting differences in the conditions of the dam.

An important question is whether these early weights in inbreds and random breds differ in variability because of characteristic differences in genotype-environment interaction. No adequately controlled data were obtained on this question and such data as are available indicate that the difference is so small that a significant difference could be demonstrated only under the most thorough controls. The following comparison (table 4.10) between random breds under very favorable conditions (1910–15), the averages within

TABLE 4.10. Statistics of litter size and average birth weight of litters.

	LITTER SIZE			BIRTH WEIGHT (gm)		
	No.	Mean	σ	Mean	σ	r
Random breds (B), 1910–15	587	3.00	1.26	83.3	18.7	−0.673
Inbreds, within strain, 1906–15	2,307	2.74	1.14	83.3	19.1	−0.665
Random breds (B), 1916–18	459	2.58	1.04	76.9	20.8	−0.656

11 inbred strains in very favorable conditions (1906–15), and the random breds under relatively unfavorable conditions (1916–18) are drawn from volume 1, table 14.15.

These comparisons are not ideal, partly because they are not between exact contemporaries and because inbreeding had not reached fixation in the inbreds (1 to 15 generations). The inbred standard deviation is also somewhat too large because of subline differentiation. It is clear, however, that there can be little if any excess variability in this character in inbreds as a consequence of their inbreeding.

The situation is essentially similar in the data on gain from birth to 33 days (table 4.11), also drawn from volume 1, table 14.15.

TABLE 4.11. Statistics of litter size and average gain to 33 days of litter within 11 inbred strains and in random breds.

	LITTER SIZE			GAIN (gm)		
	No.	Mean	σ	Mean	σ	r
Random breds (B), 1910–15	513	3.07	1.20	152.5	49.6	-0.438
Inbreds, within strains, 1906–15	2,123	2.77	1.11	157.5	45.0	-0.347
Random breds (B), 1916–18	373	2.57	0.97	137.1	43.1	-0.520

Size of Litter

Size of litter is largely a maternal character (amount of ovulation, as indicated by the delayed effect of crossing and absence of serious losses after ovulation on comparing corpora lutea and observed litter size). It is a character of the dam only at a particular moment, as indicated by the smallness of the correlation in the random bred stock (1911–18) betweeen either consecutive litters (-0.011 ± 0.034) or nonconsecutive ones $(+0.068 \pm 0.021)$. The latter includes not only heredity but effects of persistent conditions of the dam.

The study of the branch of strain 35 (1916–24), descended from a single mating in generation 12, yielded correlations with year of birth, month, and age of the dam. The squares of these give the variance components (table 4.12).

The residual category includes the condition of the dam, but the correlation found by Eaton between the weight of the dam and litter size was only 0.257 in inbreds. This is not independent of the above fractions and can account for only 7% of the variance altogether. There is some evidence that the length of the interval since the preceding litter plays a rather important role (small

TABLE 4.12. Components of the variance of size of litter of strain 35 (1916–24) due to year, month, and age of dam.

	Litter Size
Year	1.5
Month	2.1
Age of dam	1.1
Residual	95.3
	100.0

if conception occurs within a day after birth of this litter, larger if there is a delay), but a quantitative evaluation is not available. For the most part, the size of any particular litter within a particular strain must be considered a chance event.

Annual Averages

The responses of strains to environmental conditions can be studied by consideration of the fluctuations in the annual averages. Those for the nine years 1916–24, discussed in the preceding chapter from the standpoint of differentiation of strain means, will be considered here from the standpoint of variability. The conditions in these years differed greatly. It should be noted, however, that the five inbred strains were, on the whole, the most vigorous of the original 23 and thus not a random sample.

The average weight at 33 days, W, was made independent of litter size by adjusting to a mean litter size, 2.5, by means of a regression formula. The logarithms of these averages in grams, minus 80, were used in order to maximize the parallelism of the fluctuations from 1916–24 (vol. 1, tables 10.15–10.17, fig. 10.7). Fecundity is analyzed into mean size of litter, I/L, and number of litters per mating year, L/Y. The viability up to weaning (33 days) is analyzed into the percentage born alive, BA, and the percentage raised to 33 days of those born alive, RBA, both made independent of litter size by weighting the percentages in litters of one to four in the ratio 1:3:3:1. The total productivity, the average number of young raised per mating year, R/Y, is the product of the unadjusted values of the four preceding quantities.

Table 4.13 shows the variance σ_y^2 of the annual averages of each of the six strains, the unweighted average of the five entries for the inbreds, that for all six entries, and, in the three bottom rows, the results of an analysis of variance by year and strain, giving the year variance (for the total array), $\sigma_{\bar y}^2$, the strain variance, σ_s^2, and the interaction component, σ_d^2.

TABLE 4.13. The variances of the annual averages for six characters of the random breds (B) and five inbred strains, during 1916–24, and the unweighted averages of these for the inbreds and for all six populations (grand averages). The lower three rows give the variance components from the analyses of variance for each character: that due to differences among the nine years, $\sigma_{\bar{y}}^2$, that due to differences among the six strains, $\sigma_{\bar{s}}^2$, and the residual due to deviations from additivity, $\sigma_{\bar{d}}^2$.

Strain	Log $(W - 80)$	Litter Size	Litters per Year	% Born Alive	% Raised of Born Alive	No. Raised per Year
B	0.00425	0.0327	0.202	7.4	9.9	1.58
2	0.00513	0.0304	0.066	16.7	27.2	1.63
13	0.00501	0.0455	0.230	23.5	22.8	2.21
32	0.00260	0.0113	0.130	29.3	53.6	1.10
35	0.00293	0.0170	0.105	15.1	34.8	1.54
39	0.00453	0.1020	0.441	30.5	79.2	1.61
Average (inbred)	0.00404	0.0412	0.195	23.0	43.5	1.62
Average (σ_y^2)	0.00407	0.0398	0.196	20.4	37.9	1.61
$\sigma_{\bar{y}}^2$	0.00373	0.0162	0.051	5.7	25.3	0.93
$\sigma_{\bar{s}}^2$	0.00672	0.0417	0.190	3.3	18.4	1.44
$\sigma_{\bar{d}}^2$	0.00034	0.0236	0.145	14.7	12.6	0.68

The close parallelism of the average annual adjusted weights is reflected in the relative smallness of the interaction component in this case. The effects of genotype and environment on weight are largely additive on the most appropriate scale. There is relatively much more genotype-environment interaction in the other cases.

The differences between the variabilities of the controls and the average inbred strain are relatively small, except in the two mathematically independent viability percentages, in which all the inbreds are enormously more variable. There are, however, also enormous differences among the inbreds in these and other respects, with strain 39 the most variable in all but weight and total productivity. The merely average variability of strain 39 in total productivity, in spite of excess variability in all of its four components, implies compensating deviations within years.

The foregoing comparisons were concerned with absolute variabilities irrespective of the means of the characters, except in the case of weight. It is fairer to compare the coefficients of variability, except for the viability percentages. A trigonometric transformation would be appropriate here, but it would be of little importance in the range of mean percentages 55% to 68%

TABLE 4.14. The mean (M) and a measure of variability for the annual averages of six characters in the random breds (B) and five inbred strains, 1916–24.

Strain	Log (W – 80)		Litter Size		Litters per Year		% Born Alive		% Raised of Born Alive		Raised per Year	
	M	σ	M	$100\sigma/(M-1)$	M	$100\sigma/M$	M	σ	M	σ	M	$100\sigma/M$
B	2.192	0.065	2.75	10.3	3.61	12.5	68.4	2.72	66.4	3.15	6.68	18.8
2	1.972	0.072	2.37	12.7	3.71	6.9	63.3	4.08	66.2	5.22	5.82	22.0
13	2.184	0.071	2.65	13.0	3.63	13.2	62.9	4.84	61.1	4.77	5.23	28.4
32	2.106	0.051	2.16	9.1	3.21	11.2	66.1	5.42	58.9	7.32	4.11	25.5
35	2.154	0.054	2.43	9.1	3.56	9.1	62.7	3.89	60.6	5.92	5.23	23.7
39	2.084	0.067	2.56	20.5	2.52	26.4	64.0	5.52	54.8	8.90	3.21	39.6
Average (inbred)	2.100	0.063	2.43	12.9	3.33	13.4	63.8	4.75	60.3	6.42	4.72	27.8

and has not been made. Table 4.14 shows the means of log $(W - 80)$, the standard deviation of log $(W - 80)$, the mean size of litter and the coefficients of variability above 1.0 (the minimum possible value), the means and coefficients of variability for litters per year and young raised per year, and the means and standard deviations of the viability percentages.

There is not much difference between the standard deviations of log $(W - 80)$ of the random breds (0.065) and the average of those for the inbreds (0.063), which is in agreement with other evidence that inbreeding causes little difference in variability in weight. There are appreciable differences among the inbreds, however, ranging from 0.051 to 0.072.

The average inbred strain has more variability in size of litter above 1.0 (12.9%) than the random breds (10.3%), but this is largely due to wide annual variations of strain 39 (20.5%), and two of the inbred strains (32 and 35, both 9.1%) are less variable than the random breds. These strains, however, showed a declining trend in several respects, strains 2 and 13 showed relatively rising trends during the period of improving conditions from 1916–24, while strain B merely went along with the trend (chap. 2).

The situation was similar with respect to frequency of litters (B, 12.5%; average inbred, 13.4%, but inbreds ranging from 6.9% to 26.4%, with the highest value that of strain 39).

In the viability percentages, the random breds have much the lowest relative variabilities and strain 39 again has the highest. These statements apply also to total productivity, with strain 39 here very much the highest in relative variability.

The deviation y of the annual average of a strain in any respect from the grand average for all strains over the whole period can be analyzed into the deviation $\bar{y}$ of the annual average for all strains and the deviations of that for the strain in question from the latter and a joint term:

$$\sigma_y^2 = \sigma_{\bar{y}}^2 + \sigma_d^2 + 2\sigma_{\bar{y}}\sigma_d r_{\bar{y}d}.$$

In terms of degrees of determination,

$$p_{y\bar{y}}^2 + p_{yd}^2 + 2p_{y\bar{y}}p_{yd}r_{\bar{y}d} = 1,$$

where $p_{y\bar{y}} = \sigma_{\bar{y}}/\sigma_y$ and $p_{yd} = \sigma_d/\sigma_y$.

Table 4.15 gives this analysis. The values for $p_{y\bar{y}}^2$ and p_{yd}^2 for the average inbred strain and for the average strain including the controls are based on the ratio of the mean variances and may differ considerably from the mean of the ratios given for the separate strains. The latter is less affected by the very large ratios where σ_y^2 of some strains are exceptionally small. It gives no joint term for the grand averages.

The uniformly small values of p_{yd}^2 in the case of weight again reflect the close approach to parallelism of the fluctuations of all strains in this respect.

TABLE 4.15. The degrees of determination of the annual averages of the same six chara[...] latter and the grand average, by the annual mean of all six strains ($p_{y\bar{y}}^2 = \sigma/\frac{2}{y}\sigma_y^2$)

Strain	Log ($W - 80$)			Size of Litter			Litters per Year		
	$p_{y\bar{y}}^2$	p_{yd}^2	Joint	$p_{y\bar{y}}^2$	p_{yd}^2	Joint	$p_{y\bar{y}}^2$	p_{yd}^2	J
B	0.879	0.050	0.091	0.496	0.317	0.187	0.251	0.597	0
2	0.727	0.092	0.181	0.534	0.206	0.260	0.765	0.301	−0
13	0.745	0.082	0.173	0.357	0.350	0.293	0.221	1.359	−0
32	1.434	0.131	−0.565	1.438	1.925	−2.363	0.392	0.146	0
35	1.275	0.093	−0.368	0.957	2.113	−2.070	0.484	0.695	−0
39	0.822	0.092	0.083	0.159	0.503	0.330	0.115	0.736	0
Average (inbred)	0.924	0.095	−0.019	0.394	0.636	−0.030	0.261	0.770	−0
Average	0.916	0.084	0	0.408	0.592	0	0.260	0.740	0

Genotypic and environmental effects are essentially additive. The percentage raised of those born alive and the total productivity come next, while size of litter, percentage born alive, and frequency of litters show markedly non-additive genotype-environment interactions.

The joint terms indicate whether there is, in general, an exaggerated response to conditions (positive) or a homeostatic tendency (negative), with the qualification that a greater upward trend than that of the average during this period tends to give a positive joint term while a smaller upward trend (or no trend at all) tends to give a negative joint term. These trends (chap. 2) may have other causes than exaggerated or homeostatic responses to fluctuations about the trends.

The control stock showed merely an average upward trend during the period. The joint terms showed no important tendency either to exaggerate or damp the fluctuations, except for the very pronounced damping of those of both viability percentages. These account for the very small absolute variabilities of the controls in those respects shown in table 4.13.

As a final character in guinea pigs, we will consider the weight at 353 days. Environmental factors have much less effect on adult weights than on birth or weaning weights (vol. 1, chap. 14). Table 4.16 shows these weights for males of the control stock and the five inbred strains during the years 1916–25 (McPhee and Eaton 1931).

Here the inbreds are much less variable than the random breds, not only absolutely but relative to the mean weights. If it be assumed that the variance in the inbreds is wholly nongenetic and that the nongenetic variance is the same in the random breds, the portion of the variance of the latter due to environment is 51%, leaving 49% of the random bred variance genetic. Fixation may not, however, have been complete in the inbred strains. As before, it should have been most nearly complete in strain 35, descended from one mating in generation 12. Using its relative variance as the measure

table 4.13, in the random breds (B) and the five inbred strains, the average of the
tion from this ($p_{yd}^2 = \sigma_d^2/\sigma_y^2$), and a joint term.

	% Born Alive		% Raised of Born Alive			Raised per Year		
	p_{yd}^2	Joint	$p_{y\bar{y}}^2$	p_{yd}^2	Joint	$p_{y\bar{y}}^2$	p_{yd}^2	Joint
8	1.988	−1.756	2.545	1.153	−2.698	0.591	0.355	0.054
0	0.960	−0.300	0.928	0.325	−0.253	0.573	0.461	−0.034
2	0.458	0.300	1.109	0.344	−0.453	0.423	0.879	−0.302
4	0.384	0.422	0.471	0.271	0.258	0.851	0.112	0.037
6	1.297	−0.673	0.726	0.253	0.021	0.608	0.651	−0.259
6	0.528	0.286	0.319	0.308	0.373	0.580	0.424	−0.004
7	0.641	0.112	0.581	0.296	0.123	0.578	0.433	−0.011
8	0.722	0	0.666	0.334	0	0.580	0.420	0

TABLE 4.16. Number, means, standard deviations, and coefficients
of variability (CV) of the weight at 353 days of male guinea pigs,
1916–25.

Strain	No.	Mean (gm)	SD (gm)	CV (%)	(CV)²
B	68	910	111.7	12.3	151
2	112	709	64.2	9.1	83
13	102	878	75.8	8.6	74
32	98	718	60.3	8.4	71
35	61	824	63.2	7.7	59
39	67	745	74.4	10.0	100
Average (inbred)		775	67.6	8.8	77

SOURCE: Data from McPhee and Eaton 1931.

of nongenetic variance, 39% of the variance of B was nongenetic and 61%
was genetic.

Maize

Patterns of variability of characters could be illustrated from inbred and
random bred strains in many other organisms. We will, however, largely
restrict consideration in the rest of this chapter to direct tests of the question
of whether inbreds are more variable than first-generation crosses.

One type of case in maize was apparent from the first. Shull (1914) noted
that the heterotic hybrids between weak selfed lines were much less susceptible to unfavorable conditions than the latter and, in consequence, showed
much less variability over the years, or from place to place, in growth and
productivity. Jones (1918) described an extreme example in which several
F_1's were grown between their parental lines on ground saturated with water

when the plants were just starting. Many in the selfed lines were stunted and remained so through the season, producing neither tassels nor ears. Variability in all characters associated with vigor was enormous. The F_1's also had a poor start but overcame this and produced a fairly uniform stand. These cases were excluded from his statistical analysis referred to in the preceding chapter.

This situation contrasts with that described by Emerson and Smith (1950) in a study of number of kernel rows in maize, discussed in volume 1, chapters 10 and 12. The standard deviation of inbred lines was approximately proportional to the mean (vol. 1, fig. 10.5), so that the coefficient of variability was approximately constant. The average coefficient of six 8-rowed lines was 8.70, for three 10-rowed lines 10.73, and for thirteen 12-rowed lines 10.55, with a grand average of 10.07. In four crosses in which the actual parent lines averaged 9.60, the average was 11.55. Thus, as in Jones' data for the same character (chap. 2), F_1 was somewhat more variable than the inbred parents. There were, however, highly significant differences in variability among inbred lines (and among crosses) with the same numbers of rows, indicating the presence of specific genotype-environment interactions that were more important than the observed excess variability of the crosses, thus throwing doubt on whether there is any consistent difference in either direction between inbreds and first crosses in variability in this respect.

There have been many other studies of cross-fertilizing species of plants in which results similar to those with maize have been observed. The general conclusion has been that inbreds are more variable than first crosses with respect to characters that exhibit heterosis, such as weight and productivity, but not with respect to characters such as number of kernel rows in maize, which do not.

Tomato

Self-fertilizing species exhibit little or no heterosis in growth and yield characters, as discussed in the preceding chapter. There is also little or no difference in variability between inbreds and their first crosses on the appropriate scale. This may be illustrated by a study by Williams (1960) of various characters of the tomato (table 4.17). He compared eight inbred lines and six F_1's. Both the interpopulation and intrapopulation variabilities of F_1 were essentially the same as in the inbreds. There were, however, significant differences in variability among inbreds with intermediacy in most cases in F_1.

The near-identity of variability of fruit weight in inbred and first-cross tomatoes on the appropriate scale applies even with a 55-fold difference

TABLE 4.17. Standard deviations of the means of characters of the tomato in ten replicates, covering a considerable range of conditions, and the standard deviation of individual plot means.

| | STANDARD DEVIATION MEANS OF 10 REPLICATE MEANS | | | (PLANT MEANS) | | |
	Least Variable	Most Variable	F_1	Least Variable	Most Variable	F_1
No. of fruits (3 weeks)	0.71	0.94	1.07	2.40	3.19	3.05
Average weight of fruits (oz.)	0.08	0.10	0.07	0.28	0.46	0.36
Weight of plant (oz.)	3.11	4.31	3.82	6.26	7.11	7.08
Flower number (first inflorescence)	0.23	0.30	0.31	1.87	2.29	2.42
Date of flowering	0.19	0.25	0.19	0.82	1.82	0.86

SOURCE: Williams 1960.

between the parent inbreds (Powers 1942), as shown in volume 1, tables 10.7 and 15.8.

Rats

In the rats recorded by Livesay (tables 3.18 and 3.19) the 90-day weight of inbred strains showed enormously more variability than their crosses. The 90-day weight of rats is comparable to the 353-day weight of guinea pigs. The average coefficient of variability of the guinea pigs was only 8.8% (table 4.16) in contrast with those of the inbred rats (22.8%) or even of the F_1 rats (13.0%). The five strains of guinea pigs were probably the most vigorous of the original 23 strains and were, perhaps, further from the threshold of failure than two of the strains of rats. The rats of the other strain (King albinos) were, however, extraordinarily vigorous, but nevertheless were very variable in Livesay's data.

According to King (1918c,d), however, the coefficients of variability of the inbreds at 90 days were only 13.8% in males and 13.9% in females; those of the random bred controls were about the same (14.8% in males and 12.5% in females). It appears that conditions in Livesay's experiments were such as to produce an enormous increase in variability of the inbreds but not in the F_1's. The inbreds seem to have been (or to have become, at least by generation 45) much more susceptible to unfavorable conditions than the crossbreds.

Mice

In mice, Chai (1957) determined the coefficients of variability of inbred strains and their crosses with respect to weight within litters (table 4.18).

TABLE 4.18. Numbers, means, and coefficients of variability (CV) of weights within litters of two inbred strains of mice, F_1, and F_2.

Strain	At Birth			3 Weeks			60 Days		
	No.	Mean (gm)	CV (%)	No.	Mean (gm)	CV (%)	No.	Mean (gm)	CV (%)
P_1	78	1.8	11.5	48	20.1	10.4	41	37.4	5.1
P_s	96	1.2	10.4	75	9.1	9.4	65	13.6	4.8
F_1	138	1.7	7.7	162	16.5	6.9	161	25.8	4.4
F_2	321	1.5	8.6	231	18.3	9.7	216	27.2	10.6

SOURCE: Chai 1957.

The inbreds were derived from strains selected for large and small size, P_1 and P_s, respectively.

The variability of the inbreds exceeded that of F_1 by 42% at birth, 43% at three weeks, but only 12% at 60 days. The coefficients at 60 days are much less than in the guinea pigs at 353 days and enormously less than those of the rats at 90 days, but they refer to variability within litters, which was not the case with the guinea pigs and rats. As with the rats, variability is greater in the inbreds than in F_1 but the difference at 60 days is not excessive.

Inbred strains of mice tend to be characterized by higher than normal frequencies of particular morphological anomalies (Grüneberg 1951, 1952), much as was found to be the case in guinea pigs with respect to otocephaly, polydactyly, microphthalmia, and others. Of special interest was the frequent (12%) absence of lower third molars in a particular strain (C57Bl), associated with the somewhat smaller size of this tooth when present and greatly increased variability (6.8% as opposed to 2.3% in a normal strain) (vol. 1, chaps. 5 and 6, fig. 6.7). Absence represented a breaking point in a developmental process that tended to go poorly at best in this inbred strain.

Grüneberg (1954) and others have questioned the value of inbred strains as sources of uniform material for studies of physiological agents. Thus Claringbold and Biggers (1955) found excessive variability of responses of inbred female mice to estrogens as compared with crossbreds. McLaren and Michie (1954, 1956) obtained similar results with respect to duration of narcosis in response to Nembutal in adult male mice of two highly inbred strains (C57Bl and C3H) in comparison with F_1 and a random bred stock. The estimates of variance (logarithmic) from an analysis of variance are shown in table 4.19.

In this case, F_1 is intermediate in duration of narcosis although somewhat above the average of the parental strains. The striking thing is the twofold greater standard deviation of both inbred strains than either F_1 or the random bred strain. In this case, there is excess variability of inbreds over

TABLE 4.19. Logarithms of minutes of narcosis from a given dosage of Nembutal, and the standard deviations and variances in a random bred strain of mice, two inbred strains, and their first cross.

	Responses		
	Log. Min.	SD	Variance
Random bred	2.10	0.133	0.0176
C57Bl	1.73	0.235	0.0552
C3H	2.21	0.279	0.0778
F_1	2.08	0.128	0.0165

SOURCE: McLaren and Michie 1956.

crossbreds and random breds without heterosis. This is the opposite of the situation with weight in guinea pigs, in which there is heterosis but no appreciable excess variability of inbreds. While the two phenomena are probably associated as a rule, both present or both absent, the association is obviously not perfect.

Drosophila

We will compare the results of studies of three different characters of *D. melanogaster*, the number of chaetae on the abdominal segments, wing length, and the number of bristles on the scutellum.

Chaeta number is a character that resembles white spotting of guinea pigs in that most of the nongenetic variance is determined by accidents of development, as indicated by the smallness of the correlation between the numbers on different sternites. Reeve and Robertson (1954) found a correlation of only 0.068 ± 0.022 within inbred lines. They found the following percentages of variance components in a random bred stock (table 4.20).

TABLE 4.20. Components of variance of number of abdominal chaetae in random bred *D. melanogaster*.

	ACTUAL VARIANCES		COMPONENTS	
	Males	Females	Males	Females
Genetic	1.69	2.11	40	39
Nongenetic				
Common to all sternites	0.13	0.08	3	1
Peculiar to single one	2.42	3.25	57	60
Total	4.24	5.44	100	100

SOURCE: Data from Reeve and Robertson 1954.

Essentially similar results (52% additive genetic, 9% nonadditive genetic, 4% true environmental, and 35% developmental accident) were obtained by Clayton et al. (1957).

A comparison of nine inbred lines and F_1 crosses by Reeve and Robertson (1954) with respect to the sum of the variances of sternite counts gave 7.7 and 7.5 for inbred and crossbred males, respectively, 13.3 and 12.2 in inbred and crossbred females, respectively.

There was no appreciable difference between inbreds and crossbreds in the variability of this character, which thus resembled white spotting of guinea pigs in this respect.

With respect to wing length, largely an index of general size, Robertson and Reeve (1952b) showed that, like weight in rats and mice, but not in guinea pigs, there is much greater variability of inbreds than of crossbreds.

The following data (table 4.21) involve six highly inbred lines, derived from unrelated wild stocks, Nettlebed (N) and Edinburgh (E). Two had been selected for long wings (LN, LE), two for short wings (SN, SE), and two were unselected (UN, UE) before sib mating began.

TABLE 4.21. The coefficients of variability of wing length for parent lines and F_1 of six crosses in *D. melanogaster*.

STRAIN CROSSES	COEFFICIENT OF VARIABILITY	
	Average (Parents)	F_1
LN × SN	1.34	1.17 (females)
LE × SE	1.20	0.72 (males)
UN × UE	1.51	1.05 (both sexes)
SN × UN	1.69	1.37 (both sexes)
SE × UE	1.33	0.74 (both sexes)
SN × SE	2.00	1.41 (both sexes)
Average	1.51	1.08

SOURCE: Data from Robertson and Reeve 1952b.

The inbreds are more variable than their crosses in every case. On the average, their coefficients of variability are 40% greater.

The same authors also made up various combinations of the three major chromosomes from inbred lines SE and UE. In table 4.22, coefficients of variability are averaged according to the number of heterozygous pairs.

In both sexes, variability declined with increasing heterozygosity. Females homozygous in all three chromosomes were 45% more variable than those heterozygous in all three, but males homozygous in the second and third

TABLE 4.22. Coefficients of variability (CV) of wing length of *D. melanogaster* according to pairs of heterozygous chromosomes derived from two inbred strains.

HETEROZYGOUS PAIRS	FEMALES		MALES	
	Genotypes Tested	CV	Genotypes Tested	CV
0	5	1.59	5	1.28
1	9	1.49	7	1.14
2	5	1.20	2	1.09
3	1	1.10	...	...

SOURCE: Robertson and Reeve 1952*b*.

chromosomes were only 17% more variable than those heterozygous in these. The average 31% was somewhat less than in the preceding comparison but it is highly significant. A similar decline was confirmed later by more extensive studies (Robertson and Reeve 1955*b*).

Sheldon, Rendel, and Finlay (1964) have compared large numbers of isogenic lines of *D. melanogaster* with each other and their random bred sources with respect to number of scutellar bristles. There were two sources, a scute stock, Oregon RC, and a mutant scute stock, selected for low variance. Number of scutellar bristles is a very different sort of character from number of abdominal chaetae. There are nearly always just four individualized bristles on the scutellum in a wild-type stock instead of the highly variable number of nonindividualized abdominal chaetae. Occasionally there are three or five bristles. Payne (1918*a*) showed that the number could be greatly increased by selection, giving rise to typical meristic distributions (see vol. 1, figs. 6.9 and 11.8 [which are erroneously stated to refer to dorso-central bristles]). As noted in volume 1, chapters 6 and 11, meristic variation implies a succession of thresholds, the intervals between which are not necessarily the same on the underlying physiological scale. If the elements are individualized, there is usually canalization, in Waddington's terminology, so that the interval between thresholds defining a certain standard number is wider than other intervals. In the case of the wild type, the wide interval for scutellar bristles is that for just four, in that of selected scute it was that for two. If a distribution overlaps two thresholds and normal variability is assumed on the underlying scale, it is possible to calculate the mean and standard deviations of populations relative to the interval, as in the case of digit number in guinea pigs, or conversely calculate the width of the interval in terms of standard deviations. Rendel, Sheldon, and Finlay (1965), adopting the latter viewpoint, showed that the interval in their scute stock could be broadened by selection until nearly all flies had just two scutellar bristles.

From the viewpoint of constant thresholds, they reduced the variability. As long as only two thresholds are involved, it is a matter of relativity, as illustrated in figure 4.4, where the mutation Px could be interpreted either as reducing variability of guinea pigs of strain I with respect to development of the little toe or as broadening the interval between the thresholds for none and perfect development. The latter interpretation was preferred in this case, on consideration of the other thresholds introduced by $PxPx$.

Sheldon, Rendel, and Finlay (1964) found that 18 isogenic strains isolated from the wild type by use of marker genes showed almost as many cases of increased canalization (wider interval between three and five bristles) as showed a decrease. There were only slight variations in the positions of the mean, relative to the middle of the interval. In the case of 11 isogenic lines derived from the canalized scute line, there was reduced canalization (narrower interval between one and three bristles) in all cases and also considerable differences in the position of the mean relative to the midinterval. All but two shifted downward. The authors concluded that heterozygosity was not significant in either series, but that dominance effects were important in the scute series.

Summary

Patterns of variability in terms of the kinds of components are illustrated in this chapter by those of the characters of the guinea pig.

The first category is that in which there is a striking appearance of uniformity within inbred strains, comparable to that of a pair of identical human twins. Each of the strains studied had its own characteristic quality and intensity of the eumelanic and phaeomelanic regions of the coat, especially at birth, going far beyond the aspects capable of analysis in terms of specifiable genes. Aspects of conformation (relative length of legs, straightness of back, shape of nose) were also highly characteristic of each strain, although measurements would no doubt have shown some variability.

The first component of variability to be considered is that due to accidents of implantation or development that cause differences among newborn littermates for which all external environmental factors, including conditions of the mother, are necessarily the same. The occurrence of otocephaly, a threshold character, in strain 13 (always 5.5% in the main line) was determined 90% to 100% by accidental factors, depending on whether or not there were pronounced seasonal differences in the conditions to which the mothers were exposed. The percentage of white in the tricolor patterns, with medians of strains varying from 20% to 98% and variability within intermediate strains from almost 0 to 100%, was determined 87% by nongenetic factors not common to littermates in a strain (no. 35) with no appreciable genetic variability. In the case of frequency of occurrence of the little toe, a trait with a lower threshold for any occurrence and an upper for perfect development, 46% of the nongenetic variance of strain 35 was due to accidents of development.

Analysis of variability of birth weight in strain 35 indicated 26% not common to littermates. The trait of being born alive is best considered to be of the threshold sort. On this basis, analysis in several inbred strains indicated 24% not common to littermates. Differences among littermates in weight at weaning (Beltsville, 33 days; Chicago, 30 days) may be due to differential action of external factors to a greater extent than at birth, but all specifiable environmental factors are still the same. The accidental components for the threshold trait of reaching weaning among those born alive was 34% in the same inbred strains, as above, and for the continuous variable, weight at weaning was 15% in strain 35. This component approaches 0 in mature weights.

All evidence indicated that size of litter was almost wholly dependent on the dam (ovulation rate) in guinea pigs but must reflect conditions of a very temporary sort in view of the absence of significant correlation between consecutive litters, and the low value (0.07) for nonconsecutive litters in the random bred strains in which genetic variability was included. Analysis in inbred strain 35 indicated some 93% as due to intangible accidental factors.

Among tangible nongenetic factors, the age of the dam played a major role (more than one-third) in the small portion of the variance common to littermates in the case of white spotting and also in the much larger portion in the case of digit number. It was a real but minor factor in the case of litter size, a real but very minor factor (1% or 2% of the variance common to littermates), in the case of birth weight, weaning weight, and viability at birth, and was hardly detectable with respect to the probability of reaching weaning.

Size of litter was the most important tangible factor in determining birth weight (33%) and weaning weight (30%) in strain 35 and still had an appreciable effect (8%) at one year of age (strain 2). It had effects on mortality at birth and between birth and weaning, but not linear effects, since the optimum litter size is usually two in inbreds, three in crossbreds. Effects on digit number and occurrence of otocephaly were slight.

The condition of the dam, reflected to a large extent in the deviation of her weight from that characteristic of her age in the strain under consideration, undoubtedly accounted for most of the remaining variability common to littermates in the neonatal characters. This was associated to some extent with differences in annual averages in the period of rather drastic differences in which most of the studies were made and, to a lesser extent, by the differences among monthly averages. The latter caused significant annual cycles with respect to digit number, occurrence of otocephaly, size of litters, birth weight, and weaning weight in litters of a given size, and in mortality percentages in the early data (Beltsville), but not to an appreciable extent under the more uniform conditions in Chicago.

There is, of course, genetic segregation with respect to the X and Y chromosomes even in the most isogenic of the inbred lines. This determined a striking difference in frequency of otocephalic monsters in all cases (a ratio of two females to one male within all branches of strain 13 and in the sporadic occurrences in other strains). The ratio was similar for microphthalmia. It also caused males of all strains to have lower percentages of white than females and to be somewhat more variable. There were differences in these and other traits among early sublines of inbred strains, presumably due to segregation, and differences later that in one case (high incidence of otocephaly in one line) were certainly mutational. In other cases, it was difficult to decide between mutation and delayed segregation.

There were important genetic differences in all respects among the inbred lines. In the random bred control stock, the genetic variance of white spotting was about 41% (at least 38% additive, perhaps 3% from dominance deviations). It differed greatly, however, in males (54%) and females (27%), another X chromosomal effect. The genetic variances of birth weight, weaning weight, and litter size in the control stock were very slight (perhaps as much as 6% for the weights, 7% for litter size on the basis of the correlations between nonconsecutive litters). At one year of age, however, 61% of the variance of the control stock was estimated to be genetic on the basis of the reduction in the most nearly isogenic strain (no. 35).

The question of whether isogenic strains are characteristically more variable than crosses (or even than random breds where the genetic variance is small) is a very important one and has been much discussed. The guinea pig data indicated no excess variance with respect to spotting or weight at any age. Otocephaly and polydactyly were too rare in the control stock for any estimates, but variabilities in the inbred strains and crossbreds in which they occurred were best interpreted as mere differences of the means relative to two thresholds at constant distances apart, with the qualification that the tendency toward return of the little toe (brought about by a mutant gene, Px), along with tendencies toward return of the thumb and big toe (all in $Pxpx$) and of many nonatavistic digits, and many other abnormal traits in $PxPx$, required at least a two-dimensional pattern of thresholds in which the distance between the thresholds was greater with $Pxpx$ than $pxpx$.

The possibility of different amounts of genotype-environment interaction in inbreds and random breds was studied for several characters by considering the fluctuations in the annual averages in a nine-year period in which conditions varied enormously. The index of weaning weight, W, log $(W - 80$ gm), of controls and five inbred strains (from litters adjusted to a size of 2.5) fluctuated in close parallelism (92% on the average determined by the annual grand average, 8% by deviations). There was no appreciable difference between the relative variabilities of controls and average inbred strain and only slight differences among the inbreds. In other respects the degree of determination by the annual grand average was much less (size of litter, 41%; litters per year, 26%; percentage born alive, 28%; percentage raised of those born alive, 58%; and total productivity [product of preceding four factors], 58%), indicating much genotype-environment interaction. The annual averages for litter size of the random breds showed no consistent difference in relative variability from four of the inbred strains, but the other inbred strain (no. 39) was by far the most variable. The annual averages for percentage born alive and percentage of those born alive of the random breds were, however, very much lower than for any of the inbreds among which strain 39 was again the most variable. The random breds showed strong homeostatic tendencies in those respects. With respect to overall productivity, the random breds again showed less relative variability than any of the inbred strains, but the differences from four of the latter were much less than that of all of these from strain 39, which, with much the lowest average, was approaching extinction.

Data from other species were considered almost exclusively for their bearing on the relative variabilities of inbreds and first crosses. The enormously greater variability of inbred lines of maize in growth and yield than crossbreds under unfavorable conditions, but their relatively slight excess variability under favorable conditions, were noted. There was little difference in such neutral characters as number of rows. No characters showed much difference in self-fertilizing species.

Certain data in rats and mice showed much more variability of inbreds than first crosses with respect to weights, in contrast with the situation in guinea pigs. In the rats, however, the excess variability of inbreds did not apply under favorable conditions.

Mice of the same inbred strain have shown much excess variability in certain physiological traits: response of females to estrogen and duration of narcosis from Nembutal.

In *Drosophila*, the absence of excess variability within inbred strains with respect to chaeta number, a trait determined largely by accidents of development, resembles the situation with respect to spotting in guinea pigs. In contrast, wing length shows much excess variability within inbred strains as compared with their crosses. In the case of the four individualized bristles of the scutellum, in which crossing of a lower threshold occasionally gives three bristles and crossing of an upper gives five, isogenic lines were about equally likely to show more or less variability, interpretable as weaker or stronger canalization, respectively, of the pattern of four bristles. In a scute strain selected for low variability (strong canalizations), isogenic lines tend to differ much more than in the wild-type strain, but in all cases exhibited more variability (less canalizations).

As often noted, there is an association among characters between the occurrence of heterosis in crosses and excess variability within inbred lines. Thus growth and yield in maize, viability and productivity in guinea pigs, weight in certain strains of rats and mice, and wing length in *Drosophila* exhibit both phenomena while percentage of white in the coat in guinea pigs and chaeta number in *Drosophila* exhibit neither. The association is not perfect, however, since weights in guinea pigs exhibit marked heterosis but no appreciable excess variability within inbred lines, and the opposite is true of duration of narcosis under Nembutal in mice.

The threshold characters—percentages of otocephaly, polydactyly, microphthalmia, and others—of guinea pigs can best be interpreted as showing constant thresholds, constant variability, and no heterosis with respect to the underlying scale. This is probably also true of most morphological anomalies in mice. In such cases, inbreds are likely to exhibit variability, random breds uniformity, merely because of the frequent fixation of deviant genotypes that cross the threshold of abnormality in inbreds.

There may, however, be fixation of genotypes that alter variability, ones that react differently to environmental or developmental factors or else alter canalization (location of the thresholds). The experiments with scutellar bristles in *Drosophila* indicated such changes. The isogenic derivatives from the wild-type stock merely show differentiation in variability (or canalization) with no systematic direction of change. Those from the scute stock, selected for low variability, show more differentiation of their means and a systematic tendency for the selected low variability (or strong canalization) to be lost, giving excess inbred variability for a somewhat different reason than that due to approach a breaking point in development of an individual or an organ.

The situation in such cases as the excessive variability of weak inbred lines is not, however, altogether different from the simple threshold case. The breaking point is a sort of threshold. In first crosses, the probable presence of a more harmonious set of genes than in either inbred parent, which function well together and hence give dominance of the gene in each pair that is more favorable in the combination, results simultaneously in heterosis and remoteness from the threshold of instability. This accounts for the cases of positive association. The cases with neither heterosis nor excess inbred variance are ones in which there are merely additive effects or, if there are thresholds, they are ones in which the transcending phenotypes are necessarily stable (0 and 100% white, no or perfect little toe). The two frequent consequences of random fixation (instability of inbreds and heterosis in crossbreds) do not, however, involve exactly the same conditions and are not necessarily both present or both absent.

CHAPTER 5

Genotypic Persistence in Inbred Lines and Clones

It is important to consider the degree of persistence of the genetic constitution in lines in which no genetic variability is expected except as a result of mutation of some sort, before discussing the effects of selection on random bred or crossbred stocks.

In general there is a high degree of persistence. The inbred strains of mice of the Jackson Laboratory, raised in enormous numbers, have been highly uniform in characters in which heritability is nearly complete. In the long run, however, many mutations have appeared, ones with little effect being recognizable because of the general uniformity, and these have contributed enormously to our knowledge of the genetics of the mouse. A similar statement can be made of inbred lines of *Drosophila*.

The interpretation of significant changes in quantitatively varying characters in long inbred lines as mutations may be uncertain because of the possibility of delayed fixation. The guinea pig colony described in previous chapters had reached a point at which fixation should have been approaching completion unless interfered with by natural selection. Theoretically 3.27 generations of sib mating are equivalent to one of self-fertilization, with 50% loss of heterozygosis (vol. 2, p. 186). Thus 13 generations of sib mating correspond to four of self-fertilization ($F = 93.75\%$) and 23 generations of sib mating correspond to seven of self-fertilization ($F = 99.2\%$). The degree of fixation indicated by the histocompatibility reactions discussed in chapter 2 is of the right order if fixation was proceeding without interference.

The modifiers of white spotting, which showed conspicuous subline differentiation in the early generations of each inbred line, were also approaching fixation as expected. Thus in the branch of strain 35 descended from a single mating in generation 12 there was no apparent subline differentiation (fig. 4.1) and no significant correlation between parent and offspring (0.024 ± 0.019, Beltsville branch). The correlation in the Chicago branch from one mating in generation 22 was 0.026 ± 0.028 but the average in this branch was, as noted before, slightly but significantly whiter, indicating that fixation had not actually been quite complete in the earlier period. A branch of strain 13 from a single mating in generation 18 showed a small but significant parent-offspring correlation (0.064 ± 0.025), but a branch of strain 2 tracing to generation 15 yielded a parent-offspring correlation of -0.027 ± 0.017, although there had been strong subline differentiation a few generations earlier.

In marked contrast with this fixation or near-fixation with respect to white spotting was the marked subline differentiation with respect to presence of the little toe in the same branch of strain 35 as referred to above. Among 21 sublines of the Beltsville branch (1,976 young) (fig. 4.3) the frequency of the polydactyly ranged from 9% to 69% ($\chi^2 = 247$ with 20 degrees of freedom). The tetrachoric correlation between parent and offspring was 0.183 (in close agreement with the squared correlation ratio, 18.5%). The average correlation within four groups of sublines, distinguished by frequency of polydactyly, was -0.008, indicating that there must have been very little segregation within them, and that the group differences were probably due to a few rapidly fixed mutations. There are six major changes after generation 13, possibly reducible to four. On grouping, the sublines M, N, O (24%) and R, S, T (19%) from two matings in generation 13 probably do not differ from each other. Group BB, DD, EE, GG (14%), tracing to T, may or may not really be lower but P, Q (42%) from O is so isolated from other high groups as to require a mutation. Groups CC (50%) and FF (39%) both traced to BB but started three generations apart. They may trace to the same mutation. Groups U and V (42%) from T, and groups W, X, and Y (62%), also from T, seem to require separate mutations, but the latter may involve two, one that is the same as that for U and V, which may carry on in the large branch Z and AA, Chicago (40%) from W, three generations after separation from U and V.

In both spotting and polydactyly, crosses between inbred strains at opposite extremes gave segregation indexes of about 4 (chap. 4), which may indicate differences at four equivalent loci, or in one major one responsible for half the differences and many minor ones, or some intermediate situation (vol. 1, table 15.6). It appears that there was neither segregation nor mutation of importance in the case of spotting, but that at least four mutations occurred in that of polydactyly. The changes were smaller in magnitude on the underlying scale than was apparent at first because of the extreme sensitivity of the phenotype in the portion of the scale in which strain 35 was located. On a scale on which the threshold for any development of the little toe is 0 and that for perfect development 1.0, with a nongenetic standard deviation of 0.80, the location of strain 2 (no little toe) was -7 on the basis of crosses with strain D (perfectly developed little toes), which was at $+3$ (vol. 1, figs. 5.14 and 5.15). The location of group R, S, T with 18.9% polydactyly

(323 individuals) and its derivative W, X, Y (61.7%, 191 individuals) are -0.71 and $+0.24$, respectively, 0.95 apart, only 9.5% of the range between strains 2 and D. As noted, this may involve two successive mutations. The other postulated mutations all have considerably smaller effects. They are thus all minor mutations, but the apparent minimum number, four, is much greater than expected at usual mutation rates, unless the total number of loci at which mutations may affect polydactyly is very great.

It may be noted that on the underlying scale for spotting, 9.5% of the range between extreme inbred lines (strains 39 and 13) would account for a change from 60% to 70% white. There was no indication of any change of this magnitude, but one-half or less might perhaps be obscured by the enormous amount of nongenetic variability. One such change of this order did apparently occur in the origin of the Chicago branch.

It has been noted that the percentage of otocephaly (fig. 4.2) in strain 13, while showing much segregation in early generations, was apparently constant (5.5%) in all clusters tracing to a single mating in generation 13, except for one abrupt shift in a mating six generations later to a much higher percentage (18%) due, quite certainly, to a mutation that raised the incidence to about 20% in males and 40% in females in heterozygotes, and probably much higher in homozygotes. Relapse to 5.5% was due merely to segregation.

Mice

Green (1953) studied derivatives of one of the inbred strains of mice of the Jackson Laboratory (3CH), started in 1920, with respect to the number of thoracic and lumbar vertebrae. He showed that two groups of branches that separated in 1930 after some 20 generations of sib mating had come to be very different. One had about 96% with five lumbar vertebrae, the other had about 96% with six. McLaren and Michie (1954, 1956) found significant differences among certain subdivisions of these two groups.

Deol et al. (1957) studied seven branches of one of the Jackson strains, C57BL/Gr, in British laboratories. Two branches (I, II) of this strain separated from the others in 1941 after at least 40 generations of sib mating. These separated from each other 18 generations later. The group of five others branched into a group of two (III, IV) and a group of three (V, VI, VII) 13 generations after the cleavage in 1941. Groups III and IV separated after one additional generation, groups V, VI, and VII after 12 additional generations. Each of these groups, as studied, consisted of four to eleven closely related matings that separated only a few generations after the last cleavage referred to.

The authors classified the mice for 30 minor skeletal variants of which 27 were considered satisfactory for wholly objective specification of percentages of occurrence. All of these seemed trivial and unlikely to be of selective significance. Thirteen of the 27 variants occurred with the same frequencies within errors of sampling in all seven groups, but the remaining 14 showed significant differences among or within groups. Altogether, 21 significant changes seem to have occurred, although there was a possibility in a few cases that associated changes in two characters were pleiotropic consequences of a single genetic change. Disregarding this, 10 of the 21 changes occurred between the two primary groups. All groups except V and VI were, however, distinguished entirely or in part from each other.

A similar study was made by Carpenter, Grüneberg, and Russell (1957) of three American branches of the same inbred strain, C57BL. These had a common ancestral mating with each other and the seven British branches in 1929 after about 20 generations of sib mating. Two of these American branches had common ancestry for some 20 additional generations. The British groups V, VI, and VII were used to represent C57BL/Gr in this study.

The number of changes between lines that separated in 1929 was greater than between the British groups that separated in 1941. An interesting point was that the 13 characters that were alike in all of the British groups all showed significant differences.

Grewal (1962) used a measure of the divergence of groups based on all 27 characters. The amount of divergence between two groups was found to be roughly proportional to the number of generations of separation. He estimated that if the 21 changes found in the British branches were assumed to be mutations, a rate of about one mutation per character per 100 generations is indicated. This may be compared with W. L. Russell's (1963) determination of the rate of spontaneous mutation in male mice at seven convenient loci. He found 28 in 531,500 mice or 0.75×10^{-5} per locus per gamete. Grewal's estimate for the skeletal variants is 1,300 times as great if each change is identified with a single locus. The factor 1,300 should, as noted, be discounted somewhat because of possible pleiotropy, but not much.

It must be discounted much more, of course, because of the likelihood that each type of skeletal change may be affected by mutations at any one of many loci. There is also the possibility that the "minor" mutations involved in the variants may occur more frequently than the "major" mutations at Russell's seven loci. Russell, however, recorded mutations with minor as well as major effects at these loci. The possibility that homozygosis had not been attained up to the time of separation, in some cases 50 generations, is not plausible in the absence of selection.

That neither mutation nor segregation was occurring frequently in the late generations was demonstrated by Deol et al. (1960) by tests of parent-offspring correlations within groups of strains C57BL/Gr, A/Gr, and CBA/Gr, in which there was no significant differentiation. In 66 tests, only five were significant at the 0.05 level where 3.3 was the expectation by chance. These inbred lines with 70 or more generations of common sib matings were thus homallelic or very nearly so. Nevertheless the preceding analysis indicates that changes were occurring much more frequently than expected from the available estimates of mutation rate unless something like 1,000 loci can be supposed to be subject to minor mutations that affect skeletal variants.

The first-order genotypic constancy of long inbred lines does not apply to the genotypes of tumors. It became evident in the early studies of tumor transplantation (Strong 1929) that different tumors from the same strain or even from the same mouse might give different results on the same F_2 or backcross individuals. A thorough study of this matter was made by Cloudman (1932), who used eight tumors from the same strain (Bagg albinos) in tests of F_1, F_2, and backcrosses to another strain (dilute browns). His application of the formula given in chapter 4 indicated that one of the tumors had about 12 genes foreign to the dilute browns, while others indicated 4, 2, or 1. The results were consistent for each tumor, except for one that shifted from 2 to 1 apparent foreign factor in the course of the investigation.

It seems clear that something tends to happen in the tumor cell that prevents most of the antigens from inducing a sufficiently hostile reaction to lead to rejection. Somatic mutation is not plausible when so many homozygous codominant genes are involved. There seems, rather, to be a mutational suppression of effects.

There are various possibilities with respect to the number of gene products affected. Cloudman's calculations were based on the assumption that the genes are either wholly effective or wholly ineffective. Other possibilities are that all are equally reduced in penetrance or that penetrance is reduced unequally (Wright 1954). Snell and Higgins (1951) have shown that there is one locus that causes the strongest reactions and an allele, $H - 2^{dk}$, was the one that remained active in the Cloudman tumors in which the apparent number was reduced to one. There clearly were differences among the others in ease of suppression.

Drosophila melanogaster

Mather and Wigan (1942) tested the effect of plus and minus selection on the number of abdominal and sternopleural chaetae in an inbred strain of *Drosophila* with at least 78 generations of common sib mating. Their selected lines were carried along by cyclic matings among the progeny of three pair matings, with rare exceptions.

In one experiment, the high and low lines, starting with about 19 sterno-pleural chaetae, gradually diverged by about one chaeta in the course of 21 generations.

In a second experiment, the number of chaetae on the fourth and fifth abdominal segments (average, about 40) showed no divergence of the plus and minus lines for 13 generations, and divergence of about one chaeta in the next four. A new high line was started from the low line at this point. After six generations, there began to be considerable divergence, which reached six chaetae by generation 17, due in part to decrease for the first time in the low line. A third high line was started from generation 34 of the low line and soon began to rise while the low line continued to decline, bringing about a divergence of four chaetae after 53 generations of selection of this low line. There was no indication in the early generations that sufficient heterozygosis was present at the beginning for effective selection. Minor mutations and recombinations of these seem to come and go during close inbreeding, and after a slight expansion of the line give a basis for selection.

Durrant and Mather (1954) made a different sort of study of the same inbred line (Oregon R) after an unbroken run of more than 300 generations of sib mating. Ten representatives of the second chromosome of their maintenance stock of Oregon R were extracted and brought together in all possible combinations by a technique involving use of a strain in which crossover inhibitors (CyL^4/Pm) for the second chromosome had been introduced into a stock in which the X and III chromosomes and probably IV were from Oregon R.

Significant differences with respect to numbers of sternopleural bristles and abdominal chaetae were demonstrated among the ten chromosomes by analysis of variance. A retest of five of them was made after isolating for five generations. Their effects on chaeta numbers were found to have changed significantly, and a lethal mutation had appeared. These experiments again indicated that mutations are occurring continually and being fixed or lost by inbreeding, rather than that heterozygosis is maintained from the beginning. Again the rate of occurrence required for the postulated mutations seems unexpectedly high.

Clayton and Robertson (1955) selected for abdominal chaetae (10/25 of each sex) in one long-inbred line, with and without exposure to X rays (1,800 r/generation). After 17 generations, there was no significant divergence between the unirradiated high and low lines (29.0 and 28.7 chaetae, respectively) in contrast with the irradiated lines (31.6 and 28.3, respectively).

Spofford (1956) produced an isogenic strain of eyeless (ey^4) of *D. melanogaster* in which a single X, a single II, and a single III chromosome had been made homozygous by use of a stock with dominant markers and crossover inhibitors. The eyes in this strain were much smaller than in the original eyeless strain, which differed little from normal. In the course of a year, the eyes gradually increased from about half the size of the wild stock on the average to eyes as large or nearly as large as those of the wild stock. The correlation between midparent and son or daughter rose gradually from 0 to +0.46 and +0.47, respectively, in the course of two years. The right-left correlation rose similarly from 0 to 0.46. The absence of parent-offspring or of right-left correlations in the strains soon after formation make it likely that initial isogenicity had actually been achieved. The rising correlations show that modifiers were appearing by mutation and the approach to the normal eye in size shows that natural selection was operating in favor of the larger eyes.

Plants

Of special significance for the question of frequency of occurrence of mutations affecting quantitatively varying characters is evidence from plants derived by doubling the chromosome number of viable monoploids. Such plants should be completely isogenic at the time of origin. East (1935) obtained such diploids in several species of *Nicotiana*.

Over a dozen different progeny rows of rustica, for example, gave a somewhat astonishing result. The plants of each progeny test were as much alike as if they had been cut with a die. . . . The individuals in each of these populations were more strikingly similar than those of *any* ordinary inbred population that I had ever examined. And now comes the feature to which I wish to draw particular attention. Several of these lines were continued from self-fertilized seed, and within 3 or 4 years they were showing approximately the same amount of variability as ordinary selfed populations, a result which could only have come about from frequent small mutations.

A quantitative study of this phenomenon, in doubled monoploid maize stocks, has been made by Sprague, Russell, and Penny (1960). They examined large numbers of plants in randomized blocks from the third to sixth generations of selfing (S3 to S6) from the doubled monoploids, and determined the means for nine characters (plant height, leaf width, number of tassel branches, number of kernel rows, ear length, ear diameter, weight of 100 kernels, weight of shelled grain per plant, and date of silking). There were 11 experiments. Altogether they found 90 significant changes in mean from generation to generation or between sibling lines of a generation that were consistent with the hierarchic structure of the experiment (from one S_2 ear, two S_3 ears, four S_4 ears, eight S_5 ears). These changes were interpreted as mutations that occurred in a total of 176 gametes. These led to an estimate of six mutations per attribute per 100 gametes, assuming all to be independent. An examination for possible cases of pleiotropy indicated that about 25% might be of this sort.

As in most of the other cases that had been considered, this seems to require a rate of mutation enormously greater than the usual estimates unless each attribute is susceptible to effects from a very large number of loci. It differs from the inbreeding experiment in the virtual certainty that there was no initial heterozygosis. The typical mutation rates in maize, as with mice or *Drosophila*, seem to be of the order of 10^{-5} or less per gamete, except (in the case of maize) for loci affected by such transposable elements as those demonstrated by McClintock (1952, 1956) and by Brink and Nilan (1952) or subject to paramutation (Brink 1960). As there seems to be no reason to suppose exceptionally high mutation rates for the loci affecting the characters studied by Sprague and his associates, it seems most probable that there are hundreds or even thousands of loci in which mutation can affect each character enough to account for the results.

On the whole, this seems more probable for most of the cases in which long-inbred lines have produced differentiated branches with unexpected frequency, than persistence of initial heterozygosis throughout the whole course of inbreeding. The subject is one, however, on which further information is desirable.

Genotypic Persistence under Vegetative Reproduction

It has been known at least since the time of Theophrastus in ancient Greece, that there is enormously more persistence of type through propagation by cuttings than by seed. The high degree of persistence of apples, peaches, citrus fruits, grapes, and others by cuttings or grafts testifies to this.

These are, however, subject to bud mutations. East (1910) found 12 bud mutations in examination of 700 named commercial varieties of potatoes. These affected color, shape, and so forth, and few if any were of commercial significance. Florists have produced many new color varieties by bud mutations.

The frequency of occurrence seems to be no greater than the rates usually ascribed to gene mutations. This, indeed, is the way in which gene mutations are most frequently observed to occur in higher plants. In some cases, however, bud mutations have seemed common enough to warrant systematic bud selection. The Boston fern, a variety of *Nephrolepis exaltata*, has produced an extraordinary diversity, some 40, of heritable forms differing in frond type (uni-, bi-, tri-, and quadripinnate), and in color, with varying degrees of stability (Benedict 1915). Stout (1915) described the heritable varieties that appeared in two clones of cultivated *Coleus* in 833 plants. The original plant had a pattern of green in the midregion, yellow in the border of the leaves (subepidermal), with red blotches in the epidermis.

As shown in table 5.1, the rate per bud of each type was about 10^{-4}. There was segregation of the progeny from seed of two of the plants, indicating that the changes were probably gene mutations.

TABLE 5.1. Frequencies of types of bud mutations in *Coleus* and rates of mutations, assuming 200 buds per plant.

Type	Plants	Frequency	Mutant Rate
Increase of yellow, decrease of green	827	27	0.00016
Decrease of yellow, increase of green	740	50	0.00034
Reversal of positions of yellow and green	450	8	0.00009
Increase of red	770	8	0.00005
Complete loss of red	815	19	0.00012
Decrease of red but not loss	815	2	0.00001
Laciniate from entire leaves	765	13	0.00008
Entire leaves from laciniate	68	1	0.00007
			0.00092

SOURCE: Stout 1915.

Hanel (1907) studied the stability of clones of *Hydra grisea*, propagated by budding, with respect to number of tentacles. She concluded that there were heritable differences among clones but she was unable to obtain persistent differentiation within clones by up to seven generations of selection. According to Hase's (1909) analysis of Hanel's data, there was no evidence of hereditary differentiation among the clones; but Pearson (1910) concluded from the same data that there was heritability both within and among them. The significant correlations between parent and offspring and the smaller ones after two generations might, however, have been due to short-time persistence of physiological states rather than true heredity.

A more thorough study was made of various characters in clones of *Hydra viridis* (especially number, general size, and color of tentacles) by Lashley (1915). He studied the relations of age, nutrition, and various environmental factors to the characters. He concluded that the differences among clones were almost certainly genetic, the only reservation being with respect to the unknown ages as clones. He found significant correlations in number of tentacles in successive generations but no persistence for seven generations.

Parthenogenesis

Whitney (1912) found a decline in vigor in the rotifer, *Hydatina senta*, under long-continued parthenogenesis. He started a line from a single fertilized egg. After 59 parthenogenetic generations he maintained two separate lines, A and B. The former was still alive in generation 503 but in a very exhausted state. In a number of tests in about generation 400, it produced an average of 2.1 daughters per 48 hours in comparison with an average of 7.2 from contemporary parthenogenetic lines from nature (or 7.7 from a cross between two such lines). Inbreds from line A at this time averaged only 1.0, but this rose to 4.0 in a second and third inbreeding. Inbreds from line B at the time of its extinction as a parthenogenetic line averaged only 0.1, rising to 2.3 on further inbreeding. There was full recovery of reproductive capacity (average 7.3) following crosses between A (parthenogenetic or inbred) and inbred B. Thus unfavorable genetic changes seem to have been accumulated under long-continued parthenogenesis. The recombination occurring under sexual reproduction within a line brought no recovery in the first generation and only slight improvement in two or three further inbreedings. That no serious decline had appeared in the first 59 parthenogenetic generations is indicated by the full recovery of vigor on crossing lines A and B, which had these generations in common.

A. F. Shull (1912) also experimented with *H. senta*, but considered the effect of continued inbreeding in lines with relatively few parthenogenetic generations. A cross was followed by 12 generations of parthenogenesis along a single line. Successive inbreedings, separated by a few parthenogenetic generations, were then carried to the sixth generation along two lines. The size of family from parthenogenetic females fell off at an increasing rate from 48.4 after one inbreeding to 22.6 and 24.8 after six inbreedings in the two lines. Size of family from sexual females fell off from 11.0 to 7.5 and 7.6, respectively. The difficulty of rearing increased with inbreeding.

These (and other) results showed merely that there is an inbreeding decline in *Hydatina*. Shull was, however, aware of the similar decline under long-continued parthenogenesis that could not be explained by the prevailing interpretation of inbreeding decline as a by-product of genetic fixation. He made the interesting suggestion that there may be a stimulus from action of a genome on a cytoplasm that has been under the influence of a different genome, a variant of the stimulus theory of heterosis, and that the inbreeding decline is due to declining stimulation as the cytoplasm comes to be more and more the product of a single genome. Under this view, the stimulus would also decline under long-continued parthenogenesis, with no change in the genome. The concept of complementary action of alleles in heterozygous genomes seems preferable, however, in the interpretation of heterosis and that of gradual accumulations of unfavorable mutational effects seems preferable in the case of decline under parthenogenesis.

Ewing (1916) carried a line of aphids (*Aphis avenae*) through 87 parthenogenetic generations in association with selection of different sorts at different times. No significant changes were made in the ratio of length of the third to fourth segment of the abdomen by 15 generations of plus selections or by 10 of minus selections. A few generations of selection for length of cornicles, plus and minus, also seemed to be without effect. Finally, the ratio of cornicle to body length was selected for 44 generations without effect. There was clearly no appreciable heritable variability during the period of study. The possibility of an accumulation of small changes over a longer period was, of course, not ruled out.

Negative results were also obtained by Agar (1914) in experiments with parthenogenetic lines of the Cladoceran, *Simocephalus vetulus*.

The most extensive studies of selection under parthenogenesis have been those of Banta (1921, 1939) with Cladocera. His first series of experiments (1921) was on selection for and against phototropism in three species. In testing, the day-old daphnids were released in the middle of a shallow black tank 40 cm × 26.6 cm × 7.2 cm and exposed to heat-screened light of 120 candle meters at one end. The criterion was the number of seconds required to reach the light end (or failure to reach it in 15 minutes, arbitrarily counted as 900 seconds). Pairs of lines selected for high and low reactivity (plus and minus, respectively) came from the same progenitors and were maintained (in most cases by selecting several females) under as nearly the same conditions as possible and by exclusive parthenogenesis.

Reactivity was only slightly affected by variations in temperature but more by other environmental factors, especially by the composition of the pond water in which they were reared. There were parallel periods of high and low reactivity that were probably due to this cause. There was a great deal of individual variation, there being a standard deviation of 300 seconds in periods in which the mean reaction time was 400 to 700 seconds.

In *Daphnia pulex* there were no significant divergences between plus and minus in the later generations of eight pairs of lines with 44 to 152 generations of selection each (average 105). Figure 5.1 shows the parallel variations in the two-month averages of the total set. No significance can be assigned to the divergence in the direction of selection in the last two averages.

There were, however, periods of up to two years during which successive two-month averages of lines that were being compared differed consistently from the general average. Whether these were due to mutational changes that were later neutralized by ones in the opposite direction or merely accidental will require consideration.

Three pairs of sister lines of *D. longispina* were tested for 20 to 75 generations (average 45) without any suggestion of differentiation.

Five pairs of sister lines of *Simocephalus exspinosus* were tested for 70 to 183 generations (average 123). In four of these, there were again no such divergencies at the end as to indicate successful selection, although again

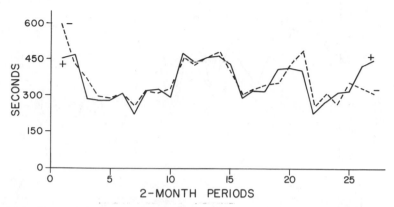

FIG. 5.1. Average numbers of seconds required to reach the light end of a tank by eight parthenogenetic lines of *Daphnia pulex*, continually selected for rapidity (+), and eight selected for slowness (−) over 54 months. From data of Banta (1921).

there were periods of up to a year in which the direction of deviations of the two-month averages were the same. In the remaining pair of lines (no. 757) (fig. 5.2) selected for 181 generations, the line of plus selection had a considerably shorter average reaction time (103 seconds) than the line of minus selection from the third to the ninth two-month record (9th to 62nd generation). There was much irregularity but an average superiority (58 seconds) of the plus line from the 10th to 16th two-month period (114 generations) followed by great average superiority (327 seconds) of the plus line to the end (27th two-month period, 181st generation). This superiority was found to be still present after 112 generations without selection. There seems no doubt that at least two favorable mutations occurred and were established by selection. Banta's analysis indicated a minimum of three, one very early in the plus line and ones in both the plus and minus lines when the period of wide divergence began. After this time, not only did the plus line show a much shorter reaction time than any of the other plus lines of *Simocephalus* but the minus line was much slower than the other minus lines.

Returning to the other lines, a rough test of whether there were more periods of consistent divergence than expected by chance can be obtained by considering the arrays of observed and expected sequence lengths with respect to the directions of difference of the paired two-month averages. These are shown for the three species and the totals in table 5.2 (excluding no. 757). Plus (+) refers to excess reaction times of the lines of plus selection

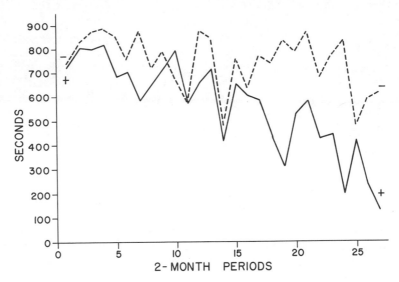

FIG. 5.2. Average numbers of seconds required to reach the light end of a tank by a parthenogenetic line of *Simocephalus exspinosus*, continuously selected for rapidity (+), and by a sister line selected for slowness (−) over 54 months. From data of Banta (1921).

and thus deviations opposite to that expected from selection, while minus (−) refers to deviations that might be due to selection. It may be seen that the distributions of + and − are very similar, indicating that they are independent of selection. As noted above, however, the individual variability was so great that selection could not have been very effective either in seizing

TABLE 5.2. Frequencies of lengths of sequences of excess (+) and defect (−) of reaction time of plus selection relative to minus selection in *D. pulex*, *D. longispina*, and *Simocephalus exspinosus*.

	D. pulex		D. longispina		S. exspinosus		Total					
	+	−	+	−	+	−	+	−	Observed	Calculated	$o - c$	$\dfrac{(o - c)^2}{c}$
1	20	23	4	4	9	11	33	38	71	63.7	+7.3	0.84
2	10	7	1	2	5	3	16	12	28	30.0	−2.0	0.13
3	2	2	1	1	1	1	4	4	8	14.1	−6.1	2.64
4	2	3	...	...	1	3	3	6	9	6.6	+2.4	0.87
5	1	1	...	...	2	...	3	1	4	3.0	+1.0	0.33
6	1	1	...	...	...	...	1	2	3	1.4⎱	+0.4	0.06
6+	...	...	...	...	...	1	...	...	...	1.2⎰		
Total	36	37	6	7	18	19	60	63	123	120.0		4.87

SOURCE: Banta 1921
* Excluding no. 757.

upon favorable mutations or in preventing the choice of unfavorable ones, if associated with favorable individual performance.

The expected lengths of sequences would form a geometric series with frequencies halving with each increment of length if taken from an indefinitely long series of observations in which it was an even chance whether one or the other line had the greater average. This is modified somewhat by the finiteness of the number of two-month averages in each pair of lines. If n is the number of such averages, the expected number of deviations preceded and followed by ones in the opposite direction is $(n + 2)/4$; of sequences of two, $(n + 1)/8$; of sequences of length m, $(n + 3 - m)/2^{m+1}$, except that the expectation that all are of the same sign, $(m = n)$, is $4/2^{m+1}$, just half that for $m = n - 1$. The expected total number of sequences is $(n + 1)/2$. The expected numbers in table 5.2 are based on the sets of expectations for all pairs of lines, excluding no. 757.

There is a slight excess of observed sequences of four, five, and six two-month averages (as well as of single ones) at the expense of sequences of two and three, but $\chi^2(= 4.87)$ is obviously of no significance. It should be said that there were a number of cases in the published graphs in which there appeared to be no divergence. These were arbitrarily assigned to $+$ and $-$ alternately. If all these signs are reversed, there is one sequence of seven minuses but χ^2 rises only to 5.97, still of no significance. There is thus no necessity to suppose that any mutations were established in any of paired lines other than no. 757 in which, after an initial minus difference followed by a plus, there was a succession of eight minuses, a plus, and finally 17 minuses, a record of a wholly different sort from that of any of the other 15 pairs of lines. On the other hand, the possibility that a minor mutation in one direction or the other became established temporarily, to be neutralized later by one in the opposite direction, is not wholly excluded.

Sex intergrades appeared in a clone of *Simocephalus*. These varied greatly in each of many secondary sexual characters and, if sufficiently extreme (above grade 30 or 40 on a scale of 70), were sterile; there were some hermaphrodites in which one gonad was an ovary, the other a testis or ovotestis. Not infrequently individuals that at first produced sperms later produced eggs (usually abortive). There was an unusual number of normal males in high grade lines. These were not counted in determining the average grade of a brood. Because of sterility, selections in the plus lines with averages of 30 to 40 were directed toward maintenance of this average rather than toward further advance. While the mode of inheritance was not tested by crossing in *Simocephalus*, there is little doubt that it was due to a dominant mutation as was proved for a similar occurrence of sex intergrades in a clone of *D. longispina*. The lines were maintained by four females per generation.

In *Simocephalus*, plus and minus selection was carried out in each of two subclones that had long been separate. Subclone XV under minus selection was derived from subclone V under plus selection. Subclone XVI under minus selection was derived from clone XIV under plus selection. Table 5.3 shows the length of sequences under selection. Note that + is in the direction of selection in this case. Tabulation was by groups of five generations.

TABLE 5.3. Sequences of higher (+) or lower (−) grades of sex intergrade in *Simocephalus exspinosus* (by numbers of five-generation units) in comparisons of a line under plus selection with one under minus selection.

Line	Sequences of Differences
V (plus) − XV (minus)	4−, 10+
XIV (plus) − XVI (minus)	1−, 1+, 1−, 2+, 1−, 8+
Totals	21+ : 7−

SOURCE: Banta 1939.

There can be no doubt from the terminal sequences of ten pluses and eight pluses that selection was effective in both cases, although there was considerable delay before a favorable mutation was seized upon in the second one. The courses of the lines under comparison are shown in figure 5.3. In both, the successful differentiation was brought about wholly by decline of the

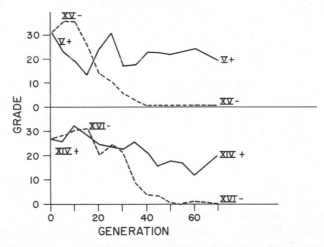

FIG. 5.3. Grades of a mutation "sex intergrade" of *Simocephalus exspinosus* in two sets of sister subclones, one in each case under plus selection (+) and one under minus selection (−). From data of Banta (1939).

minus line, as expected, because the plus line was already about as high at the beginning as it could go and remain fertile. In neither case was the basic gene lost by reverse mutation.

A similar mutation, causing sex intergrades with varying penetrance and expressivity in different subclones, was later found in *D. longispina.* In this case three lines of plus selection (I, IV, and VI) and three lines of minus selection (III, V, and VIII) were started from one line. Table 5.4 shows the

TABLE 5.4. Sequences of higher (+) or lower (−) grades of sex intergrade of *Daphnia longispina* (by numbers of five-generation units) in comparisons of three primary lines under plus selection with three primary lines under minus selection.

Lines	III (minus)	V (minus)	VIII (minus)	Total
I (plus)	1−, 16+	4+, 1−, 12+	8+, 1−, 14+, 1−, 1+, 2−	55+ :6−
IV (plus)	1−, 4+, 2−, 4+	4+, 1−, 6+	5+, 1−, 1+, 1−, 1+, 1−, 1+	26+ :7−
VI (plus)	17+	4+, 1−, 12+	8+, 2−, 13+, 4−	54+ :7−
Total	41+ :4−	42+ :3−	52+ :13−	135+ :20−

SOURCE: Banta 1939.

lengths of sequences of + and − differences in five-generation averages for all possible comparisons. There can be no doubt that minus lines III and V declined relative to the plus lines (terminal sequences 16+, 4+, 17+ and 12+, 6+, 12+, respectively). In the case of minus line VIII, there were no such records of terminal success of selection, although there were intermediate periods in which sequences of 14+ and 13+ relative to I and VI, respectively, could not be due to chance, and the total record in comparison with the plus strains (52+, 13−) seems to indicate a prevailing tendency for differentiation in the direction favored by selection.

Lines of reverse selection and reversals of the reversed lines were taken off from time to time: XII(minus) from I(plus), and XXII(plus) from the former; X(plus) and later XIX(plus) from III(minus), and XIV(minus) from X(plus); XI(plus) and later XX(plus) from V(minus), and XV(minus) from XI(plus); XIII(minus) from VI(plus) and XXI(plus) from VIII(minus).

In most of these ten cases shown in table 5.5, in which the sign is that of the difference of the line of plus selection from that of the minus line, irrespective of which was the new one, selection (indicated by +) was immediately successful by this criterion; in several of them, however, the lengths of the sequences were too short for confidence.

In these reverse selections, there were, however, 88 differences in the direction expected from selection and only 4 in the opposite direction. It is evident from the records of both the primary selections in both species and the reverse selections in one of them that selection was very much more

TABLE 5.5. Sequences of higher (+) or lower (−) grades of sex intergrade of *Daphnia longispina* (by numbers of five-generations units) in comparisons of a line under plus selection, derived from one under minus selection or the reverse.

Lines	Sequences of Differences (Plus Line) − (Minus Line)
I (plus) → XII (minus)	20+
XII (minus) → XXII (plus)	11+
III (minus) → X (plus)	8+
X (plus) → XIV (minus)	3+
III (minus) → XIX (plus)	4+
V (minus) → XI (plus)	5+
XI (plus) → XV (minus)	1+, 1−, 1+
V (minus) → XX (plus)	4+
VI (plus) → XIII (minus)	20+
VIII (minus) → XXI (plus)	10+, 3−, 1+
Total	88+ : 4−

SOURCE: Banta 1939.

generally successful in modifying the degree of the sex intergrade mutations than in changing the degree of phototropism. Again, the basic mutation was never lost.

The courses followed by the various lines in *D. longispina* are shown in figure 5.4, in which the unweighted averages of successive five-generation averages are plotted in order to smooth out some of the more violent fluctuations. The interpretation of changes is somewhat confused by the fact that broods from four selected females were graded in each generation. It is stated that selections in the high selected clones were made when possible from a group of sibs averaging high and vice versa, and that whenever feasible an individual was selected whose progeny, as indicated by preliminary observations, tended to be of the desired sort, but employment of these criteria was limited. It appears that any mutations that occurred would, in general, not be established in all individuals of a generation or even after several generations, and that a mutation present in one but not all of the four females of a line might later be lost.

Nevertheless it appears from the graphs that while progress up or down sometimes occurred with little or no delay, there were cases in which there was considerable delay or even no progress at all; others in which there was progress by widely separated steps; and several others in which an unfavorable mutation seems to have been present in one or more of the females

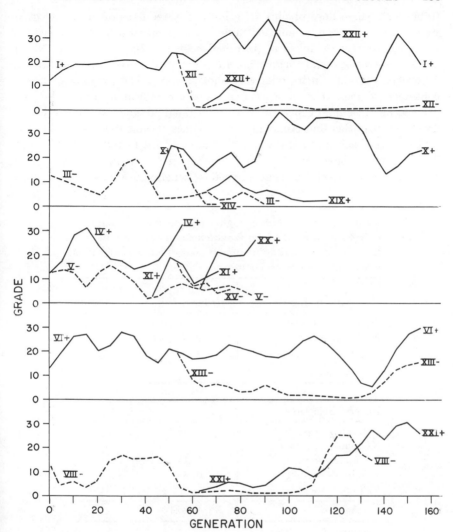

FIG. 5.4. Grades of a mutation "sex intergrade" of *Daphnia longispina* in sub-clones under plus (+) or minus (−) selection as indicated. From data of Banta (1939).

to be carried several generations before being lost. As in the case of photo-tropism, there was so much nongenetic individual variability that selection could not have been a very effective process.

Another mutation in *D. longispina*, excavated head, also showed irregular penetrance and wide variation in expressivity within the clone of origin (table 5.6). Three lines of plus selection and three lines of minus selection were carried through a sufficient number of generations to demonstrate persistent divergence in pairs (terminal sequence 10+, 18+, and 21+). The courses are shown in figure 5.5. Line I (plus) however, showed little or no progress upward in its relatively short history. The progress of line V continued so long that at least two steps are indicated. Line XI (plus) delayed more than 30 generations before a rapid advance to a much higher level than any other line. Among the minus lines, decline to a low level (never by loss of the basic mutation) was rather rapid in line II and most delayed in line IV, which, however, at a later generation (ca. 115), seemed to have acquired a second mutation that reduced its penetrance almost but not quite to zero.

TABLE 5.6. Sequences of higher (+) or lower (−) grades of excavated head of *Daphnia longispina* (by numbers of five-generation units) in comparisons of a line under plus selection with one under minus selection.

Primary Lines	Sequences of Difference (Plus Line) − (Minus Line)
I (plus) − II (minus)	0, 10+
XI (plus) − IV (minus)	0, 3−, 18+
V (plus) − VI (minus)	1−, 21+
Total	49+ :4−

Reverse Selection	
II (minus → plus)	0, 18+
XI (plus → minus)	0, 4+, 1−
IV (minus → plus)	0, 1−, 1+, 0, 14+
V (plus → minus)	0, 5+
VI (minus → plus)	0, 0, 2+, 0, 1+, 12−
Total	45+ :14−

SOURCE: Banta 1939.

Reverse selection was conducted in all cases except I (plus). Minus selection from V (plus) and XI (plus) resulted in small but consistent declines. Plus selection from II (minus) and from IV (minus) resulted in similarly small but consistent gains. Banta suggested that all of these changes might be due merely to a maternal cytoplasmic effect for which he found evidence from crosses, rather than from selection of mutations. In the case of plus selection from VI (minus), there was a decline after a few generations to very low

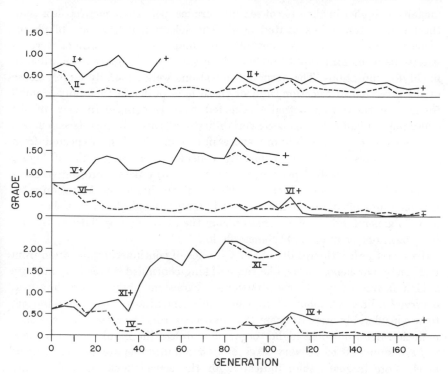

FIG. 5.5. Grades of a mutation "excavated head" of *Daphnia longispina* in sub-clones under plus (+) or minus (−) selection as indicated. From data of Banta (1939).

penetrance, indicating that a minus mutation had been inadvertently established. The absence of important mutation in the desired direction in the many generations of reverse selection is in marked contrast with the fairly rapid success of selection in all of the primary lines (except I [plus]). There seems to be no explanation but chance.

Reviewing all of these experiments with Cladocera, it appears that only two or three mutations, all in one line, were established in some 3,200 generations of selection of phototropism in these species. More than two dozen were probably established in some 2,550 generations of selection of sex intergrades in two species. At least eight, not counting the small changes that may have been maternal effects, were established in some 1,080 generations of selection of excavated head. Between 50,000 and 60,000 individuals were tested in the experiments with each of the three characters. Thus the rate seems to have been no more than 1 per 1,000 generations, 1 per 20,000 individuals tested, in the case of phototropism, but was at least an order of magnitude higher in the case of sex intergrades, perhaps somewhat less than the latter in that of excavated head. The selection in the case of phototropism were, however, of normal individuals, while the others were of variations of mutant types that may be expected to be more susceptible to modifying mutations. The observed mutations were naturally largely those in the direction of selection, and of these it is probable that only a small fraction of those that actually occurred were recognized in view of the enormous amounts of nongenetic variability in all cases. It must be concluded that selection was much more successful than would be expected from mutation rates of the order usually considered characteristic of higher organisms (10^{-5} to 10^{-6} per locus per generation) unless there were a great many loci at which mutations had effects on the characters. Mutation rates for such minor effects as modification of a highly susceptible mutant character may, in general, however, be greater than the above figures. There may also have been one or more highly unstable loci.

To a first order, the results of most studies of continued propagation from buds or by parthenogenesis, like those of long-continued inbreeding, indicate a high degree of genotypic persistence. Persistent changes are, however, observed if lines are followed up over many generations. They seem, indeed, to be considerably more frequent than expected from the rates of conspicuous mutation determined in higher organisms, unless the numbers of pertinent loci are supposed to be very large or minor mutations are supposed to be much more frequent than indicated for the conspicuous ones. There are some cases in which the frequencies of change indicate highly unstable loci.

We have not considered here eversporting variegations due to cytoplasmic elements. Those considered were due to gene mutations as far as tested. They are of a sort that could be of evolutionary significance.

Protozoa

Apparently similar phenomena have been observed in protozoa, reproducing by fission in the absence of conjugation or autogamy: first-order persistence, second-order changes that accumulate under selection more rapidly than expected from accepted mutation rates in higher organisms. Those observed in ciliates, however, can have no direct evolutionary significance. Ciliates, in general, have two kinds of nuclei: micronuclei, corresponding to those of the germ cells of higher organisms, and macronuclei, corresponding to those of somatic cells. In *Paramecium aurelia*, the most intensively studied species, there are (in most varieties) two micronuclei and one macronucleus. The former divide by mitosis and the latter apparently amitotically before apportionment at fission. In conjugation or autogamy, combined cytologic and genetic evidence (Sonneborn 1947, 1957) indicates that the micronuclei (diploid) go through the two meiotic divisions, but all but one of the products are absorbed. This one divides a third time to produce two identical haploid gamete nuclei. One from each conjugant usually enters the other conjugant. The resulting diploid genotypes of the exconjugants are thus identical and may be homozygous or heterozygous at loci, according to the genotypes of the conjugants. The fertilization nucleus divides twice. Two of the products become micronuclei, which undergo mitosis and apportionment at the first fission. The other two products become macronuclei, segregate at the first fission, and replace the macronuclei of the conjugant, which have meanwhile been absorbed. Similar processes occur at autogamy, except that the two identical products of the division following meiosis unite with each other. The fusion nucleus divides. One product replaces the macronucleus, which in this case also has been absorbed, and the other divides to produce two new micronuclei. Thus the micronuclei and the macronuclei all have the same homozygous genotype. These can normally be heterozygous only in the usually small number of generations between conjugation and the first autogamy.

Sonneborn (1947) found, however, that under certain conditions a small fragment of the macronucleus persisted through conjugation or autogamy and gave rise, by division and macronuclear regeneration, to the new macronuclei. In cases in which the genotype of this macronucleus differed from that of the new micronuclei, it was the one that determined the characters of the organisms through the following fissions (insofar as not determined by persistent properties of the cytoplasm of the conjugant). This and the discovery of clones that flourished through hundreds of fissions without micronuclei indicated that physiological control is restricted to the macronucleus (and cytoplasm). The regeneration indicates that the macronucleus consists of many equivalent components and that the apparently amitotic division at fission is really an apportionment of many complete nuclei. It follows that persistent changes observed in the course of fission must be due to dominant macronuclear mutations where they are not due merely to decay of persisting cytoplasmic properties, as in cytoplasmic lag. Because of the regular disintegration of the macronucleus at the next autogamy or conjugation, such mutations can have no direct evolutionary significance. Micronuclear mutations may, of course, occur but these will not be manifested during the succession of fissions under observation. They may, however, become manifest as properties of the macronucleus formed at the next autogamy or conjugation. The degeneration of ciliate clones in the absence of autogamy may be due to the accumulation of macronuclear changes (as well as persistent cytoplasmic damage), but the characteristic life cycle of ciliates, leading, after many generations of fission, to senescence and death in spite of temporary and decreasingly effective rejuvenescences from autogamy, also seems to imply the accumulation of micronuclear mutations.

Jennings (1908) found that members of the same clone of *Paramecium aurelia* or *P. caudatum*, varied greatly (up to twofold) in length, depending on growth stage, or nutrition, etc. The averages for different clones differed consistently under the same conditions, again by a factor of two. He was unable, by long-continued selection, to change the mean of a clone. There was thus a high degree of persistence.

Middleton (1915), on the other hand, was able to bring about great changes in rate of fission of *Stylonychia pustulata* by continually selecting the more rapidly dividing offspring and the more slowly dividing one for 150 generations of fission. The differences between the lines persisted for 50 days or more without selection.

Jollos (1913, 1934; cf. Sonneborn 1947) found persistence of characters under selection, or attempts at acclimatization to chemicals or to high temperatures, in many cases, but he also found cases in which such attempts, especially when combined with selection, were successful in bringing about profound changes. These "Dauermodifikationen" might persist for hundreds or thousands of generations but ultimately disappeared. In most but not all cases, they tended to disappear after autogamy or conjugation as well as change of environment. The phenomena are too varied and complex for detailed consideration here. Jollos considered them to be due to environmental effects on the cytoplasm. It seems probable that macronuclear mutations were also often involved, and in some cases micronuclear ones. I cite them to illustrate first-order persistence of the genetic material whatever its nature, but second-order capacity to change.

The genetics of rhizopods is much less well understood than that of ciliates, but no such distinction is known as that between the two kinds of nuclei of ciliates. As far as is known, changes that accumulate under fission can be of direct evolutionary significance.

Jennings (1916) studied a number of characters of the rhizopod *Difflugia corona*. This is a roughly spherical organism with a shell made of cemented sand, a varying number of teeth surrounding the mouth, and spines of varying number and length. The protoplasm extrudes from the mouth at fission and forms a new shell with spines about the pseudopodia and teeth in juxtaposition to the old ones. Jennings found that clones differed greatly in size and in the above respects and tended strongly to persist in these differences, but that long-continued selection brought about differentiation within clones comparable to that among the clones in nature. In the case of number of teeth, the correlation between parent and daughter was almost perfect, suggesting that the "heredity" was of a mechanical sort, the old teeth determining the formation of adjacent new ones. Experiments in which some of the teeth were removed indicated that this was an important factor but not the total explanation, since clones tended to return to their characteristic tooth pattern after a few fissions. The distributions of some of these characters, including clonal and subclonal differences, are illustrated in volume 1, figures 6.8, 11.3, and 11.9.

Similar results have been demonstrated by Root (1918) with *Centropyxis aculiata* and by Hegner (1919) with *Arcella dentata*.

Mutability

Abundant evidence, discussed in volume 1, chapter 15, has demonstrated that quantitative variability in a great variety of animals and plants depends on segregation in multiple Mendelian loci. The material for this segregation is presumably provided ultimately by mutations and immediately by alleles, which are maintained by one or another sort of balance among the following factors: recurrent mutation, selection, and immigration.

The rate of recurrence of mutations with effects of sufficient magnitude to be easily scored has been found to be of the order 10^{-4} to 10^{-6} per generation in organisms as diverse as *Drosophila*, maize, and the mouse. It tends to be higher with longer generation times. Data on persistence of genotypes in long inbred lines, and under budding or parthenogenesis, discussed in this chapter, indicate the expected persistence to the first order, but genotypic changes seem to occur with considerably more frequency than expected from observed mutation rates and very much more frequently in some cases. This is probably due to higher rates of occurrence of mutations with effects too small to be readily scored. There are, however, undoubtedly particular mutable loci at which changes occur with very great frequency, and such loci may be more numerous than supposed. The nature of these mutable loci and the question of whether they may not indicate modes of quantitative variability that approach in importance that of ordinary segregation of multiple factors require some consideration.

Mather (1941) reopened the whole question with his hypothesis that quantitative variability depends on a wholly different sort of genetic unit, a "polygene," from that concerned with major qualitative differences, an "oligogene." He suggested that the former might be heterochromatic and the latter euchromatic. While these are both chromosomal and subject to Mendelian heredity, there are certain possible qualifications of the latter in the case of heterochromatic differences. Mather (1941) supported his view in part by the results of experiments in which he demonstrated that the heterochromatic Y chromosome of *D. melanogaster* is the site of differences that contribute to quantitative variability of chaeta number.

Selection experiments (including ones of Mather and Harrison [1949] discussed later) showed that polygenes must be widely distributed throughout the chromosomes of *Drosophila*, and thus that there must be a similarly wide distribution of heterochromatin if Mather's hypothesis held. This made it difficult to make any clear pragmatic distinction from the conventional multiple factor hypothesis. When it was found that multiple allelic series may include not only oligogenic differences but polygenic ones, grading down to isoallelic ones, and that the same gene difference may have oligogenic effects on one character but pleiotropic polygenic effects on other characters, a sharp distinction became impossible. Mather himself seems to have abandoned it. The term *polygene* has, however, continued to be widely used for a gene responsible for a minor difference from type in a character under consideration. This is convenient but somewhat objectionable because of its implication of a special class of genes rather than a kind of effect. The term *polygenic* as a synonym of *multifactorial* does not suffer from this objection and has the advantage of referring specifically to genes rather than to genetic and environmental factors collectively, when the former is intended.

It still holds, however, that while euchromatic genes may have effects of all magnitudes, differences in visibly heterochromatic regions of chromosomes rarely have major effects but may contribute frequently to quantitative variability.

There are deviations from orderly Mendelian segregation associated with heterochromatin in one way or another. Thus accessory chromosomes, such as those of maize, are distributed largely at random at meiosis (Randolph 1928) and cause some reduction in vigor and fertility when too abundant (Randolph 1941). There is here a component of certain kinds of quantitative variability that do not come under the strictly Mendelian formulation of multiple factor theory.

The easily transposable elements of maize discovered by McClintock (1956) (Dissociation, Activator), may be related to the nonspecific synaptic attractions characteristic of heterochromatin. They tend to inhibit the effects of neighboring genes. The effects described are oligogenic and responsible for such long-known mutable loci as that of pericarp variegation (Emerson 1917; Brink and Nilan 1952). It would seem likely, however, that genes with minor effects would also be inhibited, and thus that there might be a significant contribution to quantitative variability.

The phenomenon of paramutation, discovered by Brink (1960), under which certain genes at the *R* locus of maize ("paramutable genes") undergo systematic and persistent changes in effect after exposure in heterozygotes to certain alleles ("paramutagens"), may also contribute to quantitative variability. The frequency of paramutation is not yet known.

As noted earlier (vol. 1, chap. 3), there is evidence that gene duplication has been a very common evolutionary phenomenon that, being followed by evolutionary differentiation of the duplicants, has led many if not all loci of higher organisms to become complex. It has also been suggested that heterochromatin differs from euchromatin in the multiplicity of replications. It has long been known that replication tends to lead to unequal crossing-over with consequent difference in the number of elements in the products of meiosis. This phenomenon was discovered in the Bar eye mutation of *D. melanogaster* where the duplication involves several loci. The "mutations" to one and three groups (normal and ultrabar, respectively) occur at the relatively high rate of 10^{-3} per generation. They occurred in Zeleny's selection experiment for modification of the number of facets in Bar eye, discussed in a later chapter, but the "mutations" were so great in comparison with the ordinary variations that he did not include them as contributing to progress by selection. If, however, unequal crossing-over should occur frequently in highly replicated heterochromatic genes with only minor quantitative effects, they might contribute to an important extent to quantitative variability. Such changes could occur in highly inbred lines. There might also be a tendency to nonequational apportionment in mitosis and thus changes from this cause under budding and parthogenesis.

The critical test of the relative importance of conventional multifactorial Mendelian segregation and of such accessory processes as listed above is the extent to which changes occur and selection is effective where no normal segregation should be occurring. Persistence under such conditions seems to be true to a first order, but the processes other than simple gene mutation may contribute to some of the unexpectedly frequent second-order changes discussed in this chapter. It should be added that there may also by cytoplasmic changes, about which little is known except that there are special cases in which quantitative differences have been shown to be due to this cause (vol. 1, chap. 4).

Summary

The degree of persistence of a number of characters of long inbred lines of animals (guinea pigs, mice, *Drosophila*) are discussed. A high degree of persistence is the rule but, nevertheless, sublines have been observed to become differentiated much more rapidly than expected from what are considered typical mutation rates (10^{-5} or -10^{-6} per generation), unless the numbers of loci at which mutations may affect the characters in question are much more numerous (hundreds or even thousands) than ordinarily supposed. Delayed segregation is sometimes a possibility but in general seems inadequate as the explanation.

This last possibility is ruled out in the cases of the persistence of monoploid plants (species of *Nicotiana*, maize), and of vegetative multiplication (potatoes, Boston fern, *Coleus*, *Hydra*), parthenogenetic reproduction in a rotifer, *Hydatina senta*, and in several species of Cladocera, fission in *Paramecium* and *Difflugia*, in all of which there is persistence to the first order but again more rapid divergence of sublines than expected from ordinary mutation rates, unless the numbers of pertinent loci are very great. The only exceptions among those referred to were the absence of any apparent genetic change in a long-continued study of *Aphis avenae* by Ewing and in a study of a Cladoceran, *Simocephalus*, by Agar.

It is thus probable that quantitatively varying characters can usually be affected by a very large number of loci. It is also likely, however, that typical gene mutation at these loci is supplemented fairly often by other types of genetic change, including release of genes from inhibition by frequently transposable heterochromatic elements (of McClintock), by paramutations (of Brink), and by unequal crossing-over between replicated genes (perhaps rather common in heterochromatin).

CHAPTER 6

The Course of Directional Selection

The immediate changes in gene frequency brought about by long-continued selection under a great variety of conditions have been discussed in volume 2. The theoretical course for a single locus may be deduced at once where the action of selection is independent of that at other loci. There rarely is such independence, however; thus it is desirable before considering the results of experiments with directional selection to consider some of the more important complications where mass selection of this sort is applied to multifactorial quantitative variability.

Assumptions

We will start from the results of the section on "Selection of Quantitative Variability" (vol. 2, chap. 5). It is assumed that the population is panmictic and continues so except for the selection of parents. Inbreeding effects will be ignored at first, recognizing that this is unrealistic in connection with the small size of most experimental populations. The effects of loci will be assumed here to be additive with respect to the selected character.

The likelihood that two or more loci have special interaction effects in actual cases, such as discussed in various chapters in volume 2, must, however, be borne in mind. These are too numerous for detailed consideration here. On the other hand, the usual type of interaction under natural selection to which most attention was paid in volume 2, that due to intermediacy of the optimum grade of quantitative variability, must always be taken into account, although it is often of only minor importance in the early generations of strong directional selection, since this shifts the optimum to the favored extreme. In the later generations, as the limit is approached, the presence of natural selection that opposes the artificial selection may, however, become very important because under such natural selection the actual parents of the next generation may be expected to deviate less than the initially selected individuals.

With respect to single loci, some heterozygotes may be favored over both homozygotes even under strong directional selection, sometimes because both homozygotes actually tend to be of lower grade, but more often because one is lower in grade and the other of such lower viability or fecundity that even strong selection leads to an equilibrium frequency instead of fixation. In this case, the empirical prediction, based on the regression coefficient of offspring on a random array of parents, wholly breaks down. The latter indicates that there should be progress where there actually is none.

Such complications led to the concept of *realized* heritability under selection (Falconer 1955). This is the ratio of the total gain under directional selection, before the approach to a plateau, to the accumulated selective deviations of the parents from the means of their generations. The realized heritabilities, up and down, often differ greatly in contrast with the common estimate from the regression of offspring on midparent in a random breeding population (assuming that the character is measured on a scale that makes the regression linear).

It will be assumed here that the most appropriate scale is used. This eliminates the common type of interaction based on a systematic relation between the effects of all factors and grade (vol. 1, chap. 10). The use of such a scale not only eliminates any initial asymmetry in progress up and down (due, for example, merely to multiplicative action of factors), but also removes an apparent approach to a plateau due merely to a physiological limit at which all factor effects approach zero (in contrast with a limit due to fixation or to equilibrium with strongly opposed natural selection).

For simplicity, only pairs of alleles will be considered. This is not a very serious restriction in dealing with strong directional selection since the primary question at issue is the increase in frequency of the most favorable allele at the expense of the others collectively, a group usually consisting mainly of the allele most favorable before the selection began. Most of the selection experiments have, indeed, started from such small foundation stocks that only two alleles are likely to be present at all loci that are heterallelic at all.

Of the modes of selection between pairs of alleles discussed in volume 2, chapter 3, most attention will be paid to the cases of exactly intermediate heterozygotes (as representative of incomplete dominance), of complete dominance, and of recessiveness of the favored allele. It does not seem necessary here to go into the common case in which selection operates differently in the sexes since, as discussed, change of gene frequency is about the same as if the selection intensity in both sexes were the average, unless the difference is very great. It also seems unnecessary to go into selective mating. The case of sex-linked heredity is usually of secondary importance in dealing with selection of multifactorial variability in most organisms. It will, however, need to be taken into account in discussing selection experiments with *Drosophila* in a later chapter.

The total phenotypic variance will be assumed here to be the sum of genetic and environmental components $(\sigma_T^2 = \sigma_H^2 + \sigma_E^2)$ without any of the diverse complications from either genotype-environment correlation or interaction. In particular, it will be assumed here that all genotypes are subject to the same nongenetic variability $(\beta_2 = \beta_1 = 0$ in the formulas of the section of vol. 2, chap. 5 on selection of quantitative variability). I will thus merely illustrate certain complications that affect the course of selection but by no means all complications.

Basic Formulas

The frequencies of the genotypes at a locus and their contributions to the character and their selection values will be represented as follows:

Genotype	Frequencies	Contributions to Mean	Recessive	Semidominant	Dominant
A_1A_1	q^2	$\alpha_1 + \alpha_2$	s	$2s$	s
A_1A_0	$2q(1-q)$	α_1	0	s	s
A_0A_0	$(1-q)^2$	0	0	0	0

The basic formula for Δq is as follows (vol. 2, eq. 5.141):

$$\Delta q = (z/p)q(1-q)[(1-q)\alpha_1 + q\alpha_2]/\sigma_T$$

Here p is the proportion at one extreme, selected as parents, and z is the ordinate of the unit normal distribution at the truncation point, so that z/p is the mean deviation of the parents in terms of the standard deviation (vol. 1, eq. 8.94). The term z/p may be replaced by $(\bar{P} - M)/\sigma_T$. This becomes important when the mean $\bar{P}$ of the parents of the next generation differs from that of the initially selected group as the result of adverse natural selection.

We will start with the highly arbitrary assumption of multiple favorable genes, all alike in effect, degree of dominance, and gene frequency, and compare the courses of change under two extreme assumptions with respect to heritability. If h^2 is very low (almost all variability environmental), σ_T in the formula may be treated as a constant, σ_E, without appreciable error. If, on the other hand, there is complete determination by heredity, σ_T $(=\sigma_H)$ is a function of the gene frequencies that approaches zero under the assumed conditions, as fixation of all of the similar genes is approached simultaneously. In this case the selection coefficients, $s_1 = (z/p)(\alpha_1/\sigma_T)$ and $s_2 = (z/p)(\alpha_2/\sigma_T)$, are frequency dependent.

The means (M) and variances (σ_T^2), rates of change of gene frequencies (Δq), the selection coefficients $(s$ and k, respectively), and the time (t) in generations to reach any gene frequency (q from $q_0 = 0.50$) are given in table 6.1 for multiple equivalent favorable semidominant genes ($\alpha_2 = \alpha_1$) and for multiple equivalent favorable recessives ($\alpha_1 = 0$, $\alpha_2 > 0$). The case of multiple equivalent favorable dominants merely requires replacement of q by $(1 - q)$ and $\alpha_2 < 0$. The results (Wright 1969b) in the cases of constant selective intensities are essentially equivalent to ones given by H. T. J. Norton (in Punnett 1915) and by Haldane (1924).

TABLE 6.1. Courses of selection under two genetic models.

Multiple Equivalent Semidominant Genes ($\alpha_2 = \alpha_1$)

$M = 2nq\alpha_1$

$$\sigma_T^2 = 2nq(1 - q)\alpha_1^2 + \sigma_E^2$$
$$\Delta q = (z/p)q(1 - q)(\alpha_1/\sigma_T)$$

If σ_T constant, ($\sigma_T \approx \sigma_E$), then $\Delta q = sq(1 - q)$ where $s = (z/p)(\alpha_1/\sigma_E)$

$$t = [\ln q - \ln (1 - q)]/s$$

If $\sigma_T = \sigma_H$, then $\Delta q = k\sqrt{[q(1 - q)]}$, where $k = (z/p)/\sqrt{(2n)}$

$$t = [\sin^{-1} (2q - 1)]/k$$

Multiple Equivalent Recessive Genes ($\alpha_1 = 0$)

$M = nq^2\alpha_2$

$$\sigma_T^2 = nq^2(1 - q^2)\alpha_2^2 + \sigma_E^2$$
$$\Delta q = (z/p)q^2(1 - q)(\alpha_2/\sigma_T)$$

If σ_T constant, ($\sigma_T \approx \sigma_E$), then $\Delta q = sq^2(1 - q)$, where $s = (z/p)(\alpha_2/\sigma_E)$

$$t = [\ln q - \ln (1 - q) - (1/q) + 2]/s$$

If $\sigma_T = \sigma_H$, then $\Delta q = kq\sqrt{[(1 - q)/(1 + q)]}$, where $k = (z/p)/\sqrt{n}$

$$t = [\sin^{-1} q + \ln q - \ln [1 + \sqrt{[(1 - q)^2}] + 0.7934]/k$$

SOURCE: Wright 1969b. Reprinted with permission. © 1969 by The University of Chicago.

Table 6.2 shows the number of generations required to pass from one gene frequency to another at given selective intensities. These intensities are chosen so that the rates of change are the same with σ_T constant and with $\sigma_T = \sigma_H$ at $q = 0.5$. Intensities for the semidominants with σ_T constant are chosen to approach the same rate as the recessives in the neighborhood of fixation, and thus the same rate as the dominants at very low frequencies. The chosen values are improbably high for constant σ_T. If selection were one-tenth as great ($s = 0.01$), the numbers of generations would be ten times as great. The intervals of gene frequency are chosen so as to make the numbers approximately uniform in the case of semidominance, constant σ_T.

Table 6.2 illustrates the extreme slowness with which favorable recessives rise from low frequencies if the selection coefficient is constant, and the

TABLE 6.2. Generations between values of q.

RANGE OF q	σ_T CONSTANT			$\sigma_T = \sigma_H$		
	Recessive ($s = 0.10$)	Semidominant 0.10	Dominant 0.10	Recessive ($s = 0.0433$)	Semidominant 0.05	Dominant 0.0433
0.0001–0.001	90,023	23	23	53	1	2
0.001–0.01	9,023	23	23	53	3	4
0.01–0.09	912	23	24	53	8	14
0.09–0.50	114	23	32	51	19	34
0.50–0.91	32	23	114	34	19	51
0.91–0.99	24	23	912	14	8	53
0.99–0.999	23	23	9,023	4	3	53
0.999–0.9999	23	23	90,023	2	1	53
Total	100,174	184	100,174	264	62	264

SOURCE: Wright 1969b. Reprinted with permission. © 1969 by The University of Chicago.

corresponding slowness with which favorable dominants approach complete fixation under this condition. The course of selection of a character determined completely by heredity is very much in contrast. Given the same rates when $q = 0.5$, it requires 379 times as many generations for a recessive or dominant to pass from $q = 0.0001$ to $q = 0.9999$ under constancy of the selection coefficient, as in the case of complete heritability. The contrast is much less striking, only threefold, in the case of semidominant genes.

The courses are compared graphically in figure 6.1 for semidominants and in figure 6.2 for recessives. There is not much difference over most of the range of q in the former, q^2 in the latter, proportional to the changes in mean in both. The slowness of the rise from low frequencies and of approach to fixation apply to the extremes of the range.

The abrupt rise from very low frequencies and the abrupt fixation where heritability is complete depend, however, on the extreme (and very unrealistic) smallness of the genetic variance and hence of σ_T in the denominator of the basic equation for Δq near the extreme values of q. Practical selection would become impossible as these values are approached.

Moreover, even a small amount of nongenetic variance would prevent the total variance from becoming as extremely small as the theoretical genetic variance in these regions. In this case there would be a small value of q below which heritability, h_q^2 $(=\sigma_{H(q)}^2/\sigma_T^2)$, would become so small that σ_T $(=\sigma_E\sqrt{[1/(1 - h_q^2)]})$ would be practically constant. In terms of h_q^2 and $\sigma_{H(q)}^2$, $\sigma_E^2 = [(1 - h_q^2)/h_q^2]\sigma_{H(q)}^2$, but in terms of $h_{0.5}^2$ and $\sigma_{H(0.5)}^2$ the same quantity is equal to $[(1 - h_{0.5}^2)/h_{0.5}^2]\sigma_{H(0.5)}^2$. Let R be the ratio of $\sigma_{H(q)}^2$ at the chosen small value of q to its value at $q = 0.5$.

$$R = [\sigma_{H(q)}^2/\sigma_E^2]/[\sigma_{H(0.5)}^2/\sigma_E^2] = h_q^2(1 - h_{0.5}^2)/[h_{0.5}^2(1 - h_q^2)].$$

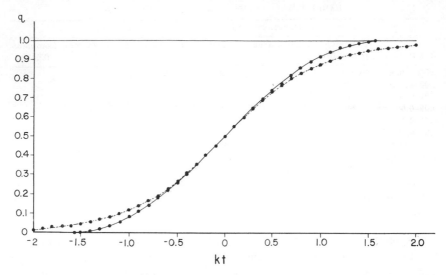

FIG. 6.1. Courses of change of frequency (q) of a semidominant gene, under directional selection, where variability is wholly genetic (*solid line*), or so much nongenetic that selection intensity may be treated as constant (*broken line*). The rates are taken as the same at $q = 0.5$ ($s = 2k$). The scale is in terms of kt, where t is number of generations. From Wright (1969b, fig. 1).

In the case of semidominance, $R = 4q(1 - q)$; in that of a recessive, $R = (16/3)q^2(1 - q^2)$. Choosing a value of h_q^2 that is sufficiently small (for example, 0.04, giving $\sigma_T \approx 1.02\,\sigma_E$), the corresponding values of R can be found for any given heritability at $q = 0.5$. It will be found that in the case of semidominance, there is substantial constancy of σ_T up to $q = 0.10$ if $h_{0.5}^2 = 10\%$, up to $q = 0.01$ if $h_{0.5}^2 = 50\%$, but only up to about 0.001 if $h_{0.5}^2 = 90\%$. For a recessive, there is substantial constancy for values of q up to 0.27 if $h_{0.5}^2 = 10\%$, up to 0.09 if $h_{0.5}^2 = 50\%$, and up to 0.03 if $h_{0.5}^2 = 90\%$.

The ratio $\sigma_E/\sigma_{T(0.5)}$ is $\sqrt{(1 - h_{0.5}^2)}$. The near-constant value of σ_T below the chosen value of q is thus $\sigma_{T(0.5)}\sqrt{(1 - h_{0.5}^2)}$. In table 6.3 the estimated numbers of generations to pass from one value of q to another in the region of near-constant σ_T for various heritabilities at $q = 0.5$ are given, with the same selection coefficients as in table 6.2.

Even with 90% heritability at $q = 0.5$, the rise of favorable recessives from very low gene frequencies is exceedingly gradual. There is still a big gap between this situation and the abrupt rise with 100% heritability. With 99% heritability at $q = 0.5$, there is approximate constancy of σ_T up to $q = 0.01$. It requires about 9,000 generations to rise from 0.0001 to 0.001, and 902 from 0.001 to 0.01. With 99.9% and even 99.99% heritability at

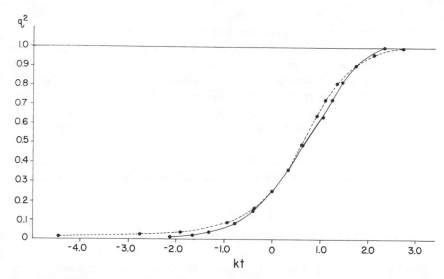

FIG. 6.2. Courses of change of frequency (q^2) of recessives, under directional selection, where variability is wholly genetic (*solid line*) or so much nongenetic that selection intensity may be treated as constant (*broken line*). The rates are taken as the same at $q = 0.50$ ($q^2 = 0.25$), so that $s = k\sqrt{(16/3)}$. The scale is in terms of kt. From Wright (1969*b*, fig. 2).

TABLE 6.3. Estimated numbers of generations to pass from one value of q to another in region of near-constant σ_T for various heritabilities at $q = 0.5$; same selective values as in table 6.2).

Range of q	Heritabilities at $q = 0.5$				
	0	10%	50%	90%	100%
	Favorable Recessive				
0.0001–0.001	90,023	85,403	63,660	28,468	53
0.001–0.01	9,023	8,560	6,380	2,853	53
0.01.0.09	912	865	645	...	53
	Favorable Semidominant				
0.0001–0.001	23	22	16	7	1
0.001–0.01	23	22	16	...	3
0.01–0.09	23	22	...	...	8

SOURCE: Wright 1969*b*. Reprinted with permission. © 1969 by The University of Chicago.

$q = 0.5$, there is approximate constancy of σ_T up to $q = 0.001$. The number of generations from 0.0001 to 0.001 are 284 and 90, respectively.

The estimates for such nearly complete determination by heredity as the preceding are again not realistic because of the practical impossibility of selection from such extremely small total variances as are implied. Thus the rise in frequency of very rare recessives must always be very slow. Similar considerations apply to the final stages in fixation of favorable dominants. The effects of recurrent mutation, moreover, are not considered here.

If there is a given genetic variance, due to an unknown number of equivalent genes with a given gene frequency, the effects, α's, and hence s and Δq, vary as $1/\sqrt{n}$, and the change in the characters in the first generation of selection, expressed as

$$\Delta M = 2n\,\Delta q[(1 - q)\alpha_1 + q\alpha_2],$$

is independent of the number of genes, irrespective of their degree of dominance.

But because Δq varies as $\sqrt{(1/n)}$, the variance is exhausted more rapidly with few than with many genes and the limit reached by selection, $M_\infty = n(\alpha_1 + \alpha_2)$, varies as $\sqrt{n}$.

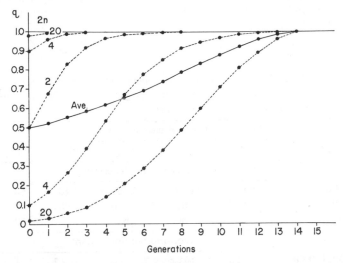

Fig. 6.3. Courses of change of the frequencies (q) of five classes of semidominant genes that wholly determine a character under directional selection, $(z/p) = 1$. The classes differ only in initial frequencies (ten at 0.02, two at 0.1, one at 0.5, two at 0.9, and ten at 0.98). The average is proportional to the changing grade of the character. From Wright (1969b, fig. 3).

As noted, the assumptions that all favorable genes are equivalent in effect, in degree of dominance, and in frequency are highly arbitrary. It is desirable to consider the consequences of mixtures of genes with different properties. These can be worked out most easily by finite steps. For simplicity, I have assumed that $z/p = 1$, corresponding to selection of the upper 38% as parents, and I have assumed complete determination by heredity. We do not run into the impossibly small values of σ_T, encountered where all genes are equivalent, if genes start at different frequencies or at frequencies as great as 0.10, and do not approach fixation simultaneously (except for the last class to be fixed).

In the first example, figure 6.3, I have started from 25 favorable semi-dominants, equivalent in effect (α), but with initial gene frequencies of 0.02 for ten, 0.10 for two, 0.50 for one, 0.90 for two, and 0.98 for ten, and thus an initial average gene frequency of 0.5, corresponding to mean, $M = \sum(2nq\alpha) = 25\alpha$. There are initially accelerating rises in gene frequency and contribution to the mean from $q = 0.02$ and $q = 0.10$, but decelerating ones from $q = 0.50$, 0.90, and 0.98, resulting in a slightly accelerating but almost linear rise of the average. The final stages of fixation should be more tapering in the presence of any nongenetic viability at all.

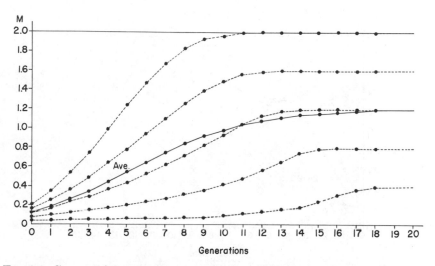

FIG. 6.4. Courses of change of the contributions (M) of five unequal semidominant genes (effects α, 0.8α, 0.6α, 0.4α, and 0.2α) to a character under directional selection, $(z/p) = 1$. It is assumed that all start from $q = 0.10$. The average (*solid line*) is one-third of the grade of the character. From Wright (1969*b*, fig. 4).

In the next example, figure 6.4, the initial gene frequencies are all 0.10, but the effects of the five genes vary: one each with contributions 0.2α, 0.4α, 0.6α, 0.8α, and α. The cumulative contributions for each gene and the average are shown. In spite of having the same initial frequencies, the courses are very different because of the differences in the factor α/σ_T in the selection coefficient. The most important, and hence most strongly selected gene, rises almost linearly until it approaches fixation. The least important, and hence least strongly selected one, rises very slowly until the more important ones have approached fixation, causing such a reduction in σ_T that its rate of increase becomes rapid. The average follows a curve that first accelerates a little more than the middle one (0.6α), but approaches its plateau more slowly because of the delayed contributions from the less important genes.

If similar genes all start from $q = 0.5$, the differences in the courses are similar but much less pronounced, and the course of the average differs less from that of the gene, with the average effect 0.6α.

In the next example, figure 6.5, all genes, all assumed to be semidominant, start from $q = 0.10$. One major gene has the effect α, ten minor ones the effect 0.1α each, thus collectively the same as the major gene. Their contributions to variance are, however, only 1% as much individually and 10%

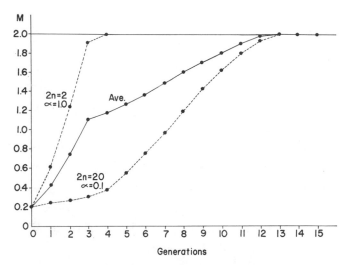

FIG. 6.5. Courses of change of the contributions (M) of a major gene (effect α), and of ten minor genes (effects 0.1α each), to a character under directional selection (z/p). All start from $q = 0.1$. The average is half the grade of the character. From Wright (1969b, fig. 5).

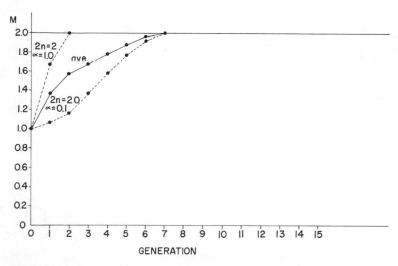

FIG. 6.6. Similar to figure 6.5, except that here all genes start from $q = 0.5$. From Wright (1969b, fig. 6).

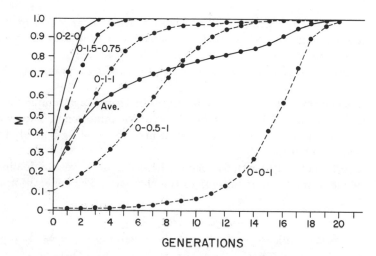

FIG. 6.7. Courses of change of the contributions of five genes that differ in degrees of dominance, to a character under directional selection, $(z/p) = 1$. The three numbers on each broken line indicate the contributions of AA, AA', and $A'A'$, respectively, where A' is the favored allele. The effects are such that the ultimate contributions are the same. The average is one-fifth of the grade of the character. All start from $q = 0.1$. Corrected from Wright (1969b, fig. 7).

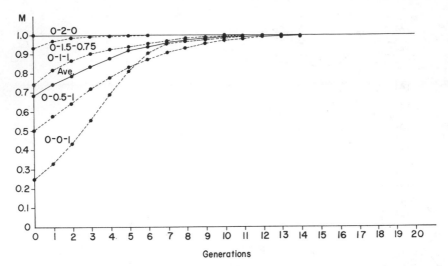

FIG. 6.8. Similar to figure 6.7, except that all genes start from $q = 0.5$. The gene with pure overdominance (0-2-0) is at equilibrium from the start. From Wright (1969b, fig. 8).

collectively. The major gene rapidly reaches fixation, after which σ_T^2 is due wholly to the minor ones, which are then enabled to rise rapidly in frequency. The average course is irregular, with an initial rapid, slightly accelerating advance until the major gene is nearly fixed, followed by an abruptly slower rate.

If all of these genes start from $q = 0.5$ (fig. 6.6), there is again rapid fixation of the major gene and delay in the rapid rise of the minor ones. The average shows a rapidly declining rate of increase at first, followed by a nearly linear rise toward the plateau.

Figures 6.7 and 6.8 deal with five genes with different degrees of dominance but the same ultimate contributions to the character. The general formulas for one locus are as follows:

$$M = 2q\alpha_1 - q^2(\alpha_1 - \alpha_2)$$
$$\sigma_T^2 = q(1 - q)[2\alpha_1^2 - q(\alpha_1 - \alpha_2)(3\alpha_1 + \alpha_2) + q^2(\alpha_1 - \alpha_2)^2].$$

The five genes dealt with are assigned the properties in table 6.4.

Gene A' tends to be fixed in the first three cases, while equilibrium is reached at $\hat{q} = 0.66$ and $\hat{q} = 0.50$ in the last two, respectively.

The initial frequencies are all 0.1 in figure 6.7, with varying contributions to the mean. The two overdominant genes rapidly reach equilibrium and make their full contribution, 1.0 in both cases. The contribution of the

TABLE 6.4. Properties of loci with diverse degrees of dominance.

Gene	α_1	α_2	M	σ^2	$\Delta q/(z/p)$
Recessive	0	1	$q^2\alpha_2$	$q^2(1-q^2)\alpha_2^2$	$q^2(1-q)(\alpha_2/\sigma_T)$
Semidominant	0.5	0.5	$2q\alpha_1$	$2q(1-q)\alpha_1^2$	$q(1-q)(\alpha_1/\sigma_T)$
Dominant	1	0	$q(2-q)\alpha_1$	$q(1-q)^2(2-q)\alpha_1^2$	$q(1-q)^2(\alpha_1/\sigma_T)$
Overdominant	1.5	−0.75	$q(2-1.5q)\alpha_1$	$q(1-q)(2-3.75q+2.25q^2)\alpha_1^2$	$q(1-q)(1-1.5q)(\alpha_1/\sigma_T)$
Pure overdominant	2	−2	$2q(1-q)\alpha_1$	$2q(1-q)[1-2q(1-q)]\alpha_1^2$	$q(1-q)(1-2q)(\alpha_1/\sigma_T)$

SOURCE: Wright 1969b. Reprinted with permission. © 1969 by The University of Chicago.

dominant gene rises rather rapidly but has an exceedingly slow tapering approach to its limit. Note that σ_T can never become small in this case because of the overdominant genes. The contribution of the recessive gene shows the expected long delay. This and the relatively rapid rise of all the others to their limits result in a rise of the mean that seems to be approaching a plateau, but then show a second rather rapid rise toward the final plateau. In this case there is more variability at the end (1.25) than at the beginning (1.133).

If all the gene frequencies start at 0.5 (fig. 6.8), the gene with pure over-dominance is already at equilibrium. The one with asymmetrical over-dominance soon reaches equilibrium. The contributions of the others all rise rapidly but approach their limits in diverse ways. The mean shows a tapering approach to its plateau. In this case the initial variance (1.887) is larger than the final variance (1.25).

Course of Change of Variance and its Components

The course of change of the total variance and of the contribution to this of each initial class of factors is of interest. Complete determination by heredity will be assumed.

In the case of many equivalent semidominant factors, starting at a very low frequency (solid line of means of fig. 6.1) the variance rises from a very low value to a maximum when all gene frequencies are 0.5, and declines symmetrically as all genes move toward fixation (fig. 6.9).

In the case of many equivalent recessive favorable factors, starting from low frequencies (solid line of means of fig. 6.2) the variance rises to a maximum when all gene frequencies are 2/3 (fig. 6.10), and then declines.

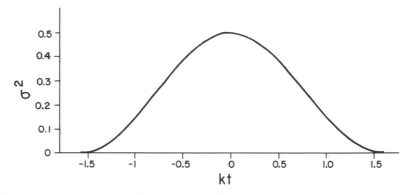

FIG. 6.9. The variances during the course of directional selection in the case of a semidominant gene with variability wholly genetic (*solid line*, fig. 6.1).

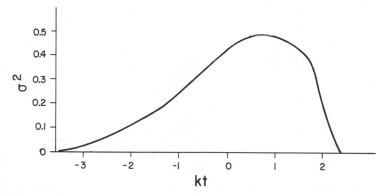

FIG. 6.10. The variances during course of directional selection in the case of a recessive with variability wholly genetic (*solid line*, fig. 6.2).

Figure 6.3 showed the changes in gene frequency of favorable semidominant genes with different initial gene frequencies, ten at 0.02, two at 0.1, one at 0.5, two at 0.9, and ten at 0.98. The changes in the total variance due to those classes and the grand total are shown in figure 6.11. The overwhelmingly greater importance of the initially rare genes as sources of variance is illustrated.

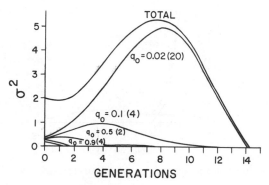

FIG. 6.11. The variances during the course of directional selection of the set of genes of figure 6.3.

Figure 6.4 showed the contributions to the changing mean of equal classes of semidominant genes with different effects (α, 0.8α, 0.6α, 0.4α, and 0.2α) all starting from frequency 0.10. The changing contributions to the variance and the grand total are shown in figure 6.12. The more important factors

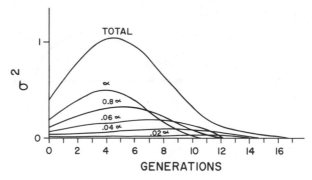

FIG. 6.12. The variances during the course of directional selection of the set of genes of figure 6.4.

contribute most in the early generations but are fixed relatively early, so that what variance there is in the later generations is due to the minor factors. This situation is shown in exaggerated form in figure 6.13 where, as in figure 6.5, one gene contributes the effect α, while ten others contribute the effect 0.1α each.

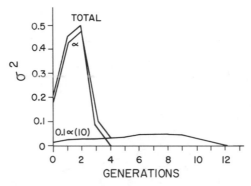

FIG 6.13. The variances during the course of directional selection of the set of genes of figure 6.5.

Finally, figure 6.14 shows the contributions to variance in the same case as figure 6.7 for contributions to the mean: five genes that differ radically in degree of dominance but that all make the same final contributions to the mean and all start from gene frequency 0.1. The variance due to the gene with pure overdominance soon reaches equilibrium and its maximum contribution (1.0) to variance. The contribution of the gene with partial over-

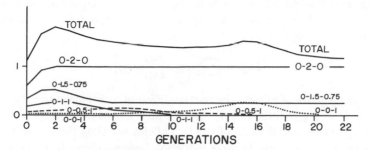

FIG. 6.14. The variances during the course of directional selection of the set of genes of figure 6.7.

dominance rises at first and then rapidly declines toward 0.25. The variances due to the dominant, the semidominant, and the recessive favorable genes are always much smaller, The contribution from the dominant gene rises to about 0.25 and then gradually approaches zero. That from the semi-dominant gene rises for a considerable number of generations before declining. That from the recessive reaches its maximum only after many generations (about 0.25 at generation 16) but ultimately approaches zero.

Nei (1963) has developed the theory for the course of change of the various components of variability at a locus with any degree of dominance (additive component σ_A^2, dominance component σ_D^2 in the terminology of Kempthorne and Cockerham) (see vol. 2) and for interacting loci (σ_{AA}^2, σ_{AD}^2, σ_{DD}^2, etc.) in the same terminology, under the assumption of negligible deviations from random combination initially and later, small changes in gene frequency in each selection cycle, and selection followed by at least one generation of random mating. The change in the genetic variance due to one cycle of selection is

$$\Delta\sigma_H^2 = \frac{d\sigma^2}{dq}\Delta q.$$

The changes in the components from pairs of alleles at the various loci (L) are as follows:

$$\Delta\sigma_A^2 = \sum_L \left[\frac{\partial\sigma_A^2}{\partial q_L}\Delta q_L\right]$$

$$\Delta\sigma_D^2 = \sum_L \left[\frac{\partial\sigma_D^2}{\partial q_L}\Delta q_L\right]$$

$$\Delta\sigma_{AA}^2 = \sum_L \left[\frac{\partial\sigma_{AA}^2}{\partial q_L}\Delta q_L\right]$$

With multiple alleles, evaluation requires use of equations of the type $\partial q_j / \partial q_i = -q_j/(1 - q_i)$ (vol. 2, eq. 3.57), as noted by Nei. The term for each allele must be weighted by its frequency.

$$\Delta\sigma_A^2 = \sum\sum_{Li} \left[q_{Li} \; \frac{\partial\sigma_A^2}{\partial q_{Li}} \Delta q_{Li} \right]$$

With only pairs of alleles, the partial differential coefficients with respect to q_L and $(1 - q_L)$ differ merely in sign and the same is true of Δq_L and $\Delta(1 - q_L)$. The contributions of the two alleles are thus the same. The sum of the weights being one, there is reduction to the simpler formulas above.

Small Populations

It has been assumed so far that the situation is such that the effect of selection is not appreciably affected by variations in gene frequency due to the accidents of sampling. This rules out favorable mutations that have only one representative in the population since (as discussed in vol. 2, chap. 5) the chance of fixation of such a mutation with selective advantage (heterozygotes) is only about $2s$. The chance that a favorable gene with several (k) representatives is lost is also not negligible (probability of fixation $= 2ks$). Thus it may be surmised that in populations as small as those usually used in selection experiments, favorable genes may have an appreciable chance of being lost. The conditions under which this inbreeding effect is important have been investigated by A. Robertson (1960).

He began with the case of a semidominant gene. In accordance with the usage in this book we will assume a selective advantage of s for the heterozygote, $2s$ for the favorable homozygote, over the unfavorable homozygote. The chance of fixation of the favored gene starting from gene frequency q is given by Kimura's (1957) formula (vol. 2, eq. 13.117):

$$u(q) = (1 - e^{-4Nsq})/(1 - e^{-4Ns}).$$

Nearly all genes for which Nsq is greater than one become fixed. The average advance, $u(q) - q$, is thus nearly $1 - q$. The ratio of this to the advance in one generation, $\Delta q = sq(1 - q)$, is $1/sq$ and independent of the size, N, of the population.

If, however, $2Ns$ (and a fortiori $2Nsq$) is less than one,

$$u(q) \approx q + 2Nsq(1 - q).$$

The total advance, $u(q) - q$, is thus $2N$ times that in the first generation. For larger $2Ns$ but very small q, the ratio is larger. With k representatives of the gene, $q = k/2N$,

$$u(q) \approx 1 - e^{-2ks} \approx 2ks.$$

The average total advance, $u(q) - q$, is thus $k[2s - 1/(2N)]$, which approaches $4N$ times the average in the first generation (about $ks/2N$).

In the case of a favorable recessive gene with frequency q and selective advantage s, Kimura's formula gives:

$$u(q) = \int_0^q e^{-2Nsq^2} \, dq \bigg/ \int_0^1 e^{-2Nsq^2} \, dq.$$

On expansion, this yields the following approximation with small Nsq:

$$u(q) \approx q + (2/3)Nsq(1 - q^2).$$

The ratio of the average total advance, $u(q) - q$, to the advance in one generation, $\Delta q = sq^2(1 - q)$, is $2N(1 + q)/3q$. This is less than $2N$ if q is greater than 0.5 but may be much greater if q is less than 0.5.

Computer studies of the effects of linkage on directional selection in small populations have been made by Hill and Robertson (1966). Latter and Novitski (1969) have discussed the limits in finite populations in which multiple alleles are present.

Interpretation of Experimental Results

For the most easily interpretable results, it is desirable to start from F_2 of a wide cross between inbred lines since this insures that initial gene frequencies are all the same ($q = 0.50$) and so removes one parameter that otherwise varies to an unknown extent. Knowledge of F_1 and F_2 variances gives, moreover, an estimate of the heritability independent of that given by offspring-parent regression, and also an estimate of the segregation index and hence information on the number of segregating loci if these are assumed to be equivalent. An attempt to isolate the leading factor or factors by repeated backcrossing gives some idea on whether major factors are present (vol. 1, chap. 15).

Comparison of the courses of change under plus and minus selection, after taking account of the nature of the scale as well as possible, gives valuable information. If all initial gene frequencies are known to be 0.50 because of derivation from a cross, asymmetry may reflect the relative frequencies of favorable dominants and recessives in the two cases. Fixation of a favorable recessive (initial zygotic frequency $0.25M_\infty$) contributes three times as much as fixation of a favorable dominant (initial zygotic frequency $0.75M_\infty$) if the effects are equal. Any asymmetries in degrees of dominance contribute to asymmetry in the rates and ultimate amounts of progress. As noted earlier, there is, however, another cause of asymmetry in the intensities of the natural selection that artificial selection opposes in each case. Artificially favorable homozygotes may suffer more from low viability or low fecundity in one direction than in the other. If the foundation stock is of random bred origin, the distribution of frequencies of favorable alleles is of primary importance with respect to the symmetry of plus and minus selection.

Degree of symmetry of plus and minus selection involves both immediate rate of change and ultimate progress. These are affected in different ways by the factors referred to above. Some inferences can be made from the forms of the curves. The most typical form is that of a rapid initial rise followed by a tapering approach to an apparent limit. Where all initial frequencies of segregating genes are known to be 0.50 and there is a secondary rise, there are several possibilities. It may imply a mutation, or it may result from the delayed breaking of close linkage between genes with opposite effects. Such delay is, of course, more likely in a small population than in a large one, but a rare crossover that occurs at once in a large population may become manifest in a late secondary rise if it releases a favorable recessive for selection. These are the most probable explanations where the foundation stock is so small that no really low initial gene frequency is possible. On starting from a large random bred stock, and maintaining large numbers, the presence of favorable recessives at very low initial frequencies is perhaps the most probable explanation of a secondary rise.

Another source of information is the persistence or exhaustion of genetic variance when a plateau seems to have been reached. The importance of trying to distinguish between fixation and the effects of physiological limit has already been emphasized. Close to a physiological limit, the absolute genetic variance tends to decrease, but as the nongenetic variance also tends to decrease the heritability remains high. Persistence of either high absolute or high relative genetic variance in spite of a failure of progress may indicate that the more extreme grades among those selected to be parents are, in general, producing fewer offspring than the less extreme ones or it may be that only favorable heterozygotes can be selected at one or more loci because the corresponding homozygotes are inviable. These are the cases in which, as noted, the offspring-parent regression as the basis for empirical prediction of the rate of progress completely breaks down. Another possibility is that heterozygotes at some loci have higher grades of the selected character than either homozygote. Segregation gives a genetic variance but there is no regression of offspring or parent.

Accessory experiments may give useful information. De Vries (1901) believed, on the basis of numerous experiments (see vol. 1, figs. 6.9c and 11.8c), that selection of quantitative variability could not be relaxed without losing all of the progress that had been made. This would obviously be expected if selection had not proceeded to complete fixation of any loci, and selection had been conducted against natural selection that favored the original grade.

If relaxation, after a plateau seems to have been reached, does not lead to any loss of the gains, it implies either that complete fixation has been reached or that there is no natural selection that is appreciable over the number of generations tested. There has certainly been no selection of heterozygotes over both homozygotes for any reason. Partial loss of the gain followed by stability implies either fixation of some loci but action of natural selection at the others or that the artificial selection may have favored heterozygotes of genes of which the homozygotes are unfixable because sterile or lethal. If relaxation at a point at which considerable progress is still being made leads to no loss, there can be no appreciable natural selection against the selected character. The occurrence of partial loss may depend on heterotic loci that relapse from forced equilibrium as above in the absence of general selection against the selected character, but it also may be due to fixation of some loci in the presence of such selection.

Reverse selection, either in a population that has reached a plateau or one that has not, plays the same role as strong antagonistic natural selection in the discussion above. Its results, in conjunction with those of relaxed selection, thus tend to narrow the range of possible interpretations. In general, its partial or complete failure implies partial or complete fixation.

CHAPTER 7

Artificial Selection Excluding Insects

A Long-continued Selection Experiment with Maize

This and the following chapter will deal with the results of unidirectional mass selection of a considerable number of kinds of characters, some varying quantitatively, some meristic, and some of the threshold sort, in widely different kinds of organisms but including principally maize, higher vertebrates, and *Drosophila*.

It is appropriate to begin with a series of selection experiments with maize that were started shortly before the rediscovery of Mendelian heredity and that have continued to the present without reaching limits. Experiments to increase and to decrease the percentages of protein and of oil in the kernel were begun in 1896 by the Illinois Agricultural Experiment Station and have continued under the successive directions of C. G. Hopkins, L. H. Smith, C. M. Woodworth, and E. D. Leng. The present account is based principally on the overall review by Leng (1962). Dudley and Lambert (1969) have analyzed variability in generation 65.

The project began with chemical analysis of 163 open-pollinated ears from a single variety, Burr White. The high protein and high oil strains began with selection of the 24 best ears in each respect, the low strains each from the lowest 12. Each strain was maintained for the first nine generations by open pollination within isolated plots and selection of 24 out of 120 analyzed ears in each case for ear-to-row planting.

At the end of this period, yield had declined considerably, even though the amount of inbreeding for loci unaffected by the selection would have been rather slight, about $1/(8N_f)$ where $N_f = 24$ per generation or about $1 - 0.995^9 = 0.046$ in nine generations. This, however, does not allow for nonrandomness in the number selected from the female parents. Only two to four of the original ears were represented in each strain by descendants in the straight female line. The decline in yield was no doubt partly due to inbreeding somewhat greater than 0.046, but it is probable that increase in the frequency of deleterious genes, linked with the selected ones or pleiotropic effects of some of the latter, were more important.

An attempt was made to counteract the decline in yield by selection and crossing within the strains. For the next 15 generations, alternate rows were detasseled and 20 ears for analysis were saved from the six highest yielding ones in each strain. As before, 24 ears (20%) were planted in rows for the next generation in each. This procedure multiplies the formal rate of inbreeding about fourfold, leading to a total value of about 0.30. By generation 18 each strain traced to only one original ear in the female line.

For the next four generations (25–28), selection for yield was abandoned and two ears were selected from each of the 12 detasseled rows.

The system was changed more drastically after generation 28. Ear-to-row planting and open pollination were abandoned. Twelve seed ears were selected each year. These were divided into two lots of six each. Plants of lot I were hand pollinated by pollen from 15 to 20 plants of lot II and vice versa. At harvest, 30 ears were chosen from each lot for analysis and as before the best 20% were chosen for seed. The formal rate of inbreeding was 0.008 per generation, leading to a total F of at least 0.49 by generation 61 for loci unaffected by the selection. There were, probably, no such loci.

These experiments are noteworthy for the long-sustained progress by selection in all strains. The courses of change under direct selection, under reverse selection after generation 48 in all strains, and under renewed direct selection beginning after seven generations of reversal are shown in figures

TABLE 7.1. Progress in altering the chemical composition of maize kernels by four lines of selection. The means are three-year averages at the end of each decade of selection with the exclusion of protein in two years and oil in four years of extreme environmental disturbance. The changes are in terms of the preceding decade.

DECADE OF SELECTION	MEAN PROTEIN (%)		CHANGE (%)		MEAN OIL (%)		CHANGE (%)	
	High	Low	High	Low	High	Low	High	Low
0	10.93				4.68			
1	14.42	8.25	+41	−15	6.98	2.82	+49	−40
2	15.08	7.87	+5	−5	8.45	2.04	+21	−28
3	18.31	7.05	+21	−10	10.42	1.43	+23	−30
4	17.95	5.95	−2	−14	12.74	1.34	+24	−7
5	19.53	5.23	+9	−12	13.77	0.96	+8	−28
6	21.79	4.85	+12	−7	14.83	0.77	+8	−20

SOURCE: Reprinted, with permission, from Leng 1962.

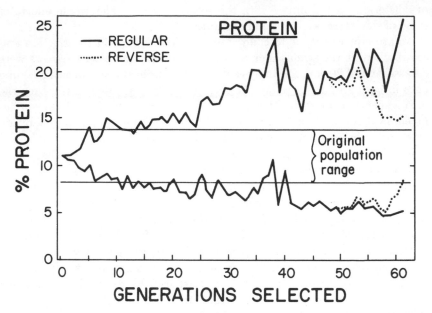

FIG. 7.1. Courses of 61 generations of directional selection and of 13 generations of reverse selection after 48 of directional selection, on protein content in maize kernels. Redrawn from Leng (1962, fig. 1); used with permission.

7.1 and 7.2. Table 7.1 shows the changes in the three-year averages at the ends of the successive decades of selection.

Oil content advanced rapidly (49%) in the first decade, about half as rapidly throughout the next three decades, but only about one-sixth the initial percentage rate in the last two decades. The astonishing threefold overall increase, 4.7% to 14.8%, suggests the likelihood of some approach toward a physiological limit. It was, indeed, shown that most of the progress after the first four or five generations was in increased germ size (the germ containing most of the oil) rather than in increased percentage of oil in the germ, which thus seems to have reached a limit rather early. The result of reverse selection after generation 48 was a drop from 13.5% to 9.7% in six generations, followed by substantial constancy (9.7%) through the next seven generations of the reversed selection. This suggests fixation or near-fixation of genes responsible for about 72% of the progress but not of the rest. There was a good response to the reversal of the reversed selection for six generations (rise to about 11.5%).

The selection for low oil content made somewhat less progress in the first decade (40%) but was still declining rapidly in the last two decades (28%,

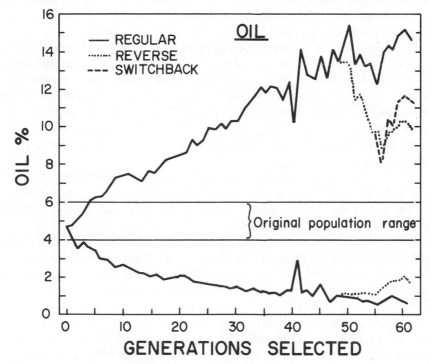

FIG. 7.2. Courses of 61 generations of selection, of the sorts indicated, for oil content of maize kernels. Redrawn from Leng (1962, fig. 2); used with permission.

20%). The total change was from 4.7% to 0.8%. In contrast with the increase in the high oil strain, most of this decline after five generations was in percentage of oil in the germ. Reverse selection after generation 48 had little effect at first, but by 13 generations it had more than doubled the oil percentage (0.8% to 1.8%), suggesting considerable fixation at loci with major effects and reverse selection based on favorable minor effects or recessive effects or both.

The total range between the results of low and high selection for oil content in generation 61 was more than 34 times the phenotypic standard deviation in the original population and about 7 times the original population range.

The selection for high protein, like that for high oil, made the most rapid progress in the first decade (41%). The average percentage rate per decade after this was only 9%. The original amount (10.9%) doubled by the end of the sixth decade (21.8%). Reverse selection after generation 48 for 13 generations reduced the content from 19.2% to 15.1% or nearly halfway back to the initial value. High protein selection resulted in small, flinty kernels with a high percentage of horny endosperm starch, in contrast with the relatively large kernels with high content of soft starch from low-protein selection.

The selection for low protein started much more slowly (15%) than in any of the other cases, but proceeded relatively uniformly through the sixth decade (average decline, 10% per decade). The end result was a decline from 10.93% to 4.85%. Reverse selection after generation 48 started slowly but brought about an advance from 5.1% to 8.1% in 13 generations. The total range between the means of high and low protein after 61 generations was 19 times the initial phenotypic standard deviation and over three times the original range.

A test of the effects of cessation of selection for eight generations was made after generation 36. Leng's analysis indicated no significant regression toward the original average in any of the four strains. Thus natural selection in favor of the origin percentages was slight.

As discussed in volume 1, chapter 15, crosses were made between the high and low oil strains by Sprague and Brimhall (1949) and carried to F_2 and to the backcrosses to the parental strains. These indicated a segregation index of about 20 in terms of the range at that time. One interpretation is near-fixation of at least 20 equivalent plus genes in the high strain and of the corresponding minus genes in the low strain. At the opposite extreme with respect to uniformity of gene effects is the differential fixation of one major pair of alleles, responsible for nearly all of the F_2 variability and for about 22% of the range, and an indefinitely large number of minor alternatives as responsible for the remaining 78%. An intermediate possibility is a geometric series of effects, each with 90% of the one next greater and the leading effect 10% of the total range. Maize, however, has only ten pairs of chromosomes. The first interpretation, if correct, could only refer to chromosome halves. Thus it is fairly certain on any interpretation that more than 20 loci are involved, perhaps hundreds even if effects are equivalent. The leading factors under the other hypotheses might well be blocks of many loci that could be gradually sorted out in later generations. In any case, it is reasonably certain that there were a great many heterallelic loci in the foundation stock, but with the large number of gene differences ultimately established and the relatively large size of the population, mutation may not have been a wholly negligible factor in supplying material for selection. Each mutation would, however, require a considerable number of generations of selection to reach the gene frequencies for rapid progress.

There is no indication that progress has approached close to its limit (on a logarithmic scale) in any of the four strains. There is indeed an indication that the high oil strain came fairly close to a physiological upper limit with respect to content of the germ (35% to 50% in the early generations), but the germ was still rising in size in the late generations. Conversely, in the low oil strain, germ size had soon approached a lower limit of 6% or 7% of kernel weight but oil content of the germ (at about 10% in the late generations) could fall indefinitely on a logarithmic scale.

The results of the experiment admirably demonstrate the possibility of producing enormous changes in a quantitatively varying character by long-continued selection from a rather small foundation stock by changes in the frequencies of a great many genes, none of which could have had very much effect by itself.

There have been far too many selection projects with cultivated plants, mostly for practical purposes, to review here. Most of them have been of a more complicated sort than in the experiments on chemical composition of maize kernels. One example has already been discussed at some length in volume 1, chapter 15: that in which Emerson and Smith (1950) nearly doubled the number of rows in maize by selection, but not by mass selection. The foundation consisted of 13 self-fertilized strains all characterized by 12 rows. Selection within these was without significant effect, ruling out mutation as a factor in a short-term experiment. All possible diallel crosses were made. Plus and minus selection was carried for several generations in all cases, and gave strains differing, on the average, by 3.6 rows (6.0 in the most extreme case). Crosses between those with the most rows, followed by plus selection for several generations, gave three strains with about 18 rows. Crosses between pairs of these, tracing to seven original inbred lines, plus selection, brought the average to 22 rows. This experiment illustrates strikingly the advantage of widening the base, but even seven inbred lines in all of which genotypes giving 12 rows had been fixed is not a very wide base. There is agreement with the protein and oil experiments in showing that a great deal of progress can be based on a rather narrow foundation.

Selection Systems

Brief accounts of patterns of selection designed to utilize overdominance at loci and specific interactions between loci will be useful here.

The most outstanding success has been that of hybrid maize, discussed briefly in chapter 2. This has involved selection on a grand scale among innumerable selfed lines for vigor, followed by selection, also on a grand scale, of pairs that produce desirable hybrids, and finally selection of pairs that produce high-yielding double crosses.

Selection among inbred lines has reached a stage at which some lines have been found that at least approach a level of seed production adequate for use as ovule parents in the production of F_1 hybrids, usable commercially in place of the somewhat cumbersome double crosses. The point has, indeed, been reached at which hybrids, which are almost single-crosses, are being produced on a large scale from ovule parents derived from crosses between somewhat divergent sublines of the same inbred strain. The resulting hybrids are much more uniform than double-cross hybrids.

Hull (1945) proposed an interesting method, which he called recurrent selection, for the production of hybrids of maximum yield. He would start from a superior but not inbred strain A. This is to be molded into complementarity to a "tester," B, which may be either a pure line or an F_1 hybrid between such lines. A very large number of plants from strain A are both to be selfed and crossed with the tester. Selection is made among these according to the merit of the hybrids, AB. The selfed progeny of those selected are interbred for the next generation of strain A. The cycle involves three generations.

Comstock, Robinson, and Harvey (1949) proposed "recurrent reciprocal selection" in which two very diverse strains are each selected on the basis of the merit of crosses between them. Selfed progenies of the selected plants, produced at the same time, are interbred within each strain to reconstitute them for initiation of another selection cycle. The cycle again requires three generations. Analytic comparison of the genetic consequence with the preceding method or a method of selection for general combining ability led the authors to the conclusion that under no circumstances would it be more than slightly inferior, and it would be definitely superior to selection for general combining ability of loci at which there is overdominance (or pseudo-overdominance as a result of strong repulsion linkage), and it would be definitely superior to Hull's recurrent selection for loci at which there is partial dominance.

These processes are necessarily rather slow because of the length of the selection cycle. They have no advantage over comparable family selection insofar as gene effects are additive. They offer the greatest possibilities where there is strong heterosis or strong interaction. In general, utilization of the idea is likely to increase with exhaustion of additive variance by direct selection. Utilization in animal breeding is possible with suitable modifications and shows some promise in connection with hybrid poultry.

Selection of the Hooded Pattern of Rats

The first extensive experimental study of selection in a mammal was conducted by Castle and Phillips (1914) on the extent of the colored areas in the hooded pattern of the laboratory rat. Their foundation stock of about a dozen hooded animals was derived from experiments by MacCurdy and Castle (1907) with black-rats (self-colored), hooded black rats, ones with the "Irish" pattern (self-black except for small amounts of white on the belly), and albinos.

The orderly gradation in the extent of black in the hooded rats made possible a linear series of grades, extending originally from −2.5 to +4 on an arbitrary scale in which each quarter grade was barely distinguishable from the preceding. The scale was later extended to −4.5, self-white except for a little black on the face, and +6, self-black.

The plus selection series started with parents of average grade +2.51,

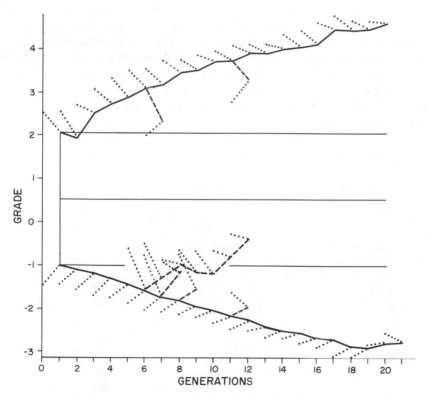

FIG. 7.3. Courses of plus and minus selection (*solid lines*) and of reverse selection (*broken lines*), for amount of color in coats of hooded rats. The dotted lines show the regressions from the grades of the selected parents (at ends). From data of Castle and Wright (1916).

offspring +2.05, and advanced rather steadily to +4.61 for the offspring of the 20th and final generation. The minus series started from parents averaging −1.46, offspring −1.00, and advanced rather steadily to an offspring average of −2.89 in generation 19. Selection was relaxed in two small subsequent generations before this line became extinct. Both lines were terminated, not because of lack of progress in grade, but because of such low vigor and fecundity that it became impossible to maintain numbers. Apart from the small first generations and those immediately preceding extinction, the average number per generation was about a thousand from dozens of selected parents in each case. The courses of progress of the offspring averages are shown in figure 7.3, together with the results of reverse selection at various times.

The difference between the offspring of the original plus and minus selections from the heterogeneous foundation stock was about 40% of that ultimately achieved and about 28% of the difference between self-white and self-black. This is in marked contrast with the slight effects of the initial selections in the experiments on protein and oil content in maize. It suggests the segregation of one or two factors with much greater effect than the rest.

Table 7.2 shows the total amounts of selection applied over intervals of six generations, as the sums of the differences between the parental means and the means of the populations from which they were selected. It also shows the corresponding amounts of response as the total gain in the offspring average during the period. The ratios show the realized heritabilities, under

TABLE 7.2. Accumulated deviations of means of parents ($\sum \Delta M_p$) and of means of offspring ($\sum \Delta M_o$) from the means of offspring of the preceding generations and the ratio $\sum \Delta M_o / \sum \Delta M_p$, as the realized heritability ($b_{op(r)}) = h_r^2$. Also the average standard deviations of the parents and offspring in each generation are given.

	Generations	$\sum \Delta M_p$	$\sum \Delta M_o$	$b_{op(r)}$	σ_p	σ_o
Plus	1–7	3.53	1.15	0.326	0.255	0.547
	7–13	1.97	0.74	0.376	0.162	0.346
	13–19	2.16	0.55	0.256	0.262	0.359
	Total	7.66	2.44	0.319	0.223	0.418
Minus	1–7	2.76	0.73	0.264	0.213	0.454
	7–13	1.72	0.67	0.390	0.208	0.308
	13–19	1.53	0.50	0.324	0.214	0.284
	Total	6.01	1.90	0.315	0.212	0.349

SOURCE: Data from Castle and Phillips 1914 and Castle and Wright 1916.

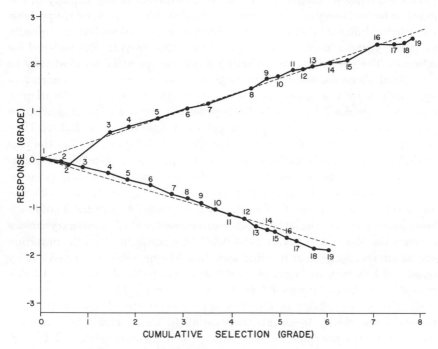

FIG. 7.4. Cumulative selection of parents and response of offspring by generation from the data indicated in figure 7.3. Trends are shown by broken lines. From data of Castle and Wright (1916).

the assumption of additive effects (semidominance, no interaction). Figure 7.4 shows the relation between cumulative selection and response by generation. While the standard deviation of the offspring was on the average 20% greater in the plus series than in minus series, the overall amount of progress was very nearly the same. Progress was considerably more rapid in the plus series in the first period but considerably less rapid in the last period. The standard deviations of the offspring were only about half a grade at first and declined to 0.32 on the average. This decline seems to have been more a matter of a damping of factor effects as the limits were approached, a scale effect, than a real reduction in genetic variability.

Six generations of return selection from generation 6 in the minus series carried the average from -1.86 to -0.39, considerably less than in the offspring of the first selected parents (-1.00). Reverse selection for single generations at various times in both series caused more loss than had been gained in two generations of forward selection.

This experiment started before the demonstration of the validity of the multiple factor theory of quantitative variability. Castle's original hypothesis was wholly different: that genes are subject to continual variation especially in heterozygotes ("contamination") and that this provides the material for selection. The possibility that selection might be operating on what may be considered allelic mutations of the original hooding factor was supported to some extent by the appearance of two animals in generation 10 of plus selection of grades 5.5 and 5.75, far beyond the average of their generation (3.73). These were born in separate pens but had the same father. They produced ten offspring of their own type (average 5.60) and six like the plus strain (average $+3.87$) and later generations gave clear Mendelian segregation. In backcrosses the mutant male produced 56 like himself (average $+5.45$) and 58 like the plus strain (average $+3.73$). In crosses with the minus strain, he produced 31 of relatively high grade ($+4.43$) and 35 of much lower grade ($+0.49$), the latter much like the results of contemporary crosses between the plus and minus series ($+0.63$). The possibility that the mutation was at another locus from hooding and its wild-type allele was ruled out by crosses of homozygous "mutant" with wild rats (self-colored). The 46 F_1's ranged from 5.25 to 6 (average 5.89) and 212 F_2's ranged from 5 to 6 (average 5.86) with no segregants of the plus type (Castle and Wright 1916). The mutation was clearly to an intermediate allele. Castle (1951) later demonstrated a minus mutation at this locus, which reduced grade $+2$ to -2. The occurrence of these allelic mutations suggests the possibility that there may be more frequent mutations that have much less striking effects but effects that nevertheless give a basis for selection.

Crosses were made between the plus and minus lines in generation 5 or 6 and 10. In both cases the offspring were about halfway between. The standard deviations in F_2 were very significantly greater than in F_1 but much less than expected from segregation of a single pair of alleles (unless there were wholesale contamination). The segregation indexes were 4.8 and 8.5, respectively. This is consistent with the accumulation of independent plus and minus modifiers in the plus and minus lines, respectively, but does not rule out the hypothesis of contamination of alleles. The results given above for the plus "mutant," however, indicated clear segregation rather than contamination.

The interpretation of crosses on the multifactorial (modifier) hypothesis is explored in table 7.3 (Wright 1917a). Systematic estimates of the potency of the largely hidden array of modifiers in strains with the type allele (S) or the above plus "mutant" (S^M) or a postulated Irish allele (S^I) can be made by assuming that the F_2 hooded segregants (or F_1 in the crosses of heterozygous S^M) have arrays with mean effects (M_O) exactly halfway between those of the hooded strain used in the cross (M_A) and those carried by the self or near-self strain in question (M_S), so that estimated $M_S = 2M_O - M_A$. Comparisons are made in table 7.3 for the Irish strain crossed with plus or minus of early generations, wild rats similarly crossed in various generations, and the plus "mutant." Weighted averages are given in the last column.

TABLE 7.3. Estimates, $(2M_O - M_A)$, of grade of hooding in genotype ss implied by the residual heredity carried by self-colored wild SS or near-self (Irish, mutant, both SS) based on the grades, M_O, of their hooded descendants from cross with plus or minus hooded, M_A. The mutant gene S^M arose in plus hooded of generation 10.

PARENTS		SELECTED HOODED ANCESTORS			F₂ HOODED OFFSPRING			RESIDUAL HEREDITY OF NEAR-SELF	
Near-self	Hooded	Genera-tion	M_A	σ_A	No.	M_O	σ_O	$2M_O - M_A$	Average
Irish	× plus	2	+1.92	0.73	239	+1.27	0.90	+0.62 ⎱	+0.51
Irish	× plus	3	+2.51	0.53	23	+0.95	0.87	−0.61 ⎰	
Irish	× minus	3.5	−1.23	0.47	90	−0.62	0.61	−0.01 ⎱	
Irish	× minus	4	−1.28	0.46	53	−0.73	0.60	−0.18 ⎬	−0.08
Irish	× minus	7.5	−1.76	0.32	66	−0.94	0.84	−0.12 ⎰	
Wild	× plus	3	+2.51	0.53	21	+2.56	0.50	+2.61 ⎱	
Wild	× plus	5.5	+3.00	0.50	38	+2.97	0.52	+2.94 ⎬	+2.67
Wild	× plus	11	+3.78	0.29	73	+3.17	0.73	+2.56 ⎰	
Wild	× minus	2.5	−1.12	0.49	62	+0.31	0.98	+1.74 ⎱	
Wild	× minus	6	−1.56	0.44	48	+0.25	0.97	+2.06 ⎬	+2.05
Wild	× minus	10	−2.01	0.24	91	+0.24	1.18	+2.49 ⎪	
Wild	× minus	16	−2.63	0.27	121	−0.38	1.25	+1.87 ⎰	
Mutant	× mutant	10	+3.73	0.36	8	+3.75	0.25	+3.77 ⎱	
Mutant	× plus	10	+3.73	0.36	58	+3.73	0.24	+3.73 ⎬	+3.79
Mutant	× plus	10	+3.73	0.36	70	+3.79	0.33	+3.85 ⎰	
Mutant	× minus	10	−2.01	0.24	35	+0.49	0.77	+2.99 ⎱	+3.15
Mutant	× minus	10	−2.01	0.24	75	+0.61	1.14	+3.23 ⎰	

SOURCE: From Wright (1917a) with a few additional cases and the standard deviations, σ_A and σ_O, from the data of Castle and Phillips (1914) and Castle and Wright (1916).

In each case, the estimate for the modifier array, based on crosses with the plus series, averages only a little more than half a grade (0.59, 0.62, 0.64) higher than that based on crosses with the minus series. This is a very small amount compared with the difference between the two series and supports the view that selection has operated almost wholly on independent modifiers.

There remains, however, the possibility that it has operated to a slight extent on mutations at the hooding locus and that this is responsible for the difference of about 0.6 referred to above. This, however, must be discounted because of the arbitrary nature of the scale. It is, as noted, fairly clear that the scale is damped somewhat toward the limits. This would tend to bring about a difference in the estimates such as observed. Use of the inverse probability transformation, in fact, reduces the difference in the estimates to 0.52 (Irish), 0.23 (wild), 0.44 (mutant). Thus, while possible, it is doubtful that the minus series has a slightly lower allele than the plus series. On comparing the estimated potencies of the modifiers of these near-self strains, it is evident that the mutant (+3.79) stands higher than the wild strain (+2.67) and both are much higher than the Irish strain (+0.51). The mutant with some white on the belly in homozygotes clearly has a lower allele than the self-colored wild. The Irish strain, in spite of its phenotypic resemblance to the mutant, has modifiers of so much less potency that it probably had the same allele as wild.

The most crucial test of the multiple factor hypothesis was by means of three cycles of backcrossing to the same, narrowly based wild strain and extraction of hooded, made with both the plus and minus strains (Castle 1919). The results are given in table 7.4. The third cycle segregants tracing to the plus strain fell back relatively little, +3.73 to +3.04, while those tracing to the minus strain advanced from −2.63 to +2.55. The slight difference between +3.04 and +2.55 in these 7/8-blood wild rats would presumably be further reduced by additional cycles of backcrossing to the wild each followed by a generation of segregation. Castle was convinced by these tests that selection had operated practically wholly on modifiers segregating independently of the hooding factor, and that the latter was subject to mutations too infrequently to give a comparable basis for selection.

It remains to consider the probable nature of these arrays of modifiers as indicated by F_1 and F_2 of the crosses between the plus and minus strains (table 7.5) and F_2 from wild × minus, attributing a potency of +2.67 to wild in two earlier crosses and +3.21 in the later ones and using the average F_1 variance, 0.529, from the plus × minus cross. The mutant strain is included with the plus strain of generation 10 from which it arose. The segregation index is taken as $S = R^2/[8(\sigma_{F_2}^2 - \sigma_{F_1}^2)]$ where R is the contemporary difference between plus and minus (vol. 1, chap. 15).

TABLE 7.4. The mean and standard deviation of hooded rats and of their hooded descendants derived by crossing and extraction to a wild strain.

	Derived from Plus				Derived from Minus		
	No.	M	σ		No.	M	σ
Plus (10)	776	+3.73	0.36	Minus (16)	1,980	−2.63	0.27
First F_2	73	+3.17	0.73	First F_2	131	−0.38	1.25
Second F_2	256	+3.34	0.50	Second F_2	49	+1.01	0.92
Third F_2	19	+3.04	0.64	Third F_2	104	+2.55	0.66

SOURCE: Castle 1919.

TABLE 7.5. Estimates of the segregation index, $S = (P_1 - P_2)^2/[8(\sigma_{F_2}^2 - \sigma_{F_1}^2)]$ from F_2 and F_1 of crosses between plus and minus rats or between wild and minus, assuming that the modifiers of wild have an average $+2.67$ and that those of F_1 have a standard deviation of 0.73 (the average of that of F_1 plus × minus).

PARENTS				OFFSPRING				
Strain	Generation	M_p	σ_p	Generation	No.	M	σ	S
Plus	5	$+2.90$	0.51	F_1	93	$+0.06$	$+0.71$ ⎫	4.8
Minus	6	-1.56	0.44	F_2	305	$+0.24$	$+1.01$ ⎭	
Plus	10	$+3.73$	0.36	F_1	49	$+0.63$	$+0.76$ ⎫	8.5
Minus	10	-2.01	0.24	F_2	148	$+0.68$	$+1.03$ ⎭	
Wild		$(+2.67)$				(0.73)		
⎧Minus	2.5	-1.12	0.49	F_2	62	$+0.31$	$+0.98$	4.2
⎪Minus	6	-1.56	0.44	F_2	48	$+0.25$	$+0.97$	5.4
⎨Minus	10	-2.01	0.24	F_2	91	$+0.24$	$+1.18$	3.9
⎩Minus	16	-2.63	0.27	F_2	121	-0.38	$+1.25$	4.1

SOURCE: Data from Castle and Phillips 1914 and Castle and Wright 1916.

The standard deviations of F_1 plus × minus (0.71, 0.76) are considerably larger than expected from those of the selected strains. The latter were, of course, far from homallelic but this should not cause F_1 of their cross to have a greater standard deviation than themselves on an additive scale. The mean of F_1, however, falls near the middle of the scale where, as noted above, the effects of factors and thus the standard deviations are expected to be maximum. The inverse probability transformation cannot indeed fully account for the large F_1 standard deviations, but perhaps an even more drastic transformation of scale is required, or perhaps the nongenetic variance of heterozygotes is greater than in near-homozygotes.

However this may be, the segregation index rose from 5 to 8 from generations 5 to 9 in crosses of plus × minus, as already noted. It was always about 4 or 5 where the plus modifiers were those of the wild strain. These values are small compared with those found in crosses between high and low oil content in maize, especially as the haploid number of chromosomes in the rat is 21 as compared with only 10 in maize.

Estimation of the rate of change of the gene frequencies may be based on the formula for truncation selection, given in chapter 6, assuming semidominance here ($\alpha_2 = \alpha_1$).

$$\Delta q = (z/p)q(1 - q)\alpha_1/\sigma$$

The term z/p ($= \Delta M_p/\sigma$) averaged 1.11 in the plus series, 0.95 in the minus series through generation 5, 1.21 in the former, 1.11 in the latter in the next five generations, and became somewhat less than 1.0 later in both series. It was never far from 1.0.

If the difference (4.31 units) between the offspring of the two series in generation 5 was due to 4.8 equivalent, nearly fixed loci (9.6 genes), the average gene effect was 0.45. With an average in offspring standard deviations of 0.50, the value of the term α_1/σ was about 0.90. A similar calculation for generation 10 gives 1.1. Thus α/σ also is never far from 1.0.

These figures imply a rate of change of the gene frequencies of about $\Delta q = q(1 - q)$, a rate so rapid that fixation should be virtually complete in the first few generations. Interpretation in terms of genes with equivalent effects is clearly untenable.

It seems probable that the initial selection brought about near-fixation of one or two major modifiers of the pattern, but that there were many other pairs of alleles with slight to very slight effects, hardly subject to selection at all at first. As the more important of the favorable genes approached fixation, others advanced to take their places with respect to strength of selection. There would continually be fresh loci for the most active selection. Linked loci that were not combined at random initially would approach randomness. In particular, closely linked pairs of loci that happened to start with repulsion, and thus of little effect, might become of major importance as the opposite association increases. In such a system, all Δq's after the initial selection could be small, because of small α/σ, as well as small $q(1 - q)$, and progress could continue a long time with no appreciable slowing of rate, until linkage of selected genes with ones deleterious in vigor or fecundity or the accumulation of similar pleiotropic effects terminate the experiment.

Selection from Mere Traces of Traits

White spotting becomes a threshold character at the extremes 0 and 100% white. The Irish rats in Castle's experiments apparently had the same allele as that giving self-color in wild rats, as noted, but an array of minus modifiers that permits white to appear on the midline of the belly. The situation is similar in guinea pigs (vol. 1, chap. 15). Genotype SS is usually self-colored, but the introduction of an extreme array of modifiers from the spotted strain (ss) with most white permits crossing of the threshold in SS. Genotype Ss with diverse background heredities ranges from invariable self to 100% low to high-grade spotted.

It is appropriate to refer here to an experiment by Goodale (1937, 1942) that originated in a self-colored strain of mice in which he noticed a single individual with a few white hairs on the forehead. This animal, a male, was mated with four females. Some of the offspring had a few white hairs. Intense selection was carried out among the descendants, at first on the basis of the number of white hairs, later on the size of a white forehead spot. The size of this was gradually increased until in 17 generations most of the face in front of the ears was white. At an intermediate stage, white had begun appearing also on the tail, feet, and midbelly.

Kyle and Goodale (1963) reported on the results many generations later (after 1959). The mean percentage of white had risen to 69.5 for males, 64.7 for females. A few of each sex were black-eyed white.

There was no indication at any point of segregation of a single locus. The first appearance was probably not due to mutation or segregation of any one factor, but to an unusual combination of several genes that had been segregating out of sight in the foundation stock. With this start, selection brought about the accumulation of ever larger numbers of favorable genes.

Castle (1906, 1907) made a number of small-scale studies of the effects of selection on characters of guinea pigs and rabbits. One of these illustrated the possibility of developing a morphological character from a single rudimentary appearance. Normal guinea pigs, like their wild ancestors, have only three toes on the hind feet. Castle started with a male in an otherwise normal strain that had a rudimentary little toe on one hind foot. By extensive mating, followed by inbreeding and selection among his descendants for five generations, he produced a strain in which all animals had perfectly developed little toes on both hind feet, associated with novel plantar pads, resembling those of species in which the little toe is normally present.

A number of years later I made crosses between this strain (D) and various inbred strains (Wright 1934d). As stated in volume 1, chapters 5 and 15, the frequencies of three-toed, poor four-toed, and good four-toed animals in an array could best be interpreted as a trichotomy of a normal distribution on an underlying additive scale, determined by lower and upper thresholds. The nongenetic variance on the underlying scale, taking the interval between the thresholds as the unit, could be estimated (as 0.80) from an inbred strain (no. 35) in which all three categories appeared, without any genetic differences within large branches. The largest F_2 segregation index ($S = 4$) came from a cross with an inbred strain (no. 2) that wholly lacked the little toe, as did F_1 of the cross. This index could be interpreted as implying a difference between strains D and 2 of four equivalent semidominant genes, but repeated backcross tests indicated an interpretation in terms of a major factor responsible for up to half of the difference (on the underlying scale) and a multiplicity of minor factors.

Selection for Body Weight in Mice

One of the favorite characters for selection experiments has naturally been body weight. Goodale (1938, 1941) attempted to breed mice as large as possible from a foundation stock consisting of only 5 males and 11 females of an albino strain obtained from a dealer. He continuously selected the heaviest males from litters, with some regard to litters as wholes, for test mating with similarly selected females, irrespective of generation, and retired them until their offspring were one month old. The parents of the largest offspring were then remated and retained as long as useful. The mean weight at 60 days was used as the criterion. Averages were found for males in successive groups of 500 in the records and for the chronologically corresponding females. He reported on 28 such groups in 1941. The groups corresponded roughly to generations after the building up of the stock in the first group, and those after 1941 will be referred to as generations.

Successive groups showed gradually increasing means. Starting from weights of 26.0 gm for males, 21.3 gm for females of the first group, the males had increased 40% and the females 37% by the 14th group. Both had increased up to about 66% by the 28th group. There was then considerable slowing down but no apparent approach to a limit up to this time.

In a brief note (included as an appendix in Falconer and King 1953) Goodale stated that he had continued the experiment to at least generation 50 but the selection had become increasingly difficult after the previous report because of much sterility and reduced litter size. Selection was suspended for a number of generations and the weight of males fell from 43.2 gm in the 28th group to a low of 39.3 gm in the 44th, but on resuming selection reached a high of 45.7 gm in generation 50. Wilson et al. (1971) reported on continued selection to generation 84. There was a plateau, after generation 35 without exhaustion of genetic variation.

Selection of mice of this strain by Falconer and King (1953) made no appreciable advance in nine generations (realized heritability 0.06 ± 0.20 gm) but reverse selection brought about a significant decline. It appeared that most loci had reached fixation or near-fixation but that an equilibrium had been reached at others because of seriously deleterious effects of favored homozygotes.

MacArthur (1944a,b, 1949) selected for both large and small size from a much more heterogeneous foundation stock than did Goodale. He started from crosses involving seven different strains of mice. His selections took account not only of individual weight and average of sibships but also of progeny, since extra litters were obtained where the young in the first litter were exceptionally large. The average generation consisted of 336 offspring from 37 pairs. As in Goodale's experiment, the average weight at 60 days was the criterion in describing progress. He attempted to apply equal intensities of selection but, because of the rapid reduction in the size of litter in the small strain relative to that in the large (about five and ten, respectively), the selection differentials of the parents in terms of standard deviation were considerably greater (by 49.4%) in the large strain.

Table 7.6 shows the means, standard deviations, and coefficients of variability of males and females of both strains over the intervals of seven generations. Selection for large size was very effective at first (gains of 50% by males, 41% by females in the first seven generations) but slowed down, the total in 21 generations being 72% in males, 77% in females. In selection for small size there were declines of 26% by males, 31% by females in the first seven generations, while in three times this period the total decline by males was 48%, by females 45%. Realized heritability declined from 0.238 in the first seven generations to 0.123 in the next seven and 0.114 in the last seven.

As illustrated above, the changes in the two directions were highly asymmetrical, but this does not take account of the inappropriateness of the scale in grams, which is obvious from the much greater uniformity of the coefficients of variability (100 C) than of the standard deviations (σ). This indicates, as MacArthur stated, that a logarithmic scale is more appropriate than the scale of actual weights. The mean M' and the standard deviation σ' on such a scale are shown in table 7.6, as given by the formulas in volume 1, chapter 10, under the assumption of normality on it:

$$M' = \log_{10} v = \log_{10} M - (1/2) \log_{10} (1 + C^2)$$
$$\sigma' = \sigma_{\log_{10} v} = \sqrt{[0.4343 \log_{10} (1 + C^2)]}$$

On this basis, progress was somewhat more rapid upward ($+0.1613$) than downward (-0.1447) at first on averaging the sexes, and was somewhat less rapid upward over the whole 21 generations ($+0.2394$ vs. -0.2729). Thus the considerably greater intensity of selection in the large strain did not produce quite as much progress as that achieved in the small strain over the whole period.

The interpretation is complicated somewhat by a negative relation between size of litter and weight at 60 days. In L21 the mean weight of males was 44.0 gm in litters of four to seven, 35.4 gm in litters of eight or nine, 39.1 gm in litters of 10 or 11, and 38.9 gm in litters of 12 to 14. Since, however, the mean for the foundation stock should be rated up about half as much as that for the large strain to put all comparisons on the basis of the small litters of the small strain of the later generations, the correction to be made is probably slight.

There was no consistent maternal effect as tested by comparing weights of young reared by foster mothers of their own and the other strain.

A branch from generation 26 of MacArthur's large size strain was continued by Falconer and King (1953) for nine generations. There was only doubtfully

TABLE 7.6. The mean (*M*), standard deviation (σ) in grams, the coefficient of variability (100 C), and the mean (*M'*) and standard deviation (σ') on a logarithmic scale of male and female mice in the foundation stock 0 and at intervals of seven generations in the selection for large size (L7, L14, and L21) and for small size (S7, S14, S21).

GENERA-TION	MALE					FEMALE				
	M (gm)	σ (gm)	100 C	*M'*	σ'	*M* (gm)	σ (gm)	100 C	*M'*	σ'
L21	39.85 ± 0.47	5.10	12.80	1.5969	0.0554	34.46 ± 0.49	5.12	14.86	1.5326	0.0642
L14	36.79 ± 0.37	3.66	9.95	1.5636	0.0431	30.71 ± 0.31	3.29	10.71	1.4850	0.0445
L7	34.69 ± 0.26	3.59	10.34	1.5379	0.0448	27.51 ± 0.39	3.80	13.81	1.4354	0.0597
0	23.16 ± 0.26	2.56	11.01	1.3621	0.0477	19.51 ± 0.25	2.65	13.52	1.2886	0.0583
S7	17.25 ± 0.30	2.61	15.13	1.2319	0.0652	13.56 ± 0.27	2.34	17.36	1.1295	0.0748
S14	13.93 ± 0.23	2.13	15.29	1.1389	0.0661	11.69 ± 0.20	1.74	14.89	1.0657	0.0644
S21	11.97 ± 0.28	1.71	14.29	1.0737	0.0617	10.79 ± 0.26	1.47	13.62	1.0313	0.0588
L21-0	16.69	2.54	1.79	0.2348	0.0077	14.95	2.47	1.34	0.2440	0.0059
0-S21	11.19	0.85	-3.28	0.2884	-0.0140	8.72	1.18	-0.10	0.2573	0.0005

SOURCE: Data from MacArthur 1949, table 2.

significant progress, realized heritability 0.33 ± 0.20. Reverse selection, however, caused significant decline, indicating that there was still some genetical variance. The situation was probably essentially similar to the late generations of Goodale's stock, but fixation or near-fixation of the plus genes that were fixable was probably not quite so far advanced.

The correlated changes observed by MacArthur are of great interest. Infertility increased in both lines. A great diversity of colors was present in the foundation stock. The small strain became exclusively black, BB, except as modified by the agouti alleles (A agouti, a^t black and tan, a black). White spotting (ss) was sometimes present but brown (bb), dilution (dd), or albinism (cc) never were. Thus its genetic composition was $CCBBDD$ $(A, a^t, a)(S, s)$. Dilute brown with one or another agouti allele but never white spotting came to be characteristic of the large strain, but albinos occasionally appeared. The genetic composition became largely $(C, c)bbdd(A, a^t, a)SS$, although it is not stated that B and D were wholly lost. The differences with respect to C, c and S, s may have been accidental, but Castle (1941) had demonstrated that B and D act pleiotropically in mice to reduce size.

The small strain came to have relatively large ears, feet, and tail. This is a familiar effect of allometric growth. These effects may be considered to be pleiotropic effects of factors for small size in general.

The mice of the small strain were much more active, wild, and nervous as compared with the docile, phlegmatic, and very fat mice of the large strain. Not surprisingly, those of the small strain had a higher rate of metabolism. These characteristics could all be explained by their larger surfaces per unit weight and consequent greater rate of heat loss. The differences may again be looked upon as pleiotropic effects of a general sort for genes with effects on growth rate.

Finally there was the striking difference in size of litter referred to above. By generation 12 and 13, the mice of the small strain were producing an average of 7.2 corpora lutea and 5.3 fetuses, in contrast with averages of 14.1 corpora lutea and 10.5 fetuses in the large strain. The difference was wholly maternal, as shown by reciprocal crosses, and presumably reflects a physiological correlation between growth rate and ovulation. Such a correlation had been observed among inbred strains of guinea pigs (Wright 1922b) and in rabbits (Gregory 1932).

Falconer and King (1953) noted a striking difference in the results of Goodale's and MacArthur's selections for large size in spite of the apparent similarity in their procedure. Goodale's mice had larger frames but were not as fat as MacArthur's. Since both strains seemed to be at or close to the upper limit attainable by selection within them, it was of interest to see whether, in a cross, each could furnish favorable genes, not present in the other, and so permit further gain in size by selection, or whether a physiological limit had been reached, or a limit at which antagonistic natural selection prevented further progress. Mice of Goodale's generation 43 were crossed with MacArthur's of generation 26. Reciprocal crossbreds were large-bodied like the former, and fat like the latter. There was no appreciable increase in variance in F_2. Nine generations of plus and minus selection yielded highly significant progress in both directions. Table 7.7 shows some of the parameters in comparison with those from nine generations of selection of the pure strains. There is no doubt of the success of selection in both directions.

TABLE 7.7. Data on nine generations of plus selection in Goodale's mice of generation 43, in MacArthur's of generation 26, and of nine generations of plus and minus selection from F_2.

| | Goodale's | MacArthur's | SELECTION FROM CROSS | | |
			Plus	Minus	Divergence
Selective differential (gm)	1.52	1.37	1.88	1.40	3.28
Realized gain per generation (gm)	$+0.06 \pm 0.20$	$+0.33 \pm 0.20$	$+0.30 \pm 0.08$	$+0.52 \pm 0.12$	0.82 ± 0.13
Realized heritability	0.038 ± 0.130	0.243 ± 0.145	0.157 ± 0.043	0.374 ± 0.088	0.26 ± 0.040

SOURCE: Falconer and King 1953.

Falconer conducted other experiments on selection for large and small body weight in mice. In one (1953, 1955), he started from crosses among inbred lines. His criterion was weight at six weeks instead of 60 days, as in Goodale's and MacArthur's experiments. He made six matings in each generation in each case, using just two mice from each of the families of the preceding generations in order to cut down the rate of inbreeding. This was, however, considerable. Using A. Robertson's formula $\Delta F = 1/[4N - (m + 1)]$ where $N = 2^m$ is the size of population, the rate of increase of the inbreeding coefficient per generation yields about 0.023. In 20 generations $0.977^{20} = 0.63$, indicating the value of F of 0.37. The initial weight, 21.6 gm (based on an average of the sexes), was 3.8 gm greater than the average of the inbred strains. A strain bred without selection in the same way as the selected lines might be expected to lose 37% of this heterosis in 20 generations (1.4 gm) and thus weigh about 20.2 gm at six weeks. The inbreeding depression in MacArthur's experiment with an average of 37 pairs per generation would have been negligible.

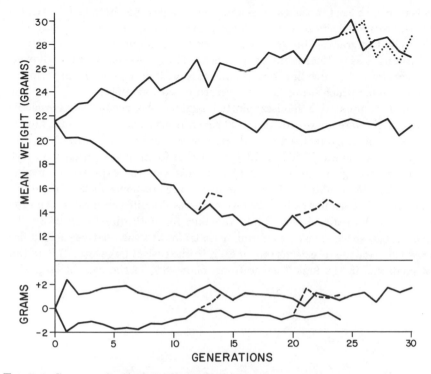

FIG. 7.5. Courses of selection of mice for high or low weight at six weeks in comparison with controls. Effects are shown of late relaxation (*dotted lines*) in the high line and of reverse selection (*broken lines*) at two times in the low line. Standard deviations are shown below. Redrawn from Falconer (1955, fig. 1), © 1955 by Cold Spring Harbor Laboratory; used with permission.

Falconer's selection based wholly on weights within two or three litters was less intense than MacArthur's and accounts for slower initial progress (fig. 7.5). Selection for large size was conducted for 30 generations, but after a rather steady rise it reached an apparent upper limit at 28 gm in generation 20. Relaxation at generation 24 caused no change, indicating absence of serious natural selection. On selection for small size, weight declined to 14 gm in 12 generations and tapered off toward 13 gm in 20 generations. Reverse selection at this time was effective, indicating that genetic variance was still present. The decline under minus selection was considerably more rapid than the rise under plus selection in the first 12 generations and had gone a little further by 20 generations in contrast with the greater absolute progress upward in MacArthur's experiment. As with MacArthur, the coefficients of variability were much more nearly uniform than the standard deviations, at least in the early generations, suggesting that the factors have multiplicative effects and that a logarithmic scale is more appropriate than the weight in grams. This greatly increases the asymmetry, instead of nearly abolishing it as in MacArthur's experiment. This, however, does not allow for the substantial inbreeding depression, expected in Falconer's experiment, allowance for which brings about a return toward symmetry.

The accumulated responses plotted against the accumulated selective differentials were almost linear for 20 generations in both directions, but the slopes, the average realized heritabilities, were very different (fig. 7.6). That for plus selection was $17.5\% \pm 1.6\%$ while that for minus selection was nearly three times as great, $51.8\% \pm 2.3\%$. If, however, the expected inbreeding depression 1.4 is subtracted from the line of zero response at 20 generations, there is little asymmetry left. MacArthur's results differed both in the rapid decline in the realized heritability (average for both directions: 23.8, 12.3, and 11.4 in successive 7-generation periods) and in the relative small difference between the directions (about 20% in the case of large size, 27% in that of small size in the first 7 generations, about 8% in the case of large size,

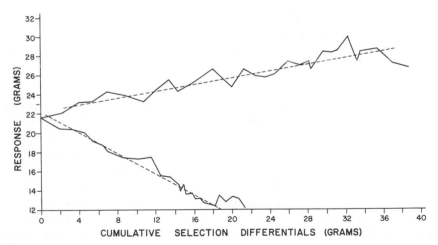

FIG. 7.6. Cumulative selection of parents and responses of offspring in the selection for weight of the mice in figure 7.5. Trends are shown by broken lines. Redrawn from Falconer (1955, fig. 2), © 1955 by Cold Spring Harbor Laboratory; used with permission.

12% in that of small size in the next 14 generations). There was, however, little inbreeding depression in his experiment.

A different aspect of the asymmetry became apparent when Falconer analyzed the six-week weight into two components that behaved very differently. The weaning weight (three weeks) was almost wholly maternal and thus not directly selected, while that from three weeks to six weeks was independent of the mother, both as demonstrated by reciprocal crosses among the large, small, and control stocks. Under selection for large size, the first component remained substantially constant at about 10 gm, while the second rose from about 11 gm to 18 gm in 20 generations. Under selection for small size, the first component declined rapidly and linearly from 10 gm to 6 gm in 20 generations and the second declined from 11 gm to 7 gm in 12 generations and thereafter changed little. It appears that selection for slow gain from three to six weeks has as a pleiotropic effect a marked decline in mothering ability, greater than can be accounted for by inbreeding depression. Selection for rapid gain may have had some pleiotropic effect in the opposite direction, which was neutralized by the inbreeding depression. The gains from three weeks to six weeks are roughly symmetrical for 12 generations but the apparent limiting values show the opposite symmetry in absolute terms from that shown by the six-week weights. If part of the inbreeding depression applies to the gains after weaning, as is probable, the asymmetry in favor of large gains is still greater.

It may be noted here that the percentage of fertile matings and the percentage of survival to six weeks both showed considerable declines, indicative of pleiotropic effects (as probable in the large-scale experiments of Goodale and MacArthur) or of inbreeding depression (probably important in Falconer's case). Size of litter rose from 8 to 12 in seven generations in the large line and then gradually declined to about 8, presumably because of the inbreeding depression. In the small line it declined to about 5 in 11 generations and remained there. This difference between the two lines indicates the same pleiotropic effect of size factors on size of litter as was apparent in the experiments of Goodale and MacArthur.

The coefficient of variability (fig. 7.7) within litters rose gradually from about 5 to 6 in 30 generations in the large line. In the small line, it followed the same course to generation 7 but in generation 9 it approximately doubled and remained high to the end. This was not merely increased environmental sensitivity when the size fell below a critical threshold, since realized heritability was not reduced. The abruptness and persistence suggested an unfixable mutation.

A study of the effect of mass selection on postweaning weight of mice by Rahnefeld et al. (1963) may be noted here as one in which the realized

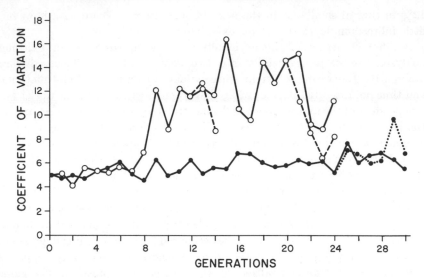

Fɪɢ. 7.7. Coefficients of variation within litters in the selected lines of mice of figure 7.5 (plus line, *solid circles*; minus line, *open circles*). Redrawn from Falconer (1955, fig. 6), © 1955 by Cold Spring Harbor Laboratory; used with permission.

heritabilities did not differ significantly from those estimated from the initial parent-offspring regressions, over 17 generations of selection for increase. There was no significant deviation from linearity and a gain of 58% relative to the controls. The fact that the foundation stock was from a cross between two highly inbred lines, so that the gene frequencies were 0.50 at all loci that were heterallelic, made this case especially favorable for prediction of the course of selection, as long as it was linear.

Selection for Gain: Relation to Ration in Mice

In an early experiment, Falconer and Latyszewski (1952) tested a thesis advanced by Hammond (1947) in connection with livestock breeding:

> Since the full capabilities of the animal for development are inhibited by undernutrition, it would appear that the selection of meat-producing animals can best be made under the optimum nutritional conditions, for only under these conditions can small differences in development be distinguished.

Hammond was doubtless referring to more than selection for rate of gain, but the efficiency of selection for rate of gain is something that can be tested readily in mice.

The character studied was the weight at six weeks. Selection for large weight was carried out on two rations, a full one and one about 25% below normal consumption. A foundation stock was derived from crosses among four strains. About 100 mice were fed from three to six weeks of age on each ration. Six pairs were selected for weight at six weeks in both cases. Thereafter each line consisted of six pairs, selected within litters (after adjusting weights for sex), just two from each parental mating to cut down inbreeding. The restricted diet was fed only from three to six weeks. The average results per generation are shown in table 7.8.

TABLE 7.8. Selective differential response and realized heritability of selected mice under two rations.

Ration	Selection Differential (gm)	Response (gm)	h_r^2
Full	1.67 ± 0.050	0.33 ± 0.10	0.20 ± 0.06
Restricted	0.88 ± 0.024	0.26 ± 0.11	0.29 ± 0.13

SOURCE: Falconer and Latyszewski 1952.

Exchange of ration were made in the fifth and seventh generations and contemporaneous comparisons were made. The mice from the line selected on the restricted ration, but themselves given the full ration, weighed very nearly as much as did mice from the line selected on this ration. On the other hand, those from the line selected on the full ration, but themselves given the restricted ration, weighed no more than unselected mice and very significantly less than the mice from the line selected on the restricted ration. Overall, it appeared that selection was most effective on a restricted ration and the results thus did not support Hammond's thesis as far as weight was concerned.

The fat content of the full ration strain was about 24% greater than that of mice from the other strain after eight generations when both were given the full ration. It is evident that selection for weight was not here selection for the same underlying character. Hammond seems to have been right with respect to fattening capacity.

Falconer came back to this problem later (1960b). He carried through two-way selection (+, −) in this case for gain from three to six weeks on both high (H) and low (L) planes of nutrition. The lines consisted of 12 pairs in each case with litters standardized at four males, four females by elimination or fostering, as far as possible. The inbreeding coefficient rose only 1.1% per generation. Selection was wholly within first litters (approximately 25%). The second litters were fed the opposite ration. Comparisons were made of the effects of both rations in all of the four lines in each generation. This program was carried out for 13 generations, after which changes were made that need not be considered here.

The nature of the results can be illustrated (table 7.9) by those of the seventh generation, for which a special control was set up (first litter, response under plane of selection; second litter on the other plane).

TABLE 7.9. Data on realized heritability in four selection experiments in mice of generation 7, and the responses (gain [gm] three to six weeks) to the same and other rations.

SELECTION (GENERATION 6)				RESPONSE (gm) (GENERATION 7)		RESPONSE (GENERAL)	
Direction	Plane	Intensity ($\times \sigma$)	h_r^2	H	L	H	L
+	H	0.83	0.26	+2.26	+0.64	+	0
+	L	0.69	0.31	+1.55	+3.07	+	+
−	H	0.80	0.42	−2.80	−2.89	−	−
−	L	0.66	0.25	−1.17	−3.23	0	−

SOURCE: Falconer 1960b.

The results in the first two lines confirm the results of the earlier study. Taking all of the data, line +H was in general only slightly superior to +L on the high plane, while +H gave no consistent response on the low plane to which +L naturally responded strongly. With respect to selection for low gain, there was in general no appreciable difference between −H and −L on the low plane, while −L gave only a slight negative response on the average on the high plane to which −H naturally gave a very strong negative response. The results were almost those indicated by exchanging directions of selection (+, −) and planes of nutrition (H, L).

With respect to weaning weights (indicating maternal capacity), −H was distinctly the poorest and +L on the whole the best, but +H and −L were only slightly poorer (in this order).

The coefficients of variability of gain on the high plane were all substantially alike and constant (about 10%). Variability was very irregular on the low plane.

Not surprisingly, +H used the food consumed most efficiently (18.7%) with respect to gain and −H the least efficiently (12.7%) on the high plane of nutrition. On the low plane, +L was much more efficient (9.3%) than +H (2.6%) and −L (1.2%). Selection −H was not tested but presumably was very low.

In agreement with the earlier study, the percentage of fat in the carcass on the high plane was considerably greater (30% at 12 weeks) in +H than +L. The percentage of water, indicative of protein, was 6% greater in +L.

The peculiar asymmetry of the results makes it impossible, as stated by Falconer, to account for the relations between the gains under high and low planes on the basis of linearly correlated heredities and environments for the two characters (as discussed in vol. 2, chap. 15).

There are probably reciprocal interactions between the processes of protein growth and fattening by which each tends to reduce the other. Either excess fattening or protein growth, and to some extent both, can occur under a high plane of nutrition, depending on the genotype. Excess protein growth can occur under a low plane with a favorable genotype, but little fattening occurs unless protein growth is severely reduced by a genotype that is normal with respect to fattening capacity (only $-L$).

Under these conditions, selection $+H$ tends to build up a genotype favorable to fattening but only at a high plane. The selection for fattening genes apparently is easier and takes precedence over selection for genes favorable to excess protein growth. Selection $+L$ can only build up a genotype favorable for protein growth. This can, however, function well under either plane, although it is perhaps offset somewhat on the high plane by reduction from normal fattening. Selection $-H$ tends to build up a genotype unfavorable to both protein growth and fattening that thus reduces gains drastically under both planes. Selection $-L$ tends to build up a genotype unfavorable to protein growth but neutral as far as fattening is concerned, except as reduced growth favors fattening. Gains are greatly reduced under a low plane, but under a high plane the reduction in protein growth is largely offset by a released enhancement of the normal fattening process.

This experiment illustrated admirably the possible complexity of the selection process (fig. 7.8).

The danger of generalizing from one or two experiments is shown by the contrary results of selection for increased gain of rats in lines on full or restricted rations obtained by Park et al. (1966). Their FF line (full feeding) was fed ad libitum on a standard ration while their LF line (low feeding) was fed three-fourths of the amount normally consumed. The character was gain from three to nine weeks of age. The lines were selected for 17 generations, after which selection was relaxed for six generations on the full ration. A control line was carried to the end. Litters were standardized at a size of six by addition or removal.

The FF line showed a regression of 0.260 ± 0.050 on midparent and the LF line showed a regression of 0.146 ± 0.050, but the realized heritabilities (ratio of total response to cumulative selection differential) was only 0.105 ± 0.009 in the former, 0.057 ± 0.010 in the latter. These may be underestimates, however, since the contemporary control stock was making considerably greater gains on both rations, as shown below, than it had at the beginning, in spite of the absence of selection. Whether this was due to environmental improvement or random genetic drift is not known.

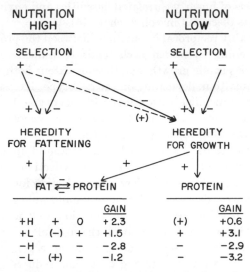

FIG. 7.8. Hypothetical effects of plus and minus selection of mice under high or low nutrition in each case, on the heredities for fattening and for growth. The actual responses (gain [gm], 3–6 weeks) in the seventh generation (Falconer 1960*b*) are shown below.

The point of most interest here is the response of each line to the heterologous ration (table 7.10).

TABLE 7.10. Gain (3–9 weeks) of control rats and rats under plus selection under two rations.

	PLANE OF NUTRITION	INITIAL CONTROL (gm)	GAIN (gm) 3–9 WEEKS GENERATIONS 13–17		
			Control	FF	LF
♂	H	168	188	203	194
	L	104	113	122	116
♀	H	109	120	134	130
	L	76	85	89	89

SOURCE: Park et al. (1966).

The FF line gained more on the full ration than did the LF line, but it also gained as much or more on the low ration, contrary to the results of Falconer and Latyszewski (1952) and Falconer (1960*b*). Selection was much less effective, but that under the full ration seems to have been acting on the same aspect of gain as that under the low ration, and doing so more effectively.

It has been noted that the weight of mice at weaning (three weeks) is largely a maternal character. Falconer (1955) described a selection experiment conducted in his laboratory by Bateman on weight at 12 days of litters made constant in size (eight) as far as possible by reduction or addition, as a measure of lactation.

Two-way selection was successful but only to a limited extent. The lines were carried apart only by about twice the original standard deviation, although the initial heritability was about 0.50. The limits were reached in five or six generations, but this was not due at all to fixation since reverse selection remained highly effective for as many generations as the lines were carried. It appears that few loci were involved and that these were essentially unfixable because of natural selection against the extremes.

This experiment illustrates a further complexity in selection for the apparently simple character of body weight.

Direct Selection for Size of Litter in Mice

As noted earlier in both MacArthur's and Falconer's experiments on the selection of size in mice, there was a marked correlative change in size of litter such that the large strain came to have litters about twice as large on the average as the small strain. It is of interest to compare this effect with that of direct selection for size of litter. This has been done by Falconer (1955, 1960a, 1964).

It is desirable to take up first the conclusions on heritability in the unselected foundation stock, which he carried through the 31 generations of the experiment as a control. This consisted of ten pairs per generation. The ten females were chosen at random, each from a different family of the preceding generation, and mated with a male similarly chosen and from a different family. The data from this line are based on 304 mothers and 977 daughters. With close to maximum avoidance of consanguinity, the inbreeding coefficient was about 0.0134 per generation and thus considerable (about 0.34) after 31 generations. There was no appreciable decline in litter size (average 7.5 throughout) in spite of an estimated decline of 0.5 young per 10% inbreeding under rapid inbreeding on the average, suggesting that natural selection, necessarily based on viability of the embryos, was strong enough to counterbalance the inbreeding depression, the amount of which, however, varied greatly in different rapidly inbred lines.

The daughters from each mating, usually three or four, were each mated with a male from a different one of the nine males chosen at random from the other matings. Each male was thus mated with several females unrelated to him or to each other. This made it possible to test for possible paternal influence on litter size. There was a probably significant influence, presumably on viability, but it was certainly responsible for less than 10% of the variance.

At first there seemed to be no heritability from the mother since the correlation between the size of litter produced in successive generations was -0.028 ± 0.040. This, however, does not take account of a correlation of -0.339 ± 0.042 between litter size and daughter's weight at six weeks and a correlation of $+0.265$ (implied but not stated) between mother's weight at six weeks and the size of litter that she produces (Falconer 1964). These would be responsible for a negative relation between litter sizes in successive generations, which would counterbalance an appreciable positive genetic correlation. The relations are represented in a path diagram in figure 7.9 in which L and L' are the sizes of litter produced by daughter and mother, respectively, and W is the daughter's weight at six weeks. The following equations can be written:

$$r_{LL'} = -0.028 = l_1 + l_2 w$$
$$r_{LW} = +0.265 = l_1 w + l_2$$
$$r_{WL'} = -0.339 = w$$

Solution yields $p_{LW} = l_2 = 0.288$ (the figure stated), $p_{LL'} = l_1 = +0.070$. This positive genetic correlation is slightly overbalanced by the negative relation through weight $p_{LWL'} = l_2 w = -0.098$.

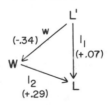

Fig. 7.9. Path diagram showing path coefficients relating size of litter (L), produced by a mouse to her weight at six weeks (W) ($l_2 = +0.29$) and to the size of the litter in which she was born (L') ($l_1 = +0.07$), and that relating W to L' ($w = -0.34$).

An independent interpretation of the correlation between successive litters can be based on the correlation between the litters produced by full sisters. Analysis of the variance of litter sizes (4.304) into the components between and within sibships, 0.460 and 3.844, respectively, yielded a correlation of $+0.107 \pm 0.035$ for full sibs (Falconer 1964). In the path diagram (fig. 7.10) the influence of genotype (G), assumed to be additive, is represented by h, the maternal influence by m, and other environmental

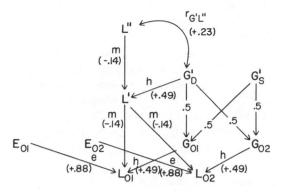

FIG. 7.10. Path diagram showing relations between litters, L_{01} and L_{02}, produced by two littermates with genotypes G_{01} and G_{02}, dam's and sire's genotypes G'_D and G'_S, and environments E_{01} and E_{02}. Litter size is represented as affected by heredity (path coefficient h), environment (path coefficient e), and maternal influence from size of litter L' in which born (path coefficient m). The relations to the size of litter L'' in which the dam was born are also shown.

influences by e. The path coefficients relating genotypes of offspring and each parent (dam G_D, sire G_S) is 0.5, treating mating as if random within the population (not strictly true). The basic equations are as follows:

$$r_{LL'} = -0.0282 = hr_{GL'} + m$$
$$r_{L_1L_2} = +0.1069 = 0.5h^2 + 2mhr_{GL'} + m^2$$
$$m^2 + h^2 + 2mhr_{GL'} + e^2 = 1.$$

These permit solution for m, h, and e if $r_{GL'}$ can be expressed in the following terms:

$$r_{GL'} = (1/2)r_{G'L'} = (1/2)r_{GL} = (1/2)(h + mr_{GL'}) = h/(2 - m)$$
$$m^2 + 2.0846m + 0.2702 = 0$$

$m = -0.1388$	$m^2 = +0.0193$
$h = +0.4864$	$h^2 = +0.2366$
$r_{GL'} = +0.2274$	$2mhr_{GL'} = -0.0307$
$e = +0.8802$	$e^2 = +0.7748$
	1.0000

The maternal influence, -0.139 by this calculation, differs only slightly from Falconer's estimate, -0.133, calculated in a different way. It is larger than that through weight at six weeks (-0.098) and would imply a genetic correlation between successive litters of $+0.111$ instead of $+0.070$. The agreement is as close as can be expected in view of the standard errors, but it is also quite possible that the relations to weight at six weeks do not fully account for the total maternal influence.

In the selection experiments, each line was maintained by ten pairs, one female and one male from each mating of the preceding generation, with avoidance of inbreeding. Selection was made between full sisters (litter-mates) all mated with the same male so that no selection was applied to males with respect to the size of litter sired. The litters were reared as born, so that the maternal influences through daughter's weight were present. The generation means were computed as means of the means of sister groups.

The response to the first generation was significantly negative, presumably reflecting an excessive maternal influence; thereafter the response was approximately linear to about generation 20, after which there was no consistent advance. The high line advanced in 20 generations from 7.56 (controls) to 9.26, while the low line declined to 5.91. The divergence (3.35) from direct selection for litter size is thus less than brought about corre-latively from selection for weight (5.2 in MacArthur's experiment, and about the same in Falconer's before being reduced by greater inbreeding depression in his high line).

While the changes in the high and low lines were about the same in absolute terms, the regressions of daughters' litters on the cumulative selection differential of mothers' litters for generations 1 to 21 were highly asymmetrical, $+0.038 \pm 0.011$ upward and $-0.125 + 0.013$ downward. These are re-gressions on only one parent, and thus should be compared with $(1/2)h^2 = 0.119$ as estimated from the control data. Even so, the selection for large litters is far below the expectation on the simple theory.

Study of the lines indicated that the increase in the high line was largely in ovulation rates, any tendency to increase viability of the embryos pre-sumably being offset by inbreeding depression, which was found to act only on this aspect under intense inbreeding. The decrease in litter size in the low line, on the contrary, was found to be due to decreased viability. As in the case of body weight, there are various reasons why it is easier to reduce the character by selection than to increase it and for failure of realized heritability to be as great as heritability based on correlations in an unselected population.

There have been many more selection experiments than it is practicable to review here. In most of them, as in some of those already discussed, the purpose was merely to produce strains deviating as much as possible from the foundation stock in one or both directions, rather than to analyze the process. The variety of characters in which selection has been tried, usually successfully, is interesting.

Carcinogenetic Action of Urethane

Falconer and Bloom (1962, 1964) found that the susceptibility of mice to the carcinogenic action of urethane varies greatly among strains and that there was strong heritability within strains. The mice were injected intraperitoneally at three and nine weeks and not mated until 12 to 14 weeks to insure elimination of the urethane. Autopsy was at 23 weeks.

Selection was conducted in a strain of mixed origin in which the mean number of visible lung tumors was 8.19 ± 0.46, with standard deviation 7.20 ± 0.33. About 49% of the variance was additive genetic, 25% nonadditive genetic, only 10% environmental, and 16% mere chance. The asymmetry and correlation of standard deviation with mean indicated the desirability of a transformation of scale. The square root of the number of tumors proved satisfactory; on it, the additive heritability was raised to 54.5%. Selection had to be retrospective. Mice were mated and offspring were obtained before the number of tumors was determined by autopsy. Already existing litters of the mice, which were then selected, were retained and the remainder discarded.

The zero generation consisted of 32 pairs. Fifteen pairs were selected to be parents of the next generation as coming from parents with the larger means. High selection was carried for nine generations on the basis of 15 pairs chosen from an average of 25 in each generation. The tumor index (mean square root of the number) rose only slightly in the first three generations (2.55 to 2.75) but it rose to 4.99 in the next six generations, with no indication that a limit was being approached. Low selection was started from the third generation and rather steadily reduced the mean to 0.83 in six generations. A control strain remained virtually unchanged. The actual mean number of tumors changed from 7.0 to 26.8 in the high strain and to 1.5 in the low strain.

Sex Ratio

Falconer (1954) conducted one selection experiment with mice in which no significant change was brought about. This was in sex ratio. The foundation stock was of very mixed origin. Selection was based on the difference between the numbers of liveborn males and females in the first litter. Fifty males and 50 females out of some 600 of each sex were chosen to produce the next generation.

The experiment was carried for four generations in each direction with no significant effect. In the last generation, the percentage of males was 53.1 in the high line and 53.6 in the low line.

There were evidently no heterallelic loci with appreciable effect on sex ratio in the foundation stock. This does not prove that there may not be such loci in other cases in mammals. We will cite first an experiment by King (1918c) in which such a difference was brought about in rats by selection in an experiment in which there had already been several generations of full-sib mating.

This experiment was started from a single litter of stock albinos, containing two males and two females. The descendants of the two matings made from those were maintained by full-sib mating for many years, as discussed in chapter 3. The first seven generations were produced by parents not selected in any way. After this, all breeding females in line A were taken from litters that contained an excess of males, while those in line B were taken from litters that contained an excess of females. After two litters had been obtained from full sibs, the females were remated to stock males and if possible two more litters were obtained. Table 7.11 gives a condensed summary for 25 generations.

TABLE 7.11. Sex ratio in two series of rats selected in opposite directions and in outcrosses.

Generations	Line A (Litters 1 and 2)		A ♀ × Stock ♂ (Litters 3 and 4)		Line B (Litters 1 and 2)		B ♀ × Stock ♂ (Litters 3 and 4)	
	No.	% ♂	No.	% ♂	No.	% ♂	No.	% ♂
1–7	930	52.5	682	54.0	932	52.1	712	49.6
8–13	1,732	55.0	1,555	53.2	1,671	46.8	1,254	48.6
14–19	2,143	55.3	1,765	54.2	1,971	43.5	1,825	47.3
20–25	2,399	55.1	1,910	53.2	2,251	44.9	1,720	47.4
Total (8–25)	6,274	55.1	5,230	53.5	5,893	45.0	4,799	47.7

SOURCE: King 1918c.

There is no indication of a difference in the percentage of males in the full-sib lines A and B in the first seven generations. There is a doubtful indication of a rise in the later generations of A and a clearly significant drop in line B. There is no indication of a change in this percentage in the outcrosses of the females in either case, but there is a clearly significant difference between the outcrosses of the two lines, although not much more than half as much as between A and B themselves. These results indicate that the character was at least partially a maternal one but probably also partially paternal. The limiting effect of selection seems to have been reached rapidly, but the close inbreeding would tend to restrict progress.

Returning to mice, Weir (1953, 1960) obtained a very marked immediate difference in percentage of males in two lines derived from a single random bred strain but not by selection for this character. He was selecting for pH of the venous blood, the effect on which will be considered first. The initial pH was 7.458 ± 0.0044. One generation of selection (7 males, 23 females selected from 280) raised the average to 7.466 while selection of 6 males and 19 females at the other extreme lowered the average to 7.420. The realized heritability was $0.008/0.073 = 0.11$ in the high line and $0.038/0.124 = 0.31$ in the low line. There was no appreciable change in two more generations with similar selective differentials (total difference 0.050). There was some inbreeding and the lines were maintained thereafter by full-sib mating without further attempt at selection. The difference between the two lines remained substantially unchanged. The interpretation, as in King's results for percentage of males, is complicated by the inbreeding, but the immediate attainment of the limit suggests that only one or two heterallelic loci were involved.

The most surprising feature of Weir's results was a rapid differentiation in sex ratio without any selection for this character. While not noted until after ten generations, most of the effect was present from the first. In the first four generations the percentage in the high line was 53.7 ± 0.7, in the low line 45.9 ± 0.7. In later generations, that in the high line was 52.5 ± 0.3, in the low line 40.5 ± 0.2. Reciprocal crosses and crosses with other strains showed that this was wholly a paternal character, in contrast with King's results.

Interpretation as a pleiotropic effect of blood pH was made difficult by finding that selection for pH in arterial instead of venous blood (Wolfe, in Weir, 1960) with avoidance of inbreeding was associated with high and low sex ratios in the opposite direction (45.0 ± 0.8 with high pH, 52.9 ± 0.8 with low pH).

Susceptibility to Dental Caries

Hunt and associates (1944, 1955, 1962) have made an intensive study of the effects of selection on susceptibility and resistance to caries in rats. This is a very complex character. The acidity of the saliva and the characteristics of the teeth (chemical composition, microstructure, fracturing, and width of sulci) are immediate factors. The presence of lactobacilli and other bacteria, mouth secretions, character of the diet (especially size of particles) are more remote. Heredity may act through more than one channel.

A cariogenic diet with much coarsely ground hulled rice was used in the earlier studies. The experiment began with 119 rats of heterogeneous origin. They were examined every two weeks for caries. Two males and five females were selected to start the susceptible line, three males and six females for the resistant line. Most matings were made between brothers and sisters, but in the seventh generation in the susceptible line and in the sixth generation in the resistant line matings were wholly between families in order to gain vigor and broaden the base. Thereafter, all matings were between full sibs except for three or four between half sibs. Altogether 9,800 albino rats were bred and observed up to 1955.

The average number of days from exposure to the cariogenic diet to the recognition of caries was determined for each sibship and these averages were averaged. The initial mean caries time was 70 days. This fell to 57 days in the susceptible rats and rose to 116 days in the resistant rats after one selection. After two selections, these means became 43 and 147 (fig. 7.11). Thereafter progress was slower and rather irregular. That for the susceptible rats had fallen to 17 days for the period from the 12th to the 19th selection. A change to a diet in which coarse particles were largely eliminated led to a rise to an average of 35 days for the next three generations and no clear changes later. The mean of the resistant line rose to 340 days for the 9th through 11th selection and, thereafter, with the change in ration, rose to 462 days.

The distributions became fairly compact in the susceptible line but continued to show enormous variability in the resistant line, ranging up to more than 700 days. Most of this was, no doubt, a matter of scale, as in the tumor frequencies in mice of Falconer and Bloom.

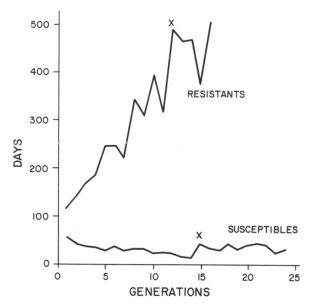

Fig. 7.11. Average number of days from exposure of rats to a cariogenic diet to recognition of caries, under selection for resistance or for susceptibility. The crosses indicate the time after change to a less cariogenic diet. Reprinted, by permission, from Hunt, Hoppert and Rosen (1955). © 1955 by the American Association for the Advancement of Science.

The rapidity of the approach to a limit indicates that the number of major heterallelic loci was small, although again the close inbreeding in most generations complicates the interpretation since this would lead to fairly rapid fixation. Whether genetic fixation was actually approached in the later generations was not, however, tested by reverse selection. Comparison of the two strains indicates that genes operate both through effects on the bacterial flora and the characteristics of the teeth.

Intelligence

There have been a great many experiments that demonstrate that differences in behavior among individuals and strains of higher vertebrates may depend to a large extent on genetic differences (Fuller and Thompson 1960; Scott and Fuller 1965). There have also been many experiments that show that differences between derivatives of a strain may be brought about by mass selection. I will cite only one example, an experiment by Thompson (1954) with a strain of rats that initially registered an error score of 200 in maze trials. After six generations of selection of "maze bright" rats, the error score of a substrain was reduced to 140 under the same conditions, while selection of "maze dull" rats raised the error score of another substrain to 280, just twice the preceding.

Miscellaneous Characters in Mice

Mice have a highly standardized number (19) of secondary vibrissae. Kindred (1963, 1967) found no extra bristles that could give a basis for plus selection. Ten generations of minus selection on the basis of the occasional absence of one bristle had little or no effect (average about 18.7). The sex-linked mutation tabby, however, reduces the average to about 16 in females and to about 9 in males, with much variability in both. Dun and Fraser (1959) were able to increase the averages to 18.7 and 13, respectively, by 22 generations of plus selection and reduce them to about 9 and 5, respectively, by the same number of generations of minus selection. Tabby and its normal allele were kept segregating in the process, making possible study of correlative effects. The plus selection of tabby produced no correlated increase in the segregating normals (still 19), but the minus selection of tabby reduced the average in the normals to about 17. Kindred (1967) showed that direct plus selection of normals after the 22 generations of plus selection of tabby, raised the average of the normals to 21, even though neither this process nor further direct plus selection of the tabbies produced any further gain in the latter. She showed that the average for the normals, after the 22 generations of minus selection of tabbies, could be further reduced by minus selection of either type of segregant (to about 15.5 in 11 generations in each case). The tabbies, after 22 generations of minus selection, could not be further reduced. The most interesting result was that the average for normal mice could be increased only by following plus selection of tabbies by plus selection of normals, and that the latter process was effective even though the tabbies had reached a plateau. The relations of the modifiers of vibrissa number to the thresholds for change in normals and tabbies are evidently complex.

Results of selections of mice for high and low resistance to X-radiation (100 r/day, 400 r/day) have been reported by Roderick (1963). All experiments were successful in producing changes in the direction selected, more markedly so at the lower dosage.

Chai (1966, 1970) selected for high and low leukocyte numbers in mice from a stock derived from six inbred strains. After a slow start, the high line reached an average of 25 in terms of thousands per cubic millimeter, in 18

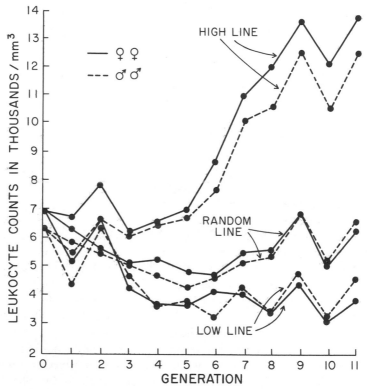

FIG. 7.12. Courses of change of leukocyte counts, in mice selected for resistance or for susceptibility, over 11 generations. From Chai (1966).

generations. The low line declined to 5.9, probably close to the physiological limit. The unselected controls at this time averaged 10. A cross between high and low made in generation 14 gave a mean of 9.3 and variance of 3.4 in F_1; mean 11.5, variance 14.0 in F_2. Using a logarithmic scale, the segregation index comes out 4.3, which could imply four or five equivalent pairs but also one or two important ones and many minor ones. An interesting feature was that the thymus became twice as large in the high line as in the low one, while the spleen and adrenal glands decreased in size by about 29% and 20%, respectively. Figure 7.12 shows results up to generation 11.

Selection Experiments with Domestic Animals

Most of the many selection experiments with domestic animals have been directed primarily toward improvement rather than testing theory. In many cases they have involved complications that are undesirable from the latter standpoint, and in many cases they have involved subjective judgments. Those with the larger animals have also often been based on undesirably small numbers because of expense. There have been experiments, however, especially with swine and poultry, comparable to those with laboratory animals.

One example with swine must suffice here. Hetzer and Harvey (1967) selected for depth of backfat in swine on the basis of a demonstration that accurate measurements are possible by means of a probe applied to the live animal.

Selections for high and low fat were made with Duroc-Jersey and Yorkshire swine for ten and eight generations, respectively, on the basis of probes made when the pigs weighed about 80 kg (fig. 7.13). Sex differences were minor and allowed for. The one or two fattest and one or two leanest of each sex were selected from each of 15 Duroc-Jersey litters and each of 17 Yorkshire litters as the starting point. Later generations were produced by about six selected boars and twice as many gilts in each breed. Unselected controls were carried along using 12 boars and 12 gilts per generation. Sib and half-sib matings were avoided, but the amount of inbreeding gradually increased; in Duroc-Jerseys to 23%, 24%, and 15% in the high, low, and control lines, respectively. The corresponding figures in the Yorkshires were 18%, 19%, and 16%.

In the Duroc-Jerseys, selection brought about a difference of 2.6 cm (68% of the initial depth of fat) between the high and low lines. In the Yorkshires the difference was 1.4 cm (44% of the initial depth). There was considerable parallelism in the annual fluctuations, dependent no doubt on environmental conditions. This justifies comparisons of progress on the basis of the deviations from the controls.

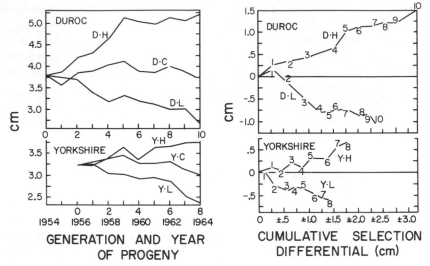

FIG. 7.13. Courses or selection of Duroc-Jersey and Yorkshire swine for high (H) or low (L) thickness of backfat, in comparison with controls (C). Redrawn from Hetzer and Harvey (1967, figs. 1 and 2); used with permission.

The most instructive comparison is that of the total selective response in each generation relative to the contemporary controls, with the cumulative selective differentials (table 7.12 and fig. 7.13). There is probably some damping of response at the lower thicknesses, but no indication of approach to a plateau in any case. The realized heritabilities were 79% of the average regression of offspring on midparental means within generations. It appears that much of the heredity was additive.

TABLE 7.12. Regression of offspring on midparent in high fat, low fat, and control lines compared with realized heritability.

	OBSERVED REGRESSION (cm)		REALIZED HERITABILITY (h_r^2)	
	Duroc-Jersey	Yorkshire	Duroc-Jersey	Yorkshire
High fat	0.43 ± 0.07	0.49 ± 0.10	0.47 ± 0.02	0.38 ± 0.03
Low fat	0.60 ± 0.08	0.70 ± 0.08	0.48 ± 0.02	0.43 ± 0.03
Control	0.62 ± 0.06	0.58 ± 0.07		

SOURCE: Hetzer and Harvey 1967.

In the high fat lines of Duroc-Jerseys, backfat depth within generations was related negatively to amount of inbreeding (0.045 cm per 1% ΔF) but there was no significant relation in the low fat line (+0.001).

The situation was largely reversed in the Yorkshires (high fat, +0.012 cm; low fat, −0.019 cm). These results seemed to reflect differences between the genetic compositions of the breeds rather than systematic effects of inbreeding.

Lerner (1958) has summarized results that he and others have obtained by selection of a great variety of traits of the domestic fowl. Selection for egg production (fig. 7.14), early growth rate, final body weight and shank length, and certain special traits were effective to varying extents, but the parameter a of the allometric growth equation, $y = bX^a$ (rate of shank growth relative to general growth), proved to be resistant to selection. This has a bearing on apparently orthogenetic evolution of certain traits of organisms in nature.

Selection for shank length (fig. 7.15) was conducted for 18 years with fairly large numbers reared per generation (average 75) from 15 to 33 selected dams and 2 to 4 selected sires. There was advance from 9.7 cm to 10.8 cm in seven years, followed by leveling off. The plateau could not have been due to fixation since suspension of selection was followed by return almost halfway

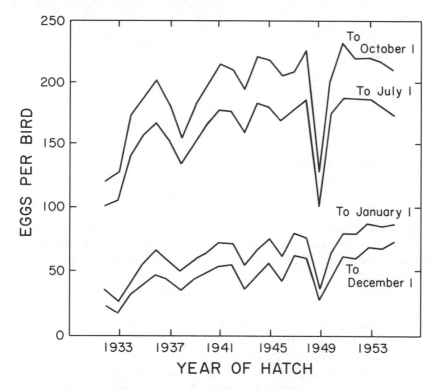

FIG. 7.14. Course or selection for increased egg production in a flock of White Leghorn fowls. The disastrous drop in 1949 was due to an epidemic of severe respiratory disease. Redrawn from Lerner (1958, fig. 7.1); used with permission.

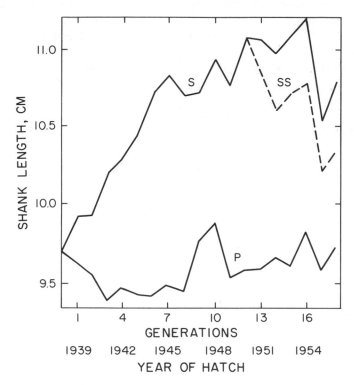

FIG. 7.15. Courses of selection (S) of White Leghorn fowls for increased shank length and suspension of selection (SS) in comparison with controls (P). Redrawn from Lerner (1958, fig. 4.10); used with permission.

to the control level in six years. The gain had clearly led to increased opposition by natural selection. As shank length increased, a fitness index declined, and as shank length decreased after suspension of selection the fitness index recovered.

The percentage of a certain abnormality, crooked toes (fig. 7.16), rose from 1% or 2% to 100% in seven years of selection. Reverse selection, begun when the incidence was 60%, rapidly reduced it to 10%, but enough fixation had occurred to make further reduction difficult.

Selection for increase in the number of comb blades in a strain in which about half had two or three instead of the usual single comb, eliminated singles in two years, duplex in four years, but apparently reached a limit in six or seven years with predominance of quadruplex but many triplex and quintuplex (fig. 7.17).

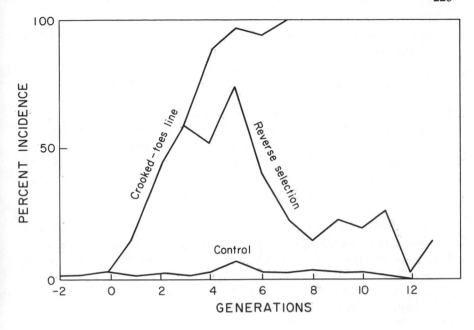

FIG. 7.16. Selection of White Leghorn fowls for crooked toes and reverse selection. Redrawn from Lerner (1958, fig. 4.9); used with permission.

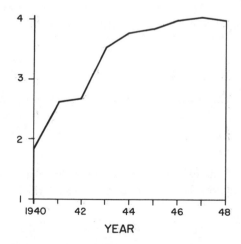

FIG. 7.17. Selection of White Leghorn fowls for number of comb blades. From data of Lerner (1958).

There have been many experiments by many workers in laboratories in which the incidences of infective or of neoplastic diseases have been reduced with varying degrees of success.

Estimates of Selective Value Based on Relatives

In the improvement of livestock, much more intense selection can be applied to males than to females because of the much smaller number required. Unfortunately, in such important cases as milk production in dairy cattle and egg production in fowls, males cannot be selected at all on the basis of their own characteristics. It is important to make estimates of their expected grades, if they had been females, on the basis of the records of tested female relatives.

Such estimates are multiple regression equations that depend on the kinds and numbers of female relatives with known grades (Wright 1932a; Lush 1947; and vol. 1, chaps. 12 and 13). Consider first estimates that can be made before any daughters have been produced. Assume random mating within the population from which prospective sires are to be chosen, additivity and semidominance of the pertinent genes, and no correlation between the occurrence of genetic and environmental factors in this population, noting that there are likely to be strong genetic-environmental correlations in a breed as a whole. Heritability, h^2, also relates to the population as defined.

Represent the hypothetical grade of the male by X. It is hardly worthwhile to go beyond the closest female relatives, dam and full sisters. Since the genetic correlations involving these relatives are the same under the assumed conditions, it is convenient to represent them by the same symbol, Y. Assume n such relatives.

$$r_{XY} = 0.5h^2$$
$$r_{Y_1Y_2} = 0.5h^2$$

Let y $(=p_{XY})$ be the standardized partial regression coefficient relating X to any Y:

$$r_{XY} = y + (n - 1)r_{Y_1Y_2}y = y[1 + 0.5(n - 1)h^2]$$
$$y = h^2/[2 + (n - 1)h^2]$$

The deviations of the estimated grades of males from the mean grade of the population can now be expressed in terms of the mean grade of the female relatives:

$$\Delta X = ny\,\Delta \overline{Y} = \{nh^2/[2 + (n - 1)h^2]\}\,\Delta \overline{Y}$$

With complete heritability, $h^2 = 1$, $\Delta X = [n/(n + 1)]\,\Delta \overline{Y}$ and thus approaches $\Delta \overline{Y}$ as n increases.

More direct estimates can be made after daughters (O) have been produced and tested. These contribute more to the estimate if they are half sisters rather than whole sisters because of their greater independence. In this case, $r_{XO} = 0.5h^2$, as before, but $r_{O_1O_2} = 0.25h^2$.

$$r_{XO} = y[1 + 0.25(n - 1)h^2] = 0.5h^2$$
$$y = 2h^2/[4 + (n - 1)h^2]$$
$$\Delta X = ny \Delta \bar{O} = \{2nh^2/[4 + (n - 1)h^2]\}h^2 \Delta \bar{O}$$

With complete heritability, $\Delta X = [2n/(n + 3)] \Delta \bar{O}$, which approaches $2 \Delta \bar{O}$ instead of $\Delta \bar{O}$, with large n. It is, however, again $0.5h^2 \Delta \bar{O}$ if $n = 1$.

More accurate estimates can be made by taking account of the deviations of the dams of these daughters, if known. Let z be the standardized partial regression coefficient relating sire to dam, D, and assume zero correlation between them and between dams of different daughters.

$$r_{XD} = z + yr_{OD} = z + 0.5h^2y = 0$$
$$z = -0.5h^2y$$
$$r_{XO} = y[1 + 0.25(n - 1)h^2] + 0.5h^2z = 0.5h^2$$
$$y = 2h^2/[4 + (n - 1)h^2 - h^4]$$
$$\Delta X = n[y \Delta \bar{O} + z \Delta \bar{D}]$$
$$= \{nh^2/[4 + (n - 1)h^2 - h^4]\}[2 \Delta \bar{O} - h^2 \Delta \bar{D}]$$

With complete heritability, $\Delta X = [n/(n + 2)][2 \Delta \bar{O} - \Delta \bar{D}]$. With only one daughter-dam pair, $\Delta X = [h^2/(4 - h^4)][2 \Delta \bar{O} - h^2 \Delta \bar{D}]$. The dam and full sisters of the prospective sire can also be taken into account with considerable increase in complexity. This is also the case with allowance for correlations assumed here to be zero.

The most effective method of large-scale improvement in milk production has been a combination of a primary choice of males on the basis of close female relatives or merely on the record of herd as a whole, with final choice based on the large number of daughters, made possible by artificial insemination, tested under standard conditions.

Egg production by domestic fowls has also been improved greatly by artificial selection, based to a large extent on standardized progeny testing of potential sires. Thus in experiments conducted by Lerner (1958) there was a rise from an initial production of about 126 eggs per year to a high of 200 in three years. This was followed by fluctuations due to disease and recovery but a slowly rising trend toward a plateau of about 230 eggs per year in the favorable years (fig. 7.14).

In the country as a whole, selection among dairy herds or flocks of poultry of demonstrated high productivity has probably played a major role in raising the general level.

Character Interactions

The greatest obstacles to evolutionary progress arise from the interactions among characters and the interactions of gene effects on each character. Successful selection must apply to the organism as a whole, but favorable complexes are continually being broken up by meiosis and thus cannot be selected as wholes under mass selection except insofar as recombination is reduced by linkage.

Similarly, the greatest source of difficulty in improving domestic animals has come from the necessity to select for vigor in all its aspects and fecundity in addition to that for special traits, themselves affected by interacting genes. Hazel and Lush (1942) made theoretical comparisons of three methods of mass selection for livestock improvement: (1) selection of traits in tandem, (2) simultaneous selection according to specified culling levels for each, or (3) simultaneous selection according to a total score. Each presents weighting difficulties. The authors concluded that if all difficulties could be solved, the third is most efficient, the first, least.

The greatest source of difficulty is the frequent occurrence of negative correlations among favorable traits. An especially illuminating discussion is that of Dickerson (1955) on the basis of his experience as a scientific advisor of a large and highly successful poultry farm. After 30 years of selection for a complex of economically important traits, there was still much additive variance (table 7.13).

TABLE 7.13. Percentages of additive variance due to heredity in a large commercial poultry farm (White Leghorns) after 30 years of selection.

Trait	%	Trait (4 Eggs, June)	%
Viability (0–72 weeks)	8	Albumen score	29
Egg production (survivors, 72 weeks)	32	Specific gravity	26
Age at first egg	26	Weight	37
Egg weight (all eggs in 4 days/week, March)	59	Shape	55
Blood spots (% May, June)	23	Average (all traits)	33

SOURCE: Dickerson 1955.

After following a particular program of selection from 1931 to 1953, viability and egg production had declined slightly, sexual maturity had improved slightly. March egg weight had increased through the first half of the period but then became static. The percentage of blood spots had been reduced to a probable minimum. Dickerson (1955) attributed much of the apparent genetic "slippage" from expected to observed results to environmental changes to which different genomes responded differently. There were successive cycles of disease incidence involving different neoplastic and respiratory diseases. Negative phenotype correlations between most of the favorable traits and the largely negative genotypic correlations deduced from them (vol. 2, pp. 442–45) were especially troublesome. Table 7.14 gives the observed phenotypic and deduced genotypic correlations as reported by Dickerson.

TABLE 7.14. Phenotypic and estimated genetic correlations between traits of White Leghorns on poultry farm after long selection.

	EGG WEIGHT		ALBUMEN		SPECIFIC GRAVITY		SHAPE	
	Pheno-typic	Geno-typic	Pheno-typic	Geno-typic	Pheno-typic	Geno-typic	Pheno-typic	Geno-typic
Egg production	−0.04	−0.39	−0.28	−0.32	−0.28	−0.24	−0.10	−0.50
March egg weight			+0.09	+0.51	+0.06	+0.44	−0.04	−0.09
Albumen quality					+0.06	+0.14	−0.12	+0.30
Specific gravity							+0.03	+0.54

SOURCE: Dickerson 1955.

A wild population presumably has a genetic complex that is close to a selective peak on the surface of fitness values relative to the prevailing environmental conditions. Selection for usefulness to man implies systematic changes throughout the set of selective values of genotypes and thus changes in the whole system of selective peaks. That which comes to be occupied tends to be at a rather low point relative to the original system, much lower if selection for certain special traits is far out of balance with that for major elements of fitness. The set of gene frequencies moves toward the new, artificial selective peak on the slope of which the foundation population happened to be located but, having reached this, further progress becomes impossible by mass selection in spite of much additive variance of each separate trait. The only possibility of significant progress is by somehow crossing a saddle leading to a fitness peak that is higher under the same conditions. In this respect the improvement of domestic animals by artificial selection does not differ from natural selection.

RNA

Finally I will note an experiment in which the organism was RNA extracted from bacteriophage QB, which duplicates outside living cells with the aid of a replicase, isolated from the same source (Mills, Peterson, and Spiegelman 1967). The intervals of synthesis between transfers were adjusted to select the first molecules completed. By the 74th transfer, 83% of the original genome had been eliminated, resulting in the smallest self-duplicating model known (molecular weight, 1.7×10^5).

Summary

This chapter has given examples of mass selection from widely diverse organisms (plants represented by maize; higher vertebrates represented by the mouse, rat, guinea pig, pig, and domestic fowl). Experiments with a great variety of characters have been illustrated: general size, morphologic variability, pigmentation, intelligence, fecundity, various egg characters, sex ratio, susceptibility to infectious and neoplastic disease and to dental caries, leukocyte count, resistance to X-radiation and to a carcinogenetic agent. Most of these are quantitatively varying characters, but a few were threshold characters. Nearly all experiments have been successful in producing significant, and usually large, changes. Further discussion will be deferred until after the results of selection experiments with *Drosophila* have been presented in chapter 8.

CHAPTER 8

Artificial Selection with Insects

There have been more selection experiments with *Drosophila* species than with any others, primarily no doubt because of the short interval between generations, the ease of handling, and the large numbers that can be produced at small expense, but to a large extent because of the greater possibilities for genetic analysis. This advantage is associated, however, with a peculiarity: the much greater restriction on recombination than in any of the organisms discussed in the preceding chapter. This may be expected to make the consequences of selection somewhat different. There is another feature of selection experiments with *Drosophila* that makes them somewhat more pertinent to the problem of evolution: the greater closeness of the foundation stocks to unmodified natural species than those of maize, mice, and rats. A final advantage of *Drosophila* is the great variety of meristic and quantitatively varying characters available for counting or measurement.

Selection for Number of Macrochaetae

MacDowell (1915, 1917, 1920) was stimulated by Castle's experiments with hooded rats to try the effect of selection for increased number of dorsocentral bristles in *D. melanogaster*. The normal number is four, but he found that about 1% in several wild stocks, and 5% in one, had one or more extra bristles.

His experiments began with a mating between a male with one extra bristle and a female with two from the stock referred to last. There were 55% with extra bristles among the offspring. On inbreeding from parents with extra bristles, the percentage rose to 90% and in the third generation to 97%. The average for generations 4 to 11 under continued selection was 98.3%.

The rapid progress suggested a single, incompletely penetrant recessive gene. This proved to be the case. Selection was continued to generation 49. Occasional normals appeared to the end. The mean number of extra bristles rose rapidly to generation 8 (1.5 to 3.1 in males, 1.5 to 4.4 in females) but with little further advance. Reverse selection from the second generation was effective but not from generations 15 or 26. Selection here depended on a major recessive gene and minor modifiers of its penetrance.

Sturtevant (1918), experimented with several strains of *D. melanogaster*, carrying Dichaet, a dominant mutation, lethal when homozygous, that tends to reduce the number of dorsocentral bristles from four to two and less regularly the number of scutellars. High and low selection produced small but significant changes. He demonstrated modifiers in different cases in the second and third chromosomes.

Payne (1918*a,b*), obtained much more striking results from selection for extra bristles on the scutellum from a culture of wild flies. He found one female with five instead of four bristles among 613 flies in the culture. This female was mated with a normal male. Two daughters with extra bristles, mated with normal brothers, produced 43 out of 978 with five or six bristles. Thereafter flies with as many bristles as possible were selected and were

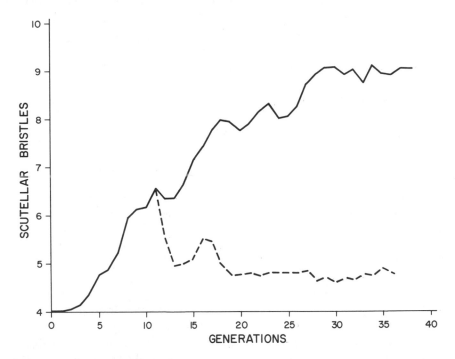

FIG. 8.1. Course of selection for number of scutellar bristles in *D. melanogaster*, starting from a single wild fly with one extra bristle. The course of reverse selection from generation 11 is shown (*broken line*). Redrawn from Payne (1918*b*, fig. 2).

usually mated brother with sister. The percentage of flies with extra bristles rose gradually to more than 50% in generation 6, to more than 90% in generation 11, and after generation 16, normals were rare or absent. The distributions of numbers of scutellars (incorrectly referred to as dorsocentrals) are shown in volume 1, chapter 10).

The average number rose to 6.5 in generation 11, but seemed to be stationary for three generations before an advance to 8 by generation 18. It then remained stationary until generation 25, rose rapidly to about 9 in generation 29, and remained there to the end (generation 38) (fig. 8.1).

There was thus the appearance of a steplike progress that has since proved to be characteristic of *Drosophila* experiments. There was, however, so much possibility of unfortunate selection of parents whose genotypes did not match their phenotypes that it is not wholly clear that more was involved in this case than fairly steady genetic progress to about generation 18 and much less thereafter. The general course may be indicated by the total responses to selection over periods of five generations (last: seven generations) in relation to the cumulative selective differentials for these periods (table 8.1).

TABLE 8.1. Gains over five generation periods ($\sum (\bar{O} - M)$), cumulative selective differentials in these periods $\sum (\bar{P} - M)$, and realized heritability h_r^2 during selection for increased number of scutellar bristles in *D. melanogaster*.

Offspring Generation	$\sum (\bar{P} - M)$	$\sum (\bar{O} - M)$	h_r^2
2–6	6.337	0.873	0.138
7–11	9.902	1.676	0.169
12–16	7.826	1.269	0.162
17–21	10.194	0.459	0.045
22–26	8.949	0.330	0.037
27–31	8.170	0.659	0.071
32–38	7.417	0.114	0.015

SOURCE: Data from Payne 1918a.

Back selections from generation 11 reduced the average from 6.5 to 5.5 at once, and to 4.7 in eight generations, but no farther by generation 35. Thus some fixation seems to have occurred.

Linkage tests demonstrated genes in the X and third chromosomes. The results differ from the preceding in the conversion of a threshold character into a typical meristic one.

A sex-linked mutation appeared in the sixth generation of plus selection that conspicuously reduced the number of bristles (4 to 0). Lines of plus and minus selection were started from matings of (4 × 4) and (0 × 0) (Payne 1920). In the minus line, selection was effective for 17 generations. It reduced the mean from 0.504 to 0.004, but it was not possible to produce a strain with no bristles even in 65 generations. In the plus line, the mean was increased from 1.95 to 3.25 in generation 18, with essentially no further change to generation 46. There were some minor changes later. Back selections in both plus and minus lines in late generations were ineffective.

Modifiers were demonstrated in the X and third chromosomes. Crosses with normal wild flies indicated that the plus modifiers were largely the same as those selected for in the original plus line in which the mutation had appeared.

In a much later experiment with the same character, Sismanides (1942) was also successful in increasing the number. His experiment was notable for the steplike advances, with responses delayed as much as 19 generations in spite of the sib mating that he practiced. Either mutation or delayed breakage of close linkages are possible explanations.

Defective Crossveins

The foregoing accounts have been concerned with thresholds in which (as in some of the cases in vertebrates discussed in chap. 7) it proved possible to make advances by selection in spite of rarity of the desired character in the foundation stock. Waddington (1955) has gone a step farther and produced strains with almost 100% of a character by selection where it was wholly absent under normal conditions in the foundation stock.

He started from a wild strain of D. melanogaster that, at normal temperatures, produced no flies with a break in the posterior crossveins. It did, however, produce some 30% of such flies on exposure of the pupae to 40°C for four hours. Selection of these crossveinless flies increased the percentage after exposure to 70% in 10 generations and to 96% in 18 generations, where it remained. Selection of flies that remained normal in spite of this treatment reduced the incidence to 15% (generations 13 to 23).

Untreated flies from the high line were examined from time to time and some crossveinless flies were found in generation 14. High selection in lines started from these ultimately raised the incidence to 67%–95% at 25°C, and 100% at 18°C. Flies from the low selection produced none at 25°C or 18°C. Crosses and chromosome studies indicated multifactorial differences (vol. 1, chap. 6).

This sort of result simulates somewhat the inheritance of acquired characters, but, as Waddington pointed out, the manifestation under special conditions is as much a result of the joint action of genome and environment on the developmental process as in any character under normal conditions.

It is merely that the character in question is ordinarily below a physiological threshold at normal temperatures. Genetic variability is nevertheless present and is capable of selection if manifestation can be induced by special conditions. Plus selection under these conditions ultimately permits crossing of the threshold even under normal conditions and further advance by selection under the latter conditions.

It is probable that the rare individuals with an extra scutellar bristle in Payne's wild *Drosophila* or the one with a rudimentary little toe in an otherwise normal guinea pig stock (Castle) or a few white hairs in an otherwise self-colored strain of mice (Goodale) showed the trait because of the conjunction of a favorable genotype with an unusual environmental condition. In guinea pigs, for example, it is known that extreme immaturity of the dam and unfavorable nutrition are conducive to appearance of the little toe in a strain with 31% incidence.

There has been a considerable number of other studies by Waddington and others of successful selection of characters that initially crossed the threshold of manifestation only under special conditions.

Bar Eye

The number of facets in the mutation Bar eye is a meristic character, but one with so many elements that it may be expected to behave much like a continuous one. Zeleny (1922) showed from the nature of the distribution and the effect of temperature that factors tend to act multiplicatively on it (vol. 1, chap. 10). He used what was essentially a logarithmic scale with "factorial units," each with 10% more facets than the preceding.

He encountered high rates of mutation back to normal and to a much more extreme type, Ultrabar, later found to result from unequal crossing-over of the short duplication that constitutes the Bar mutation. He disregarded these in his selection experiment.

The control lines were maintained by sib mating. Minus selection brought about rapid decrease in the number of factorial units in the first three generations ($h^2 = 0.31$) with considerable reduction in variability. No further change took place until generation 12, but then there was a rapid drop for two generations with no further change to the end (generation 42).

The situation was more complicated in the plus line. A marked rise (5 units) occurred in the first five generations ($h^2 = 0.32$) and then there was essential constancy to generation 21, at which time a marked rise occurred, but only in females, clearly due to a sex-linked lethal mutation (2:1 sex ratio). This mutation was lost after generation 28, but meanwhile a smaller rise had occurred in the males at about generation 25, which became recognizable also in the females after the loss of the lethal.

As in the preceding cases, the histories indicate rapid fixation of one or a few major modifiers, heterallelic in the foundation stock. Advantage was taken later of minor favorable mutations or recombinants. Second-order modifiers in the foundation stock were probably fixed largely at random in the early generations by the sib mating so that the long-continued steady progress, noted in the experiments with maize, rats, mice, with broader bases, was not to be expected.

Microchaetae

A new cycle of selection experiments with *D. melanogaster* was initiated by Mather (1941 and later). He used the number of microchaetae on the ventral surfaces of the fourth and fifth abdominal segments. This is a meristic character with so many elements (means ranging typically from 30 to 40 in males, 31 to 45 in females) that it behaves essentially like a continuously varying one.

Mather made crosses between different more or less inbred stocks and began selecting from F_2. In two experiments, the lines of high and low selection were maintained by two single-pair matings. Counts of chaetae were made on 20 of each sex from each culture when possible and the two most extreme males and females were selected to continue them, usually avoiding sib mating. In a third experiment, there were two males and two females in each of two cultures. The experiments were carried 8 or 12 generations. Selection brought about changes in the means in all lines in accordance with the directions of selection but the course and apparent limit was unique in each case. The changes appeared to occur by abrupt steps. Wigan (1941) obtained similar results from cultures derived from wild *D. melanogaster*.

The most extensive series of selection experiments from Mather's laboratory was that of Mather and Harrison (1949) (fig. 8.2). These were from F_2 of a cross between two inbred strains with about 34 and 40 abdominal chaetae, respectively. Each line consisted of two cultures each with two males and two females. The two most extreme of each sex were selected from each culture and mated with those from the other. With eight parents, the expected rate of loss of heterozygosis would be somewhat more than 0.06, depending on the numbers of sterile flies.

The low line declined somewhat irregularly from an initial 36 chaetae to 25 by generation 35, when the line became extinct from sterility. A mass culture (line 1) had, however, been taken off at generation 20 after only four chaetae had been lost. It stayed at about the same level for more than 90 generations before it was dropped. It repeatedly resisted change by selection in branch lines, in either direction, by the occurrence of sterility in a few

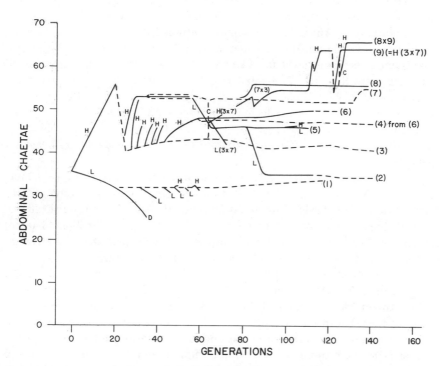

Fig. 8.2. Courses of selection for number of abdominal chaetae (high, H; low, L) along branches of a strain of *D. melanogaster*. Lines maintained without selection are indicated by dashes. Crosses (C) were made at two points. From data of Mather and Harrison (1949).

generations. The authors concluded that the low selection had created a nearly balanced situation early and that this mass line soon acquired a heterozygous genome in which both homozygotes (for some chromosome) tended toward sterility.

The history of the high line was more complex. It advanced rapidly and smoothly from 36 to about 56 chaetae at generation 20. Here, however, loss of fertility was so great that there had to be recourse to mass culture (line 3). This line rapidly lost about three-fourths of the gain by selection. Thereafter there were fluctuations of long period between means of 41 and 43 chaetae for 110 generations. This line was at first highly responsive to higher selection in branch lines. The first new high line (no. 8) started in generation 24, four generations after the relaxation. In the course of four generations it returned almost to the point previously reached but this time without the drastic loss of fertility. It advanced no farther, however, in 50 generations of high selection. There was then a rapid rise from 53 to 56 chaetae, which persisted to generation 138 when the line was discontinued.

A mass line, taken off six generations after plateauing at 53 chaetae, maintained the same level for 14 generations and was then discontinued. Another mass line (no. 7) started a generation later than the preceding, behaved similarly for about 32 generations (generation 80 from the beginning of the experiment), and then declined slightly (to 52 chaetae) at the same time that the parent line (no. 8) rose to 56 chaetae. It continued at 52 chaetae to generation 131. It then rose abruptly and averaged 55 chaetae in its last six generations.

A low selection line (no. 2) was taken off from the preceding (no. 7) at generation 56 and declined from 52 to 46 chaetae in the next nine generations. It stayed at this level for about 15 generations and then went through a second rapid decline for about nine generations to reach a level (35 chaetae) slightly below that of the foundation stock. This persisted for over 20 generations of low selection, followed by 20 of mass culture.

High selection (line 5) was initiated in the preceding line (no. 2) at the point (generation 81) at which it began its final decline. Line 5 merely maintained the previous level (46 chaetae). A low line was taken off at generation 100 but there had been no appreciable change in either it or the parent high line when they were terminated ten and six generations later, respectively. The striking point in the histories of lines 8, 7, 2, and 5 was the resistance to high selection at two levels, those of 8 and 5 (apart from a slight, long-delayed response in the former), and the rapid decline of line 2 under low selection from the level of mass line 7 to an intermediate level and the occurrence after a long delay of a second decline to the original level but no further.

Returning to mass line 3 at about 41 chaetae, seven more tests were made of its responsiveness to high selection. A line begun three generations after line 8 advanced to 52 chaetae and was discontinued. Four lines started in the next dozen generations rose to only about 46 chaetae and then stopped and were soon discontinued. One started in generation 43 (no. 6), behaved similarly, but was continued. It rose very gradually to about 50 chaetae (generation 112) and was carried for some dozen generations more without change as a mass line. A mass line (no. 4) that had been taken off early (generation 60) remained at 46 to 48 chaetae for nearly 80 generations.

Crosses between two static lines, no. 3 (about 42 chaetae) and no. 7 (about 53 chaetae), gave intermediates (48 chaetae) that on high and low selection reached the parental levels rapidly and the high line (no. 9) went beyond.

This line, after about 20 generations of near-constancy at 54 chaetae (including a brief period of relaxation from the high selection), rose in two rapid steps separated by a generation of relaxation, to a higher level than had hitherto been attained (64 chaetae) and stayed there in spite of continued high selection (except for another brief relaxation that resulted in a severe but temporary decline).

A cross between this line and the long-persistent line no. 8 (56 chaetae) permitted high selection to reach a still higher level (66 chaetae) before it in turn became static. Maximum progress thus came about through a combination of selection, relaxation, crossing of sublines, and renewed selection rather than from pure high selection.

Chromosome assays of many of the lines demonstrated important differences in the effects of all of the three major chromosomes.

Correlated responses were found in various characters in addition to the frequently encountered sterility. The most conspicuous was variation in the number of spermathecae, 0 to 5, in contrast with the normal 2 of the original stock. Other differences that appeared among the lines were in numbers of sternopleurals and of coxal chaetae, in body pigmentation, and in eye form. As noted by the authors, correlated selection of linked genes probably played a greater role here than pleiotropic effects, although contributions from the latter were not excluded.

Some striking examples of delayed responses to selection were encountered by Thoday and Boam (1961) in experiments with number of sternopleural chaetae (fig. 8.3). Their lines were maintained by cyclic mating of four pairs, selected in each generation from 20 of each sex from each pair. The initial chaeta number was about 20. The histories of lines and sublines under selection were, as in preceding cases, marked by long delays and abrupt steps. In three closely related lines, each long static at 23 or 24 chaetae under high selection, there came rapid advances to new plateaus at about 30 chaetae. These histories were so similar that a common basis seemed indicated. After weighing various possibilities the authors concluded that a specific rare recombination between linked genes (+ −/− + giving a + + gamete) was probably responsible. By crossing high selection lines, including others than those referred to above, and renewing high selection, a level of 46 chaetae was ultimately attained. The general pattern resembles that observed by Mather and Harrison in their experiments with abdominal microchaetae.

Another important series of selection experiment has been conducted with abdominal microchaetae by A. Robertson and associates. The primary purpose of the first two papers (Clayton, Morris, and Robertson 1957; Clayton and Robertson 1957) was an experimental testing of the possibility of predicting the course of selection.

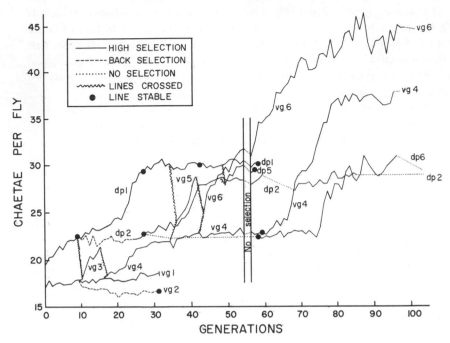

Fɪɢ. 8.3. Courses of selection for number of sternopleural chaetae along lines of *D. melanogaster*, as indicated. Redrawn from Thoday and Boam (1961, fig. 1); used with permission.

The material came from a stock derived from a few wild females and maintained for some 60 generations in a population cage (some 5,000 individuals per generation). There should have been a fairly close approach to random combination among all but rather closely linked loci since two loci with 5% crossing-over in females (none in males) go halfway toward random combination in about 28 generations.

TABLE 8.2. Estimates of heritability with respect to abdominal chaetae of *D. melanogaster*.

Source of Estimate	Estimated h^2
Daughter-dam regression	0.54 ± 0.11
Son-dam regression	0.48 ± 0.11
Half-sib correlation	0.48 ± 0.11
Full-sib correlation	0.53 ± 0.07
Average	0.52 ± 0.06

Sᴏᴜʀᴄᴇ: Clayton, Morris, and Robertson 1957.

The initial average number of abdominal chaetae (35.3) was about the same as in Mather and Harrison's experiment. The variability was not excessive (σ about 3.3). The estimates of heritability obtained are shown in table 8.2.

TABLE 8.3. Components of variance of number of abdominal chaetae of *D. melanogaster*.

Contributions to Variance	
Genetic	
Additive	0.52
Other	0.09
Environment	0.04
Accidents of development	0.35
	1.00

SOURCE: Clayton, Morris, and Robertson 1957.

The analysis of the variance into components (table 8.3) indicated that those due to dominance and interaction were relatively small. There was not much of a true environmental contribution but much from accidents of development.

Individual selection and various modes of family selection were tested. Individual selection of 20 males and 20 females was conducted at four levels: 20% (20/100), 26.7% (20/75), 40% (20/50), and 80% (20/25), with five replicates in the first and three in the others (fig. 8.4). Five control lines were maintained in lines of 20 of each sex. There was considerably more differentiation of these than expected from accidents of sampling, probably implying that the effective number of parents in each case was considerably less than 40 (fig. 8.5).

The expectation from the regression of offspring on midparent in the foundation stock is $\Delta M_O = b_{O\bar{P}} \Delta M_P$ (vol. 2, eq. 5.123) where ΔM_O and $\Delta M_{\bar{P}}$ are the deviations of the means of the offspring and parents from population mean. The regression coefficient $b_{O\bar{P}}$ is an estimate of the heritability, and the deviation of the parental mean if there is a truncated selection of the extreme portion p of an assumed normal distribution, is $(z/p)\sigma$ (vol. 2, eq. 5.126), giving $\Delta M_O = h^2(z/p)\sigma$ or $h^2 i \sigma$ if $i = (z/p)$ is written for the intensity. The aspect of the theory to be tested is whether the realized heritability (the ratio of the gain under selection to the cumulative selective differentials of the parents) agrees with the heritability, h^2, estimated from the offspring-midparent regression, or the proportion that the additive variance forms of the total, from data from the foundation stock.

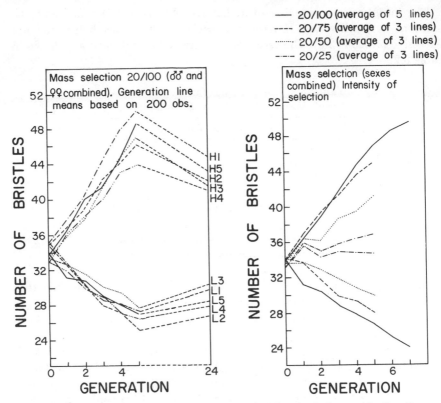

FIG. 8.4. Courses of selection for number of abdominal chaetae in five lines of *D. melanogaster*, in each direction, followed by 19 generations of relaxation (*left*). Courses of selection (high and low) at different intensities (*right*). Redrawn from Clayton, Morris, and Robertson (1957, figs. 1 and 3).

The observed changes under selection in the first five generations proceeded more regularly than in the preceding experiments, as expected from the larger numbers of parents, even allowing for effective numbers less than 40 (table 8.4).

The observed average response under high selection agreed fairly well with expectation except at the lowest intensity, but was systematically much lower for low selection, and actually reversed at the lowest intensity. There was considerable differentiation of replicates.

This asymmetry may be accounted for in various ways; to a large extent it may be a matter of scale since the coefficient of variability was more nearly constant than the standard derivation in this period. There is also the possibility that low selection was opposed more by natural selection than

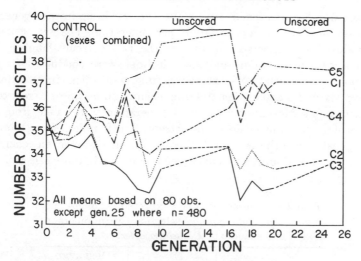

FIG. 8.5. Differentiation among five unselected lines of *D. melanogaster* with respect to number of abdominal chaetae. Redrawn from Clayton, Morris, and Robertson (1957, fig. 8).

was high selection, a matter not allowed for in the formulation above. This was tested by suspending selection for 19 generations after the fifth generation of selection. The high line lost 36% of the gain under strong (20/100) selection and the low line lost only 28%. The fact of loss in both cases indicates that natural selection of some sort was operating against the artificial selection in both directions but not more against low than high selection. There were smaller losses of gains after weak selection. The causes of the asymmetry were thus not wholly clear.

Experiments were also made with family selection of various sorts, in which case $\Delta M_O = h_f^2 i \sigma_f$, where f refers to the family treated as a unit. Ten

TABLE 8.4. Observed and expected changes per generation from individual selection of number of abdominal chaetae in *D. melanogaster*.

		CHANGE PER GENERATION		
SELECTION	INTENSITY		Observed	
p	$i = (z/p)$	Expected	High	Low
20/100	1.40	2.42	2.62	1.48
20/75	1.24	2.14	2.20	1.26
20/50	0.97	1.68	1.46	0.79
20/25	0.35	0.61	0.28	−0.08

SOURCE: Clayton, Morris, and Robertson 1957.

single-sire families, ten females per male, were made up in each generation, scoring ten of each sex in each family per generation. The next generation was produced by unscored members of the two extreme families. Three replicates were made. High and low selections were conducted for seven generations. In a second series, there was full-sib selection. In each line, 240 offspring (six males and six females from each of 20 matings) were scored in each generation. Matings for the next generation were made from equal numbers of unscored members of the extreme four families. Again there were three replicates in each case and the lines were carried for seven generations, followed by relaxation. A small experiment combined family selections with sib mating.

TABLE 8.5. Observed and expected changes per generation from family selection of numbers of abdominal chaetae in *D. melanogaster*.

| | | OBSERVED | |
	EXPECTED	High	Low
Half-sib families	1.33	1.38	0.94
Full-sib families	2.02	1.62	1.36

SOURCE: Clayton, Morris, and Robertson 1957.

Again there was better agreement in the case of the high than low selection (table 8.5). Crosses between high and low lines gave results close to the base populations, and backcrosses to the latter were close to the intermediate results expected from additive effects.

On the whole, the early generations of selection agreed moderately well with expectation from the base population, after allowing for scale effect and for antagonistic natural selection.

In later generations of the initial experiment the rate of gain slowed down and in most cases reached a plateau by generations 20 to 30 (fig. 8.6) (Clayton and Robertson 1957).

This was to be expected if fixation was occurring, but actually there was persistence and, indeed, great increase of relative variability, much of it demonstrably genetic. The situation in each line was, moreover, unique. Much of the lack of progress in spite of strong genetic variability was clearly due to greatly increased frequency of lethals of which the heterozygote was more extreme in the direction selected than the viable homozygote. Two high lines were found to carry lethals in the second chromosome and these and a third carried ones in the third. Two low lines carried ones in the second and two different ones in the third. In one case in a high line, the heterozygotes had 22 more bristles than the viable homozygote, a major effect that could

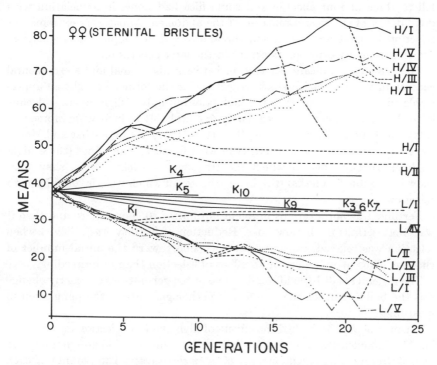

FIG. 8.6. Courses of selection, high, H; low, L; in five lines each, for number of abdominal chaetae (sternital bristles) in females of *D. melanogaster*, continuing the selection of figure 8.4. The effects of relaxation are shown by broken lines. The courses in a number of unselected lines, *K*, are also shown. Redrawn from Clayton, Morris, and Robertson (1953).

not have been present in the base stock with standard deviation only 3.7. This gene, or an enhancing modifier, may have arisen by mutation or by breaking of a close repulsion linkage and then have been selected in the heterozygotes.

The high line that rose most slowly in the first 20 generations became equal or superior in generation 30 to the line that had risen most rapidly, the other two having been discontinued after plateauing at relatively low levels after generation 20. There was a striking example of delayed effect in one of the lines of low selection. This declined more slowly than the others up to generation 22 but then declined precipitately to a much lower level. One remarkable feature of all five lines of low selection was that at some point the females dropped well below the males instead of being well above them as in the foundation stock and all of the high lines. The average for females fell to three or four chaetae and some flies had none, in association with gross defect of the sclerotinization of the abdomen. Apparently extreme low selection brought about a serious breakdown of developmental regulation. There were many other peculiarities in the later generation.

There was a significant correlation between abdominal and sternopleural chaetae in the foundation stock. Selection for the former brought an appreciable advance in the latter in the high lines but only a slight average decline in the low lines, with much irregularity due probably to increasing inbreeding (Clayton, Knight, Morris, and A. Robertson 1957). As in Mather and Harrison's experiment, extra spermathecae were found in some lines (three high, no low). In two lines in which the frequencies reached 40%, they were shown to be due to a lethal, the heterozygotes of which were favored by high selection.

Fertility was not a serious problem until after 20 generations, and then it was more serious in the low lines. Reduction of fertility was relieved when selection was relaxed, causing some reversion toward the initial number of chaetae. Thus, it was more of an effect of selection than of inbreeding.

The general result beyond the first five or ten generations of orderly change was the building up of one or another pathologic state of the population in most lines.

Rasmussen (1955) studied the effects of high and low selection on the number of sternopleurals and in other lines on the number of abdominal chaetae in four freshly caught wild strains of *D. melanogaster*. The parents of each generation consisted of two extreme males and two extreme females, selected from 20 or 25. Two long inbred laboratory strains did not respond appreciably to either high or low selection.

With respect to sternopleurals, high selection caused much more percentage change than low selection in three of the strains but not the others. The amounts varied enormously. There were pronounced steps in two cases. The decline under low selection tended to be rapid but to reach a level much sooner (8 to 12 generations) than the high selection.

With respect to abdominal chaetae, it was the low selection that declined most in three of the four strains in contrast with the results of Clayton, Morris, and Robertson (1957). In all cases, there was rapid decline under low selection in the early generations. In three of them a constant level was soon reached but not in the others, which declined rapidly for 11 generations. Relaxation for a generation because of low fertility caused a 50% reversion, but renewed low selection brought renewed long-continued decline. Seriously low fertility was encountered in five lines of low selection and two of high selection.

The effect of relaxation was investigated at the end of all of the selection experiments. There was little or no immediate effect. Ten of the lines were then maintained for a year without scoring. The two high sternopleural lines returned almost to the original average, while the three high abdominal lines changed little. The two low sternopleural lines showed slight but significant increase and the same was true of the three low abdominal lines. There seemed to have been less maintenance than usual of favorable heterozygotes of genes that were unfavorable when homozygous.

Crosses between high and low lines indicated that genes favoring an extreme were more nearly recessive than dominant.

The determination of the major portion of the nongenetic variance by accidents of development has been noted. Thoday (1958a) made experiments to determine whether selection in either direction tends to disturb developmental homeostasis as indicated by right-left asymmetry in sternopleural counts. He started from a stock derived from a single wild female. Replicated high, low, and control lines each consisted of four single-pair cultures from flies selected from 20 of each sex. Mating was cyclic to avoid close inbreeding.

Progress in the selected lines was unusually regular, resulting in a gain of five or six chaetae in the high lines in ten generations, losses of three and five, respectively, in the two low lines, and little change (from 17) in the two control lines. The selected lines all showed increases in the amounts of asymmetry, while there were no significant changes in the controls. It appears that selection does tend to disturb developmental homeostasis.

Body Size

The effect of selection on a different sort of character, body size, as indicated by the lengths of wing and thorax, has been investigated by F. W. Robertson and E. C. R. Reeve (1952a; Reeve and Robertson 1953). They started from decidedly narrow bases of potential variability, stocks derived from single wild females, carried for ten generations in mass cultures.

In their first series of experiments, selection was for long or short wing length, or for long or short thorax length in a stock from a fly caught at Nettlebed, Scotland. The lines at first consisted of only three cyclically mated pairs in each generation. The selected flies were the most extreme from 20 of each sex measured in each culture. After 12 generations, three of each sex were selected from each of three cultures and thus potentially 18 parents per generation instead of six.

Similar experiments were started later from a stock tracing to a single fly caught in Edinburgh. Unselected controls were measured in the Nettlebed lines in a few early generations, and regularly for later generations. Controls were measured from the first in the Edinburgh lines.

An analysis was made of the squared coefficient of variability with respect to wing length in the unselected Nettlebed stock on the basis of the regression of offspring on midparent and in comparison with variability in inbred lines (table 8.6).

TABLE 8.6. Variance components of wing length in *D. melanogaster*.

	$(CV)^2$	%
Genetic, additive	1.02	32
Genetic, other	0.46	14
Environmental	1.72	54
Total	3.20	100

SOURCE: Reeve and Robertson 1953.

There is a striking difference from the situation with respect to abdominal chaetae in the nature of the nongenetic component. Wing length was greatly affected by conditions common to cultures, in contrast with chaeta number, for which the same authors found only 2% of this sort but 58% from nongenetic factors peculiar to individuals.

Selection brought about substantial changes before plateaus were reached. There were great differences in the amounts of direct change, of correlated changes, and in the courses followed (table 8.7).

Selection for short wing was three times as effective as for long wing in the Nettlebed stock, only slightly more effective in the Edinburgh stock, while

TABLE 8.7. Percentages of change in wing and thorax length of *D. melanogaster*, achieved by generations 48–50 Nettlebed and 28–30 Edinburgh, averaging the sexes. The ratios of final to initial variance are also given. Correlative results are shown in parentheses.

SELECTED CHARACTER	Stock	Wing	Thorax	VARIANCE FINAL/INITIAL Wing	Thorax
Long wing	Nettlebed	+7.2	(+7.3)	1.5	(1.3)
Short wing	Nettlebed	−23.5	(−11.4)	3.8	(3.0)
Long thorax	Nettlebed	(+5.4)	+10.8	(0.9)	1.0
Short thorax	Nettlebed	(−9.0)	−8.2	(1.8)	2.2
Long wing	Edinburgh	+11.5	(+10.5)	1.7	(1.6)
Short wing	Edinburgh	−14.2	(−10.8)	1.0	(1.3)

SOURCE: Robertson and Reeve 1952a.

selection for long thorax was slightly more effective than for short thorax (Nettlebed). Wing and thorax length were strongly correlated (0.75) in the unselected stock, so that strong correlative effects of selection are not surprising. Selection seems to have been almost wholly for general size in all cases except for short wing, Nettlebed.

The striking increase in the variance in most cases, in spite of cessation of progress, parallels the results in most of the other cases and poses a similar problem.

The selection for long wing, Nettlebed, was studied intensively (Reeve and Robertson 1953). A first plateau was reached at about generation 20. A second rise began ten generations later and at generation 40 reached the level that persisted to the end (generation 76). There was much heterozygosis at the first plateau, as shown by rapid loss of all of the gain on back selection. Relaxation of selection at generation 27 also brought rapid loss of all of the gain, indicating not only much heterozygosis but also that the gain had been made against rather strong natural selection. Later (generations 37, 43, and 66), after the final level had been nearly or quite attained, relaxation brought about loss of only about half of the total gain and back selection of one of these lines carried it no closer to the original level, indicating that the delayed gain had been fixed.

The phenotypic variance rose from the first and became much greater than that of the unselected stock by generation 20, but declined to somewhat less on either relaxation or back selection. Parent-offspring regression indicated that at the final level this additive genetic variance had risen from 1.02 units to 2.50. This fell to 1.80 on relaxation.

The viability of eggs after attainment of the final level fluctuated about 25%, much less than the 56% of the contemporary controls.

Tests with marker genes indicated the curious situation that some of the flies were homozygous for a third chromosomal gene that was lethal in the background of the tests. Detailed analysis indicated, indeed, that two linked lethals or near lethals with favorable heterozygous effects on size were being carried along, and that there was at least one other gene (or block) with heterozygous superiority in size over both homozygotes. It appears that the first effect of the strong selection had been to increase the frequency of heterozygotes of these sorts. Only after many generations was there either fixation of genes with minor positive effects on size, or more probably in view of the delay, release from repulsion linkage and fixation of a gene (or block) with a major positive effect.

There was also intensive study of lines selected for short wing. The Nettlebed line fell off in size gradually for some 15 generations and then rapidly for the next 15 generations, followed by attainment of a limit at about 40 generations. In contrast with selection for long wing, relaxations (generations 27 and 37) brought no change, and back selection (generations 23 and 41) brought little. It appears that homozygosis was being approached. In spite of this, the variance rose above that of the unselected stock from the first, and to an enormous extent after generation 20, and was not much reduced either by relaxation or back selection in the late generations. It must have been due largely to greatly increased susceptibility of homozygotes of small size to environmental fluctuations. A strong genotype-environment interaction is thus indicated.

Robertson and Reeve (1953) and Robertson (1954) studied the net effects of the major chromosomes in all combinations in comparison with those of the unselected stock. A line taken off from the Nettlebed short-wing line after progress had ceased was carried through 40 generations of sib mating. A line from the unselected stock had been carried through 100 generations of such mating. The former had wings 23.2% shorter and thorax 9.5% shorter.

Crosses were made with a strain carrying dominant markers that largely prevented crossing-over in each of the major chromosomes, and by suitable matings and tests, strains were produced that were homozygous in all eight combinations, including the two already at hand. All 27 female genotypes and all 18 male ones were produced and measurements made under similar conditions. Using subscript s for the chromosomes from the short-winged Nettlebed line and no subscript for those from the unselected line, the differences between X_sY and XY were closely similar in all cases to those between X_sX_s and XX. There were no maternal effects. Evidence indicated that there were no important differences in the fourth chromosome. Table 8.8 shows the average differences from the unselected stock.

TABLE 8.8. Average differences (0.01 mm) in wing length in *D. melanogaster* between all combinations of the major chromosomes derived from the short-winged Nettlebed line, relative to the unselected line.

	XX, XY			XX_s			X_sX_s, X_sY		
	II II	II II_s	II_s II_s	II II	II II_s	II_s II_s	II II	II II_s	II_s II_s
III III	0	+2.2	−1.9	−3.7	−3.7	−8.3	−23.5	−25.3	−31.2
III III_s	−0.6	−0.7	−5.2	−4.1	−5.1	−10.1	−27.0	−25.9	−26.6
III_s III_s	−10.0	−10.5	−16.1	−14.0	−15.1	−22.3	−32.5	−26.6	−44.0

SOURCE: Robertson 1954.

The chromosomes of the unselected line tend to be dominant, but the degrees of dominance vary in different combinations. There are some rather strong interaction effects.

Assuming that X_s, II_s, and III_s were comparably rare in the foundation stock, selection would have tended to increase the frequency of X_s most rapidly, III_s next, and II_s most slowly. The near-recessiveness of X_s and II_s would account for the slowness of progress for 15 generations, followed by fairly rapid progress in the next 15 generations, leading to near-fixation of both. The fixation of III_s would be considerably delayed by the peculiar interaction effects in X_sX_s.

Selection for long thorax (Nettlebed) started very slowly, but after generation 20 proceeded rather rapidly to a plateau at generation 40. Selection for short thorax had gone farther by generation 20, but seems to have reached its plateau five or six generations later.

High and low selection for wing length in the Edinburgh stock proceeded more uniformly, with the rate somewhat greater down than up. Plateaus were probably reached four or five generations before discontinuance at 29 generations. There was no loss from relaxation in the high line at generation 14, while there was complete return of the low line to the unselected level on relaxation at this time, indicating opposition of strong selections in this case. Relaxation at generation 20, on the other hand, caused only slight reversion, indicating fixation in spite of the antagonistic natural selection.

Robertson (1954) made a chromosome study of the Edinburgh short-wing line similar to that in the Nettlebed short-wing line. There had been much less progress, $X_sX_sII_sII_sIII_sIII_s$ scored only -13.9 relative to the unselected Edinburgh line in comparison with -44.0 in the corresponding Nettlebed line. The only important effect was that of III_sIII_s (-10.3 by itself), much as in the Nettlebed case. There were differential effects of the other chromosome pairs but those were negligible, except in $X_sX_sII_sII_s$ (-4.7 with III III, -4.6 with III III_s, and -3.6 with III_sIII_s). While percentage of viability of eggs remained high (about 70%) throughout high selection, it fell to only 15% in generation 17 of the low selection line. Selection must have largely operated up to this time to increase the frequency of III_s, with largely deleterious effects on egg viability. There was no systematic rise in egg viability under continued selection but, as noted earlier, fixation must have been reached by generation 20 to account for the failure of relaxation at this time to lead to reversion in either wing size or egg viability. There was not much progress after generation 20 and what there was could have been due to the very slow increase in frequencies of X_s and II_s.

A more consistent picture of the effects of selection on body size in *D. melanogaster* emerged from a number of later experiments by Robertson and Reeve (Robertson 1955). As before, each started from a single impregnated female, one from Ischia, Italy, one from Renfrew, Scotland, and one from Crianlarich, Scotland. The Nettlebed experiment was repeated. In all of these, thorax length was used as the indicator of size. The populations consisted of 20 pairs per generation, selected in each case from 100 of each sex, with selection equally distributed among five cultures. With a lower selection intensity (0.20 instead of 0.05 or 0.15), there was less pressure to increase the frequencies of genes favorable to size in opposition to very strong natural selection.

Even in the earlier experiments, moreover, the course of selection for thorax length had been somewhat more orderly than for wing length, which was probably more subject to the effects of special factors, other than those for general size. As shown by figure 8.7, for the three new stocks, there was fairly regular progress of both high and low selection and fairly regular

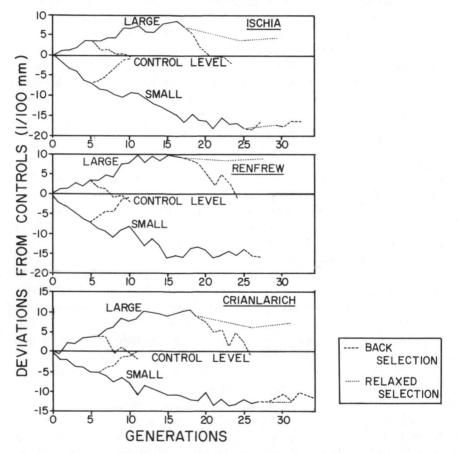

FIG. 8.7. Courses of direct, back, and relaxed selection for large or small thorax in three lines of *D. melanogaster*, each tracing to a single wild female. Redrawn from F. W. Robertson (1955, fig. 1), © 1955 by Cold Spring Harbor Laboratory; used with permission.

attainment of plateaus in 10 to 15 generations. The advances were about 8% to 10% upward under high selection and 14% to 17% downward under low selection. The new Nettlebed lines behaved similarly. In each of the experiments shown in figure 8.7, back selection at generation 5 led to rapid return to at least the level of the control line. This was also true of back selection from the high lines in generations 15–17, indicating that there was still abundant heterozygosis after progress had ceased. Relaxation at this time led to partial loss of the previous gains, indicating considerable antagonistic natural selection. In the low lines, on the other hand, neither back selection nor relaxation after progress had ceased caused appreciable change, indicating fixation.

The variability of the high lines remained roughly constant, reflecting permanent heterozygosis, while variability increased in the low lines in spite of fixation, which again must be interpreted as due to greatly increased sensitivity of the homozygous small-sized flies to environmental differences (fig. 8.8).

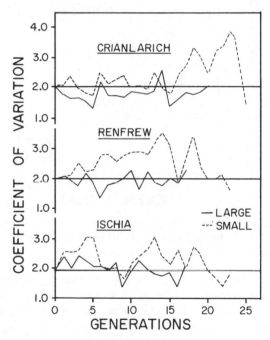

FIG. 8.8. Coefficients of variation during the selection experiments of figure 8.7. Redrawn from F. W. Robertson (1955, fig. 3), © 1955 by Cold Spring Harbor Laboratory; used with permission.

In all of these respects there was agreement with the earlier results, except for the greater regularity of the new series.

The percentage of eggs that produced adults declined considerably (70% to 40%) in the Renfrew and Crianlarich lines, but not much (55% to 45%) in the Ischia lines (fig. 8.9). In the two former instances, decline in the low line preceded that in the high line but the end result was about the same. There were no consistent differences between the Ischia lines.

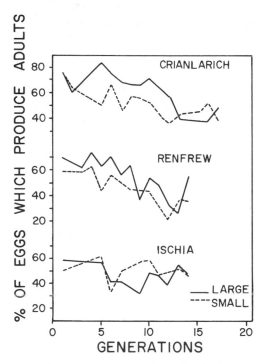

FIG. 8.9. Percentages of eggs that produced adults during the selection experiments of figure 8.7. Redrawn from F. W. Robertson (1955, fig. 4), © 1955 by Cold Spring Harbor Laboratory; used with permission.

The authors interpret the greater amount of change under low selection as due mainly to the greater number of ways of weakening a developmental process than of strengthening it. The asymmetry with respect to persistent heterozygosis upward in contrast with ultimate fixation downward is probably due to the tendency toward positive heterosis often associated with lethality of the homozygotes of the favored gene. Such heterozygotes tend to be favored in high selection but opposed in downward selection.

The chromosomal analysis suggests that heterosis here relates more to chromosomal blocks than to single loci. Assuming that there is a prevailing tendency for genes with a positive metabolic effect to have a dosage-effect curve in which effect tends to fall off as dosage increases (vol. 1, chap. 5), and that such genes tend more to favorable than unfavorable effects on growth, coupling of genes with like effects merely simulates genes of double effectiveness but repulsion linkage tends to give heterosis. Thus if *aa, Aa*, and *AA* have plus effects in the ratio 0:2:3, and *bb, Bb, BB*, similarly, coupling gives blocks with effects in the ratio 0:4:6 but repulsion gives blocks with effects in the ratio 3:4:3. In stocks based on single flies with little subsequent opportunity for random combination of linked loci, and with severe selection, apparent near-dominance and overdominance of chromosomal blocks would be expected to be characteristic, whereas in a large natural population with a closer approach to random combination even of closely linked loci, and with only relatively weak selection, incomplete dominance would be expected to be the rule.

Genotype-Environment Interactions in Directional Selection

F. W. Robertson (1960*a,b*) selected *D. melanogaster* for large body size on different media, one much more favorable than the other, and tested both lines and unselected controls on both media. There was less deviation from the controls in both cases on the strange medium than on that on which the line had been selected and presumably become better adapted.

Druger (1962), using *D. pseudoobscura*, selected lines for short and long wings at 16°C (favorable) and 25°C (unduly high) and tested the response in each of 28 generations at 16°C, 19°C, and 25°C. Plus selection at 25°C led to responses that did not differ significantly at the three temperatures and the same was true of minus selection at 16°C. On the other hand, minus selection at 25°C and plus selection at 16°C both gave lines that showed significantly greater dispersion at the temperatures of their selection than to the other temperature. Selection brought about marked changes in the frequencies of inversion patterns of chromosomes with respect to which *D. pseudoobscura* is strongly polymorphic.

Frahm and Kojima (1966) selected lines of *D. pseudoobscura* for large and small body weights at two densities: 50 eggs per bottle and 200 eggs per bottle. Plus selection at density 50 and minus selection at density 200 performed better at both densities than the opposite combinations in the early generations, but there were no significant differences later.

Thus there is no universal rule. Lines selected under given conditions tend to respond similarly under other conditions, but if there is a difference a line tends to show the effect of selection more under the same condition as those of the selection than under different conditions.

Selective Differences in Behavior

De Souza, da Cunha, and dos Santos (1968) observed marked differences in the percentages of pupation outside the food cups in a number of cage experiments with *D. willistoni*. Selection according to site was effective in each of 40 lines. After six generations, more than 90% of the pupations were in the selected site in all cases. There had clearly been an adaptive polymorphism in the foundation strains.

Carson (1958) studied the response of strains of *D. robusta* to selection for phototaxy and motility. Ten strains traced to matings of single pairs of wild flies from the periphery of the range (northwest, Nebraska), in which nearly all wild flies (and all in the 10 strains) were homokaryotic, and 15 strains from the Missouri Ozarks, where so many were heterokaryotic that he estimated that only 67% of the euchromatin was available for crossing-over.

The selection consisted in giving a group of 50 flies of the same sex the opportunity to move horizontally toward the light from one half-pint bottle to another, during 15 seconds, and similarly from the second to third bottle and so on, until only two flies moved. Two such selections were made of each sex to obtain eight parents for the next generation. Each generation of the 25 lines had an unselected control.

The controls for the Nebraska lines made, on the average, 4.7% of the total moves per generation made by flies of the strain (controls and selected) in the six generations. The corresponding figure for the Missouri controls was 5.6%. There was highly significant progress by selection in both groups, but the Nebraska lines went further (15.5% ± 1.8%) than the Missouri lines (12.3% ± 2.0%). The author attributed the greater success in the homokaryotic lines to the greater amount of free crossing-over.

Hirsch (1962) devised a maze for studying geotaxis. It was arranged so that flies put into a tube at the middle height would make 15 successive choices of upward or downward movement and ultimately emerge into one or the other of 16 tubes of which no. 1 was uppermost and no. 16 lowermost. The number defined the grade of geotaxis. He showed that selection of *D. melanogaster* was effective in either direction and found that at least three of the four pairs of chromosomes were involved.

Dobzhansky and Spassky (1962, 1967a) made use of the same type of maze to study selection for geotaxis in *D. pseudoobscura*. They studied two homokaryotic lines, CH/CH and AR/AR, and a heterokaryotic line with 50% of each arrangement. These were derived from 12 strains of each sort. In each generation, 25 flies were selected from 250 of each sex according to their grades (table 8.9 and fig. 8.10).

TABLE 8.9. Grades of geotaxis, under selection in each direction, of strains of *D. pseudoobscura.*

	Plus Selection			Minus Selection		
Generation	AR/AR	CH/CH	AR+CH	AR/AR	CH/CH	AR+CH
0	8.7	9.6	9.2	8.7	9.6	9.2
9	10.6	10.6	11.6	5.9	6.1	4.9
18	11.9	11.9	12.8	3.9	5.0	...
(Relaxation, 11 generations)	9.9	11.6	11.6	4.2	5.9	...
(Reversal, 7 generations)	10.4	7.2	9.3	8.3	7.0	...

SOURCE: Dobzhansky and B. Spassky 1962.

After generation 8, the chromosomes of 150 larvae of each line of the heterokaryote populations were examined. The plus line was found to be 66% AR. The minus line had become 100% AR. It was discarded and a new minus line was started from a cross between the others, giving a grade of 10.0, which

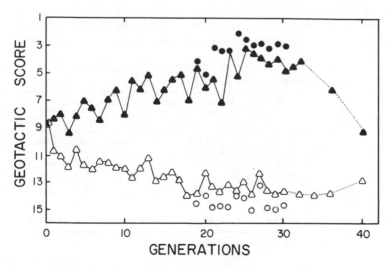

FIG. 8.10. Courses of selection in each direction for geotaxis in *D. pseudoobscura.* Mean scores of retests of 100 "best" flies (*solid circles,* minus; *open circles,* plus). Relaxation of selection, *dotted lines.* Reprinted, by permission, from Dobzhansky and Spassky (1969).

fell to 7.1 in seven generations of minus selection. It had again become 100% AR. At this time, the plus line, after 17 generations, was found to be 53% AR. It appeared that the AR chromosome strongly favored negative geotaxis while AR/CH was most favorable for positive geotaxis.

Reverse selection for 7 generations went far toward restoring neutral geotaxis and even relaxation carried the grade back significantly. In a later study, Dobzhansky and Spassky (1969) found that relaxation for 10 generations after 30 generations of selection in both directions reduced the divergence to 25% of its value. Relaxation of the plus line for 11 generations did not greatly change the chromosome composition (50% AR, 50% CH) but after 7 generations of back selection (after 18 of plus selection) there were about equal numbers of AR/CH and CH/CH but no AR/AR, indicating that the AR chromosome had become lethal in this case.

Selection was slightly more effective in the heterokaryotic lines than in the homokaryotic ones, contrary to Carson's results in *D. robusta* with chromosomes from different sources.

Hadler (1964) constructed a maze for grading phototaxis, analogous to that designed by Hirsch for geotaxis, and demonstrated the efficiency of selection of *D. melanogaster* in either direction.

Dobzhansky and Spassky (1967*a,b*) studied selection for this trait in *D. pseudoobscura* in a maze of this type, in comparison with the selection experiments on geotaxis. In this case, the 25 best flies of each sex were selected from 300 of each. All populations were started with 50% AR, 50% CH, tracing to the same wild strains as before (table 8.10 and fig. 8.11).

TABLE 8.10. Comparison of results of geotactic and phototactic selection in strains of *D. pseudoobscura*. Realized heritability h_r^2.

	GEOTACTIC SELECTION				PHOTOTACTIC SELECTION			
	Plus		Minus		Plus		Minus	
	Mean	σ^2	Mean	σ^2	Mean	σ^2	Mean	σ^2
0	8.5	16.6	8.5	16.6	8.7	8.3	8.7	8.3
9	11.9	10.5	5.6	10.5	12.9	3.7	5.4	7.0
17	12.7	9.0	4.5	13.3	14.1	2.6	3.2	4.9
h_r^2								
(15 generations)	0.026		0.029		0.100		0.084	

SOURCE: Dobzhansky and Spassky 1967*b*.

The results for the sexes are here (as before) averaged, although in both respects males made slightly lower scores than females.

With respect to chromosome arrangements, unselected populations were in equilibrium at about 75% AR, 25% CH. After 15 generations of geotactic

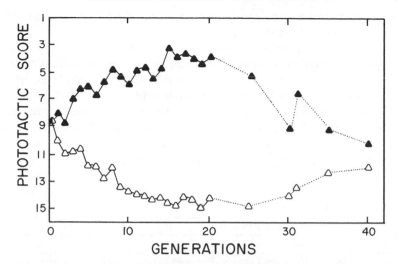

FIG. 8.11. Courses of selection (*solid triangles*, minus; *open triangles*, plus) in each direction for phototaxis in *D. pseudoobscura*. Relaxation of selection, *dotted lines*. Reprinted, by permission, from Dobzhansky and Spassky (1969).

selection, the plus line showed 85% AR, the minus only 32%. These results differ greatly from previous results, where minus selection favored AR. There was little differential effect of selection for phototaxis (plus 86%, minus 74%).

Dobzhansky, Spassky, and Sved (1969) studied the joint effects of selection and migration on these characters in comparison with uncomplicated selection on each, of the same intensity as before (25/300 of each sex). The realized heritability in the latter case was calculated for generations 1–10 and 11–20 (table 8.11). Geotactic selection was more effective than before, phototactic about the same as before.

The joint effects of selection and migration on the recipient population were found by taking the 30 out of 300 tested of each sex closest to the mode and adding the 10 flies of each sex from the current donor population that

TABLE 8.11. Comparison of realized heritabilities in uncomplicated geotactic and phototactic selection of *D. pseudoobscura*.

	GEOTACTIC SELECTION		PHOTOTACTIC SELECTION	
GENERATIONS	Plus	Minus	Plus	Minus
1–10	0.069	0.065	0.108	0.098
11–20	0.002	0.040	0.044	0.060

SOURCE: Data from Dobzhansky, Spassky, and Sved 1969.

were most extreme in the opposite direction from the 25 selected to continue the donor line (table 8.12).

In both cases the recipient populations progressed in the same direction as the donor populations, in spite of the merely average characters of the 60 parents selected from them and the opposite character of the 20 migrant flies. Because of the low heritabilities, the heredity transmitted by migrants was, however, largely that built up in the donor strains; the heredity transmitted by average flies of a recipient strain was largely that accumulated in them by the migrants of preceding generations. The mean grades estimated in this basis agreed fairly well with those observed.

TABLE 8.12. Comparison of courses of geotactic and phototactic selection, complicated by migration.

	GEOTACTIC SELECTION				PHOTOTACTIC SELECTION			
	Plus		Minus		Plus		Minus	
GENERATION	Donor	Recipient	Donor	Recipient	Donor	Recipient	Donor	Recipient
0	9.3	9.3	9.3	9.3	10.0	10.0	10.0	10.0
10	13.2	12.4	6.4	8.9	13.5	12.0	4.9	7.7
20	13.3	12.9	5.9	7.7	14.3	9.9	2.9	4.1

SOURCE: Data from Dobzhansky, Spassky, and Sved 1969.

Selection Systems

Bell, Moore, and Warren (1955) made tests of different systems of selection with *D. melanogaster*. They started from eight wild-type strains and conducted family selection (FS) ("closed population," fig. 8.12) with a combination of these. Another system used was Hull's recurrent cross selection, RCS, between this combination and an inbred tester line. Comstock's reciprocal recurrent selection (RRS) (fig. 8.12) was carried out between a combination of four of the strains with one of the other four. Finally two independent sets of F_1 crosses, taking two lines from each of the initial stocks, were used in extensive terminal comparisons.

The effects of these systems on egg size, a trait with high heritability in mixed stocks (0.30 to 0.60) and only slight heterosis, were compared over 16 cycles of selection of the first three sorts. Adaptation to a tester (RCS) began with a remarkably high average but went somewhat further only in the later generations and ended only very slightly above the family selection (FS), which had risen almost linearly. Reciprocal recurrent selection (RRS) was less than half as effective per cycle as family selection. In the terminal tests, it was about the same as the two sets of F_1 hybrids. Neither of these last systems is expected to be as effective as family selections on a highly heritable

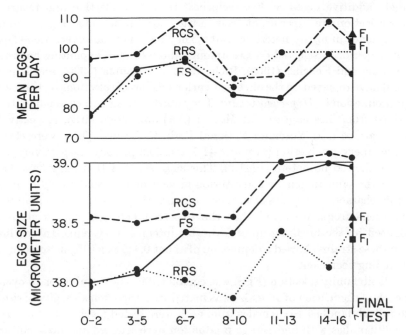

Fɪɢ. 8.12. Comparison of courses of different systems of selection of *D. melanogaster* for fecundity and egg size, and a replicated terminal test of F_1 of cross. Redrawn from Bell, Moore, and Warren (1955, fig. 1), © 1955 by Cold Spring Harbor Laboratory; used with permission.

nonheterotic trait. The high score of RCS was somewhat surprising and probably due to the exceptional merit of the particular tester strain used.

The effects on fecundity (eggs laid in four days), a trait with low heritability (5% to 15%) and strong heterosis, was obtained in the same experiment. Adaptation to the tester (RCS) again started surprisingly high, but in this case progressed no farther. In the terminal tests it averaged about the same as the F_1 hybrids and slightly above reciprocal recurrent selection (RRS), which was above family selection (FS). The reversal of the relation of the F_1 crosses and RRS to FS are as expected for a hcterotic trait.

The tests with fecundity were repeated in an experiment carried to 39 selection cycles. In this case RRS and the F_1 hybrids came out on top with no significant difference between them, and both were significantly superior to RCS, which was slightly but significantly superior to FS, which had started with much the highest score but soon reached a plateau.

With respect to application, the authors concluded that combined individual and family selection is superior for improving quantitative traits with largely additive heredity, but reciprocal recurrent selection and recurrent cross selection are superior, at least in the long run, in obtaining maximum performance in highly heterotic traits. They have not, however, been shown to be superior to selected crosses among inbred lines in exploiting heterosis.

Kojima and Kelleher (1963), Richardson and Kojima (1965), and Kojima (1969) have reported on the effects of various modes of selections on three-day egg production in *D. pseudoobscura*. They started from two cage populations derived from flies captured at Mather (AA) and Mono (BB), respectively, two locations near Yosemite National Park, California. These were closely similar in egg production (averages 41.7 and 39.2 per day, respectively) and crosses showed no heterosis (41.2). This suggests that they may have been almost the same in genetic constitution in view of the usual strong heterosis of this character. It is possible, however, that they differed considerably in some gene frequencies and that the expected heterosis happened to be balanced by the breaking up of favorable interaction systems. Inbred lines from these strains showed a depression of about 0.4% per 10% increase in the inbreeding coefficient.

Full-sib family selection (FFS) was conducted in both strains for 13 cycles. In each of them, each of 40 males was mated with three females, giving rise to 120 full-sib families. Six females from each were tested (each in two bottles). The 20 families with highest egg production were selected and mass cultures made. Two males and six females tracing to each selected family were mated, in such a way as to minimize inbreeding, to start the next cycle.

The control (C) for each cycle in these and other experiments consisted of 72 F_1 hybrid females from symmetrical interlocality crosses among nine closely inbred lines, three from Mather, three from Mono, and three from Bryce, Utah. The control average was close to that of Mather, Mono, and their F_1.

Under high FFS (fig. 8.13), average egg production rose at once to about 14% (about six eggs per day) above the controls, but showed no significant trend thereafter through 13 cycles (realized heritabilities after two cycles, Mather 0.03, Mono 0.04, on allowing for probable slight inbreeding depressions). This is illustrated in figure 8.14 in which eggs per day is plotted against the cumulative selection. Under five cycles of low selection, however, begun after cycle 10, most of the superiority over the controls was lost. The two strains behaved much alike in these respects.

Apparently there was immediate adjustment of certain gene (or chromosome) frequencies with major effects on egg production, probably at the expense of other aspects of fitness, since the strains were presumably close to equilibrium in general fitness under natural selection. The immediacy of the rise can be explained better by shifts to new equilibriums at overdominant

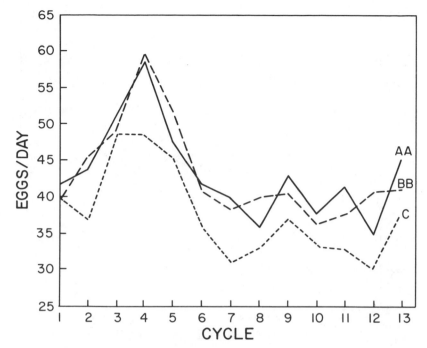

F<small>IG</small>. 8.13. Observed performance of lines of *D. pseudoobscura*, in full-sib family selection (FFS), from Mather (*solid line AA*) and from Mono (*broken line BB*), in comparison with control *C*, described in the text. Redrawn from Kojima and Kelleher (1963, fig. 3); used with permission.

loci than at fixable ones. There was apparently little or no additive variance in the strains with respect to minor genes with favorable, more or less dominant effects in egg productions.

So far there had been no clear evidence for any appreciable differences between the Mather and Mono populations. Evidence was provided by carrying the cross between them for several cycles (to F_6 and F_4 in replications) without selection, made so as to minimize inbreeding. Egg productions increased in F_2 on the average, and still more in F_3, rising to about 20% above F_1 and the controls, and then declined somewhat. Apparently the intermixing of the strains led to diverse trends among loci with respect to heterosis, recombination and the breaking up and formation of interaction systems.

Eight cycles of high half-sib family selection after six of no selection gave no further significant increase (average about 17% above the controls). Low half-sib family selection, conducted for eight cycles in parallel, caused a

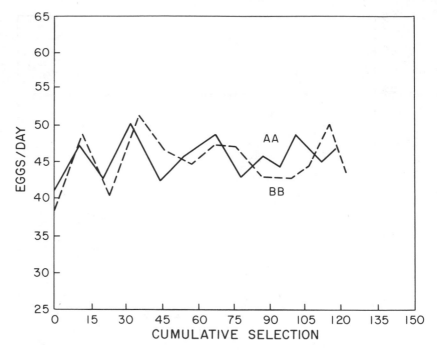

Fig. 8.14. Cumulative responses to full-family selection (FFS), relative to the controls in the experiments of figure 8.13. Redrawn from Kojima and Kelleher (1963, fig. 4); used with permission.

drastic decline to some 20% below the controls. It is evident that the cross between the two strains provided material for considerable segregation, capable of exploitation by selection.

Reciprocal recurrent selection (RRS) was tried on a scale comparable to FFS. Each of 60 males from each population was mated with two females from the other and later with three females of his own population. Six cross-bred daughters were tested as before in each full-sib family. Ten of the 60 purebred populations were selected on the basis of the tests of their half sisters, again in such a way as to minimize inbreeding.

Egg production of the RRS crossbreds rose linearly in relation to the cumulative selection (fig. 8.15) for the first 10 cycles with realized heritability 0.16. It then tapered off to a plateau in cycles 15 to 21, about 28% above the controls (fitted lines Y_1 and Y_2 in fig. 8.16). This was the greatest gain obtained by any method.

Selected families of cycle 19 were intercrossed. The crossbreds were mated at random, except for precautions to minimize inbreeding, and carried to F_4.

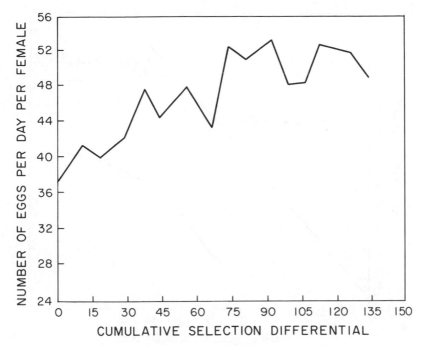

FIG. 8.15. Responses to reciprocal recurrent selection (RRS) of the Mather and Mono lines of *D. pseudoobscura*, relative to the controls. Redrawn from Kojima and Kelleher (1963, fig. 6); used with permission.

Egg production remained practically constant at about 22% above the controls. After this, high and low half-sib family selection was practiced for eight cycles.

Divergence was less than half of that which followed crossing of the original strains, random mating for several generations, and similar selection. There was no appreciable change in the line of high selection (about 22% above control) while the line of low selection declined in six cycles to the control level. It appears that the amount of variability in the pure strains, available for this sort of selection, had been greatly reduced by RRS.

It was desirable to test directly what had happened to these strains themselves because of the selections for crossbred performance. There was an insignificant decline on the average to cycle 5, followed by a sigmoid rise, slight at cycle 10, then a rapid approach to a plateau by cycle 21 at about 17% above the controls and thus at about 60% of the level of the RRS crossbreds (lower line in fig. 8.16).

In view of the failure of FFS to produce any cumulative gains after the

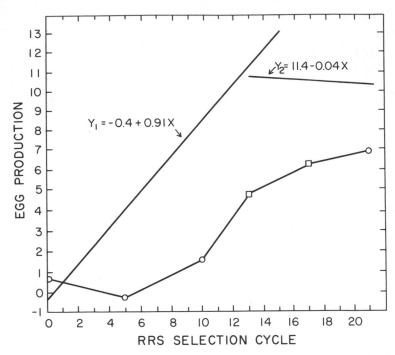

FIG. 8.16. Average deviation from control values of purebred egg production of *D. pseudoobscura*, under reciprocal recurrent selection (RRS), in comparison with results from crossbred RRS, for cycles 0–15 (Y_1) and cycles 15–21 (Y_2). Redrawn from Richardson and Kojima (1965, fig. 2); used with permission.

first cycle, it is unlikely, as already noted, that additive loci without overdominance played any appreciable role in the progress of either RRS crossbreds or RRS purebreds. Overdominant loci on the other hand, probably played a major role. If allele A_1 was relatively more frequent in the Mather population than in the Mono population and the reverse was true of allele A_2, selection on the basis of crossbred performance, assuming genotype A_1A_2 to have maximum effect, would mean a tendency to select, and ultimately fix, A_1A_1 in the Mather population and A_2A_2 in Mono. This should depress the productivity of the purebreds, unless there were a countervailing tendency.

Such a tendency could be introduced by selection based on the productivity of crossbreds with superior interaction effects. There are various possibilities. Assume that B_x–C_x– is a favorable interaction system with maximum effect when homozygous. If Mather carried B_x at a low frequency but not C_x and the reverse was true of Mono, selection on the basis of the crossbreds B_x–C_x–

would increase the frequency of B_x in Mather, leading ultimately to $B_x B_x CC$, and would increase that of C_x in Mono, leading ultimately to $BBC_x C_x$ and thus to $BB_x CC_x$ in the crossbreds. There should be some decline, as in the preceding case, in purebred Mather and Mono, with no subsequent rise. If however, Mather carried both B_x and C_x but at such low frequencies that favorable selection for the very rare combination B_x–C_x– was overbalanced by the assumed selection against B_x and C_x by themselves, and if Mono carried both B_x and C_x at fairly high frequencies, selection of Mather on the basis of crossbred performance would slowly increase both B_x and C_x, not necessarily in parallel, in Mather and also in Mono, leading ultimately to establishment of $B_x B_x C_x C_x$ in both pure strains and their cross. Another possibility is that Mather carried both B_x and C_x at low frequencies as above but Mono carried one of these at a fairly high frequency, the other not at all. The selection of Mather on the basis of crossbred performance would lead gradually to fixation of the best genotype $B_x B_x C_x C_x$ in it, but only of $B_x B_x CC$ with no interaction effect at all in Mono and $B_x B_x CC_x$ with submaximal effect in the crossbreds. At other pairs of loci, Mather and Mono may exchange roles. The net effect would be the somewhat greater average production in the crossbreds.

There are, of course, many other possibilities. A combination of those cited is enough to account for the observed results.

Under selection for egg production alone, instead of for fitness as a whole, the system of selective peak is modified. Full-sib family selection carries each population to a new peak, determined in each case by a shift in the equilibrium frequency of heterotic loci, but not by changes with respect to possible interaction effects. Material seems to be lacking for changes with respect to loci with additive effect and no overdominance. Selection from mixtures of the two strains leads to a crossbred population at another selective peak than those of the parent strains. Overdominant loci give the basis for a new equilibrium and some advantage may be taken of interaction effects. Reciprocal recurrent selection produces crossbreds that utilize overdominance as much as possible and interaction effects to a major extent, although not necessarily to quite the maximum. It produces pure strains that utilize the interaction effect, for which they themselves carry the genes, to a major extent if assisted by the presence of the necessary interacting genes in the other strain, but these strains tend toward lower productivity insofar as alternative homozygotes at overdominant loci become established in them.

In this case, the best results were obtained in RRS crossbreds in spite of the fact that the strains that were used did not apparently differ very much in their genetic constitutions.

Selection for Parthenogenesis

A case of great interest from the evolutionary standpoint is selective change in mode of reproduction. Stalker (1954, 1956) discovered the occurrence of thelytokous parthenogenesis in a *Drosophila* species, *D. parthenogenetica*, and later found it in two other species after a survey of 28 species. He was able to raise the incidence from 0.08% to 1.5% by 17 generations of selection, but there was no further increase in 45 generations. By backcrossing to the original strain and extracting twice the incidence was raised to 3.1%.

Carson (1967) found low rates in strains of *D. mercatorum* and raised the incidence in one from 0.1% to 1.0% by alternating unisexual and bisexual reproduction and from 0.004% to 0.2% in another strain. Crossing followed by selection enabled him to reach 6.4%.

Stabilizing and Disruptive Selection

All of the selection experiments considered so far have been directed toward changing the mean in one direction as far as possible. This is an extreme pattern of selection, not likely to occur often in nature since natural selection may be expected to be conservative as a rule, operating against both extremes. The optimum at any time is, however, likely to be to one side of the mean, so that the selection has to some extent a directional effect, and this may be of great importance if consistent over a long period.

In a diversified environment or in a species that has become genetically differentiated among localities, the type of selection may differ in different places. The total pattern of selection of a species is thus a much more complicated matter than that in one locality or in a simple experiment. Simpson (1944) proposed terms for three types of selection that may be combined in various ways in natural selection: *centripetal* (centrally directed), *centrifugal* (directed differently in different individuals or demes), and *linear* (unidirectional). Mather (1955) proposed terms that correspond largely to Simpson's: *stabilizing* (with *centripetal*), *disruptive* (with *centrifugal*), and *directional* (with *linear*). There was a difference in aim, however, in that Simpson's terms were purely descriptive, as best suited to the complexities of natural selection, while Mather's refer to the expected consequences in relatively simple laboratory experiments.

Centripetal selection is expected to reduce variability and thus be stabilizing in a single homogeneous population. In the broader context of an array of partially isolated demes, even though all are subject to the same selection, it tends to bring about genotypic diversity by stabilizing different gene systems of similar effect on the selected character, in different places, and thus phenotypic diversity in unselected pleiotropic effects. In other words, centripetal selection is stabilizing locally but tends to be disruptive for a species.

Centrifugal selection is, of course, expected to increase variability locally and may often be genuinely disruptive as applied to different localities. How far it may actually be disruptive locally is a matter on which some of the experiments to be discussed here give evidence.

Selection may be centripetal locally but centrifugal among localities, causing very conspicuous diversification if not actual disruption. It may, on the other hand, be centrifugal locally but not among localities, because the same sort of environmental heterogeneity extends throughout the species. In this case no disruption is to be expected.

It is evident that terms based on expected consequences are highly ambiguous when applied to natural selection. They are, however, the ones generally used in laboratory studies and it seems least confusing to use them in our discussion of these. They are sufficiently clear in meaning in this context.

There have been a number of studies of stabilizing selection of chaeta numbers with *Drosophila*. Falconer (1957) compared a line subject to such selection with an unselected control. Both were maintained by use of 20 pairs, chosen from 60 of each sex in one experiment (14 generations) and 40 of each sex in another (13 generations with two replications). The variance σ_S^2 of the sum of the bristle counts on the fourth and fifth abdominal segments and that of the difference, σ_D^2, were obtained for each sex.

The between-fly variance ($\sigma_S^2 - \sigma_D^2$) is almost wholly genetic within cultures. It might be expected to fall off under stabilizing selections with any increase in homozygosis or any balancing of genes with plus and minus effects along the chromosomes, as suggested by Mather. Falconer, however, found no difference from the controls that could be attributed to the selection. There was also no significant change in developmental stability as measured by the within-fly variance, σ_D^2. The pressures toward fixation are, however, expected to be very weak. More drastic selection of this sort seem to be required for positive results.

The most extensive studies of stabilizing and disruptive selection with *Drosophila* have been those of Thoday (1958a–c and later), using number of sternopleural chaetae. In most cases, the selection lines have consisted of only four single-pair cultures bound together by some pattern of cyclic mating.

In one series (Thoday 1959) a line of stabilizing selection, S, was maintained by scoring 20 flies of each sex from each of the four cultures and choosing the male and the female closest to the mean of each to be parents (mated cyclically) of the next generation. The mean remained essentially unchanged, but in this case there was a significant decline in variance in the course of 44 generations. There was again, however, no appreciable change in the amount of asymmetry.

He conducted two types of disruptive selection. In one, D−, the male and female at opposite extremes were selected out of 20 of each sex from each of the four cultures to be parents (cyclically mated) of the next generation. There was an unexpected increase in the mean (from 18 to 21 chaetae). Of more interest was an increase of variance (of somewhat doubtful significance, however) and a strongly significant increase in asymmetry. This line was distinctly more fertile in the later generations than the stabilizing line. The most important difference was a significantly greater responsiveness to directional selection, indicating a greater amount of genetic variance.

It is convenient here to digress to a later comparison (Thoday 1960) of stabilizing, S, and disruptive selection, D−, derived from a cross involving markers for the second and third chromosomes:

$$\frac{vg\ +++}{vg\ +++} \times \frac{+\ se\ cp\ e}{+\ se\ cp\ e}.$$

Under S, variance declined as before and this line approached the constitution

$$\frac{+\ se\ ++}{+\ se\ ++},$$

involving recombination in the third chromosome. Under D−, there was again a delayed rise in the mean and persistence of high variability. There was approach to a state in which matings were of the type

$$\frac{vg\ +\ +\ +}{vg\ se\ cp\ e}\ (\text{high}) \times \frac{vg\ se\ cp\ e}{vg\ se\ cp\ e}\ (\text{low}).$$

Thus there came to be a striking difference between populations derived from the same gene pool.

Going back to the earlier series, instructive results were obtained from a second type of disruptive selection, D+, from the same stock as S and D−. In this case, two of the four matings were between high extremes and the other two between low extremes, but the high females were selected from low × low and the low females from high × high. Using H and L for high and low, respectively, two of the matings were ♀H(LL) × ♂H(HH) and the other two ♀L(HH) × ♂L(LL). There were alternating sublines in each case. There was little change in mean or in symmetry but there was significant increase in variability. This was found (Thoday and Boam 1958) to be due to an increase in chaeta count in one H × H line (H1) and a decrease in an L × L line (L4). Chromosome analysis showed that a third chromosome with H effect relative to the others was being carried along in heterozygous state in the straight male line of H1 and a second chromosome, with L effect, relative to the others, was being carried along similarly in the straight male line of L4. Further analysis (Gibson and Thoday 1962) of the second chromosome of this line showed that there were two loci (at 27.5 and 47.5 in the chromosome map) that were generally in repulsion linkage $+ -$ and $- +$, being about equally frequent in D+, although the former prevailed in the foundation stock in the ratio 46:2. It was a recombinant, $- -$, lethal when homozygous, that was being carried along by selection in L4. The type $+ +$ was dominant lethal. It was evident that the very strong disruptive selection was able to maintain polymorphism of the second chromosome, $- -/+ -$, against the strong natural selection inherent in a recessive lethal, $- -$. Presumably a similarly vulnerable polymorphism of the third chromosome was being maintained in H1, in spite of the 50% gene flow.

An experiment was devised later (Millicent and Thoday 1961) that involved only 25% gene flow. Two of the lines alternated between H(HH) × H(HH) and H(HH) × H(LL) and two between L(LL) × L(LL) and L(LL) × L(HH) in a cyclic pattern. The results were compared with a line with 50% gene flow like D+ except that continuity of H and L was along the straight female lines.

The H × H and L × L lines with 50% gene flow diverged relatively little (by one chaeta) in replicates carried 26 and 28 generations. The isolated H × H and L × L lines (0 gene flow) had already diverged by three chaetae in generation 7 in both of their replicates and reached deviations of four and six chaetae at the end. The lines with 25% gene flow diverged more slowly than the isolated lines (one or two chaetae in generation 7), but their average divergence in the final generation was about the same (five chaetae). This experiment illustrated the great effectiveness of very strong disruptive selection (1 in 20) with 25% gene flow between high and low lines, in contrast with borderline effectiveness with 50% gene flow. The within-culture variance rose markedly with 25% gene flow, slightly with 50%, and not at all with directional selection.

Another test of the balancing of selection and immigration was made by Streams and Pimentel (1961), who used number of sternopleural chaetae of *D. melanogaster* as the character. Starting from 25 wild flies, they compared a control line A1 (four pairs) and a line B1 (four pairs), both selected for chaetae numbers (4/40), and a line C1, similarly selected but with the addition of a male and a female from A1 (five pairs, 20% from immigration), and a line D, similarly selected but with the addition of four males and four females

from A1 (eight pairs, 50% from immigration). The controls remained substantially constant at 18–19 chaetae. Pure high selection brought about an
increase to 25 chaetae in seven generations, after which the line became
extinct from low fertility. With 20% from immigration, selection produced
about as much progress as with no immigration and went farther (to 28
chaetae) because of normal fertility. One-third of the gain was lost on
relaxation. With 50% immigration, there was little progress (about one
chaeta).

Astonishing evidence for the possibility of actual splitting of a line under
disruptive selection was reported by Thoday and Gibson (1962). They started
from a stock derived from four wild females collected from the same garbage
can. In each generation assays were made of 20 virgin females and 20 males
from each of four cultures. The eight of each sex with the highest and the
eight of each sex with the lowest chaeta numbers were selected regardless
of culture. These 32 flies were put together for 24 hours, after which the
males were discarded and two groups of high females and two of low females
were isolated. Again 20 offspring of each sex were assayed from each of the
four cultures and selections made as above, and so on generation after
generation. In spite of the equal exposure of the females to both types of
males, the high and low lines diverged so rapidly that by generation 12 there
was a mean difference of nine chaetae and no overlap (low 11–18, high 20–34).
Comparisons were made of low × low, low × high, high × low, and
high × high in separate vials with four of each sex in each. The crossbreds
were intermediates 15–25, while low × low (12–17) and high × high (21–34)
gave essentially the same results as the low and high branches of the line of
disruptive selection.

Since further tests became impracticable because of much infertility, a new
experiment was started from the same base population (Thoday 1965). There
was even more rapid divergence under disruptive selection. Tests of prefer-
ential mating were made in generations 7–10 and two in 19. Extra flies were
scored, selected, and mated as in maintaining the line, but after mating the
16 females were put in separate vials to determine which type of male these
had mated with. The results were closely similar in all tests (table 8.13).

There was thus strong assortative mating but not strong enough to account

TABLE 8.13. Test of assortative mating under disruptive selection in *D. melanogaster.*

	H × H	H × L	L × H	L × L	Fertile	Sterile	Total
No.	71	15	27	62	175	17	192
%	40.6	8.6	15.4	35.4	100.0		

SOURCE: Thoday 1965.

for the complete separation between the high and low groups of offspring in the main experiment. Other tests, however, indicated that the crossbred larvae competed rather poorly and probably developed more slowly than the purebreds and so did not get into the samples collected over a standard five-day period for assay. The complete separation was thus attributed to a combination of strong assortative mating and low competitive ability of the crossbred larvae.

This experiment has been repeated by Scharloo, Boer, and Hoogmoed (1967) and Scharloo, Hoogmoed, and ter Kuile (1967) with the important qualification that their base populations were laboratory stocks of long standing instead of ones derived from wild females caught in a single garbage can. In both experiments the high and low lines diverged by about three chaetae at once and by about five or six ultimately, but never separated as in the experiments of Thoday and Gibson. The authors suggest that the complete separation found by the latter was probably the consequence of a special situation in which there happened to be a genetic correlation between chaeta number and an isolating tendency. The rather extreme conditions required for complete separation have been discussed by Thoday and Gibson (1970).

The experiment has also been repeated with four strains of diverse origin by Chabora (1968). She found significant divergence after 5 generations in one strain, 10 in another, and only after 27 in two others, but never obtained such a separation as found by Thoday and Gibson. There have been other negative results listed by Thoday and Gibson (1970).

Tests of stabilizing and disruptive selection have been made with a different character, time of development of *D. melanogaster*, by Prout (1962). Five lines of stabilizing selection, 11 of disruptive selection, and 2 control lines were compared over 40 generations. Under stabilizing selection, the variance ultimately decreased 32%, while under disruptive selection it increased 67%.

Scharloo (1964); Scharloo, Boer, and Hoogmoed (1967); and Scharloo, Hoogmoed, and ter Kuile (1967) have also studied stabilizing and disruptive selection relative to the expression of the mutant gene cubitus interruptus dominant of Gloor of *D. melanogaster*. The character used was the sum of the percentages for both wings of the length of the fourth vein beyond the cross-vein relative to that of the third vein. Four selection methods were used: (1) high and low directional (lines of four flies each), (2) stabilizing selection (S) (four flies closest to the mean), (3) disruptive selection with random mating (D^R) (two highest and two lowest flies in single bottle), and (4) disruptive selection with disassortative mating, D^- (high females with low males and vice versa in separate vials). In this case, after 24 hours, males were discarded and both high and low females put in one bottle.

High and low directional selection diverged rapidly with considerable decrease in variance in two low lines and one of two high lines. In a comparison of stabilizing selection, S1, and disruptive selection with disassortative mating D_1, the variance of the former decreased and that of the latter increased greatly in the course of 13 generations. In a second experiment, stabilizing selection, disassortative, disruptive selection, D^-2 and disruptive selection with random mating, D^R2, were compared for 35–38 generations. Table 8.14 compares the results at midcourse and near the end.

TABLE 8.14. Heritability and variance under diverse mating systems with *D. melanogaster.*

	Base	D^R2		D^-2		S2	
Generation	18	18	37	18	35	19	36
No. of pairs	40	34	23	46	19	36	31
Heritability	0.68	0.94	0.85	0.54	0.73	0.47	0.30
Additive genetic variance	157	1,154	1,398	484	1,557	25	11
Within-fly variance	26	28	32	82	100	19	19
Residual variance	48	45	213	330	476	9	7
Total variance	231	1,227	1,643	896	2,133	53	37

SOURCE: Scharloo, Hoogmoed, and ter Kuile 1967.

Under stabilizing selection, the additive genetic variance and the residual (largely environmental) variance were enormously reduced and even the within-fly variance fell off considerably, but there was still some heritability after 36 generations (0.30 ± 0.08). There was clearly much buffering. With disruptive selection but random mating, D^R2, the additive genetic and residual variances increased enormously, the within-fly variance slightly. Under disassortative disruptive selection, all three components increased enormously, but there was no disruption of the population.

Tribolium

The flour beetles of the genus *Tribolium* have notable advantages as experimental material, as demonstrated especially in their very extensive use by Thomas Park in the analysis of the regulation of population size and of the results of interspecific competition. Their ease of culturing and handling, short reproductive cycle (between one and two months), high reproductive rate, longevity (life expectancy of half a year), and conveniently measured metric traits strongly recommend them for studies in population genetics. With ten pairs of chromosomes (including XY) and typical amounts of recombination in both sexes, the genetics of *T. castaneum* is much less dominated by linkage phenomena than is that of *D. melanogaster* with only three major pairs of chromosomes and one minor, and no crossing-over in males.

We will begin with an experiment by Enfield, Comstock, and Braskerud (1966) in which pupal weight of *T. castaneum* was selected for 12 generations for increase in two lines from a cross between two highly inbred lines. Both gained about 28%. The realized heritabilities (0.37 and 0.34) were in excellent agreement with the initial sire-offspring regression (0.35) and with the estimates based on variance components (0.34). There was no indication that a plateau was being approached; the average level of dominance was estimated at 0.57 in females and 0.75 in males. With heterallelic gene frequencies all 0.50 and only partial dominance on the average, the situation was especially favorable for successful prediction of the gains for a considerable number of generations.

Bell and his associates have made an extensive series of experiments (Bell 1968) concerned with the predictability of the results from selection under different environmental conditions and from different systems of mating.

Bray, Bell, and King (1962) compared the results from eight generations of mass selection for increased or decreased pupal weight (a rather strongly inherited character) with various types of controls, intended to show which sort is most adequate. These included different ways of maintaining a foundation stock (derived in 1954 by combining eight strains from widely diverse sources). In certain controls, various precautions were taken to avoid sampling drift. Other controls consisted in repeating the sets of parental matings in the following generations. One control was subject to stabilizing selection of the mean. In other lines, selection was relaxed. There were also two lines derived independently from long-continued sib mating, and the first cross between these. In all cases, lines were maintained by means of 50 pair matings in order to make sampling drift a small factor.

Humidity was found to be an important factor but one with strong genotype-environment interaction. The pupae of the foundation stock were 10% heavier under 70% relative humidity than under 40%. The inbred families, on the other hand, responded in the opposite direction. In all lines in this experiment, two generations at 70% were alternated with two at 40%.

The changes up and down under selection were both approximately linear and symmetrical but much greater in the dry than the wet environment. There was some irregularity among families from the first in their responses to the two environments, and in the later generations the large line showed more growth in the dry environment, contrary to the average response in the foundation stock. It was concluded that, because of the genotype-environment interactions, the controls should be closely related to the selected lines in origin and time.

In a later experiment, replicated selections for increased pupal weight were conducted separately in each environment for nine generations (Bell and McNary 1963). Heritability was initially 0.58 and 0.55 in the wet and dry lines, respectively. Initial pupal weight was in this case 5% higher in the wet environment, but the variance was nearly twice as great in the dry environment as in the wet. The initial genetic correlation (vol. 2, p. 442) between the responses to the two environments was almost perfect (0.98), indicating additivity at this time. The direct responses for both replicates in both environments were in good agreement with those predicted from the initial regressions. The correlated response of the dry selected line to the wet environment was about the same as the direct response, as expected from the nearly perfect genetic correlation, but the correlated response of the wet selected line to the dry environment was much less than its direct response, and the realized genetic correlation between the responses had fallen to 0.66. It was again evident that the environmental responses were complex characters in which selection might introduce changed genotype-environment relations.

Experiments, analogous to those of Falconer with mice, were made on the responses of lines selected for larger or smaller 13-day larval weight, or randomly selected weights, under different levels of nutrition and rearing at the same or opposite level (Hardin and Bell 1967). Replicated lines were carried for eight generations in each case. The mean weights were about twice as great under the good ration as under the poor one, but the variances were about the same. Heritability was 0.21 ± 0.06 under the good ration, 0.19 ± 0.05 under the poor one, both much less than for pupal weight. The initial genetic correlation between the responses to the two rations was 0.60 ± 0.21 and thus far from perfect in this experiment.

The four replications in each series over the eight generations were consistent but their averages revealed serious deviations from the values predicted from the initial parameters: the actual changes up and down under selection on the good ration were very far from symmetrical (table 8.15) but the realized heritabilities were roughly symmetrical (high 0.31, low 0.35) because of differences in the selective differentials. Conversely, the actual changes on the poor ration were roughly symmetrical but the realized heritabilities were very different (high 0.14, low 0.40). The correlated responses of the both good and poor lines to the opposite ration were again very far from symmetrical. The randomly selected lines showed no appreciable changes in response to either ration. The genetic correlation between the environmental responses of both the good and poor lines after high selection was relatively low (0.44 and 0.50, respectively), while after low selection the correlations were somewhat higher than at first (0.75 and 0.79, respectively). This explains in part the extreme asymmetry of the correlated responses.

TABLE 8.15. Change in 13-day larval weight in populations *T. castaneum* under selection for large size, at random, or for small size, under good or poor nutrition and tested under the same or opposite ration. (The latter is indicated by parentheses.)

SELECTION		TEST RATIONS		
Direction	Ration	Good	Poor	Average
Large	Good	+ 8.4	(+ 0.9)	+ 4.6
Large	Poor	(+ 3.6)	+ 8.8	+ 6.2
Random	Good	− 0.2	(− 1.8)	− 1.0
Random	Poor	(− 1.2)	− 1.2	− 1.2
Small	Good	− 18.1	(− 8.6)	− 13.3
Small	Poor	(− 13.8)	− 10.7	− 12.2

SOURCE: Data from Hardin and Bell 1967.

A later experiment was carried for 16 generations with the same character (larval weight) and the same two nutritive levels (Yamada and Bell 1969). In addition to high and low selection at the two levels separately, lines of high and low selection were based on the average response to both levels, high $\overline{\text{GP}}$ and low $\overline{\text{GP}}$, and others on the responses to good and poor levels in alternate generations, high (GP alt) or low (GP alt). The character used was the mean of five full sibs instead of single individuals, with the consequence that the initial heritabilities appeared higher (good 0.30, 0.44 in two replications; poor 0.35, 0.51 in two replications). The initial correlations between the response to the two rations were 0.82 and 0.78 in the two replications.

Progress was approximately linear in all cases for the first eight generations, with realized heritabilities of 0.40 and 0.46 under high and low, respectively, but fell off in the next eight generations with realized heritabilities of 0.24 and 0.31 in high and low, respectively.

The results for the first and last eight generations are shown separately in table 8.16. There is some irregularity, but this is much less than in the previous results. With respect to the average results, low selection goes only slightly farther in the first eight generations and somewhat less on the whole in the last eight than high selection. There seems to be no advantage in selecting for the best average directly or in selecting for responses to the two rations alternatively, over continued selection under one ration. There is perhaps a slight advantage (first eight generations) in selecting under the poor ration.

TABLE 8.16. Change in 13-day larval weight in populations of *T. castaneum* under selection for large or small size and under good or poor nutrition or selection for the average response to these rations ($\overline{\text{GP}}$) or selection for responses to these conditions in alternate generations (GP alt). The responses to these conditions and the averages are given.

Selection		FIRST 8 GENERATIONS			LAST 8 GENERATIONS		
		Test Ration			Test Ration		
Direction	Ration	Good	Poor	Average	Good	Poor	Average
Large	Good	+ 9.1	(+ 8.6)	+ 8.9	+ 5.7	(+ 2.9)	+ 4.3
Large	Poor	(+ 7.4)	+ 13.2	+ 10.3	(+ 3.5)	+ 6.0	+ 4.8
Large	$\overline{\text{GP}}$	+ 6.9	+ 9.8	+ 8.4	+ 4.4	+ 5.3	+ 4.9
Large	GP alt	+ 7.2	+ 9.9	+ 8.5	+ 5.6	+ 4.9	+ 5.3
Small	Good	− 13.4	(− 8.7)	− 11.1	− 7.3	(− 1.9)	− 4.6
Small	Poor	(− 13.7)	− 11.4	− 12.5	(− 3.5)	− 2.1	− 2.8
Small	$\overline{\text{GP}}$	− 4.5	− 6.7	− 8.9	− 6.7	− 3.0	− 4.9
Small	GP alt	− 4.9	− 7.4	− 9.9	− 5.0	− 2.7	− 3.8

SOURCE: Data from Yamada and Bell 1969.

One of the populations in this experiment gave the peculiar result that larval weight became larger on the poor ration. It was found on further investigations (Constantino, Bell, and Rogler 1967) that a recessive gene cos (corn oil sensitive) had become fixed in this line and that unsaturated fatty acids present in the corn oil (which was increased in the good ration) sharply depressed growth. Here was a genotype-environment interaction that could be attributed to a particular gene.

In one of the earlier studies (Bell and Moore 1958), replicated comparisons were made of reciprocal recurrent selection, within-line selection, and selection of hybrids between inbred lines for increased pupal weight (moderately high heritability but only slight heterosis). The experiment was designed to allot equal facilities to the first two. Actual progress was about three times as great under selection within lines as under reciprocal recurrent selection, although there was no appreciable difference in terms of realized heritability. The genetic conditions were indeed not ones that indicated that reciprocal recurrent selection would be a useful procedure.

The effects of other mating systems on selection on pupal weight were investigated by Wilson, Kyle, and Bell (1965) and on 14-day larvae weight by

Wilson et al. (1968). The former showed heritability of 0.46, the latter only 0.15. The systems, compared over six and eight generations, respectively, of plus and random selection, were random mating of selected individuals, inbreeding, outbreeding, assortative mating, and disassortative mating. The inbreeding had a slight depressive effect. Progress in pupal weight was greatest with assortative mating, but the difference was not statistically significant. There was no indication of an advantage for assortative mating in the case of larval weight. These results are quite in line with the theory of the effect of assortative mating on variance, great with very strong heritability, but otherwise slight.

Summary

The most important general conclusion from review of experiments on directional selection in mixed populations is that these are nearly always successful. This applies even to populations derived from an exceedingly narrow base, ones derived, for example, from single impregnated wild female *Drosophila* (Robertson and Reeve). No doubt isolated experiments that fail to show a change in mean in the desired direction are less likely to be reported than ones that do, but this hardly applies to extensive series of reports from the same laboratory in which absence of change in an experiment conducted on the usual scale would be of special interest. The only case described here in which an attempt at directional change of a phenotypically varying character proved ineffective is Falconer's experiment with sex ratio in mice. This is not, a priori, a favorable character for such an attempt, although genetic differences have been demonstrated between certain stocks (Weir) and selection was effective in a population of rats (King).

There are, indeed, populations so far below the threshold for manifestation of a character, variable in other populations, that there is no basis for selection. There are also ones in which the number of units of a meristic character is so standardized that again there is no handle for selection. Several cases have been cited, however, in which selection from a rare variant has led to extensive change. These included production of a strain of guinea pigs with 100% perfect development of the little toes from selection based on a single animal with a rudimentary little toe in a strain of normals (Castle), production of a strain of *D. melanogaster* with an average of nine scutellar bristles from the descendants of a single wild fly with one more than the normal four (Payne), and production of a strain of mice with more than 65% white in the coat from a start provided by a single mouse with a few white hairs in its forehead in an otherwise self-colored strain (Goodale). Somewhat similar were the results of experiments by de Vries (vol. 1, p. 133) in increasing the modal number of petals in *Ranunculus bulbosus* to 9 from a wild population in which 91% had only 5 petals. Waddington found no basis at all for selection of defective crossveins in a population of *D. melanogaster*, except in flies exposed briefly as pupae to a very high temperature, but successful selection among such flies gave rise to a strain in which 100% showed this trait at 18°C. Kindred was unable to modify the almost invariant standard number, 19, of secondary vibrissae, in a strain of mice by direct selection but succeeded after preliminary selection in a variable mutant.

From these results, it is probably safe to conclude that in a wild population of many thousands, nearly all loci are significantly heterallelic. This agrees with conclusions based on mutation rates and the theoretical expectations based on the many possible ways in which heterallelism may be maintained (vol. 2, chap. 3).

The fact that long-continued close inbreeding, or derivation of a population from a monoploid mutation, or from parthenogenetic or vegetative reproduction, does not prevent genetic differentiation in the long run (chap. 5) reinforces the conclusion that the basis for extensive change by selection is always present in wild species. Moreover, such a basis seems to be usually present even for characters that are almost invariable because the effects of genetic differences are below the threshold for phenotypic manifestation except under exceptional conditions. These considerations seem to imply the omnipotence of natural selection. This implication is no doubt warranted in the very long run, but before accepting the power of natural selection to bring about changes in periods comparable to those of experiments there are important qualifications, to be taken up later.

An important question in connection with experimental directional selection is how far the course of change is predictable. The most generally available basis for prediction is the estimate of the amount of additive heritability in the foundation stock, as given in the regression of offspring on midparent or by comparisons of variability in this stock with that in closely inbred lines or their first crosses. Some experiments have been cited in which there was excellent agreement between the realized and estimated heritability (postweaning weight of mice, 17 generations [Rahnefeld et al.]; pupal weight of *T. castaneum*, 12 generations [Enfield, Comstock, and Braskerud]). Clayton, Morris, and Robertson found fairly good agreement in the case of abdominal chaetae of *D. melanogaster* under high selection for 5 generations, not so good for low selection. Most of the experiments of Falconer and his associates on body weight and other characters of mice, of Robertson and Reeve with body size of *D. melanogaster*, and of Bell and associates with larval and pupal weights of *T. castaneum* showed less realized than estimated heritability and, in many cases, marked asymmetry of high and low responses.

Several explanations of these discrepancies have been noted. Change in one direction may be restricted by approach to a mathematical limit (0 or 100% of a character) or to a physiological limit. There is usually damping of the effects of factors with approach to either sort of limit, implying that the character should be measured on a transformed scale (vol. 1, chaps. 10 and 11). Such transformations (often logarithmic) have in some cases made realized heritabilities symmetrical and in accord with expectations (at least for a considerable number of generations), but not in all cases.

The estimated heritabilities referred to above were those based on the assumption that the ordinary statistical parameters of a population tend to remain the same as long as conditions are constant. But even one generation of selection changes the situation. How much depends on the genetic composition of the populations, on the set of gene frequencies and the properties of the allelic genes with respect to contributions of each, dominance, interaction with genes at other loci, linkage relations and genotype-environment interactions, quantities that usually cannot be determined for quantitative variability. It is not surprising that even the early course of selection can often be predicted only very roughly on the basis of available knowledge of the foundation stock. The most successful estimates have been in populations derived from a cross between two inbred lines, and thus with a gene frequency of 0.50 at all loci that are heterallelic at all.

The changes brought about by selection, after the roughly predictable ones of the first few generations, may follow very diverse courses in different cases. Perhaps most typical is a gradually decreasing rate of change with asymptotic approach to a limit. Many of the experiments described here have had this sort of result. In some cases, the deceleration is very gradual from the first and there is still demonstrable progress after a great many generations (for example in low selection of oil and of protein content of maize kernels during more than 60 generations [Leng 1962]). In other cases, very rapid progress for a few generations is followed by many generations of slow progress. High selection of oil and of protein content of maize kernels differed from low selection of both in this way. A more extreme example was given by the divergence of high and low selection of the hooded pattern of rats in which about 40% of that attained after 19 generations was reached in one generation. Another pattern is that in which there is almost linear progress for some 20 or 30 generations, at which time there is abrupt cessation of all progress. Some of the courses described by Falconer for weaning weight of mice, by Clayton and Robertson for abdominal microchaetae of *D. melanogaster*, and by Robertson and Reeve for body size in the same species, appeared to be of this type although it is impossible to be sure of the abruptness with which the limit is reached because of the irregular fluctuations from generation to generation. Occasionally selection follows a sigmoid course, with a slow start, accelerated progress after a few generations, and finally a tapering off to an upper limit. Low selection for wing length (Robertson and Reeve) and high selection for number of lung tumors induced in mice by urethane (Falconer and Bloom) and for leukocyte count in mice (Chai) show accelerated responses in the earlier generations. Possible interpretations of various orderly courses in terms of different genetic conditions with respect to gene frequencies, effects of genes, and states of dominance have been given in chapter 6.

Finally, the course may be highly irregular, with progress by a series of irregularly spaced steps. There have been many illustrations of this in selection experiments with *Drosophila* but especially in the very long-continued experiments of Mather and Harrison. This sort of history is expected especially in experiments with *Drosophila*, in which the number of parents per generation is small, because of the major role played by linkage. The differences within the three major chromosomes tend to behave like three multiple allelic series over a number of generations until a crossover from repulsion linkage $(-+/+-)$ introduces chromosomes $++$ and $--$ and makes possible rapid steplike advance to more extreme levels (Mather).

Another complication in the course of selection relates to the compound nature of most of the characters selected. An apparently smooth course, such as that followed by selection for high oil content of kernels in maize, may break up into a rather rapid approach of the percentage of oil in the germ to its physiological limit and a slow, long-continued increase in the proportionate size of the germ. The changes in the six-week weight of mice observed by Falconer were found to depend largely on maternal influences up to weaning at three weeks, and growth capacity of the young mice in the second three-week period. Susceptibility to dental caries was obviously a very complex character, but others differ only in degree.

Evidence for antagonistic natural selection was obtained in some cases from the loss, under relaxation of selection, of previous gains (mouse weight, [Goodale]; long but not short wing length in one strain of *Drosophila*, the opposite in another [Robertson and Reeve]).

Reverse selection tended to cause partial to complete loss of previous gains when applied before progress had ceased, as expected in populations in which there was still additive genetic variance. In some cases both relaxation and reverse mutations caused loss of previous gains even after progress had ceased, indicating a state of balance between the directional selection and natural selection in a population in which fixation was not complete (*Drosophila* [Robertson and Reeve]).

The natural selection in such cases might relate either to secondary physiological consequences of extreme variations of the characters or to peculiarities of particular loci. The latter is the case where the homozygote of a favored gene is less extreme than the heterozygote or is infertile or lethal.

Variability was not usually reduced by attainment of a plateau, and in some cases was greatly increased. This could be due in part to persistent additive genetic variance, as in the states of balance referred to above, but in some cases it was shown to be due wholly or in part to increased sensitivity of extreme genotypes to environmental variations (Falconer; Robertson and Reeve). The complications from changing genotype-environment interaction have been noted in several cases (response of growing mice [Falconer et al.] and of *Tribolium* larvae [Bell et al.] to good and poor rations; responses of *Tribolium* strains to different humidities [Bell et al.]; responses of body size to temperature in *D. pseudoobscura* [Druger]).

One of the most important conclusions from experiments with directional selection is the prevalence of low fertility or viability as selection proceeds. This no doubt is in part mere inbreeding depression resulting from the smallness of parental populations in many of the experiments. This was not the whole explanation, however, as shown by the frequent improvement in these respects during relaxation or reversal of selection associated with loss of previous gains. It appears that much of the decline in fertility and viability is due either to deleterious pleiotropic effects of genes favored by the selection or to linked genes, the latter no doubt especially likely in *Drosophila*. It is these secondary effects of directional selection that make permanent improvement of species by this means enormously more difficult than apparent at first sight.

An important question after high and low selection have reached plateaus is the segregation index in crosses between them. The value for the 19-fold difference in oil content in maize kernels was about 20, probably indicating more or less equivalent genes in both halves of all of the ten chromosomes. In the hooded rats the index rose from 5 in generation 5 to 9 in generation 10, probably because of one or two quickly fixed major gene differences and delayed fixation of a great many minor ones. In other cases segregation indexes of about 4 could indicate four equivalent pairs or a major one, determining half the difference, and many minor ones. Experiments with the effect of stabilizing (centripetal) and disruptive (centrifugal) selection on *Drosophila* populations have been considered. The former has been shown (Thoday et al.; Prout; Scharloo) to reduce variability. This was not due wholly to the tendency to fix some genotype, but partly to reduced responsiveness to external environmental and developmental accident by selection for greater homeostasis with respect to the selected character.

Disruptive selection has been shown to have the opposite effects: increased genetic variance, sometimes involving marked polymorphism, and also increased nongenetic variance. It has been shown that these effects of disruptive selection can occur in *Drosophila* with 20%–25% gene flow (Thoday et al.; Streams and Pimentel; Scharloo et al.; Chabora) and occasionally with 50% gene flow (Thoday et al.). Complete disruption by this means, reported by Thoday and Gibson but not confirmed by Scharloo et al., Chabora, and others, probably required a prior association of the character with an isolating tendency. The great effectiveness of disruptive selection in causing divergence in *Drosophila* undoubtedly depends on the major role of linkage; much less effectiveness is to be expected in most other organisms.

There has been discussion of experiments (by A. Robertson, Bell, and Kojima and their associates) in which there was more than individual selection. These include various forms of family selection, recurrent selection of crosses to a tester strain, reciprocal cross selection between two strains, selection among first crosses of inbred strains, and accompaniment of selection by assortative and disassortative mating. The results do not permit summarization beyond the statement that in general they agreed, with some qualifications, with those expected by theory.

CHAPTER 9

Natural Selection in the Laboratory

L'Héritier and Teissier (1933) devised a cage for maintaining large *Drosophila* populations over long periods of time. It was a metal box 50 cm × 30 cm × 15 cm covered with glass, with 21 holes closed by corks which carried food cups. These were replaced in daily succession with fresh food cups. A cage maintained a population of some 3,000 to 4,000 flies. About 200 adults emerged daily, replacing a roughly equal number of deaths. Because of the severe larval competition, this was only about 1% or 2% of the eggs laid daily.

They (1934) followed the course of a population with an initial frequency of the Bar gene (sex-linked, semidominant) of 33.1% in competition with its wild-type allele leading to a frequency of about 10% at 150 days. In two later experiments (1937) with the same gene, the observed frequencies

TABLE 9.1. Percentages of the sex-linked gene Bar (B) of *D. melanogaster*.

I		II		III				IV			
					%B						
Day	%B	Day	%B	Day	♀	♂	Average	Day	%B	Day	%B
0	99.9	0	99.9	0	52.0	91.2	71.6	0	99.5	217	12.2
28	96.0	216	41.5	36	44.5	48.9	46.0	27	98.6	248	10.6
235	27.7	243	40.5	108	17.5	21.7	18.9	40	94.3	297	3.0
263	18.7	290	27.0	162	7.6	9.6	8.3	72	88.1	332	0.9
328	10.5	323	23.1	278	2.2	6.4	3.6	94	74.7	370	0.5
426	1.40	415	6.89					119	61.1	421	0.3
489	1.25	524	3.35					149	43.7	452	0.0
534	0.55	551	4.06					180	36.1	592	0.0
565	0.94	593	1.05								
604	0.37										

SOURCE: I and II, cage data from L'Héritier and Teissier 1937; III, cage data from Petit 1951; IV (from bottle populations), data averaged after angular transformation from Merrell 1965. Frequencies in males and females were reported separately by Petit. The average is $\bar{q} = (2/3)q_F + (1/3)q_M$.

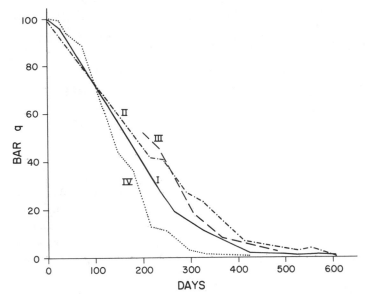

FIG. 9.1. Courses of natural selection in the laboratory of Bar eye of *D. melano-gaster* in competition with normal eye in experiments by L'Héritier and Teissier (1937) (I and II), Petit (1951) (III), and Merrell (1965) (IV).

declined from 99.9% Bar to 0.37% in 604 days and to 1.05% in 593 days, respectively. These latter results and other data on Bar are shown in table 9.1 and figure 9.1.

They (1937) made a similar study of the third chromosomal recessive gene ebony (*e*) in competition with its wild-type allele (table 9.2). The frequency of ebony flies declined from about 99% to 27.7% in 238 days, was nearly the same (26.3%) on day 369, but then declined to 14.1% on day 610.

Wiedemann (1936) reported on the results of crosses between 11 mutations of *D. melanogaster* and wild type that he followed through ten generations. Three sex-linked mutations, semidominant Bar (*B*) and Beadex (*Bx*) and recessive white (*w*), declined fairly rapidly in frequency. Among the autosomal mutations, the recessive vestigial (*vg*) declined rapidly, recessive jaunty (*j*) and dominant Dichaet (*D*) less rapidly, in spite of the lethality of the latter when homozygous. The recessive dumpy (*dy*) declined still less rapidly. All of the preceding seemed clearly destined for complete elimination. The autosomal recessives, black (*b*) and sepia (*se*), declined only slightly, and Wiedemann was doubtful of their ultimate fate. He found the autosomal recessive brown (*bw*) to be at no significant disadvantage relative to type. The domi-

TABLE 9.2. Percentages of Curly ($Cy/+$, cy/vg) in two populations of *D. melano-gaster* and of ebony in a third population. In one population Curly is segregating from its normal allele, and in the other, from the gene vestigial (vg). Curly is associated with an inversion in chromosome II that prevents crossing-over with other genes of this chromosome, including vg. Cy/Cy is lethal (Teissier 1942). Percentages of ebony and the estimated gene frequencies are given under III.

	I	II		III				
Day	% $Cy/+$	% Cy/vg	Day	% ee	q	Day	% ee	q
2	67.0	74.5	0	99+	0.995+	401	20.0	0.447
16	36.8	65.1	238	27.7	0.526	441	18.4	0.429
31	16.5	65.2	270	31.6	0.562	506	19.2	0.438
45	6.5	66.2	309	23.5	0.485	531	16.5	0.406
59	3.0	69.2	351	21.9	0.468	610	14.1	0.375
72	1.6	66.8	369	26.3	0.513			
87	0.8	69.0						
100	0.3	64.7						

SOURCE: Data from L'Héritier and Teissier 1937.

nant autosomal mutation Stubble (Sb) also seemed to be maintaining itself in $Sb/+$ in spite of the lethality of Sb/Sb.

In a cage experiment by L'Héritier, Neefs, and Teissier (1937), it was shown that conditions could be arranged in which vestigial actually tended to displace wild type instead of rapidly declining in frequency as in Wiedemann's experiment. The flies were reared in the open, exposed to the wind, but

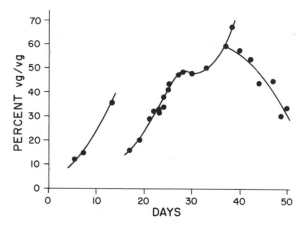

FIG. 9.2. Course of natural selection of vestigial (vg/vg) of *D. melanogaster* in competition with wild type (percentages) in open cages exposed at times to wind. From L'Héritier, Neefs, and Teissier (1937).

remained some one or two thousand in number. Eggs were taken out at frequent intervals during 50 days and the percentage of vestigial winged flies was determined with the results shown in figure 9.2.

The first arc gives the percentages from eggs laid on the specified days during the period before eclosion of adults from eggs laid on the first day. Eggs laid on the fifth day yielded 12.5% vestigials but the percentage rose to 32% from eggs laid on the 14th day, owing to the escape of a large proportion of the winged flies before this day. Most of the adults that emerged on the 15th and 16th days were, however, normals, so that the percentage of vestigials from eggs laid on the 16th day fell precipitously to 16%. Thereafter, the percentage of vestigials rose irregularly, according to the state of the wind, to 67% from eggs laid on the 38th day. Shortly before this, the cage was put in a room not exposed to the wind. The percentage of vestigials from eggs laid during this period declined to about 32%.

This was a case in which a selective disadvantage of genotypes $+/+$ and $+/vg$, due to a propensity to "emigrate," overbalanced the usual very strong advantage of these genotypes. This is a component of selective value that is wholly absent in ordinary laboratory populations but may be significant in nature. The authors began with a reference to Darwin's interpretation of the frequent winglessness of endemic species of insects on small oceanic islands as due to natural selection based on the likelihood of winged insects being blown away.

Teissier (1942) followed the declining frequencies of the second chromosomal gene Curly (Cy), lethal when homozygous in competition with normal, from 33.5% to 0.15% in 100 days (table 9.2). In a parallel experiment, the

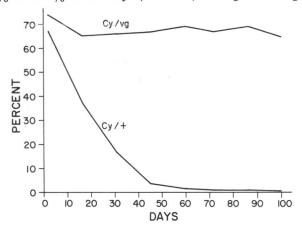

FIG. 9.3. Courses of natural selection of Cy/vg of $D.$ *melanogaster* in competition with vg/vg and of $Cy/+$ in competition with $+/+$ (percentages). From Teissier (1942).

chromosome that did not carry Cy carried vestigial (vg). Recombination was prevented by the inversion always associated with Cy. In this case there was no significant change in frequency throughout the period (table 9.2 and fig. 9.3).

Theoretical Courses of Change in Allelic Frequencies

Before further discussion, it is desirable to consider the theoretical courses under various conditions. Those for simple dominants, semidominants, and recessives were presented in chapter 6 as a preliminary to consideration of selection of multifactorial systems. We need here a somewhat broader treatment of unifactorial systems extending the discussion of rates in volume 2, chapter 3, to consider the implied course of change. It is convenient here to deal with momentary rates dq/dt as appropriate in continuous populations, although in some of the cases considered later the term $\bar{w}$ in the denominator of the expressions for Δq, change per generation, has been used. Random mating is assumed.

Genotype	f	w
A_1A_1	q^2	$1 - s$
A_1A_2	$2q(1 - q)$	$1 - hs$
A_2A_2	$(1 - q)^2$	1

$$\frac{dq}{dt} = -sq(1 - q)[h + (1 - 2h)q]$$

$$\frac{1}{h} \log_e q - \frac{1}{1 - h} \log_e (1 - q) - \frac{1 - 2h}{h(1 - h)} \log_e [h + (1 - 2h)q] = C - st$$

If $h = 0.5$, $2 \log_e q/(1 - q) = C - st$ (semidominance).

The cases of complete dominance ($h = 1$) and complete recessiveness ($h = 0$) require special treatment:

If $h = 0$, $\log_e \dfrac{q}{1 - q} - \dfrac{1}{q} = C - st$ (complete recessiveness).

If $h = 1$, $\log_e \dfrac{q}{1 - q} + \dfrac{1}{1 - q} = C - st$ (complete dominance).

These and the intermediate cases, $h = 0.25$, 0.50, and 0.75, are illustrated in figure 9.4 for populations that start with 95% of the less favorable allele, taking $s = 0.1$ per unit time in all. It would not be practicable to make these comparisons for a much larger common initial gene frequency than 95% because of the extreme slowness of the decline in the case of dominants. The rates of decline are the same in all at $q = 0.5$.

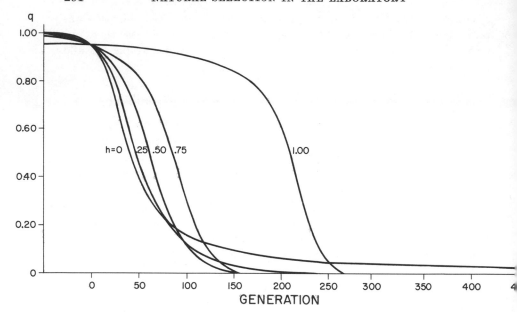

Fig. 9.4. Theoretical courses of selection with selective values, 1 for A_2A_2, $1 - hs$ for A_1A_2, and $1 - s$ for A_1A_1, starting from $q_1 = 0.95$ for values 0, 0.25, 0.50, 0.75, and 1.00 of h.

In figure 9.5, $\log_e [q/(1 - q)]$ is plotted against time. There is a linear relation in the case of semidominance ($h = 0.5$). In other cases, there is curvature without any point of inflection. The slope, $-s[h + (1 - 2h)q]$, continually decreases with time in absolute value if h is between 0 and 0.5; the reverse is true if h is between 0.5 and 1, but there is very strong curvature only with close approach to 0 (complete recessiveness) or 1 (complete dominance).

If there is overdominance, $h < 0$, gene frequency approaches an equilibrium $\hat{q} = -h/(1 - 2h)$. It is more convenient to take the selective value of the heterozygote as the standard (vol. 2, eq. 3.32):

Genotype	f	w
A_1A_1	q^2	$1 - s_1$
A_1A_2	$2q(1 - q)$	1
A_2A_2	$(1 - q)^2$	$1 - s_2$

$$\frac{dq}{dt} = -(s_1 + s_2)q(1 - q)(q - \hat{q}), \qquad \hat{q} = s_2/(s_1 + s_2)$$

$$\frac{1}{\hat{q}(1 - \hat{q})}[\log_e |q - \hat{q}| - (1 - \hat{q})\log_e q - \hat{q}\log_e (1 - q)] = C - (s_1 + s_2)t$$

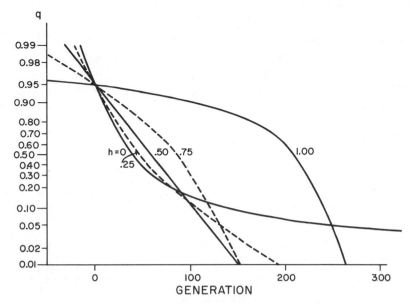

Fig. 9.5. Similar to figure 9.4, except on scale $\log_e [q/(1 - q)]$ which rectifies in the case of exact semidominance, $h = 0.5$.

The case of $s_1 = 0.7$, $s_2 = 0.3$, $\hat{q} = 0.3$ is illustrated in figure 9.6. If h is greater than 1, or s_1 and s_2 are negative, there is a point of unstable equilibrium.

We will consider sex linkage here only under the assumption that the gene frequencies of females and males do not differ so much as to invalidate the use of the average $\bar{q} = (2/3)q_F + (1/3)q_M$ and the zygotic frequencies of random mating (vol. 2, eqs. 3.102, 3.103).

Females		
Genotype	f	w
A_1A_1	q^2	$1 - s$
A_1A_2	$2q(1 - q)$	$1 - hs$
A_2A_2	$(1 - q)^2$	1

Males		
Genotype	f	w
A_1	q	$1 - ks$
A_2	$(1 - q)$	1

$$\frac{dq}{dt} = -\frac{s}{3} q(1 - q)[2h + k - (4h - 2)q]$$

$$\left(\frac{1}{2h + k}\right) \log_e q - \left(\frac{1}{2 - 2h + k}\right) \log_e (1 - q) +$$

$$\left(\frac{4h - 2}{(2h + k)(2 - 2h + k)}\right) \log_e [(2h + k) - (4h - 2)q] = C - \tfrac{1}{3}st$$

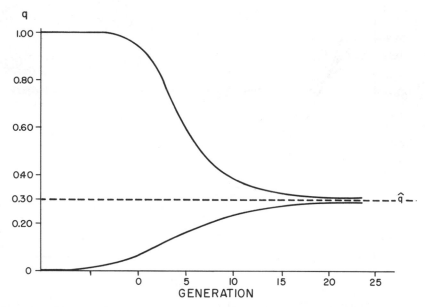

Fig. 9.6. Theoretical courses of selection for an overdominant gene, A_1 with $w_{11} = 0.3$, $w_{12} = 1.0$; $w_{22} = 0.7$; $\hat{q} = 0.3$.

The simplest special cases are the following, all with $k = 1$:

If $h = 0$, $3 \log_e q - \log_e (1 - q) - 2 \log_e (1 + 2q) = C' - st$
                                                              (A_1 recessive)

If $h = 0.5$, $1.5 \log_e [q/(1 - q)] = C' - st$ (semidominant)

If $h = 1$, $\log_e q - 3 \log_e (1 - q) + 2 \log_e (3 - 2q) = C' - st$
                                                              (A_1 dominant)

With overdominance in females, h is negative. There may be equilibrium at $\hat{q} = (2h + k)/(4h - 2)$. For stable equilibrium with $k = 1$, $h < -0.5$. This should occur relatively infrequently.

The decline in frequency of a sex-linked gene on starting from $q = 0.95$ is illustrated in figure 9.7 for the cases $h = 0$, 0.25, 0.50, 0.75, and 1. The differences are much less than with autosomal genes. This is further illustrated by plotting $\log_e [q/(1 - q)]$ against time, under which there is again linearity if $h = 0.5$ (fig. 9.8).

We consider next the simplest sort of pure frequency dependence: linearly varying selective values relative to that of the heterozygote, and the latter always exactly intermediate (vol. 2, eq. 5.78).

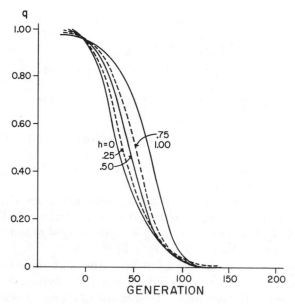

Fɪɢ. 9.7. Theoretical courses of selection of a sex-linked gene, otherwise similar to the case of the autosomal gene of figure 9.4.

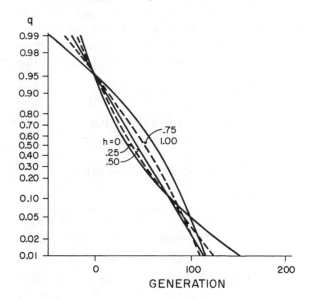

Fɪɢ. 9.8. Similar to figure 9.7, except on scale $\log_e [q/(1 - q)]$, which again rectifies the case of exact semidominance, $h = 0.5$.

Genotype	f	w
A_1A_1	q^2	$1 + a - bq$
A_1A_2	$2q(1 - q)$	1
A_2A_2	$(1 - q)^2$	$1 - a + bq$

$$\frac{dq}{dt} = -bq(1 - q)(q - \hat{q}), \qquad \hat{q} = a/b$$

$$\frac{1}{\hat{q}(1 - \hat{q})} [\log_e |q - \hat{q}| - (1 - \hat{q}) \log_e q - \hat{q} \log_e (1 - q)] = C - bt$$

This follows the same course as with overdominance if $a = s_2$, $b = s_1 + s_2$, ignoring $\bar{w}$ in the denominator of Δq.

If there is jointly overdominance and minority advantage of the above sort, represented by subtracting s_1 and s_2 from the relative selective values of A_1A_1 and A_2A_2, respectively, $\hat{q} = (s_2 + a)/(s_1 + s_2 + b)$ and the coefficient of t in the equation above is $(s_1 + s_2 + b)$.

Frequency dependence due to selective mating may cause wide deviations from the binominal square distribution of zygotic frequencies. This would complicate greatly the prediction of the course of change of gene frequencies.

Another simple but very important sort of frequency dependent selection is that in which the selective values are modified, favorably or unfavorably, by their ecologic relations with other genotypes and thus according to the frequencies of the latter instead of by their own frequencies. Since we will assume linearity, there is no essential mathematical difference from the set above but the parameters have more obvious meanings in this connection.

Genotype	f	w
A_1	q	$1 + b_1(1 - q)$
A_2	$(1 - q)$	$1 + b_2q$
$\bar{w} = 1 + (b_1 + b_2)q(1 - q)$		

$$\frac{dq}{dt} = q(1 - q)[b_1 - (b_1 + b_2)q]$$

Genotype	f	w
A_1A_1	q^2	$1 + 2b_1(1 - q) - s_1$
A_1A_2	$2q(1 - q)$	$1 + b_1 + (b_2 - b_1)q$
A_2A_2	$(1 - q)^2$	$1 + 2b_2q - s_2$
$\bar{w} = 1 + 2(b_1 + b_2)q(1 - q) - s_1q^2 - s_2(1 - q)^2$		

$$\frac{dq}{dt} = q(1 - q)[(b_1 + s_2) - (b_1 + b_2 + s_1 + s_2)q]$$

If both b_1 and b_2 are positive, there is mutual facilitation. There is stable equilibrium at $\hat{q} = b_1/(b_1 + b_2)$ or if combined with overdominance, as in the genotypic selective values above, $\hat{q} = (b_1 + s_2)/(b_1 + b_2 + s_1 + s_2)$. Figure 9.6 represents the course of change with appropriate change of parameters. Other cases may, of course, be represented by adding other constants to w.

If b_1 is positive but b_2 is negative (or if b_2 is replaced by $-b_2'$, with b_2' positive), the model may represent competition in which A_1 benefits but A_2 suffers.

$$\frac{dq}{dt} = q(1 - q)[b_1 - (b_1 - b_2')q]$$

A_1 here tends toward fixation in the pattern of genic selection above. This, of course, could be reversed by subtracting a sufficient amount from the selective value of A_1 (or adding it to that of A_2). The combination of over-dominance with competition represented in the genotypic selective values gives equilibrium $\hat{q} = (s_2 + b_1)/(s_1 + s_2 + b_1 + b_2')$ if $s_1 > b_2'$.

There is an analogous case if b_1 is negative and b_2 positive.

Finally, there may be mutual inhibition, represented by reversing the signs of both b_1 and b_2, with metastable equilibrium at $b_1'/(b_1' + b_2')$, which may be changed to stable equilibrium by sufficient overdominance, $\hat{q} = (s_2 - b_1')/(s_1 + s_2 - b_1' - b_2')$, $s_1 > b_2'$.

Comparisons of Observed and Theoretical Curves

The fact that the course of selection may be the same under patterns of selection as different as overdominance, minority advantage, and mutual facilitation precludes interpretation on this basis alone. There must be subsidiary information to be discussed later.

If there is reason to believe that relative selective values are constant, a rough judgment of degree of dominance in this respect may be made by considering the form of the curves for q (or better, $\log_e [q/(1 - q)]$), plotted against time.

In the case of L'Héritier and Teissier's data on Curly (table 9.2), the plotting of $\log_e [q/(1 - q)]$ against time yields a rather close fit to a straight line compatible with semidominance (fig. 9.9).

Interpretation must, however, be discounted in all cases where based on adult frequencies unless preadult selection of the genotypes A_1A_1, A_1A_2, and A_2A_2 is in a ratio w^2: w:1, which leaves the adult frequencies still in a binomial square distribution, if there had been such a distribution in the fertilized eggs. This implies multiplicative semidominance in this component

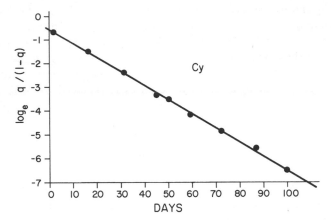

FIG. 9.9. Course of natural selection of $Cy/+$ of $D.$ $melanogaster$ in competition with $+/+$ (fig. 9.3) plotted against $\log_e[q/(1-q)]$.

of selection, which differs markedly from additive semidominance $(1-s:1-0.5s:1)$ if s is more than about 0.5.

The estimates of gene frequency of a recessive as the square roots of frequencies of recessive individuals (as in the case of ebony, table 9.2) may be seriously in error if the preadult selective values deviate from the above ratio. In the case of sex linkage (males XY), the gene frequency in adult males is, indeed, given directly, but that in females suffers from the same unreliability as in autosomal recessives.

Where preadult selective values can be determined by direct experiment under conditions comparable to those of the cage experiment under consideration, the observed adult frequencies can be divided by them to give estimates of the corresponding zygote frequencies.

In the case of a lethal such as Cy/Cy, interpretation can be based most reliably on direct determination of relative selective values of the two observed genotypes in successive generations. The frequencies of $Cy/+$ and $+/+$ are obviously $2q$ and $1-2q$ in terms of the gene frequencies of Cy in adults. If it be supposed that $Cy/+$ and $+/+$ differ only in preadult viability, ratio $w_v:1$, the frequency q_1 of $Cy/+$ in the next generation is

$$2q_1 = \frac{2q(1-q)w_v}{2q(1-q)w_v + (1-q)^2} = \frac{2qw_v}{1-q+2qw_v}$$

$$w_v = \frac{q_1(1-q)}{q(1-2q_1)}$$

If, on the other hand, $Cy/+$ and $+/+$ differ only in productivity, ratio $w_P:1$, the frequencies of Cy and $+$ gametes are in the ratio $qw_P:(1 - 2q + w_P)$ and the frequency of $Cy/+$ in the next generation is

$$2q_1 = \frac{2qw_p}{1 - 2q + 3qw_p}$$

$$w_p = \frac{q_1(1 - 2q)}{q(1 - 3q_1)}$$

TABLE 9.3. Estimates of w_v or w_p for the data in table 9.2 I under the assumption that the decline in frequency of Cy is due wholly to w_v or w_p.

Days	q	w_v	w_p	Days	q	w_v	w_p
2	0.335	...	...	59	0.0150	0.46	0.45
16	0.184	0.58	0.40	72	0.0080	0.53	0.53
31	0.0825	0.44	0.38	87	0.0040	0.50	0.50
45	0.0325	0.39	0.36	100	0.0015	0.37	0.37
Average		0.47	0.38	Average		0.47	0.46

The data can be interpreted almost equally well (table 9.3) under either assumption. The average productivity of $Cy/+$ per interval of 14 or 15 days is about 0.38 for the first three intervals and 0.46 for the less reliable later four intervals under the assumption that this is the only component of selective value. The average preadult viability is about 0.47 for either the first three or the later four intervals under the assumption that only this component varies. In either case or in any combination, there is additive semidominance to a sufficiently close approximation to account for the near-linearity of $\log_e [q/(1 - q)]$ in time.

Returning now to the data on competition between Cy and vg (table 9.2 II), it may be seen that there is apparent equilibrium from the 16th to the 100th day at almost exactly 2/3 Curly, 1/3 vestigial. This is what would be expected

TABLE 9.4. Expected frequencies of progeny from indicated parents.

Parents		Progeny		
♂ ♀	Cy/Cy	Cy/vg	vg/vg	
$Cy/vg \times Cy/vg$	0	2/9	$(1/9)w_v$	
$Cy/vg \times vg/vg$	...	$(1/9)w_{PF}$	$(1/9)w_{PF}w_v$	
$vg/vg \times Cy/vg$	...	$(1/9)w_{PM}$	$(1/9)w_{PM}w_v$	
$vg/vg \times vg/vg$	...		$(1/9)w_{PF}w_{PM}w_v$	
Total	0	$(1/9)(2 + w_{PF} + w_{PM})$	$(1/9)(1 + w_{PF})(1 + w_{PM})w_v$	

if vg/vg had normal preadult viability but were sterile in both sexes. This, however, cannot be the explanation since vestigial stocks can easily be maintained and vestigials may even outbreed normals under certain conditions, as already noted.

A population that remains in equilibrium with the above frequencies would show the following progeny array w_{PF} and w_{PM} for productivities of female and male vg/vg, respectively, relative to Cy/vg (table 9.4).

The proportion of Cy/vg at equilibrium at 2/3 $Cy/+$ would thus be

$$(2 + w_{PF} + w_{PM})/[(2 + w_{PF} + w_{PM}) + (1 + w_{PF})(1 + w_{PM})w_v] = 2/3$$

$$(2 + w_{PF} + w_{PM}) = 2(1 + w_{PF})(1 + w_{PM})w_v$$

Assume that $w_{PF} = w_{PM} = w_P$ to reduce to one degree of indeterminacy $(1 + w_P) = 1/w_v$.

Thus, if $w_v = 1$, $w_P = 0$. This is contrary to the observed productivity of vg/vg. If $w_P = 1$, $w_v = 0.5$. This is improbable in view of the great differences between the adult phenotypes. Some intermediate situation such as $w_v = 0.8$, $w_P = 0.25$ or $w_v = 0.6$, $w_P = 0.67$ is not unreasonable.

We will add some other population experiments reared in much smaller containers than those used by L'Héritier and Teissier. Reed and Reed (1948) devised a system consisting of two half-pint milk bottles, held mouth to mouth by rubber tubing, with a stoppered hole for ventilation. Each bottle had a paper cap perforated to permit flies, but not debris, to move across. Sets of 10 units were kept in boxes, painted so that one bottle of each pair was in the dark, the others in the light. The flies would go to the latter. Each bottle was replaced by one with fresh food after it had been in the light for two months, and in the dark for two months. The flies were counted two or three days after moving to the new bottles, at which time the population number was at its smallest, always less than 600 and averaging less than 200.

Merrell (1965) reported on populations of several mutations of *D. melanogaster* in competition with wild type, reared in Reed bottles. The results from individual bottles were irregular, as expected from the rather small effective numbers, but several replications were made in most cases. Averages by day of count are distorted by the differences in rates of change near 50% and at the extremes. To overcome this, angular transformations (vol. 1, pp. 255, 258) were averaged and transformed back to percentages. We consider here Merrell's results for Bar (B), Beadex[3] (Bx^3) (like Bar a sex-linked semidominant), Lobe[3] (L^3) (an autosomal semidominant), and brown (bw) (an autosomal recessive). Six replications of B, Bx^3, and L^3 are averaged in each case and five of bw. The values of $\log_e[q/(1 - q)]$ are shown in figure 9.10. All three semidominants differ markedly from linearity and thus behave somewhat like recessives with respect to selective value. In the case of bw there is

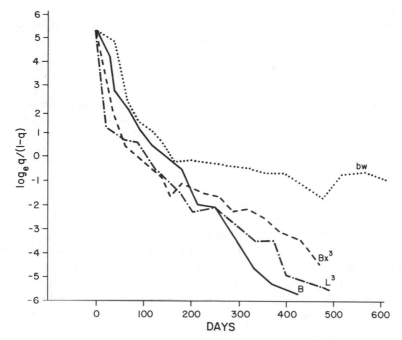

FIG. 9.10. Courses of natural selection of mutations *bw*, *Bx³*, *L³*, and *B* of
D. melanogaster, each competing with its wild-type allele, plotted against
$\log_e [q/(1 - q)]$. From data of Merrell (1965).

an approach to constancy in the later records, which suggests overdominance
with equilibrium at about $q = 0.30$.

Merrell and Underhill (1956) studied the course of decline of the autosomal
recessives vestigial (*vg*), frequently referred to above, and glass (*gl*). The
frequency of vestigial flies declined from more than 99% to only about 21%
in 35 days and to only about 1% in 70 days. Flies with the glass phenotype
behaved rather similarly, declining from more than 99% to about 36% in 35
days and to only about 1% in 70 days. In both cases a few recessives appeared for many generations thereafter. There was a marked contrast with the
slow reduction of brown.

The same authors also studied the competition between sex-linked white
(*w*) and its alleles eosin (*wᵉ*), apricot (*wᵃ*), coral (*wᶜᵒ*), and satsuma (*wˢᵃᵗ*),
each starting from gene frequencies of 50%. The changes during more than
300 days were rather small and varied considerably in replications. On the
average, white declined from 50% to about 20% in competition with eosin;
it hardly changed at all with satsuma, rose to about 70% with apricot, and

to 85% with coral. This indicates that these allelic mutations are almost neutral in relation to each other. Thomson (1961), on the other hand, found that white declined rapidly in competition with satsuma but there were threefold differences in rate among strains. He also found that the decline was much more rapid in either continuous light or continuous darkness than where the flies had a choice of light or darkness. In observations of mating, white males were much more handicapped relative to satsuma males in the light than in the dark, but Thomson doubted whether this had much significance for cage populations. His experiments showed that selection coefficients are highly dependent on both genetic background and environmental conditions.

Components of Selective Value

One of the major premises in the deductive aspect of population genetics is that of equational segregation based on equational disjunction of chromosomes in meiosis. As discussed earlier (vol. 1, chap. 3) this assumption is not without exceptions. The opportunity for asymmetrical segregation is obviously present in oogenesis and often occurs with asymmetrical chromosome aberrations. It is much less likely where there is merely a pair of segregating genes.

There would seem to be less likelihood of asymmetry in spermatogenesis, but cases are known, for example, the casting out of all paternal chromosomes in *Sciara* in a monopolar spindle (Metz 1938).

An important example of asymmetrical segregation in spermatogenesis occurs in the so-called sex-ratio genes of several species of *Drosophila* (*D. affinis* [Morgan, Bridges, and Sturtevant 1925], *D. obscura* [Gershenson 1928], *D. pseudoobscura* [Sturtevant and Dobzhansky 1936]). In the latter, the effect is associated with a particular inversion in the X chromosome. Males carrying it transmit it to more than 90% of their offspring, which thus become females. This gives a great selective advantage that, however, is balanced by disadvantages to be discussed later. The dynamics, including the conditions for equilibrium (at up to 30%) in some natural populations were discussed earlier (vol. 2, pp. 62–64).

Another example of "meiotic drive," as unequal segregation has been termed by Sandler and Novitski (1957), is provided by "segregation distorter," a second chromosomal condition in *D. melanogaster*, discovered by Hiraizumi (Sandler, Hiraizumi, and Sandler 1959) and found to be fairly common in nature. Heterozygous flies transmitted the condition to more than 95% of their offspring as indicated by linked genes (vol. 2, pp. 59–61).

A third much studied example of apparent meiotic drive is that of the compound t locus in the mouse (Dunn 1956). Segregation of up to 95% of certain alleles occurs in the progeny of males heterozygous for the normal allele. This again gives an advantage (balanced by lethality of homozygotes) that causes such alleles to be common in nature (vol. 1, p. 37; vol. 2, pp. 61–62).

Such cases seem to be rare, however. Those referred to above all involve more drastic chromosome differences than mere gene replacement. Moreover most visibly recognizable chromosome differences, restricted to one pair of homologues (duplications, deficiencies, inversions), show equational segregation.

The causes of differences in selective value other than meiotic drive may be classified into differential viability preceding the stage at which ratios are observed, subsequent differences in viability and productivity, differences in age and duration of sexual maturity, differences in mating success, either in general or in relation to particular genotypes of the other sex, differences in fertility, fecundity, and finally in tendency to leave the population, whether by death or by emigration, before the end of the reproduction period.

One of the earliest attempts to estimate one of the components of selective value was made by Timofeeff-Ressovsky (1934). He tested the preadult viability of several mutations of *D. funebris*, relative to wild type, by obtaining the ratios in the progeny of backcrossed heterozygous females, or by putting different total numbers of eggs, equally apportioned between two segregating genotypes, in a vial.

Most mutations were definitely less viable than wild type. Two, however, (eversae, heterozygous Abnormal abdomen) were significantly superior under particular conditions. In general the coefficients varied greatly with temperature and amount of crowding. Combinations of mutations generally showed cumulative affects, but there were striking interaction effects in some cases.

Experiments by Teissier (1942) and Helman (1949) bear directly on some of the population experiments already discussed. Their comparisons were based in each case on eggs laid by some 300 heterozygous females. Some were kept in such small amounts of nutrient that only about one-tenth gave rise to adults while others were put with a superabundance of nutrient.

It may be seen from table 9.5 that $Cy/+$ and Cy/vg differed little in preadult viability from $+/+$ where larval competition was weak, but have greatly reduced relative selective values under strong competition. On the other hand, Cy/vg has some advantage over vg/vg even under weak competition and nearly a twofold advantage under strong competition. The superiority of wild type over vestigial was not determined directly in these experiments but was presumably much greater, nearly fourfold if the effects

TABLE 9.5. Numbers of segregants (*D. melanogaster*) under weak and strong larval competition.

	Weak Competition			Strong Competition		
	Cy	Non-*Cy*	$1 - s$	*Cy*	Non-*Cy*	$1 - s$
Cy/+ × +/+	8,326	7,450	1.12	3,188	5,539	0.57
Cy/+ × *Cy*/+	4,332	2,136	1.01	1,140	1,107	0.51
Cy/*vg* × *vg*/*vg*	3,424	2,712	1.26	4,157	2,253	1.84
Cy/*vg* × *Cy*/*vg*	4,807	1,917	1.25	1,721	429	2.01
Cy/+ × *vg*/*vg*	6,094	6,540	0.93	2,954	6,761	0.44
Cy/*vg* × +/+	4,276	4,776	0.89	1,723	4,125	0.42
	D	Non-*D*	$1 - s$	*D*	Non-*D*	$1 - s$
D/+ × +/+	1,389	2,937	0.47	388	2,698	0.14
D/*e* × *e*/*e*	1,428	1,962	0.73	380	3,830	0.10
D/+ × *e*/*e*	1,363	3,360	0.41	414	2,016	0.20
	Sb	Non-*Sb*	$1 - s$	*Sb*	Non-*Sb*	$1 - s$
♀+/+ × ♂*Sb*/+	577	529	1.09	1,605	597	2.6
♀*Sb*/+ × ♂+/+	682	616	1.1	1,619	658	2.6
♀*Sb*/+ × ♂*Sb*/+	816	377	1.08	141	46	1.53

SOURCE: Data from Teissier 1942 (cases involving *Cy* and *D*) and Helman 1949 (cases involving *Sb*).

are multiplicative. Unfortunately, these results cannot be applied quantitatively to the cage experiments in the same laboratory, although these certainly involved severe competition. There can be little doubt, however, that differences in preadult viability were an important factor.

Dichaet (*D*/+, *D*/*e*) was shown to be at a serious disadvantage relative to +/+ and *ee* even under weak competition, and much more under strong competition (relative selective value of *D*/*e* only 0.10 to 0.20). This fits in with the rapid decline of the frequency of *D* in competition with its wild-type allele in Wiedemann's experiments.

Stubble (*Sb*/+) seems to be at a slight advantage over wild type under weak competition and at a very great advantage (2.6:1) under strong competition. This fits in with the persistence of Stubble in Wiedemann's experiments, in spite of its lethality when homozygous.

Diederich (1941) made experiments in which flies (*D. melanogaster*) from a strain carrying two sex-linked recessives, yellow (*y*) and white (*w*), competed with wild type. Groups of females were exposed to one-fourth as many males for limited periods, followed by isolation of each female. Altogether some 6,000 females were tested. In some experiments one type of female was exposed to equal numbers of two types of males. In others, two types of females in equal numbers were exposed to one type of male so briefly that only a portion were fertilized. In a third series, there were equal numbers of both sexes. She obtained the following percentages:

♂ \ ♀ ++/++	
50% ++	91
50% yw	9

♂ \ ♀ yw/yw	
50% ++	79
50% yw	21

♂ \ ♀	50% ++/++	50% yw/yw
++	50	50

♂ \ ♀	++/++	yw/yw
yw	20	80

♂ \ ♀	50% ++/++	50% yw/yw	Total
50% ++	47	33	80
50% yw	3	17	20
Total	50	50	100

With both genotypes present in equal numbers in both sexes, the component of selective value of *yw* males from differential mating success, relative to the wild type, was 0.25. It is also apparent that wild-type females discriminated against *yw* males more than did *yw/yw* females. With only wild-type females present, the component of relative selective value for *yw* males was 0.10; with only *yw/yw* females present it was 0.27. There were, on the other hand, no differences between the two types of females when exposed briefly to wild-type males. When exposed briefly to *yw* males, the greater acceptance of these by the *yw/yw* females was shown strongly.

We have so far treated mathematically only cases in which there was no selective mating. In the present case, selective values differ in recessive males

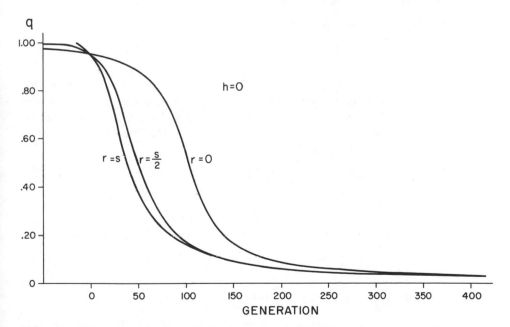

FIG. 9.11. Theoretical courses of selection where recessive males are at a disadvantage, $s = 0.1$ with *AA* and *Aa* females, and either half of this or at no disadvantage with *aa* females.

according to the kinds of females. The resulting nonrandom union of kinds of gametes greatly complicates the matter. We will merely go into an approximate theory for the course of change applicable where there is dominance and selective differences are small (rates: vol. 2, eq. 5.121 for the autosomal case and eq. 5.122 for the sex-linked case, with r substituted for t in both).

$$w_1 \text{ Autosomal Case, } (qa + [1 - q]A)$$

♂\♀	$A-$	aa
$A-$	1	1
aa	$1 - s$	$1 - r$

$$\frac{dq}{dt} = -\frac{s}{2}q^2(1 - q)(1 - k^2q^2), \qquad k^2 = (s - r)/s$$

$$\log_e q - \left(\frac{1}{1 - k^2}\right)\log_e (1 - q) + \frac{k^2}{2(1 - k)}\log_e (1 - kq)$$

$$+ \frac{k^2}{2(1 + k)}\log_e (1 + kq) - \frac{1}{q} = C - \frac{st}{2}$$

If $r = 0$,

$$\log_e q - (5/4)\log_e (1 - q) + (1/4)\log_e (1 + q)$$

$$- 1/q + 1/[2(1 - q)] = C - \frac{st}{2}$$

The cases $r = s/2$ and $r = 0$ are illustrated in figure 9.11. There is little difference between the former and the case of a constant average selective disadvantage of recessives, $r = s$. If recessive males are at a disadvantage only with dominant females ($r = 0$), there is much difference at high values of q. In the case of sex linkage:

♂\♀	$A-$	aa
A	1	1
a	$1 - s$	$1 - r$

$$\frac{dq}{dt} = -\frac{s}{3}q(1 - q)(1 - k^2q^2), \qquad k^2 = (s - r)/s$$

$$\log_e q - \left(\frac{1}{1 - k^2}\right)\log_e (1 - q) + \frac{k}{2(1 - k)}\log_e (1 - kq)$$

$$- \frac{k}{2(1 + k)}\log_e (1 + kq) = C - \frac{st}{3}$$

If $r = 0$,

$$\log_e q - \left(\frac{3}{4}\right) \log_e (1 - q) - (1/4) \log_e (1 + q) - \frac{1}{2(1 - q)} = C - \frac{st}{3}$$

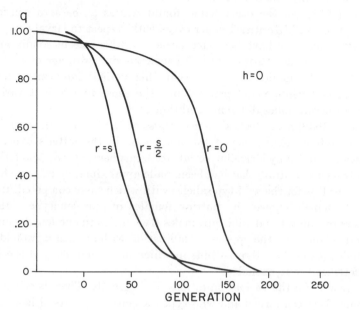

FIG. 9.12. Similar to figure 9.11, except for sex-linked instead of autosomal recessive.

The situation is closely similar to the autosomal case (fig. 9.12). While these theoretical values cannot be applied quantitatively to such strong selective differences as in Diederich's data, they indicate qualitatively how selective mating modifies the results expected for a simple recessive.

Reed and Reed (1950) followed the decline in the percentage of the white gene in competition with wild type from 67% to extinction over 25 generations. Accessory tests in which one $+/w$ female, one w/w female, one $+$ male, and one w male were kept together for 24 hours, followed by isolation of the females, indicated that the red-eyed males mated at random with the two types of females but the white-eyed males, responsible for 43% of the 359 offspring, mated more freely with the white-eyed females than with the red-eyed females, indicating components of selective value relative to that of the red-eyed males of 0.90 and 0.61, respectively. The difference is similar but less than Diederich found with yw. Reed and Reed also made tests of prenatal viability. They found no significant deviation from expected equality in the progeny of $+/w$ females by w males, at either 20°C or 24°C or with a fivefold difference in density (grand totals: 12,979 red-eyed, 12,927 white-eyed). They showed that the course of evolution over the 25 generations was in rough agreement with expectations on the basis of selective mating of males as the only differential factor.

Petit (1958), on the other hand, found whites to be at a considerable disadvantage in L'Héritier-Teissier cages with respect to larval competition. The difference may have been due either to difference in conditions or in genetic background. Petit confirmed a selective disadvantage of white males in mating. She found, moreover, that this varied significantly with the percentage of white males present, but the sense in which it varied was wholly different under different conditions.

Merrell (1949) made tests similar to those of Diederich with several mutations including yellow, one of the mutations in the latter's experiments. The others studied by Merrell were cut[6] (ct^6), raspberry[2] (ras^2), and forked (f). These traced to a strain that had been made approximately isogenic for all of them as well as for the wild-type alleles with which these competed. He made tests of "female choice" by putting females of one genotype with equal numbers of mutant and wild-type males (two males to one female) and tests of "male choice" by the opposite combination with at least a twofold excess of females (table 9.6). Experiments in which more than 80% of the females were fertilized were excluded.

The results for the various mutations differ greatly. There is no significant difference between forked and wild-type flies either in tests of female choice or male choice. Mating is substantially at random.

TABLE 9.6. Female and male choice experiments. In the former, one type of female (indicated) was put with equal numbers of the same type of male and wild type. In the latter, the situation was the reverse.

FEMALE CHOICE			MATING RATIO	MALE CHOICE			MATING RATIO
Female	No.	% Fertilized	Mutant ♂ : + ♂	Male	No.	% Fertilized	Mutant ♀ : + ♀
y/y	150	84.7	0.12	y	156	43.6	2.40
$+/y$	126	92.1	0.06	$y+$	158	79.7	1.07
ct^6/ct^6	132	48.5	0.28	ct^6	132	70.5	0.58
$+/ct^6$	107	90.7	0.28	ct^+	122	67.2	0.44
ras^2/ras^2	76	80.3	0.54	ras^2	112	59.8	1.16
$+/ras^2$	68	83.8	0.46	ras^+	60	60.0	1.00
f/f	87	70.1	1.18	f	80	65.0	0.93
$+/f$	75	72.0	1.25	f^+	86	64.0	0.83

SOURCE: Data from Merrell 1949.

There is marked but apparently equal discrimination against raspberry males by the two types of females (w for males about 0.5 with respect to mating success). There is no significant discrimination by either type of male against the two types of female (w for females about 1.0).

There is very marked but equal discrimination against cut[6] males by both types of females (w for males 0.3). The percentages of cut[6] females fertilized is low (48.5%) in comparison with all of the other kinds of females tested. This low fertility of cut[6] females is also reflected in the low ratio of cut[6] to wild-type females fertilized by either type of male. The selective value of cut[6] females has a component of about 0.5 relative to wild type because of this.

The yellow males show the most severe selective disadvantage of any of these mutations in competition with wild-type males (0.12 with yellow females, 0.06 with wild-type females). The low mating capacity of yellow males is also reflected in the percentage (43.6%) of females that they succeeded in fertilizing under the conditions of the experiment, much lower than in any other case. Among those fertilized, the high rate of yellow to wild-type females illustrates more strongly than above the discrimination against them by wild-type females. Thus Diederich associated two genes, y and w, that share properties such that males with these genes are not only at a severe disadvantage with either mutant or wild females but are at a greater disadvantage with the latter than with the former.

It may be added that population experiments by Ludwin (1951) with the same four mutations as above were in general agreement with those expected from the relative selective values with respect to mating found by Merrell. Yellow was rapidly eliminated and cut[6] even more rapidly, while raspberry and forked did not decline decisively. Merrell (1953) carried population experiments with raspberry and forked much longer. The frequency of raspberry males was reduced from 0.50 to about 0.05 after 500 days but persisted at low frequencies (at 1,351 days). The frequency of forked males fell from 0.50 to about 0.10 in 500 days but fluctuated about 0.06 between days 1,000 and 1,200. From his previous studies he estimated the mating capacities of yellow, cut[6], raspberry, and forked, relative to their wild-type alleles, as 0.1:0.15:0.5:1.0, and the capacity of female cut[6] for successful mating as 0.5. He showed that the observed frequencies of both males and females were fitted rather well by using these selective values in the cases of yellow and raspberry, less well with cut[6]. Differential preadult viability and differential fecundity of females could have played little role in the first two, perhaps some in the case of cut[6], and must be involved to some extent in the case of forked.

Barker (1962) made an especially thorough study of the role of selective mating in the case of yellow, in competition with its wild-type allele. In all cases, he compared the results for young and old flies. The types of matings and the results are indicated in table 9.7 by the percentage of successful fertilization of the indicated females in each case. The numbers of males and females in each set and the numbers of sets are shown.

TABLE 9.7. Percentages of the indicated types of females fertilized by the indicated types of males in each of six types of experiments. Experiments with immediately emerged flies and older flies are distinguished. Only one type of male but three types of females were present in the male choice experiments. Only one type of female but two types of males were present in the female choice experiments. Three types of female and both types of males were present in the multiple choice experiments.

	Young Flies				Old Flies			
	♂ ＼ ♀	+/+	+/y	y/y	♂ ＼ ♀	+/+	+/y	y/y
Pair mating (1♂:1♀) (100–170 pairs)	+	71	85	79	+	91	96	94
	y	1	8	50	y	22	40	82
Male choice (3♂:3♀)	46+	85	96	89	62+	98	98	100
	49y	8	6	82	62y	32	29	98
Male choice (5♂:15♀)	50+	52	72	78	70+	99	99	100
	75y	3	12	52	85y	22	21	97
Female choice (2♂:2♀)		146	194	104		120	128	106
	+	60	78	40	+	88	79	49
	y	3	1	39	y	5	12	45
Female choice (10♂:5♀)		143	136	90		170	135	142
	+	90	89	66	+	94	99	51
	y	3	6	26	y	4	5	47
Multiple choice (6♂:6♀)	+	90	91	72	+	79	83	57
	y	1	3	25	y	19	15	40

SOURCE: Data from Barker 1962.

There is qualitative agreement with the previous results with yellow. In all types of experiment, yellow males are highly unsuccessful in fertilizing wild-type females. They are especially so if young. They are more successful with yellow females and, in the case of the old flies, do almost as well as the wild-type males, except in the multiple choice experiment. Again there is no clear evidence of any differences among the three types of females in total mating success. In the multiple choice experiments, which are most comparable with population experiments, the relative selective values of yellow males in mating with type females is only 0.02 in young flies, 0.21 in old flies, while in the mating with yellow females the corresponding figures are 0.35 and 0.70.

Petit (1951) studied the selective disadvantage of Bar eye relative to wild type intensively. She estimated the relative viability during larval competition by introducing about 1,000 heterozygous females and an equal number of Bar or of wild-type males into a L'Héritier-Teissier cage (table 9.8).

TABLE 9.8. Numbers of Flies of Each Genotype from Mating $B\male \times B/+\female$ and $+\male \times B/+\female$.

	Female			Male	
	B/B	$B/+$	$+/+$	B	$+$
$B\male \times B/+\female$	153	216	...	171	244
$+\male \times B/+\female$	...	864	922	708	1,022
				879	1,266

SOURCE: Data from Petit 1951.

The viability of B/B relative to $B/+$ is 0.708 ± 0.075 and thus is significantly less than 1. That of B males relative to $+$ males is essentially the same (0.694 ± 0.030). On the other hand, $B/+$ does not differ significantly from $+/+$ (0.937 ± 0.044).

Petit also tested the relative productivity of B males. She made 13 such experiments in populations covering the range from 7% to 94% Bar among males. There were none in which there were significant differences in mating capacity among genotypes of females. On the other hand, Bar males were at a disadvantage in every case. The average relative selective value in Bar males from this cause was 0.497. The results were similar in eight other experiments in which the females were isolated before mating and left with the males for 16 hours. Again there were no significant differences among genotypes of females, while Bar males were at a great disadvantage in all cases ($w = 0.499$ in Bar males from this cause).

There was a strong indication of frequency dependence of this component of selective value since the average for the six cases in which frequency of Bar males was above 70% was 0.37, in contrast with 0.55 in the other 15 cases.

Similar experiments made two and three years later, however, showed that the coefficient could vary drastically under what seemed the same conditions. One series showed an average selective value of Bar of only 0.30 in six experiments while another showed no significant disadvantage (average 1.0 in five experiments). In both, however, the regression of the coefficient on the percentage of Bar males was strongly negative, confirming the type of frequency dependence of the first series (relative advantage of Bar in rarity).

Merrell (1965) studied preadult viability and mating capacities of males and females with respect to B and its wild-type allele. Among 4,357 offspring from $+\delta \times B/+\female$ the relative viability of B was 0.99. Presumably competition was less severe than in Petit's experiments.

Merrell's tests of mating success agreed with Petit's results in the absence of any significant difference among female genotypes and drastic discrimination against Bar males. The relative selective value for Bar males from this cause was indeed only 0.11 ($=23/209$). His results contrasted with hers in that the values of w in Bar males was higher with 35 B/B females (0.17) than with 85 $B/+$ females (0.13) or 94 $+/+$ females (0.07). As with Petit's results with white eye, the type of frequency dependence seems to vary with the conditions.

Neither Petit nor Merrell studied fecundity as a component of selective value. The indication from the population experiments that B behaves as recessive (in females) with respect to selective value seems to require more than the disadvantage in larval competition in Petit's (but not Merrell's) data. Strong selection against B/B but not $B/+$ in fecundity is suggested.

Merrell (1965) made tests of the preadult viability and of mating success of males for the mutations Bx^3, L^3, and bw referred to earlier, and L^2, which behaved much like L, and bw^D, which declined rapidly, in his population experiments (table 9.9). These were made at the same time as the tests of B.

TABLE 9.9. Tests of preadult viability from progeny of heterozygous females.

Genotypes	No.	w
$B/+$	4,354	0.99
$Bx^3/+$	2,434	0.94
$L/+$	1,672	0.99
$L/^2+$	2,948	0.90
$bw^D/+$	2,875	0.89
$+/bw$	2,602	0.94

SOURCE: Data from Merrell 1965.

None of these showed evidence of strong disadvantage in larval competition with their type alleles under his conditions.

Tests of differential selective mating of males were made in all cases. As already noted, Bar males were only 11% as successful as wild-type males, L^2 males were relatively unsuccessful with both $L^2/+$ and $+/+$ females (30%

and 40%, respectively). Brown males showed only 51% as much success with bw/bw females, 68% with $+/bw$ females as with $+/+$. The other male mutants were less handicapped (B_x^3, 82%; L, 91%; and bw^D, 93%), with no important differences according to type of female.

Lewontin (1955) studied hatchability of eggs and larval viability in 22 strains of *D. melanogaster*, all wild type except one white, and studied equal mixtures of white and each of the others at densities ranging from 1 or 2 to 40 per vial. In a later paper (1963) he made similar studies in *D. buskii* at densities ranging from 2 to 512 larva per vial using a wild strain, four mutations, and mixtures.

There was usually an intermediate optimal density at which the relative viability in mixtures tended to agree with those in pure cultures, but there were marked deviations at both higher and lower densities. In some cases, both strains did better in mixtures than in the pure cultures (mutual facilitation), which he showed could lead to equilibrium if the strains were not too unequal. In other cases, competition benefited one strain and depressed the other.

While laboratory studies of natural selection in higher animals have been largely limited to ones with *Drosophila* species, there are many more studies with other species of components of fitness than it is practicable to review here.

Especially noteworthy, however, are a number of studies by Sokal and his associates on the house fly (Sokal and Sullivan 1963; Sullivan and Sokal 1963; Bhalla and Sokal 1964) and on the flour beetle, *Tribolium castaneum* (Sokal and Huber 1963; Sokal and Karten 1964). The principal characters studied were viability from egg to emergence, time of emergence, and dry weight. Each genotype was studied by itself and at diverse frequencies in mixtures. Effects of different densities were studied systematically.

The general conclusion, like that from Lewontin's *Drosophila* experiments, was that the results in mixture were not predictable from those in pure cultures. Again, there might be improved performance of some or all genotypes in an association with others, in other cases lowered performance. The frequency dependence might also take the form of a selective advantage in rarity or in some cases the reverse. Selective values, whether frequency dependent or not, were functions of density so that the ranking in the same mixture might be different at different densities.

The studies of the components of selection in insects show that hatchability of eggs, larval viability, mating success (especially of males), fertility, and probably, though less studied, fecundity, all play important roles. The degrees of importance vary greatly among mutations and according to conditions. Differences in density, temperature, and in some cases lighting have important effects. The overall selection coefficients, both absolute and relative, are very far from being constants.

Competition between Varieties

It is not practicable to go far into the extensive literature on components of selection in domestic animals or cultivated plants. It is, however, desirable to touch on the subject of competition between clones of self-fertilizing plants since the contrast between reproduction capacity of these in pure stands and competitive ability in mixture are here especially clear.

Montgomery (1912) noted that "when left in competition, the variety (of cereal) which is the best yielder when placed alone may not always dominate but, on the other hand, a less productive type may be best able to survive competition." He also noted that "in almost every case with both wheat and oats, two varieties in competition have grown a greater number of plants at harvest and a greater yield than when either variety was sown alone."

Sukatschew (1928, discussed by Dobzhansky 1941) compared three clones of dandelions (*Taraxacum officinale*) collected in the same meadow near Leningrad with respect to percentage survival and number of flowers in pure and mixed stands at both high and low densities. There were wide differences among the varieties in all cases. One of the strains was highest in both respects in mixed stands, irrespective of density, although in pure stands it was the lowest in time of flowering, again irrespective of density, and the lowest in viability at high density. It was indeed first in viability at low density, showing that the two characters were not perfectly correlated in pure stands. The most important point for our purpose is that competitive ability, in which this strain clearly excelled, was far from perfectly correlated with rank in pure stands in either respect.

Harlan and Martini (1938) planted a mixture of 11 commercial varieties of barley (*Hordeum vulgare*) at each of 10 experiment stations, widely distributed throughout the United States. New crops were planted each year, with no artificial selection, from thoroughly mixed seed, from which a sample of 500 was taken each year to determine the relative frequencies of the varieties. The experiment was continued in each locality until the crop had come to consist largely of one variety. This point was reached in four years in some places but hardly in 12 years in others. The successful variety varied from station to station and that at a station was not necessarily the one that did best there in pure stands. Similar results with both barley and wheat varieties were obtained by Suneson and Wiebe (1942).

Sakai (1955, 1961) has made extensive studies of various aspects of competition among clones of crop plants, such as the effects of spacing and of soil fertility. Effects on vegetative and propagative characters were in general strongly correlated, for example, 0.90 between straw weight and yield of rice varieties. He found that competitive ability was to a large extent independent of vigor in pure stands, including that due to heterosis.

Interpretations can be made quantitative by path analysis. Data on average plant weights of seven varieties of rice (Sakai 1955) will be used as an illustration. Table 9.10 gives the grades in pure stands in the major diagonal (taken from a graph on an arbitrary scale). The deviations of the strains in equal mixtures with the other varieties are shown vertically. The competitive effects of each strain on the others are thus shown horizontally.

TABLE 9.10. Average plant weight for seven varieties of rice in pure stands (major diagonal on arbitrary scale) and deviations from this in equal mixtures with the other varities (vertical columns).

	I_2	I_1	I_3	J_4	J_2	J_1	J_3	Average
I_2	(9.9)	+1.4	+0.4	−1.0	−1.4	−1.5	−0.3	−0.4
I_1	−3.2	(9.4)	−3.2	−3.7	−3.2	−3.0	−1.0	−2.9
I_3	−1.8	+6.3	(8.5)	−3.9	−1.7	−1.4	+0.3	−0.4
J_4	−0.2	+7.4	+0.5	(7.8)	−0.3	−0.5	+1.4	+1.4
J_2	+0.9	+4.1	+2.7	−1.3	(6.7)	−0.3	−0.6	+0.9
J_1	+0.9	+5.7	+0.8	−1.3	−1.2	(6.2)	−0.7	+0.7
J_3	+0.2	+4.8	+2.2	−0.9	−2.2	−0.4	(5.9)	+0.6
Average	−0.5	+5.0	+0.6	−2.0	−1.7	−1.2	−0.1	−0.0

SOURCE: Data from Sakai 1955.

Two subspecies are represented. Strains I_1, I_2, and I_3 are of *Oryza sativa indica* while strains J_1–J_4 are of subspecies *japonica*. The former have greater plant weights and are on the whole the stronger competitors, but the relation is far from perfect. The lightest of all of the strains, J_3, was a somewhat stronger competitor than I_2, the heaviest strain. Moreover, I_1 was outstandingly the strongest competitor although lighter than I_2.

It may be seen from table 9.11 that there was no appreciable difference between the mean for the pure stands, P, and mixtures, M. The mixtures were, however, 2.4 times as variable and the differences ($D = M - P$) between weight in mixture and pure stand were 1.8 times as variable as the latter. The correlations among P, M, and D for given strains and for competitors (primes) are as shown in the table.

It is instructive to compare the results from various standpoints. Attention may be focused on the factors determining weight in mixture with differences treated as peripheral or the reverse. The factors may be restricted to weight

TABLE 9.11. Means, standard deviations, and product moment correlations for characters of seven rice varieties. P is the average plant weight in pure stands, M is that of the same variety, where mixed with another, and D ($= M - P$) is the excess in the mixture. The cross correlations between these, characters of the variety in question, and those of the competitive varieties (P', M', D') are also given.

	Mean	σ	Direct Correlations			Cross Correlations		
			P	M	D	P'	M'	D'
P	7.771	1.473	...	0.738	0.425	0	−0.326	−0.343
M	7.764	3.501	0.738	...	0.925	−0.326	−0.624	−0.652
D	−0.007	2.610	0.425	0.925	...	−0.343	−0.652	−0.681

SOURCE: Calculated from data from Sakai 1955.

in pure stand P, P' (as measures of innate growth potentials) and residuals U, U', or reciprocal interactions may be postulated and weight in pure stand P' of the competitor excluded as a direct factor.

We consider first the pattern in which M is represented as determined directly only by P, P', and U (fig. 9.13a).

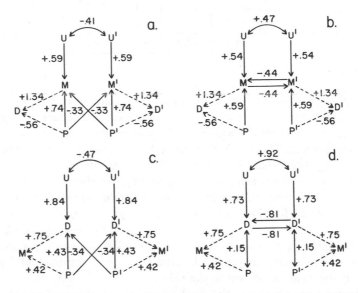

FIG. 9.13. Path coefficients calculated from the correlations in competition experiments with seven rice clones. Plant weights in pure stands (P), in equal mixtures with each other strain (M), and the differences D($= M - P$). Derived from data of Sakai (1955).

<table>
<tr><th colspan="2" align="center">Equations</th><th align="center">Path
Coefficients</th><th align="center">Determination</th></tr>
</table>

Equations	Path Coefficients	Determination
$r_{MP} = p_{MP} = 0.738$	$p_{MP} = 0.738$	$p^2_{MP} = 0.545$
$r_{MP'} = p_{MP'} = -0.326$	$p_{MP'} = -0.326$	$p^2_{MP'} = 0.106$
$r_{MM'} = 2p_{MP}r_{MP'} + p^2_{MU}r_{UU'} = 0.624$	$p_{MU} = 0.591$	$p^2_{MU} = \underline{0.349}$
$r_{MM} = p^2_{MP} + p^2_{MP'} + p^2_{MU} = 1$	$r_{UU'} = -0.410$	1.000

Not surprisingly, the most important factor in determining weight in mixture is the capacity to grow in pure stand (54.5%). The determination by the capacity of the competitor to grow in pure stand is relatively low (10.6%), leaving 34.9% determination by residual factors (specific effects of competitors, environment).

If reciprocal interactions between competitors, measured by $p_{MM'}$, is introduced and $p_{MP'}$ omitted, the following equations can be written, noting that $r_{MP'} = r_{M'P'}$ and similarly.

$$\text{Equations}$$

$$r_{MP} = p_{MP} + p_{MM'}r_{M'P} = r_{MP}/(1 - p^2_{MM'}) = \quad 0.738$$
$$r_{MP'} = p_{MM'}r_{MP} - r_{MP}r_{MM'}/(1 - p^2_{MM'}) = -0.326$$
$$r_{MM'} = p_{MP}r_{M'P} + p_{MM'} + p_{MU}r_{M'U} = -0.624$$
$$r_{MM} = p_{MP}r_{MP} + p_{MM'}r_{MM'} + p_{MU}r_{MU} = \quad 1$$

Path Coefficients		Determination	
$p_{MP} = +0.594$	$p^2_{MP} =$		0.353
$p_{MM'} = -0.442$	$p^2_{MM'} =$		0.195
$p_{MU} = +0.539$	$p^2_{MU} =$		0.291
$r_{UU'} = +0.473$	$2p_{MP}p_{MM'}r_{M'P} =$		0.171
	$2p_{MU}p_{MM'}r_{M'U} =$		-0.010
$r_{MU} = \quad 0.530$			1.000
$r_{MU'} = \quad 0.021$			

The replacement of a direct effect of P' ($p_{MP'} = -0.326$) by reciprocal interaction of the M's ($p_{MM'} = -0.442$) has reduced both the effects of P and U (fig. 9.13b). The negative sign of $p_{MM'}$ shows that competitive effects outweigh mutually stimulatory or mutually inhibitory effects. The most striking change is that in the sign of $r_{UU'}$. It appears that the residual term p_{MU} included nearly as much of the competitive effect as did that of P' in the first pattern, but that the introduction of a term $p_{MM'}$, representing reciprocal interaction, has absorbed not only the aspect of competition measured by P' but also that involved in the residual term, U', leaving the U's indicators of environmental facts that act similarly on both strains in the mixtures.

The correlations involving the changes due to mixture (apart from the common environmental effects just referred to) have not been used. D and D' can be introduced into the path diagram peripherally as determined by M and P. Since the path regressions $c_{DM} = 1$ and $c_{DP} = -1$, the path coefficients are given by the proper ratios of standard deviations $p_{DM} = \sigma_M/\sigma_D = 1.341$, $p_{DP} = -\sigma_P/\sigma_D = -0.565$. The same values can be obtained by solving the equations

$$r_{DM} = p_{DM} + r_{MP}p_{DP} = +0.925 \qquad p_{DM} = 1.341$$

$$r_{DP} = r_{MP}p_{DM} + p_{DP} = +0.425 \qquad p_{DP} = -0.565$$

The expected values of the three unused correlations can now be found.

	Calculated	Observed	Observed − Calculated
$r_{DP'} = p_{DM}r_{MP'} =$	-0.437	-0.343	$+0.094$
$r_{DM'} = p_{DM}r_{MM'} + p_{DP}r_{PM'} =$	-0.653	-0.652	$+0.001$
$r_{DD'} = p_{DM}r_{MD'} + p_{DP}r_{PD'} =$	-0.629	-0.681	-0.052

The changes in weight due to mixture may be considered to be measures of competitive ability. The analysis may be repeated, exchanging D and M in all equations. This yields the following results in the first pattern (fig. 9.13c).

Path Coefficients	Determination
$p_{DP} = 0.425$	0.180
$p_{DP'} = -0.343$	0.118
$p_{DU} = 0.838$	0.702
$r_{UU'} = -0.474$	1.000

There is relatively little determination (11.8%) by the growth potential of the competitor as measured by P' and not very much (18.0%) by that of the strain itself, leaving 70.2% hidden in the residual factor U, which shows a somewhat stronger negative correlation, -0.474, with residual factor of the competitor than where the total weights in the mixture were considered. On replacing P' by D' in the array of factors, the results were as follows (fig. 9.13d).

Path Coefficients	Determination	
$p_{DP} = +0.148$	p_{DP}^2	0.022
$p_{DD'} = -0.807$	$p_{DD'}^2$	0.651
$p_{DU} = +0.728$	p_{DU}^2	0.530
$r_{UU'} = +0.923$	$2p_{DP}p_{DD'}r_{D'P}$	0.082
$r_{DU} = +0.532$	$2p_{DU}p_{DD'}r_{D'U}$	-0.285
$r_{DU'} = +0.243$		1.000

The change in weight due to mixture is determined largely by the reciprocal interaction, 65.1%, and by the residual factors, 53.0%, and a negative contribution from the correlation between the residual factor and the change in weight of the competitor (-28.5%). The contributions from the weight in pure stand and its correlation with the change in the competitor are small. The most striking result is the near-identity, $r_{UU'} = 0.923$, of the residual factors of the competitors. They represent almost wholly variations in the common conditions, the effects of which were, as indicated by the standard deviations, very much greater than in the pure stands.

The path coefficients relating to M and M' as peripheral variables and comparison of calculated and observed values of the correlations involving M are as follows:

	Calculated	Observed	Observed − Calculated
p_{MD}	+0.746		
p_{MP}	+0.421		
$r_{MP'}$	−0.256	−0.326	−0.070
$r_{MD'}$	−0.652	−0.652	0
$r_{MM'}$	−0.591	−0.624	−0.033

There is again fairly good agreement of the calculated values for the correlations, not used previously, and the observed values.

Apparent Overdominance of Mutations

A number of cases have been referred to in which certain well-known genes of D. melanogaster have seemed to reach stability at a fairly high frequency in competition with wild type in population cages in spite of being so rare in nature that it has seemed probable that they only occur because of recurrent mutation. These include ebony in L'Héritier and Teissier's early experiments; brown in Merrell's experiments; and brown, Stubble, and perhaps black and sepia in Wiedemann's experiments. None of these were, however, carried far enough for confidence.

Nozawa (1958) described experiments in which brown actually rose in frequency in competition with its wild-type allele in a particular inbred strain, and was decidedly higher in egg viability (73% vs. 50%) and in speed of development. Rasmussen (1958) found apparent equilibrium for several mutations (among others ebony, sepia, and brown).

Buzzati-Traverso (1952) found that six populations starting with 12.5%, 50%, or 87.5% of the gene light, in competition with wild type, all converged toward 60%. This gene was found more frequently in certain wild stocks than most mutations but at a very much lower frequency than 60%. Other cases of apparent equilibrium at a rather high frequency could be cited. Some for which this was first reported seemed headed for extinction on being followed up for longer periods.

The results of Teissier's follow-up (1947) of the frequency of ebony in the population reported by L'Héritier and Teissier (1937) are typical. As noted, ebony's frequency had declined from 99.9% to 14% in 610 days with such a slow rate of change in the second year as to suggest approach to equilibrium.

The frequency continued at this level to the 28th month, at which time the population was divided into four parts. One was still at the same level (14.3%) in the 32nd month, but then fell to 11% in the course of seven months. In the second branch, the frequency declined for four months and then stayed at about 8% for seven months. In the third branch, the average remained at 14% to its accidental loss in the 43rd month, a total period of stability of about 60 generations. In the fourth branch there was stability at 8% between the 30th and 43rd months.

Three other populations, independent of the above and of each other, also reached apparent stability, one at about 10%, another at about 16%, and the third at 7%. Similar results were obtained with sepia.

The history of one of these lines is thus one of long periods of apparent stability at one level, followed by a fairly rapid shift to another level. There is no one equilibrium characteristic of a gene but rather diverse temporary equilibriums, presumably dependent on the genetic background and capable of change by recombination.

Susman and Carson (1958) introduced a single haploid set of autosomes from an inbred wild-type stock into a stock of sepia (s), spineless (ss), and rough (ro) (third chromosome recessives). There was a rapid decline in the frequency of the mutants but an apparent equilibrium was established at 16% rough ($q = 0.40$), 4% sepia ($q = 0.20$) that lasted 70 generations. Spineless declined to 1.5% but was not eliminated. The results at this stage seemed to indicate single-locus heterosis, at least for ro and se, under the conditions of the vials. Later tests by Smathers (1961) at the same laboratory demonstrated, however, the great importance of the genetic background, and indicated that apparent stability at one level or another depended on heterosis of persistent chromosome blocks rather than on single gene heterosis.

Similarly, Mukai and Burdick (1959) reported what seemed a case of single gene heterosis, associated with a second chromosome recessive lethal (also in D. melanogaster), and not associated with any crossover inhibitors. Two populations were started with 50% heterozygotes and two with only 5% heterozygotes. In spite of differences within each pair in genetic background, all seemed to approach the same equilibrium. Later, however, McAlpin,

Mukai, and Burdick (1960) found that the level could be altered greatly by changing the genetic background. Moreover, in the course of 72 generations, marked reductions in the gene frequency occurred in each line.

Frydenberg made thorough studies of a number of apparent polymorphisms in *D. melanogaster* using a type of cage devised by Bennett (1956) that maintains some 500 flies. In the first paper (1962) of a series, he described three populations that started from crosses between cinnabar (cn^m/cn^m) and a particular wild-type strain. In all three, the proportions of cinnabar flies declined rather uniformly from 25% to about 1% in the course of two years. There was no indication here of anything but competition of a slightly deleterious recessive with wild type. In six crosses with a different inbred wild-type strain, however, the proportions of cinnabar flies, after declining in ten generations from 25% to 10% or less, all rose, reaching an average of about 18% at 600 to 800 days.

The author's interpretation was that both the chromosome carrying cn^m and that from the inbred wild-type strain probably carried one or more deleterious genes, each with a sufficiently dominant allele in the other to give chromosomal heterosis and equilibrium at a certain frequency of cn^m ($\hat{q} = \sum s_b/[s_{cn} + \sum s_a + \sum s_b]$) where s_{cn} is the selective disadvantage of cn^m, $\sum s_a$ is that of associated deleterious genes, and $\sum s_b$ is that of deleterious genes in the wild-type chromosome. The randomization resulting from crossing-over of these in the wild-type chromosome lowers $\sum s_b$ and hence $\hat{q}$, while randomization of these linked with cn lowers $\sum s_a$ and hence increases q, and this may occur first if one of the loci is the furthest from cn^m. Both processes tend to reduce heterosis, which sooner or later is lost, leading to elimination of cn^m. The long-term trend of the frequency of cn^m is thus down, but there may be a short-term trend upward as in these six populations.

In his next paper (1963) he studied a Stubble mutation, Sb^W, lethal when homozygous, found in a wild population. This was linked with a short inversion, *In (3R) MO*, that interfered with, but did not wholly prevent, crossing-over. Ten cage populations started at gene frequency 0.50, declined in frequency less slowly than expected of a recessive lethal; some reached apparent stability, but one dropped to gene frequency 0.02 in generation 14. Ten populations that started at gene frequency 0.025 rose to levels ranging from 0.07 to 0.31, followed in four cases by gradual decline to lower levels and in one case by a rapid drop toward extinction. There was clearly no one equilibrium value.

In a further study (1964a), 35 cage populations, all started from gene frequency 0.50, fell into two groups. In one there was decline in frequency at a rate expected from a recessive lethal with 10% depression of the heterozygote. The other group maintained a much higher frequency of Sb^W for 12 to 60 generations, followed by decline and ultimate loss. In this second group, Sb^W was found to be coupled with the inversion as long as there was apparent equilibrium at a high frequency. In the first, coupling and repulsion were soon at random. The inversion tended to persist in all cases because of favorable effects when heterozygous, irrespective of the frequency of Sb^W, but female homozygotes proved to be sterile.

In a fourth paper (1964b), he observed three cage populations in which ebony competed with wild type. As with the other studies involving this gene, there were periods of temporary stability at various levels, but in the long run a trend toward elimination.

The general conclusion of these studies of populations in which the frequency of one of the conspicuous *Drosophila* mutations reaches an apparent equilibrium is that this is the result of the chromosome heterosis that is expected to occur often where only a few chromosomes carrying the mutation and its wild-type allele are brought together at the beginning. The very low frequency of such mutations in nature indicates their prevailing deleteriousness when in a population in which they are combined at random with linked loci.

Experiments with Chromosome Arrangements

So far, all of the experiments with laboratory populations that have been discussed have dealt with the frequencies of mutations that are very rare in nature. While purportedly dealing with allelic genes, it has become abundantly clear that the properties of unknown alleles at linked loci have played a major role in the results. In the case of those involving the mutation Curly, the alternatives were indeed whole chromosomes because of the associated inversion.

We will now consider laboratory studies of alternative chromosome arrangements that are known to be abundant in nature and thus necessarily balanced in one way or another. We are concerned here with the dynamics of their changing frequencies and will reserve for volume 4 discussion of their contributions to variability within and among natural populations.

The first such studies were conducted by Dobzhansky with natural inversion of *D. pseudoobscura*: Standard, *ST*; Arrowhead, *AR*; and Chiricahua, *CH*, and analyzed mathematically by the present author (Wright and Dobzhansky 1946). The flies all traced to ones collected at a single locality, Piñon Flats, California. Eleven experiments were made with all three arrangements in cages of the type devised by L'Héritier and Teissier. Six involved only *ST* and *CH*, one only *ST* and *AR*, and one only *AR* and *CH*. Two other experiments with *ST* and *CH* and another with *AR* and *CH* were analyzed similarly in a report the following year (Dobzhansky 1947) and will be considered here in conjunction with those reported in 1946.

Conditions were varied so that no two experiments were exact replications. Variations in lighting and nutrition had little effect in this case, but variations in temperature were important. Thus, changes in frequency were slight in five experiments at 16.5°C while striking at higher temperatures, usually about 25°C. The sample of each arrangement, introduced into each cage, was drawn from some half-dozen different laboratory cultures. The initial populations, drawn from crosses among the arrangements, consisted of several hundred to several thousand flies at known frequencies, checked after a short interval by sampling. Monthly egg samples were taken in six daily subsamples, and put in regular culture bottles. The salivary chromosomes of 25 of the developing larvae in each subsample (300 chromosomes altogether) were examined. Chi-square tests indicated substantial homogeneity.

Analysis began with determination of the change in gene frequency in each sample from the preceding one and rating of them to a constant length of interval of about a generation (3.5 weeks at 25°C, 5.2 weeks at 16.5°). Seven experiments, 30 intervals, involving all three arrangements at 25°C were deemed sufficiently comparable for grouping. Three experiments (eight intervals) involved all three at 16.5°C. Six experiments (21 intervals) involved ST and CH at 25°C and two involved these at 16.5°C. In addition, one involving all three arrangements at 21°C and one involving ST and AR at 25°C were not dealt with mathematically and two involving AR and CH

TABLE 9.12. Percentages of the Standard (ST), Arrowhead (AR), and Chiricahua (CH) arrangements at successive dates in experiment 18 and of the Standard and Chiricahua arrangements at successive dates in experiment 19. Both conducted at room temperature. Percentages are based on 300 chromosomes.

Time	Experiment 18			Experiment 19	
	ST	AR	CH	ST	CH
Initial	19.9	43.6	36.5	38.3	61.7
Mid-November 1944	33.3	27.3	39.3	...	...
Mid-December 1944	37.7	28.7	33.7	53.0	47.0
Mid-January 1945	39.3	30.0	30.7	63.3	36.7
Late February 1945	44.3	30.0	25.7	60.3	39.7
Late March 1945	42.0	39.0	19.0	65.3	34.7
Late April 1945	46.7	30.3	23.0	65.3	34.7
Early June 1945	56.4	27.3	16.3	70.4	29.6
Late July 1945	50.3	31.7	18.0	72.0	28.0

SOURCE: Data from Wright and Dobzhansky 1946.

TABLE 9.13. Mean gene frequency ($\bar{q}$), change per generation (Δq), regression of $\overline{\Delta q}$ on q, and variance of Δq for a given q, and tests of significance in each case.

	Two Types at 25°C	Three Types at 25°C			Two Types at 16.5°C	Three Types at 16.5°C		
	ST vs. CH	ST	AR	CH	ST vs. CH	ST	AR	CH
N	21	30	30	30	6	8	8	8
$\bar{q}$	0.5627	0.4201	0.3182	0.2617	0.7040	0.4927	0.2169	0.2904
$\overline{\Delta q}$	0.0434	+0.0391	−0.0122	−0.0268	−0.0125	−0.0156	+0.0167	−0.0111
$\mathrm{SE}_{\overline{\Delta q}}$	0.0110	0.0076	0.0068	0.0072	0.0162	0.0190	0.0110	0.0102
t	3.92	5.17	1.80	3.74	0.77	0.82	1.53	0.11
P	<0.001	<0.001	0.05–0.10	<0.001	0.40–0.50	0.40–0.50	0.10–0.20	>0.90
$b_{(\Delta q \cdot q)}$	−0.228	−0.387	−0.394	−0.261	+0.036	−0.237	−0.614	−0.055
SE_b	0.029	0.074	0.116	0.079	0.092	0.119	0.176	0.095
t	7.93	5.21	3.40	3.31	0.39	1.99	3.48	0.58
P	<0.001	<0.001	0.001–0.01	0.001–0.01	0.70–0.80	0.05–0.10	0.01–0.02	0.50–0.60
$\sigma^2_{(\Delta q \cdot q)}$	0.00063	0.00172	0.00139	0.00154	0.00163	0.00290	0.00096	0.00083
$\bar{q}(1-\bar{q})/300$	0.00082	0.00078	0.00071	0.00062	0.00059	0.00075	0.00055	0.00065
F	0.77	2.21	1.95	2.51	2.77	3.88	1.74	1.28
P	>0.20	0.001–0.01	0.01–0.05	0.001–0.10	0.05–0.20	0.01–0.05	>0.20	>0.20

SOURCE: Data from Wright and Dobzhansky 1946 and Dobzhansky 1947.

at 25°C are considered later for a special purpose. The type of history is illustrated in table 9.12 by two examples.

Table 9.13 gives the number of intervals, N; the mean gene frequency, $\bar{q}$; the mean change of gene frequency per generation, $\overline{\Delta q}$, and its standard error, $SE_{\overline{\Delta q}}$; the regression of Δq on q, $b_{(\Delta q)q}$, and its standard error, SE_b, and the variance of q for a given gene frequency, $\sigma^2_{\Delta q \cdot q}$, in comparison with that expected from sampling errors, $\bar{q}(1 - \bar{q})/300$, the ratios, t, of the last three quantities to their standard errors, and the probabilities that those ratios may arise from sampling (P).

There were no significant changes in gene frequency at 16.5°C, so that no further analyses seemed warranted. There were, however, highly significant changes in gene frequencies and their regressions at 25°C. There was even less variability of Δq for a given q than expected from sampling in the case of ST vs. CH. There was significantly more such variability in the experiments involving all three arrangements, but this was largely from differences in rates rather than in the equilibrium frequency. This may be illustrated by comparing the initial (control) and final frequencies in table 9.14.

TABLE 9.14. Initial and final frequencies of three chromosome arrangements in seven cage experiments with D. *pseudoobscura*.

Cage	Initial %			Months	Final %		
	ST	*AR*	*CH*		*ST*	*AR*	*CH*
1	27.1	35.0	38.0	6	52.9	27.1	20.1
2	34.5	39.9	25.6	6	53.7	33.0	13.3
5	44.2	37.1	18.7	5	57.3	30.7	12.0
6	29.0	28.3	42.7	3	44.3	21.0	34.7
10	35.7	24.7	39.7	3	48.3	22.3	29.3
12	22.0	38.3	39.7	3	50.7	30.3	19.0
18	33.3	27.3	39.3	9	50.3	31.7	18.0
Average (all cages)					51.1	28.0	20.9
Averages (excluding cages 6 and 10)					53.0	30.5	16.5

SOURCE: Data from Wright and Dobzhansky 1946.

The final frequencies do not differ very much and the most seriously divergent are two (cages 6 and 10) that were of such short duration (three months) that they had probably not yet approached close to their final frequencies. The other three-month experiment (cage 12) is fairly close to those of longer duration.

The data on competition between ST and CH were analyzed first on the hypotheses of constant relative selective values of the genotypes with the equilibrium due to heterozygous advantage. While generations overlapped in the cages, analysis was made in terms of the changes in frequency from one sample to the next, rated according to average length of generation as noted earlier.

With constant relative selective values:

$$\Delta q = -q(1 - q)[s_1 - s_2(1 - q)]/\bar{w}, \qquad \bar{w} = 1 - s_1 q^2 - s_2(1 - q)^2$$

Estimates were made of s_1 and s_2 by the method of least squares, letting $y = \Delta q$, $x_1 = q^2(1 - q)/\bar{w}$, $x_2 = q(1 - q)^2/\bar{w}$, using trial values of $\bar{w}$ and iterating.

$$-(\textstyle\sum x_1)^2 s_1 + (\textstyle\sum x_1 x_2) s_2 = \textstyle\sum x_1 y$$
$$-(\textstyle\sum x_1 x_2) s_1 + (\textstyle\sum x_2)^2 s_2 = \textstyle\sum x_2 y$$

Solution for the 13 intervals available in 1946 yielded $s_1 = 0.304$, $s_2 = 0.695$, $\hat{q} = s_2/(s_1 + s_2) = 0.696$. This value of $\hat{q}$ differed little from an estimated $\hat{q} = 0.685$ based on the regression, which was almost linear over the observed range.

The least squares estimate was revised in the 1947 paper on the basis of 21 available intervals. There was a little reduction of the selective disadvantages of the homozygotes $s_1 = 0.238$ for ST/ST and $s_2 = 0.621$ for CH/CH, but the equilibrium frequency changed only slightly ($\hat{q} = 0.723$).

Solution of the 1946 data was also made on the hypothesis of a linear selective advantage of rarity, with no heterozygous advantage (see vol. 2, pp. 136–38), accepting the approximate equilibrium frequency $\hat{q} = 0.70$.

$$\Delta q = -bq(1 - q)(q - \hat{q})/\bar{w}$$

where $q = a/b$ and $\bar{w} = 1 - (a - bq)(1 - 2q)$

Solution yielded $a = 0.90$, $b = 1.29$.

The expected relation of Δq to q with constant but different selective values in the sexes was also found, taking $\bar{s}_1 = 0.30$, $\bar{s}_2 = 0.70$, and $q = 0.70$. Cases of moderate and extreme sex differences were considered (subscript F = female, M = male). The expectations with a moderate difference did not differ appreciably from those with no difference. There was much difference in the extreme case below $q = 0.25$ but not above.

Genotype	Moderate Difference			Extreme Difference		
	w_F	w_M	$\bar{w}$	w_F	w_M	$\bar{w}$
ST/ST	0.8	0.6	0.7	1.0	0.4	0.7
ST/CH	1.0	1.0	1.0	1.0	1.0	1.0
CH/CH	0.4	0.2	0.3	0.6	0	0.3

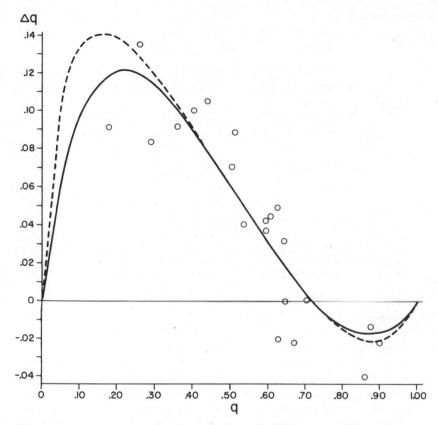

FIG. 9.14. Observed changes (*open circles*) in frequency of the *ST* arrangement of *D. pseudoobscura* in competition with *CH* in population cages, per 3.5 weeks, in comparison with theoretical curves for heterozygote advantage (*solid line*) or linear frequency dependence with the same equilibrium value ($\hat{q} = 0.70$) (*broken line*). From data of Wright and Dobzhansky (1946) and Dobzhansky (1947).

Figure 9.14 compares the expected relations of Δq to q under heterosis (*solid line*) (1946 data) and frequency dependence (*broken line*), including the 1947 data.

It is evident that the data fit all the preceding theories about equally well. There is, indeed, no difference whatever between the formulas for momentary rates (dq/dt) which differ from those for changes per generation only by omission of $\bar{w}$ in the denominators.

The experiments in which all three arrangements were present were analyzed only under the hypothesis of constant selective values relative to that of ST/CH. The theory for multiple alleles has been discussed earlier (vol. 2, pp. 38–47). Evaluation was made by the method of least squares with iteration for trial values of $\bar{w}$ in six normal equations, analogous to the two for the two-allele case. The solution is given in table 9.15.

TABLE 9.15. Estimated relative selective values of genotypes (*D. pseudoobscura*) involving *ST*, *AR*, and *CH* under the hypothesis of constancy.

GENOTYPE	INITIAL FREQUENCY	w $w_{13} = 1$	EQUILIBRIUM Genotype	Gene
ST/ST	q_1^2	0.43	0.152	$\hat{q}_1 = 0.531$
ST/AR	$2q_1q_2$	1.30	0.586	
ST/CH	$2q_1q_3$	1.00	0.173	
AR/AR	q_2^2	0.05	0.007	$\hat{q}_2 = 0.339$
AR/CH	$2q_1q_3$	0.71	0.078	
CH/CH	q_3^2	0.21	0.004	$\hat{q}_3 = 0.130$
			1.000	1.000

SOURCE: Data from Wright and Dobzhansky 1946.

The considerable differences between the relative selective values of *ST/ST* and *CH/CH* relative to *ST/CH* where *AR* is present and those where absent (*ST/ST* 0.43 vs. 0.76; *CH/CH* 0.21 vs. 0.38) indicate that they are not actually constants, as assumed in these estimates, but are functions of the frequencies of the arrangements present. In the case of *ST* vs. *CH* there are, of course, an infinite number of hypotheses intermediate between that of constant w's with heterosis and that of linear frequency dependence with no heterosis, which combine heterosis and frequency dependence in all proportions. The data from the three arrangements rule out constancy of the w's but are indeterminate over a wide range of hypotheses.

Direct Tests of Heterosis

An attempt was made by Dobzhansky (1947) to test the hypothesis of heterosis by comparing the genotypic frequencies in larvae from egg samples taken monthly from the cages but reared under optimal conditions and in adults of both sexes taken from the cages (after severe larval competition) and tested by crossing to homozygotes and examining six larvae. There were 17 samples of *ST* vs. *CH*, four of *AR* vs. *CH*.

The deviation under optimal conditions from Hardy-Weinberg distributions were relatively small in both sets, but the frequencies of heterozygotes were somewhat in excess of expectation in 19 of the 21 cases. If this is assumed to be due wholly to overdominance, the estimated selective values are as given in table 9.16.

TABLE 9.16. Estimates of relative selective values of genotypes of larvae (*D. pseudoobscura*) reared under optimal conditions.

Genotypes	w	Gene	w
ST/ST	0.91	*AR/AR*	0.92
ST/CH	1.00	*AR/CH*	1.00
CH/CH	0.78	*CH/CH*	0.80

SOURCE: Data from Dobzhansky 1947.

The apparent genotypic frequencies of the flies drawn off from the cages were corrected for incorrect classification of heterozygotes *ST/CH*, which happened to produce six homozygous larvae (1/64 classified as *ST/ST*, 1/64 classified as *CH/CH*, and similarly with *AR/AR* and *CH/CH*). The distributions for the eggs from the same cages, given optimal conditions, are also shown in table 9.17.

TABLE 9.17. Genotypic frequencies of genotypes (*D. pseudoobscura*) from cages in comparison with those from larvae reared under optimal conditions.

	% Frequencies: Cage 22				% Frequencies: Cage 23			
	No.	*ST/ST*	*ST/CH*	*CH/CH*	No.	*AR/AR*	*AR/CH*	*CH/CH*
Old males	125	20.8	68.8	10.4	100	22.0	63.0	15.0
Young males	...	...	...	...	134	25.4	52.2	22.4
Young females	130	23.9	63.8	12.3	100	24.0	63.0	13.0
Total	255	22.3	66.3	11.4	334	23.9	58.7	17.4
Optimal	150	27.3	54.7	18.0	150	32.0	49.3	18.7
Difference		−5.0	+11.6	−6.6		−8.1	+9.4	−1.3

SOURCE: Data from Dobzhansky 1947.

The heterozygote frequencies after severe competition are far above those of the binomial square distribution, under which they can never exceed 50% (found only if $q = 0.50$). They are well above the heterozygote frequencies in the larvae from eggs from the same cages reared under optimal conditions.

An excess of heterozygotes does not, of course, necessarily imply overdominance. The excess, at least in the case *ST/CH*, is, however, so great that some overdominance is virtually certain. Thus if it is assumed that *ST/ST* and *ST/CH* have the same selective value (mere dominance of *ST*) and that mating is random, the implied zygote frequencies are in the ratio 22.3 *ST/ST* : 66.3 *ST/CH* : 49.3 *CH/CH*, taking the first two unchanged from table 9.17. The implied relative selective value of *CH/CH* is thus 0.231 = 11.4/49.3. The percentage of *ST* has risen from 40.2% in the zygotes to

55.5% in the adults on this basis. The population was still well below the equilibrium frequency of 70% ST but the increase in ST is far above expectation. The evidence for heterosis is less convincing in the case of AR/CH.

Studies of larvae taken from a single locality in nature have indicated only a slight deviation from panmixia (Dobzhansky and Levine 1948). An analysis of 66 adult male samples (5 to 279 individuals) showed that in 56, including all with more than 50 individuals, the frequencies of the homozygotes were less than expected under panmixia and that the overall deficiency had a probability of less than 10^{-6} of origin by sampling errors. The overall deficiency of homozygotes was again so great, about 15%, that some overdominance is virtually certain. It should be noted that these were tests of heterosis only with respect to preadult viability.

The particular hypothesis of frequency dependence considered as a possible alternative to heterosis by Wright and Dobzhansky (1946) is ruled out by the excess frequencies of heterozygotes. On the other hand, a combination of some sort of frequency dependence and heterosis is quite possible. Spiess (1957) suggested this on the basis of irregularities in the courses of selection of chromosome arrangements of *D. persimilis* collected in Yosemite, California. In the case of Whitney (WT) vs. Klamath (KL) he interpreted the course as indicating a minority disadvantage, while in that of Whitney vs. Mendocino (MO), he suggested minority advantage.

There are so many other causes of irregularities (selection among chromosomes with the same arrangement but different selective properties, unknown changes in conditions, and accidents of sampling) that conclusions based on irregularities are apt to be inconclusive. To demonstrate a mode of frequency dependence conclusively requires direct tests of selective value at different frequencies (as in male and female choice experiments referred to earlier).

Tests of Frequency Dependence

Ehrman (1967) has made such tests of frequency dependent mating with the Chiricahua and Arrowhead arrangements of *D. pseudoobscura* using two different strains, A and B. She observed mating in groups of 20 males and 20 females. The ratios of $CH/CH:AR/AR$ (distinguished by wing notches) varied from 2:18 to 18:2 at intervals of 10% in both sexes. She observed about 1,000 matings in each strain. Table 9.18 gives the ratios of mating success per fly of CH/CH to that of AR/AR.

It is of interest to test how well these data may be fitted by assuming linear frequency dependence of the selective values of CH/CH and AR/AR on the frequency of CH. $w_{CHCH} = 1 + a - bq$, $w_{ARAR} = 1 - a + bq$, $R = w_{CHCH}/w_{ARAR} = (1 + a - bq)/(1 - a + bq)$. Nine normal equations of the

TABLE 9.18. Ratios, R, of mating success per fly of CH/CH to that of AR/AR in observations made with different frequencies of CH.

% CH	Average						
	$A\male$	$B\male$	$\male$	$A\female$	$B\female$	$\female$	$(\male + \female)$
0.10	1.45	3.71	2.58	1.13	1.07	1.10	1.84
0.20	1.69	2.88	2.28	1.02	1.64	1.33	1.81
0.30	1.83	1.22	1.52	1.18	1.48	1.33	1.43
0.40	1.82	1.36	1.59	0.94	0.94	0.94	1.26
0.50	0.96	0.74	0.85	0.93	0.95	0.94	0.90
0.60	0.47	0.37	0.42	0.88	0.96	0.92	0.67
0.70	0.64	0.59	0.62	0.69	0.83	0.76	0.69
0.80	0.32	0.44	0.38	0.97	0.98	0.97	0.68
0.90	0.13	0.32	0.23	0.83	0.86	0.85	0.54

SOURCE: Data from Ehrman 1967.

type $a - bq = (R - 1)/(R + 1)$ can be written from the entries in any column of table 9.18 and solved by the method of least squares. The results from five of the columns are given in table 9.19.

TABLE 9.19. Coefficients of the selective values of CH/CH and AR/AR ($D.$ *pseudoobscura*) under the hypothesis of linear frequency dependence.

	a	b
$A\male$	0.553	1.295
$B\male$	0.641	1.385
$\male$	0.634	1.388
$\female$	0.128	0.258
$\male$ and $\female$	0.399	0.811

SOURCE: Calculated from data of Ehrman 1967.

The relation of R to q is curvilinear in contrast with the w's. The curve for males and the observed values in strains A and B and their average are shown in figure 9.15 (left). The observed values of A and B diverge considerably at low values of q, especially at $q = 0.1$, but the curve fits the average reasonably well. The same is true of the slightly declining curve for females in figure 9.15 (right). The curve for the unweighted average of the males and females is shown in figure 9.16. Anderson (1969) suggested use of w's with hyperbolic frequency dependence. The observed curvature of R

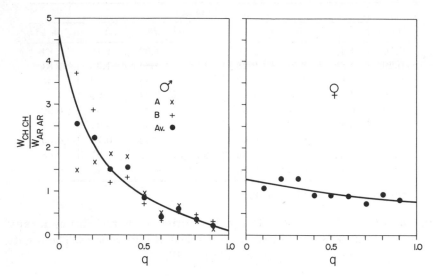

FIG. 9.15. Ratio of mating success of CH/CH to that of AR/AR of *D. pseudoobscura* in observations at different frequencies of CH. Curves based on hypothesis of linear frequency dependence. From data of Ehrman (1967).

does not, however, differ much from that expected from linear frequency dependence.

Ehrman (1969) showed by an ingenious series of experiments that the cue by which the females recognized the relative frequencies of the two types of males is odor. Sound and touch were excluded.

Spiess (1968) made similar observations of the success of AR relative to the Pikes Peak (PP) arrangement of *D. pseudoobscura*. He found minority advantage among males, but it was much less striking than in Ehrman's studies of CH and AR. Where the proportions of the sexes varied together (as in Ehrman's studies), the success of PP relative to AR varied only from 0.47 for 18 PP:2 AR to 1.00 for the opposite proportions, a twofold difference in contrast with Ehrman's elevenfold difference. There was somewhat more minority advantage of males if the females were always in the proportions 10:10, the corresponding values being 0.47 to 1.23. If the proportions varied only in females (males 10:10), the relative success of PP males varied only slightly (0.42 with 18 PP:2 AR in females), 0.65 with 2 PP:18 AR in females. In no case was there appreciable minority advantage in females. In both sexes, however, PP was only about 60% as successful as AR.

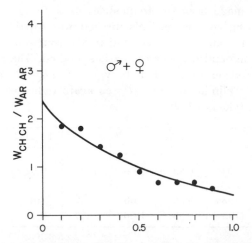

FIG. 9.16. Unweighted averages of males and females of figure 9.15.

Spiess also found minority advantage of males (but not females) in similarly varying proportions of heterozygotes (AR/PP) and homozygotes (AR/AR) of males (females always 10:10). The range was from 1.25 (2 Het:18 Hom) to 0.78 (18 Het:2 Hom), average 1.00. In females, AR/AR possibly had a slight advantage (1.21) over AR/PP with no significant differences according to proportions in the male.

Variation of the Same Chromosome Arrangement

Levine, Pavlovsky, and Dobzhansky (1954) conducted cage experiments (at 25°C) with the same three arrangements as Wright and Dobzhansky (1946) from the same locality, but a later collection. The results of two cage experiments, started from 25% ST, 25% AR, 50% CH, agreed well with each other but differed widely from the earlier estimates with deduced equilibrium frequencies 53.1% ST, 33.5% AR, 13.0% CH.

Cage	Initial			Months	Final		
	ST	AR	CH		ST	AR	CH
A	25.0	25.0	50.0	12	80.0	16.7	4.3
B	25.0	25.0	50.0	12	81.7	12.7	5.7

Least squares estimates of relative selective values on the hypothesis of constant w's indicated ultimate elimination of CH and an impossible slightly negative value for AR/AR. Levine devised an iteration method in which deductions were made from the frequencies in their actual order, on the basis of minimizing chi square. This method was capable in principle of taking account of changes that occurred in the course of the experiments that would be confounded under the earlier method. The results were not very different, however, except that the value of AR/AR came out slightly positive. Ultimate elimination of CH was again indicated. The value for ST/AR was here taken as 1.00.

		ST/ST	ST/AR	ST/CH	AR/AR	AR/CH	CH/CH
From Δq by least squares	(1946)	0.33	1.00	0.77	0.04	0.55	0.16
From Δq by least squares	(1954)	0.83	1.00	0.67	(0)	0.60	0.43
Minimum χ^2	(1954)	0.83	1.00	0.77	0.15	0.62	0.36

In the new sample, the relative selective value of ST/ST was apparently enormously superior to that of the earlier sample. The same authors (1958), however, made another experiment with the same arrangements from the same strains but with only 5% ST initially instead of 50% and discrete generations. Replications agreed well with each other but not at all with the predictions based on the previous results.

Since there can be only one triple equilibrium under the hypothesis (or under linear frequency dependence), this seems to imply that there were two or more genotypes for at least one of the arrangements and that selection reduced the number of genotypes present according to the initial frequencies. A role of frequency dependence was thus indicated.

Pavlovsky and Dobzhansky (1966) conducted cage experiments with various combinations of the arrangements ST, AR, CH, and two others, Pikes peak (PP) and Tree Line (TL), all derived from strains collected at Mather, California. In some cases, the Δq's clustered about a linear regression line of Δq on q as with ST vs. CH in the 1946 and 1947 data, but in others there was so much scatter at each value of q that no analysis could be made. In these it is again indicated that the same gene arrangement included chromosomes with widely different properties among which there was selection in the course of the experiments, with consequent changes in the set of relative selective values of the arrangements. In most cases, the heterozygotes came out with higher selective values than the homozygotes under the hypothesis of constant sets, but not in all, for example in AR vs. PP, the latter a rare arrangement at Mather until a few years before the samples used here were taken. As in the preceding cases, wide differences between the apparent selective values in experiments with two and those with more arrangements tended to indicate dependence of selective values on the composition of the array in the cage.

A series of eight experimental populations of *D. pseudoobscura*, all involving *ST*, *AR*, *CH*, and *PP* from the same strains used in the preceding study, gave surprisingly different results where the initial arrangement frequencies were different, although excellent agreement of replications (A, B) (table 9.20).

TABLE 9.20. Initial and final frequencies of arrangements of *D. pseudoobscura*. There are two replicates of each initial set.

	Initial					Final			
	ST	*AR*	*CH*	*PP*	Months	*ST*	*AR*	*CH*	*PP*
IA	30.0	5.0	50.0	15.0	12	82.7	15.7	1.3	0.3
IB	30.0	5.0	50.0	15.0	12	78.0	20.3	1.0	0.7
IIA	15.0	5.0	30.0	50.0	12	78.7	18.0	2.7	0.7
IIB	15.0	5.0	30.0	50.0	12	72.3	25.0	2.7	0
IIIA	5.0	50.0	15.0	30.0	12	36.7	63.0	0	0.3
IIIB	5.0	50.0	15.0	30.0	12	34.0	65.0	0.7	0.3
IVA	25.0	25.0	25.0	25.0	12	63.0	36.3	0	0.7
IVB	25.0	25.0	25.0	25.0	12	64.3	35.0	0.3	0.3

SOURCE: Data from Watanabe et al. 1970.

The similarity of replicates in all four cases precludes interpretation in terms of merely accidental differences in the constitutions of different experiments. In all cases *CH* and *PP* appeared to be tending toward elimination, but the frequencies of *ST* and *AR* had reached very different frequencies after 12 months, depending on whether the initial frequence of *AR* was low, *CH* high (I, II), or the reverse, III, with IV intermediate. The frequency of *AR* rose markedly even when its initial frequency was 50% (in III) and *ST* rose to such high frequencies in I, II, and IV that the equilibrium frequency of *AR* in these must have been far below 50%. The conclusion again indicated that there were differences within the *ST* and *AR* samples and that different ones became established by frequency dependent selection if the initial frequencies differed sufficiently.

Sex Ratio Genes

The tendency of male *D. pseudoobscura* with a certain type of X chromosome to produce almost exclusively X-bearing sperms has been referred to earlier as a condition that behaves like meiotic drive. This X chromosome has been shown to carry three closely linked inversions and to exhibit very little crossing-over with the standard chromosome ST in the right arm. Wallace (1948) studied the course of change in its frequency in cages of the type devised by L'Héritier and Teissier and also determined relative values of the probable major components of selective value of the various genotypes.

Two of Wallace's cage experiments were maintained at 25°C, two at 16.5°C. At each temperature, one cage started with 500 SR males, 500 ST males, 500 ST/SR females, 500 ST/ST females, while the other started from a similar array of males but 500 SR/SR females, 500 ST/SR females. The course of change in the four cages in terms of the frequency of SR are shown in table 9.21.

TABLE 9.21. History of four cage experiments with respect to frequency of the sex ratio X chromosome (vs. the standard X chromosome) of *D. pseudoobscura*.

	25°C					16.5°C			
	Cage 10		Cage 11			Cage 12		Cage 13	
Day	♂	♀	♂	♀	Day	♂	♀	♂	♀
0	0.500	0.250	0.500	0.750	0	0.500	0.250	0.500	0.750
39	0.226	0.266	0.273	0.421	20	0.216	0.323	0.361	0.421
61	0.091	0.184	0.167	0.301	77	0.244	0.352	0.350	0.349
96	0.040	0.143	0.023	0.189	112	0.279	0.243	0.346	0.363
142	0.034	0.049	0.059	0.043	162	0.151	0.155	0.200	0.270
188	0.036	0.016	0	0.012	202	0.100	0.167	0.220	0.363
212	0	0	0	0	272	0.078	0.141	0.089	0.154
					303	0.090	0.151	0.054	0.138
					434	0.087	0.062	0.060	0.048

SOURCE: Data from Wallace 1948.

It may be seen that SR declined in frequency at both temperatures in spite of the excess number of females it produced. Its frequency declined at least three times as rapidly at 25°C as at 16.5°C. In the former it was almost eliminated at 188 days and wholly at 212 days, while at the lower temperature SR was still fairly common at 434 days and may have been approaching an equilibrium. On making allowances, however, for the different average intervals between generations (about 25 days at 25°C, 40–45 days at 16.5°C) the rate of decline was only about twice as great as 25°C as at 16.5°C.

It is evident that the advantage of SR from the excess production of females (as long as an adequate number of males is present) must be offset by disadvantages to bring out its reduction to low frequencies in nature, and its probable elimination in the cage experiments. Wallace estimated the various components of selective value of each genotype by suitable experiments (taking those of ST and SR/ST as 1.000). Table 9.22 shows his results, using averages of two slightly different estimates for both mating success of males and fecundity of females. The total average selective value is given by the product in each case.

TABLE 9.22. Components of selective value in *D. pseudoobscura* in genotypes involving the sex ratio (SR) and standard (ST) chromosome arrangements. Two closely similar estimates of sexual activity of males and two closely similar estimates of fecundity are each averaged.

	At 25°C			At 16.5°C		
	♂ ($ST = 1$)	♀ ($SR/ST = 1$)		♂ ($ST = 1$)	♀ ($SR/ST = 1$)	
	SR	SR/SR	ST/ST	SR	SR/SR	ST/ST
Larval competition	0.407	0.152	0.847	0.740	0.622	0.940
Longevity	1.086	0.645	1.000	1.113	0.819	1.000
Fecundity	...	0.603	0.579	...	0.548	0.842
Sexual activity	0.804	...	...	0.882	...	...
Egg hatchability	...	0.234	0.603	...	1.000	1.000
Product	0.355	0.014	0.296	0.726	0.279	0.791

SOURCE: Data from Wallace 1948.

The conditions under which these estimates were made differed from those of the cage populations so that no quantitative explanation of the courses of change in the cages would be expected. It is, however, interesting to compare the results.

The equilibrium frequencies (if any) of SR in eggs, $\hat{q}_e$, and sperms, $\hat{q}_s$, have been given in volume 2, equations 3.142 and 3.140, respectively:

$$\hat{q}_e = (w_1 + 0.5 - w_{22})/[2w_1(1 - w_{11}) + 1 - \dot{w}_{22}]$$
$$\hat{q}_s = w_1\hat{q}_e/[1 + (w_1 - 1)\hat{q}_e]$$

	w_1	w_{11}	w_{22}	$\hat{q}_e$	$\hat{q}_s$
25°C	0.355	0.014	0.296	0.398	0.190
16.5°C	0.726	0.279	0.791	0.346	0.277

The equilibrium frequencies of SR, implied by the estimated components, may seem unexpectedly high and it may seem surprising that the frequency in eggs at 25°C is actually higher than at 16.5°C in view of the much lower values of w_1 and w_{11} at this higher temperature. The selective value (w_{11}) of SR/SR plays, however, only a minor role if this genotype is uncommon and the relatively low value of w_{22} (for ST/ST) offsets the effect of the low value of w_1. The elimination of SR at 25°C and its near-elimination at 16.5°C in the cage populations seem to indicate that w_1 was so small and w_{22} so large in these in nature that the latter was in excess by 0.5, at least at 25°C and probably also at 16.5°C.

The persistence of SR in nature at moderate frequencies (11% SR males among 144 tested in the collection at Piñon Flats, California, to which the experimental material traced) indicates that conditions in nature were intermediate between those of the cage experiments and those under which the components of selective value were estimated.

Differentiation of Isolated Cage Populations

Vetukhiv (1954, 1956) found significant heterosis in F_1 of crosses between local populations of *D. pseudoobscura*. He later started a series of experiments with six populations of *D. pseudoobscura*, all of common mixed origin. They were maintained in population cages; two at 16°C, two at 25°C, and two at 27°C. These were maintained by others after his death in 1959. Ehrman (1964) found that after $4\frac{1}{2}$ years of isolation there was a statistically significant greater tendency toward mating of flies from the same strain (54.7%) than from different ones (45.3%) when ten males of one sort were put with ten females of the same sort and ten of another sort. The difference was in the same direction, with minor exceptions, irrespective of the temperatures at which the strains had been maintained or the temperatures at which the tests were made. It is remarkable that incipient sexual isolation should have arisen in such a short time.

Mourad (1965) studied the longevity of flies from four of the same six strains as above and F_1 and F_2 hybrids, all at 16°C and 27°C, again after $4\frac{1}{2}$ years of isolation. There were significant tendencies toward heterosis in F_1 and breakdown of vigor in F_2 at both temperatures. Thus different interaction systems seem to have been arrived at in the course of $4\frac{1}{2}$ years.

Anderson (1966) studied body weight and wing length and developmental time in the same six populations as above. He found that significant divergence in wing length (0.005 level) and developmental time (0.05 level) had occurred after six years. The flies kept at 16°C had become larger and developed more rapidly than those kept at 25°C or 27°C when tested under the same conditions. Significant heterosis was found in most cages, most of which was lost in F_2.

Selection for Resistance to Injurious Substances

Resistance to injurious effects of substances in the environment is necessarily an important class of adaptive characters for all organisms. Natural selection for such resistance was observed in the spreading resistance of red scale insects to cyanide poisioning in the orange groves of California (Quayle 1938). Dickson (1941) found that laboratory data could be accounted for by a single sex-linked resistance factor.

Since this time the increasing resistance of insect pests to DDT, and of bacteria to antibiotics, testifies to the importance of natural selection of this sort. A few examples of the many laboratory studies must suffice.

Crow (1954, 1957) exposed a large, highly heterogeneous cage population of *D. melanogaster* to increasing amounts of DDT from generation to generation. A relatively resistant strain was produced. A dose that killed 90% of the controls killed only 5% of the selected population. There was no short-term tendency to return to susceptibility on relaxation.

He made crosses and backcrosses to a susceptible line that carried markers for the three major chromosomes with crossing-over prevented by exclusive use of males as the crossbred parents. The percentages of survival ranged from about 5% to 70% according to chromosome composition. It was evident that each major chromosome from the resistant line contributed significantly to survival when heterozygous and that the effects were roughly additive, except that there was dominance of the resistance factor of the X chromosome. Further analysis indicated that there were at least two regions in chromosomes II and III that contributed independently. Oshima (1958) obtained somewhat similar results.

King (1955*a*,*b*) also developed resistance in *D. melanogaster* by selection, but the genetics was rather different. Most of his selection lines were derived from a mass culture, tracing to a collection of a few wild flies, that was much more homogeneous than Crow's foundation stock. He exposed large batches of flies to an aerosol carrying DDT at doses that would kill about 50% in some experiments, 95% or 99% in others.

Several strains developed appreciable resistance after a dozen generations and more in later generations. Reciprocal crosses with an untreated susceptible control strain gave intermediate resistance in F_1 that declined significantly in F_2 but returned to nearly the F_1 level in F_3. Crosses between different resistance lines were also intermediate, also declined in resistance in F_2 and returned roughly to the F_1 level in F_3, and stayed at this level in one line carried another generation. Representative results in terms of the estimated number of minutes to 50% mortality from a particular aerosol are given in table 9.23.

The genetic difference seems to have consisted of multiple semidominant genes as in Crow's case, but the decline in F_2 seems to imply that interaction systems had been built up and that these differed in different lines (selective advance up different selective peaks). Each system would be intact at F_1 at least in heterozygous state, either from the cross to the susceptible strains or between different resistant ones, but would be broken up in F_2. This,

TABLE 9.23. Estimated number of minutes to
50% mortality from a particular aerosol containing DDT, in crossbred populations of *D.
melanogaster*. R = resistant, C = control.

	F_1	F_2	F_3
R♀ × C♂	7.8	4.8	6.4
C♀ × R♂	7.4	4.8	7.3
R_1 × R_2	13.6	7.4	11.4
R_2 × R_1	12.5	7.1	11.7

SOURCE: King 1956a.

however, leaves unexplained the unselected recovery of resistance in F_3, which
is expected to have the same genetic composition under random mating as
F_2, except for further breaking up of linkages and hence of interaction
systems.

Later King and Sømne (1958) made extensive studies of two of the resistant
lines with results much more like those of Crow. The differences between
these lines and a nonresistant strain from the stock from which the resistant
lines had been developed by selection were subjected to chromosomal
analysis. In each case, the LD_{50} was obtained for each of the 27 possible
combinations from the three pairs of major chromosomes.

There were highly significant differences in each pair of chromosomes, but
no significant interaction effects among chromosomes and, with one exception,
no deviation from exact intermediacy on the log-dose-probit scale of
mortality that was used. The exception consisted of near-dominance of
resistance in the case of the X chromosome of one of the lines. The contributions of the chromosomes from the two resistant lines were clearly not the
same in at least this case. In one line the contributions on the above scale
were 0.17:0.30:0.22 for X, second, and third chromosome respectively; in
the other, 0.31:0.24:0.21 in the same order.

Sokal and Hunter (1954) also successfully selected a strain of *D. melanogaster* in which the larvae were resistant to DDT. They noted that the resistant larvae tended to pupate at the edge of the medium and found that
this tendency was strongly heritable (as found later by de Souza, da Cunha,
and dos Santos in *D. willistoni*). They further showed that selection of the
control stock for peripheral pupation not only increased this tendency but
automatically increased larval DDT resistance.

Summary

Aspects of the theory of gene frequencies, discussed in volume 2, have been
checked by numerous experiments, mainly with species of *Drosophila*, in

which cages have been populated by an array of individuals with known frequencies of particular sets of alleles or of chromosome arrangements, usually far from those found in nature, and these individuals and their descendants have been allowed to reproduce without any artificial selection. A considerable number of such experiments have been described in this chapter. The mathematical expectations from different situations discussed in volume 2 were expressed in terms of the immediate change in gene frequency. For use in interpretation of experiments conducted over many generations, it has been necessary to extend the theoretical results by integration of the immediate rates.

Experiments involving a single pair of alleles with easily distinguishable affects have usually shown general agreement with some simple theoretical interpretation, but there has also usually been more or less complication, indicative of the complexity of natural selection.

Observed genotypic frequencies often differ considerably from the expected random union of gametes merely because there has been some selection prior to the stage at which observations have been made and the relative selective values have not been in the ratio $W^2 A_1 A_1 : W A_1 A_2 : 1 A_2 A_2$ that leaves genotypes in a binomial square distribution if a distribution of this type had been produced initially.

Genetic unions may not, however, be random. It has been found that wild-type females often reject mutant males to a greater extent than do mutant females.

There is often differential selection of the sexes in one or more respects, such as mortality at various stages and mating success, especially of males. Differences in fecundity of females of different genotypes are often important. Fortunately for simplicity of analyses, the course of change of frequencies in populations is given approximately by averaging the selective values of the sexes even where the selective differences are moderately great.

Factor interaction has long played a major role in the theory of natural selection, but it has rarely seemed necessary to invoke it in the type of experiment referred to above. Frequency dependence has also been stressed as a complicating factor in selection in nature, but until rather recently it was rarely invoked in interpreting laboratory results. It now appears to be common. In particular this seems to be true of minority advantage with respect to success of males in mating. Where there is approach to an equilibrium, it requires special experiments to decide how far it is due to overdominance or to minority advantage, or to certain other processes.

The difference between the fitness function of a population that determines the course of genotypic change within it, and its mean selective value that determines its success as a whole in isolation, has often been emphasized here. The success of a population in competition with another involves both its mean selective value and its competitive ability, which may be high or low generally, or high or low specifically in particular cases. The contrast between degree of success of strains by themselves and in mixtures is brought out especially clearly where there are clones that cross rarely if ever. The factors involved in success of plants in pure stands and in mixtures differ considerably in different cases. Path analysis of a case involving a number of varieties of rice is gone into in some detail.

In experiments with *Drosophila* mutations, it has often been found that cage populations reach an apparent equilibrium with a wild-type allele at a frequency far above that found in nature. These have sometime been interpreted as indicating single-locus heterosis but it has usually been found that if such experiments are continued for many generations after attainment of apparent stability there is ultimately a rapid drop in frequency to a lower level, more rarely a rise, followed in the course of time by similar abrupt changes, until the mutation is ultimately eliminated. The interpretation in such cases is that the alleles under observation were not in linkage equilibrium with other gene differences between the strains, often inbred, that had been used in making up the populations. It is likely that each of the chromosomes that carry the alleles in question carry deleterious genes, absent in the other, and that complementation in the heterozygote gives a chromosomal heterosis that may be abolished only after considerably delayed recombination where linkage is strong.

Alternative chromosome arrangements, found abundantly in nature in many species of *Drosophila*, have been the subject of many cage experiments, some of which have been discussed here. Cage experiments with samples of two or three arrangements from a locality have usually led to a stable set of frequencies indicating one or more mechanisms of equilibrium such as chromosomal heterosis or minority advantage or a combination. In some cases, the frequencies in all cages started from many different representatives of the same arrangements have come to essentially the same set of frequencies, indicating that all representatives of each arrangement had about the same properties. In other cases, however, this has not been the case, indicating either initial differences among representatives, or differences that arose during the experiment by a rare recombination or mutation.

Equational segregation is indicated for most alleles, but a number of cases have been studied involving "meiotic drive." These include the sex ratio gene (or chromosome) in several species of *Drosophila*, the condition "segregation distorter" of *D. melanogaster*, and the *t* locus of mice, all widely distributed in nature. Cage experiments with "sex ratio" were associated with a systematic attempt to determine the various components of the selective values. In this, as in other cases, the strong dependence of these components on conditions, especially of crowding, has prevented more than qualitative interpretation of the course of change in cage experiments.

Natural selection has been shown to be rapidly effective for resistance of insects to poisons and antibiotic resistance of bacteria.

The most general conclusion is that natural selection is effective in meeting challenges of the most diverse sorts.

CHAPTER 10

Experimental Stochastic Distributions of
Gene Frequencies

The effects of inbreeding, discussed in chapters 2 to 5, have been treated as due primarily to the rapid random drift from accidents of sampling in lines that consist of only one individual per generation in the case of self-fertilization, two in the case of brother-sister mating, four in double first cousin mating, and so on. The complications of the results of artificial selection in small populations have been noted in the chapters dealing with this subject. We consider first in this chapter some experiments in which comparisons have been made between the courses of natural selection in population cages containing thousands of *Drosophila*, according to whether they pass through bottlenecks of small size or not. It has been noted earlier that the effective size of a population depends theoretically much more on the number at population minimum than on the average number. Where there are recurrent bottlenecks, the effective size is the harmonic mean, not the arithmetic mean of the numbers in the successive generations (vol. 2, pp. 210, 215). Later in this chapter, experiments will be discussed in which the actual distributions of gene frequencies were determined in populations maintained at very small sizes, generation after generation.

Natural Selection Following One or More Bottlenecks
of Population Size

Dobzhansky and Pavlovsky (1957) started from 12 strains of *D. pseudo-obscura*, collected near Austin, Texas, in each of which a particular third chromosome carrying the Pikes Peak (PP) inversion had been fixed, and 10 strains collected in Mather, California, in which a particular third chromosome carrying the Arrowhead (AR) inversion had been fixed. Crosses were made, involving all of these, and carried to F_2. Ten population cages were started from foundation stocks of about 4,000 flies, each with approximately equal representation of all of the PP and AR strains. These were their "large" populations. Ten other cage populations were started, each from 20 flies

(10 males, 10 females) representing all F_2 cultures. These were their "small" populations. They were smaller than the large populations only in having passed through a bottleneck of 20 flies, instead of 4,000 at the start. All cages started with 50% PP.

After four months (about four generations), egg samples were taken and the pairs of third chromosomes of the salivary glands of 150 larvae, which had hatched from the eggs, were classified. The percentage of PP varied from 25% to 42% in the small populations and from 29% to 42% in the large ones, with significant heterogeneity in both. The small populations showed a greater variance (26.7) than the large ones (15.3), but the difference was not significant.

A second test, made after 17 months from the start (19 generations), showed much more differentiation of the small populations (range 16% to 47% PP), in contrast with 21% to 35% in the large population. The difference in variance (small 119, large 27) was now clearly significant at the 2.5% level. An average selective advantage of AR was shown by the decline of the mean frequency of PP from 50% to 33% in the small populations, to 27% in the large populations (not significantly different). Polymorphism was maintained as usual, probably by some combination of heterozygous advantage and minority advantage as discussed in chapter 9.

It is noteworthy that the ten large populations became significantly differentiated in spite of population numbers in the thousands from the start with equal representatives of all of the foundation chromosomes under controlled environmental conditions. It appears that they responded to random processes of some sort, perhaps chance differences in the recombinants formed in all chromosome pairs, AR/AR and PP/PP, where different AR and PP chromosomes were involved. The sampling drift in the frequencies of particular AR and PP chromosomes should have been rather small.

The effect of accidents of sampling for small numbers is, however, demonstrated by the greater differentiation of the small populations. It is noteworthy that this reached high significance only after five or more generations, long after the actual bottlenecks of small size. It required time for the train of events initiated at the time of the bottleneck to develop its full consequences. This can be interpreted as the result of selection directed toward different selective peaks in the ten replicates, following the crossing of different saddles in the surface of selective peaks at the time of the bottleneck.

Dobzhansky and Spassky (1962) carried through an experiment with crosses between the same group of strains, in this case ten PP third chromosomes from near Austin, Texas, and ten AR chromosomes from Mather, California. Five cages were derived, each from 20 F_2 flies taken so that all 20 ancestral strains were represented ("multichromosomal" lines). In a second set of five cages each came from the 20 F_2 flies, involving only one of the PP chromosomes and one of the AR chromosomes ("bichromosomal" lines). The population were allowed to breed freely for four months, when the frequencies of PP in 300 sample chromosomes were determined in each line. A new cycle was started in each case from 20 flies. The process was repeated through nine cycles in the multichromosomal lines, except that no samples were taken of the seventh and eighth cycles. The mean frequency of PP fell at once from 50% to 25% and remained there (average 23%). The variance rose to 100 in five cycles (as compared with 119 at 19 months in the small experiment) and to 140 in nine cycles. In the bichromosomal lines, the percentage of PP rose slightly (average 57%). There was no significant rise in variance during the four cycles through which it was carried. A second set of five bichromosomal lines was then started from different chromosomes and carried for five cycles. It maintained an average of 38% PP and reached a variance of about 50 in the fifth cycle, only half that of the multichromosomal lines.

The increase in variance in the multichromosomal lines confirms the main result of the previous study, although the variance did not rise quite as rapidly in spite of its recurrent bottlenecks. The absence of any significant rise in the first bichromosomal set and the relatively small increase in the other probably reflects the relative poverty of sources of variability where there were only two third chromosomes, between which crossing-over is prevented, and less genetic diversity in other chromosomes in the two strains used than in the 20 used in the multichromosomal lines.

It may be added that in the multichromosomal lines there was a significant excess of heterozygotes in only one of the eight cycles, while in the bichromosomal sets there was significant excess (at the 5% level) in all eight of the cycles in which the chromosomes of larvae were examined. This difference was presumably due to the fact that in the former AR/AR and PP/PP usually involved different AR or PP chromosomes, respectively, while in the latter these homozygotes were homozygotes in their whole genetic composition.

A third investigation with the same material was conducted by Simmons (1966). She started from the two most extreme of the multichromosomal lines of the ninth cycle. One of these had 37% PP, the other 6%. She crossed these reciprocally using ten males and ten females in each case and pooled the F_1 populations. In F_2, five groups of ten pairs were taken at random and 1,000 offspring served as the foundation of each of five replicant "hybrid" cage populations. Five groups of ten pairs were taken from one of the lines and served without outcrossing to start five "nonhybrid" cage populations.

Each set was carried through nine cycles (43 months). Samples of 300 third chromosomes (from 150 larvae) were classified after each cycle in each population. There was considerable variation in the percentage of PP chromosomes in the five hybrid populations (range, 16% to 24%) but not among the five nonhybrid populations (28% to 30%). The variance of the five hybrid populations rose from an average of 34 for the first three cycles, to 43 for the next three, and to 50 for the last three, all highly significant. The variance of the five nonhybrid populations averaged only an insignificant 4 for the nine cycles.

It was concluded that the genetic situation had been so stabilized within Dobzhansky and Spassky's multichromosomal lines after nine cycles, each initiated by a random selection of 20 flies, that there was no basis for further selective differentiation, but that the different lines had become so differentiated from each other that they behaved almost like different geographic strains in providing a basis for renewed differentiation of lines under recurrent random drift and natural selection.

Stochastic Distributions of Gene Frequencies

A number of experiments have been made with $D.$ $melanogaster$ that illustrate the development of stochastic distributions of gene frequencies in small populations of constant size.

The sex-linked mutation forked (f) is at only a very slight selective disadvantage relative to its type allele, as noted in chapter 9. In an experiment conducted by Kerr (Kerr and Wright 1954a), 96 lines were started in vials, each from four females $(1 +/+, 2 +/f, 1 f/f)$ and four males $(2 +/0, 2 f/0)$ (thus with $q = 0.50$), and continued under controlled conditions either to fixation or to the 16th generation, by random selection of four of each sex in each generation to be parents of the next generation. The numbers of forked flies among these were recorded. Table 10.1 shows the history with respect to the number not fixed and the numbers fixed one way or the other.

It may be seen that the rate of fixation of unfixed lines approached constancy after three generations, with an average of 8.6% for generations 4 to 8, 9.3% for generations 9 to 16, or 8.9% combined.

After 16 generations, 41 lines had become homallelic wild type, 29 homallelic forked, and 26 were still heterallelic. There was no suggestion of any selective difference in the early generation (fixation of 17 wild type: 23 forked) but forked was at a disadvantage later (24 wild type: 6 forked fixed in generations 9 to 16).

The exact theoretical formula for the panmictic index, P ($= 1 - F$), for sex-linked alleles was given in volume 2, equation 7.85:

$P = P' - C_1(2P' - P'') + C_2(2P'' - P''')$, where
$C_1 = (N_f + 1)/8N_f = 5/32$ with four females
$C_2 = (N_m - 1)(N_f - 1)/8N_m N_f = 9/128$ with four males and four females.

TABLE 10.1. The frequencies and amount of fixation of forked and its type allele in 96 lines, each consisting of four males and four females in each generation and carried 16 generations unless fixed earlier.

Generation	Type Newly Fixed	Unfixed	Forked Newly Fixed	Total	Newly Fixed No.	%
1	1	94	1	96	2	2.1
2	0	92	2	94	2	2.1
3	1	87	4	92	5	5.4
4	5	79	3	87	8	9.2
5	3	70	6	79	9	11.4
6	1	66	3	70	4	5.7
7	5	59	2	66	7	10.6
8	1	56	2	59	3	5.1
9	3	52	1	56	4	7.1
10	4	47	1	52	5	9.6
11	5	39	3	47	8	17.0
12	2	37	0	39	2	5.1
13	3	34	0	37	3	8.1
14	3	30	1	34	4	11.8
15	1	29	0	30	1	3.3
16	3	26	0	29	3	10.3
1–3	2	273	7	282	9	3.2
4–8	15	330	16	361	31	8.6
9–16	24	294	6	324	30	9.3

SOURCE: Kerr and Wright 1954a.

Starting from $P' = P'' = P''' = 1$ (no inbreeding for three generations), P theoretically becomes 0.9141 in the next generation. Thereafter P/P' oscillates about, and rapidly approaches, its limiting value 0.9261. The proportional rate of decrease of P, and therefore of heterozygosis, thus rapidly approaches $\Delta P/P' = 0.0739$. The observed rate, 0.0891, is 20.5% larger and implies a correspondingly smaller effective population size than the theoretical effective size 6.76 with four males, four females, and sex linkage. This can be accounted for by unequal productivity, including failure of one or two flies to reproduce.

The actual distribution of gene frequencies in successive generations could not be observed in this case because of the indistinguishability of $+/f$ and $+/+$.

Similar experiments were conducted with 108 lines, starting from equal frequencies of the semidominant sex-linked gene Bar (strictly a duplication) and its type allele (Wright and Kerr 1954). This experiment had the advantage that gene frequencies could always be observed directly, but involved complications from the rather strong and variable selective differences.

The lines started with four $B/+$ females and two $B/0$, two $+/0$ males ($q = 0.50$). They were maintained by random selection of four females and four males as parents of each generation and carried either to fixation or to the tenth generation.

Maximum likelihood estimates indicated no significant differences in viability among the genotypes of either sex under the conditions of these experiments. There were, however, clearly differences in productivity that had to be taken into account. That for B/B females was about 37% of that for $B/+$ or $+/+$ females, and that for $B/0$ males was about 59% of that for $+/0$ males in the presence of at least one B/B female but only 26% in the absence of any such females, indicating selective mating. It should be noted that these estimates are derived from data by generations, not from side experiments, as in the studies of natural selection with these genes discussed in chapter 9. They do not agree wholly, but the conditions differed.

TABLE 10.2. The distribution of numbers of Bar genes, in each generation among 108 lines consisting of four males and four females each. Newly fixed lines are distinguished from those fixed in previous generations.

No. of Bar Genes		0	1	2	3	4	5	6	7	8	9	10	4–10	%
Old	0				15	30	43	55	65	78	88	91		
New	0			15	15	13	12	10	13	10	3	4	65	22.0
	1		8	12	10	17	10	11	7	5	3	4	57	19.3
	2		7	13	16	14	14	6	3	1	2	2	42	14.2
	3		24	22	10	5	9	8	6	2	4	1	35	11.9
	4		23	20	16	10	6	4	3	2	1	1	27	9.2
	5		21	11	13	4	2	3	4	4	2	1	20	6.8
	6	108	17	8	5	7	4	5	2	1	…	…	19	6.4
	7		3	3	3	3	3	2	2	1	1	1	13	4.4
	8		3	2	3	2	…	2	1	…	…	…	5	1.7
	9		2	…	1	2	3			1	1	…	7	2.4
	10		1				1			…	…	…	1	0.3
	11		…							1	1	…	2	0.7
New	12		1				1					1	2	0.7
Old	12				1	1	1	1	2	2	2	2		
Total		108	108	108	108	108	108	108	108	108	108	108	295	100.0

SOURCE: Reprinted, with permission, from Wright and Kerr 1954.

The distribution of Bar genes in the ten generations and their percentage in the offspring of heterallelic parents in generations 4 to 10 are shown in table 10.2.

It may be seen that after ten generations, 95 of the 108 lines had become homallelic type, 3 homallelic Bar, and 10 were still heterallelic.

Stability of the form of the distribution of heterallelic classes was approached by the fourth (or even third) generation. In the total for generations 4 to 10, the chance of fixation of wild type per generation averaged 22.0%, of Bar 0.7%, and the distribution of unfixed classes was markedly asymmetrical, declining from 19.3% with one Bar gene to 0.7% with 11.

Table 10.3 shows for each gene frequency (q) in the total for parents of offspring generations 2 to 10, the observed change (Δq) to the next generation, the expected change (calc) on the basis of the exact composition of each parental group, and the above selective values. It may be seen there is fair agreement where the numbers are at all adequate.

TABLE 10.3. Bar genes in 495 parents of generations 2–10, gene frequency q, observed change in gene frequency in offspring Δq, calculated values on the basis of the composition of the parents and by the empirical formula $\Delta q = -0.34 \times q(1 - q)$, ratio $\Delta q/[q(1 - q)]$; variance $\sigma^2_{\Delta q}$ among offspring, and estimate of N_e from ratio $q(1 - q)/\sigma^2_{\Delta q}$.

No. of Bar Genes	No. of Parents	q	Δq Observed	Calculated	$-0.34q(1-q)$	$\dfrac{\Delta q}{q(1-q)}$	$\sigma^2_{\Delta q}$	$\dfrac{q(1-q)}{\sigma^2_{\Delta q}}$
1	83	0.083	-0.025	-0.021	-0.026	-0.325	0.0063	12.1
2	76	0.167	-0.031	-0.049	-0.047	-0.226	0.0141	9.9
3	90	0.250	-0.081	-0.074	-0.064	-0.434	0.0200	9.4
4	85	0.333	-0.072	-0.075	-0.076	-0.326	0.0226	9.9
5	64	0.417	-0.094	-0.088	-0.083	-0.386	0.0259	9.4
6	49	0.500	-0.088	-0.101	-0.085	-0.354	0.0348	7.2
7	21	0.583	-0.079	-0.106	-0.083	-0.326	0.0399	6.1
8	13	0.667	-0.090	-0.101	-0.076	-0.406	0.0119	18.6
9	10	0.750	-0.150	-0.095	-0.064	-0.800	0.0321	5.8
10	2	0.833	0	-0.112	-0.047	0	0.0556	2.5
11	2	0.917	-0.042	-0.056	-0.026	-0.546	0.0313	2.4
Average						-0.351		9.74

SOURCE: Wright and Kerr 1954.

The theoretical formula for the selection pressure under the conditions would be hardly manageable but it turns out that the simplest possible formula, $\Delta q = sq(1 - q)$ with s constant, gives fairly good empirical agreement. The regression of the observed value of $\Delta q/[q(1 - q)]$ on q is $-0.344 - 0.024q$. Column 6, table 10.3, gives estimates of $\Delta q = -0.34q(1 - q)$. The curve is shown in figure 10.1, together with the observed values (*open circles*) and the direct estimates (*solid circles*) referred to above.

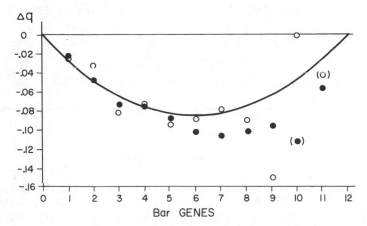

FIG. 10.1. Observed values of Δq (*open circles*), estimates from parental genotypes (*solid circles*), and theoretical values (*curve*) by the empirical formula $\Delta q = -0.34q(1 - q)$, according to the number of Bar genes (*D. melanogaster*) in the parents. Redrawn from Wright and Kerr (1954, fig. 1); used with permission.

The observed variances, $\sigma_{\Delta q}^2$, of gene frequencies in the array of offspring of parental groups with each gene frequency, q, are given in table 10.3. Theoretically these should be of the type $q(1 - q)/2N_e$. The last column of the table (the values of $q(1 - q)/\sigma_{\Delta q}^2$) should thus permit estimates of $2N_e$. The weighted average for $2N_e$ is 9.74, a reduction from expected 13.5, presumably due to unequal productivity.

The condition for stability of form of the stochastic distribution $f(q)$, of gene frequencies under fixation at the rate k per generation can be represented as follows, writing q_1 for the frequency of the Bar gene in a given offspring class that receives contributions from all parent classes (gene frequency q) and letting $p = 1 - q$, $p_1 = 1 - q_1$ (as in vol. 2, eq. 13.6 for the case of no selection).

$$\frac{(2N)!}{(2Np_1)! \, (2Nq_1)!} \sum_0^1 (q + \Delta q)^{2Nq_1}(p - \Delta q^{2Np_1}f(q) = (1 - k)f(q_1)$$

In the limiting case in which N is indefinitely large but $2Ns$ has the specified value, leaving Δq unchanged, the condition for stability of form becomes

$$\frac{\Gamma(2N)}{p_1q_1\Gamma(2Np_1)\Gamma(2Nq_1)} \int_0^1 (q + \Delta q)^{2Nq_1}(p - \Delta q)^{2Np_1}\phi(q) \, dq = (1 - k)\phi(q_1).$$

Starting from a population with gene frequency 0.50, the estimated offspring distributions for the indicated generations, including newly fixed classes, rated to a total of 1.000 are given in table 10.4 for $2N = 10$. The observed percentage of newly fixed type (22.0%) agrees reasonably well with the limiting theoretical rate 20.19 and that for observed newly fixed Bar, 0.7%, agrees well with theoretical 0.75%.

TABLE 10.4. Estimated offspring distributions for the indicated generations as fractions of 1.000.

Bar Genes		Generations $2N = 10$ $2Ns = -3.4$							$2N = \infty$ $2Ns = -3.4$
	0	1	2	4	6	8	10	∞	
New 0		0.005	0.060	0.161	0.191	0.200	0.202	0.2019	0.2066
1		0.033	0.128	0.206	0.225	0.231	0.232	0.2311	0.2602
2		0.106	0.174	0.178	0.175	0.174	0.174	0.1731	0.1704
3		0.201	0.185	0.140	0.126	0.122	0.121	0.1204	0.1142
4		0.250	0.165	0.105	0.091	0.086	0.085	0.0847	0.0783
5	1.000	0.212	0.126	0.077	0.065	0.062	0.061	0.0610	0.0551
6		0.126	0.083	0.054	0.047	0.045	0.044	0.0448	0.0397
7		0.051	0.047	0.036	0.034	0.033	0.033	0.0335	0.0293
8		0.014	0.022	0.024	0.024	0.024	0.024	0.0250	0.0222
9		0.002	0.008	0.014	0.015	0.016	0.017	0.0170	0.0171
New 10		0.000	0.002	0.005	0.007	0.007	0.007	0.0075	0.0069

SOURCE: Reprinted, with permission, from Wright and Kerr 1954.
NOTE: Estimates made generation after generation for $2N = 10$, $2Ns = -3.4$, except for the last column, which is for the limiting curve ($2N = \infty$) for $2Ns = -3.4$ at intervals of $q = 0.10$.

The observed frequencies of the unfixed classes, $q = 1/12$ to $11/12$, cannot, however, be compared directly with the theoretical frequencies for $q = 1/10$ to $9/10$. We may, however, make the same sort of calculations as above for $N = 12$ instead of $N = 10$. The theoretical rates of fixation are, of course, too low, but the relative frequencies of the unfixed classes may be compared with the observed frequencies if $2Ns$ ($= -3.4$) is used without change. This is because the form of the distributions depends largely on $2Ns$. This holds approximately even for indefinitely large N but unchanged $2Ns$. The form in this limiting case was discussed in volume 2, chapter 13 (eq. 13.125):

$$\varphi(q) = Ce^{2Nsq}(1 + C_1 pq + C_2(pq)^2 + C_3(pq)^3 \ldots), \quad \text{where}$$

$$C_1 = 1 - 2Nk$$

$$C_2 = 1/3[(6 - 2Nk)C_1 + 2N^2s^2]$$

$$C_n = \frac{2}{n(n + 1)} \{[n(2n - 1) - 2Nk]C_{n-1} + 2N^2s^2C_{n-2}\}.$$

Using the method discussed in volume 2, chapter 13, $2Nk$ was found to be 2.135368. The limiting values for $2N = \infty$, $2Ns = -3.4$ are given for intervals of $q = 0.10$ in the last column of table 10.4.

The second column in table 10.5 repeats from table 10.2 the observed percentage frequencies among the offspring of generations 4 to 10 in each frequency class of the Bar gene, 0 ($=$ newly fixed $+$) to 12 ($=$ newly fixed B). These may be compared in the next two columns with the theoretical values for $2N = \infty$ but $2Ns = -3.4$, and $2N = 12$ also with $2Ns = -3.4$. These do not differ much from each other but do not fit the observed frequencies well. Column 5 gives the theoretical frequencies for the newly fixed classes with $2N = 10$ (from table 10.4), but estimates for the unfixed classes obtained by multiplying the entries for $2N = 12$ by 0.9584 so as to make the total, including the newly fixed classes, equal to 100. These numbers agree fairly well with the observed frequencies. The last two columns give the actual observed frequencies and the theoretical ones for $2N = 10$, $2Ns = -3.4$, rated up to the actual total 295 instead of 1.000.

TABLE 10.5. Theoretical distributions of frequencies of Bar gene.

Bar Genes		Percentages			No. Observed	Calculated $(2N = 10)$
	Observed	Calculated $(2Ns = -3.4)$				
		$2N = \infty$	$2N = 12$	$2N = 10$		
New 0	22.0	17.22	16.89	(20.19)	65	59.6
1	19.3	23.46	20.59	19.73	57	58.2
2	14.2	16.42	16.47	15.79	42	46.6
3	11.9	11.67	12.14	11.63	35	34.3
4	9.2	8.42	8.94	8.57	27	25.3
5	6.8	6.19	6.68	6.40	20	18.9
6	6.4	4.62	5.07	4.86	19	14.3
7	4.4	3.51	3.91	3.75	13	11.1
8	1.7	2.71	3.06	2.93	5	8.6
9	2.4	2.13	2.41	2.31	7	6.8
10	0.3	1.70	1.88	1.80	1	5.3
11	0.7	1.38	1.34	1.29	2	3.8
New 12	0.7	0.57	0.62	(0.75)	2	2.2
Total	100.0	100.00	100.00	100.00	295	295.0

SOURCE: Wright and Kerr 1954.

NOTE: Includes newly fixed populations after attainment of stability of form. Column 2 gives the observed percentages (as in table 10.2); columns 3–5 give ordinates of the limiting curve for $2Ns = -3.4$ (column 3 for $2N = \infty$, column 4 for $2N = 12$). Column 5 gives estimates for the newly fixed classes with $2N = 10$, but the others equal 0.9584 $\times$ column 4). Column 6 gives the observed number, generations 4–10. Column 7 gives the calculated numbers, 2.95 $\times$ column 5.

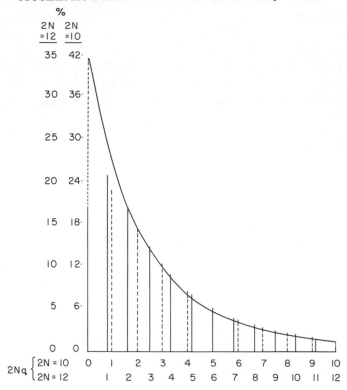

Fig. 10.2. The theoretical distribution of gene frequencies, including the classes of newly fixed genes after attainment of stability of form among populations in which $2Ns = -3.4$, $2N = 12$ (*solid vertical lines*), $2N = 10$ (*broken lines* except at 0.5 and 1.0, on which there is agreement with the values for $2N = 10$ on the scales used). The scale unit for frequencies is 20% greater for $2N = 12$ than for $2N = 10$ to bring out the similarity in form. The distribution for $2N = \infty$ is shown as a smooth curve of unit area. Redrawn from Wright and Kerr (1954, fig. 2); used with permission.

The theoretical distributions are shown graphically in figure 10.2 in terms of two scales of gene frequencies, one for $2N = 12$ (*solid lines*) and one for $2N = 10$ (*broken lines*) intended to bring out the close similarity in form. The theoretical distribution for $2N = \infty$, $2Ns = -3.4$ is shown as a smooth curve. Even it agrees fairly well with the others, except near the extremes.

Figure 10.3 compares the observed percentage frequencies with the theoretical curve.

A third series of experiments was made by Kerr (Kerr and Wright 1954*b*). This involved two allelic autosomal recessive mutations, spineless (*ss*) and

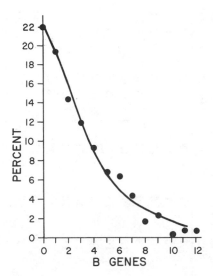

Fig. 10.3. The observed percentages of populations with 1 to 11 Bar genes or newly fixed type (0) or Bar (12) in generations 4–10 (*circles*). The expected numbers are shown by the curve, calculated as described in the text. Redrawn from Wright and Kerr (1954, fig. 3); used with permission.

aristapedia (ss^a). The heterozygote of these morphologically very different mutations is almost of wild type. In this case, 113 lines, each started from four ss^a/ss of each sex, were carried either to fixation or to ten generations by random selection of four females and four males in each generation (table 10.6). A stochastic distribution was reached in two generations that remained essentially unchanged except for fixation of spineless in eight lines. Aristapedia reached a maximum gene frequency of 0.875 (tables 10.6 and 10.7). A test for homogeneity of generations 2 to 10 gave $\chi^2 = 61.5$, $n = 80$, $P = 0.94$. The equilibrium frequency for aristapedia was about $\hat{q} = 0.3882 \pm 0.0049$. The standard deviation of frequencies was 0.1536 ± 0.0035.

Analyses of the data indicated viabilities of the three genotypes ss^a/ss^a, ss^a/ss, ss/ss in the ratio of about 0.50:1:0.75, largely irrespective of sex. The productivities of the females of the genotypes in the same order were approximately as 0.40:1:0.75, and of males 0:1:0.25. On averaging the product for the sexes, the overall selective values were approximately in the ratio 0.10:1:0.375 and thus with very strong overdominance, although not enough to prevent occasional fixation of ss.

From the rarity of fixation, 8 in 986, the distribution may be treated without serious error as if it were in equilibrium, in contrast with the cases of

TABLE 10.6. The distribution of numbers of aristapedia genes in each generation among 113 lines consisting of four males, four females each. Newly fixed *ss* is distinguished from previously fixed *ss*. The latter is not included in the totals or in the calculations of q and σ_q^2. The observed frequencies are shown for generations 2–10. The percentages are compared with the estimations described in the text.

No. ss^a	Generations												%	
	0	1	2	3	4	5	6	7	8	9	10	2–10	Observed	Calculated
Old					(1)	(4)	(4)	(4)	(5)	(6)	(7)	Observed		
New 0				1	3	2	2	1	1	1	1	8	0.81	1.06
1			2	3	1	2	2	3	1	3	1	18	1.83	2.12
2		1	5	5	4	8	4	0	3	3	5	37	3.75	3.94
3		1	6	9	9	10	6	8	6	8	10	72	7.30	6.99
4		10	10	8	17	12	13	11	15	12	7	105	10.65	10.72
5		11	15	17	13	19	19	17	15	10	13	138	14.00	14.08
6		11	17	18	18	18	19	24	14	14	19	161	16.33	15.85
7		22	22	21	18	16	16	14	17	18	12	154	15.62	15.31
8	113	22	17	12	16	14	12	12	14	14	16	127	12.88	12.64
9		20	8	10	6	7	12	9	7	10	10	79	8.01	8.81
10		7	3	5	2	1	4	7	9	10	6	47	4.77	5.08
11		6	6	3	3	2	1	2	2	2	3	24	2.43	2.35
12		1	1	1	0			1	2	1	1	8	0.81	0.82
13		1	0		1		1		2	1	2	6	0.61	0.20
14			1		1							2	0.20	0.03
Total	113	113	113	113	112	109	109	109	108	107	106	986	100.0	100.00
$\bar{q}$	0.500	0.463	0.404	0.383	0.371	0.354	0.383	0.389	0.405	0.403	0.402	0.388		
σ_q^2	0	0.0171	0.0235	0.0233	0.0252	0.0190	0.0193	0.0210	0.0260	0.0268	0.0269	0.0235		

SOURCE: Kerr and Wright 1954b.

TABLE 10.7. The observed changes in gene frequency per generation (Δq), in relation to gene frequencies of aristapedia (parents of generations 2–10), and the calculated values described in the text, the variance of changes, and the estimate of $2N_e$ based on the ratio $q(1 - q)/\sigma_{\Delta q}^2$.

| No. ss^a | No. of Parents | q | Δq | | σ_q^2 | $\dfrac{q(1 - q)}{\sigma_{\Delta q}^2}$ |
			Observed	Calculated		
1	17	0.0625	+ 0.088	+ 0.064	0.0127	4.6
2	33	0.1250	+ 0.051	+ 0.087	0.0106	10.3
3	63	0.1875	+ 0.097	+ 0.087	0.0165	9.2
4	108	0.2500	+ 0.069	+ 0.072	0.0192	9.8
5	136	0.3125	+ 0.053	+ 0.049	0.0131	16.5
6	153	0.3750	+ 0.005	+ 0.019	0.0174	13.5
7	164	0.4375	− 0.031	− 0.015	0.0165	14.9
8	133	0.5000	− 0.054	− 0.051	0.0162	15.4
9	89	0.5625	− 0.109	− 0.089	0.0237	10.4
10	48	0.6250	− 0.095	− 0.126	0.0233	10.1
11	27	0.6875	− 0.157	− 0.162	0.0151	14.2
12	8	0.7500	− 0.297	− 0.194	0.0131	14.3
13	5	0.8125	− 0.175	− 0.217	0.0147	10.4
14	2	0.8750	− 0.250	− 0.223	0.0078	14.0
15	0	0.9375	. . .	− 0.186	. . .	. . .
	986					13.03

SOURCE: Kerr and Wright 1954b.

forked and Bar. There are complications in deriving the theoretical formula, however, because of lack of knowledge of the frequencies of the zygotes. The observed frequencies are those of imagoes, following strong larval selection, but preceding strong selection in productivity. As in the case of Bar, an empirical approach seems to be indicated, based on the relation of the observed changes (Δq) to the corresponding gene frequencies.

The weighted average of the estimates of $2N_e$ (from $[q(1 - q)]/\sigma_{\Delta q}^2$) (table 10.7) is 13.03 instead of 16, but this is undoubtedly too high because it is based on the variability after the frequencies of the homozygote have been cut down severely by selective viability.

If gene frequencies were based on zygotes instead of imagoes, with s as the selective disadvantage of ss^a/ss^a and t that of ss/ss, relative to the heterozygotes we would have (vol. 2, eq. 3.32, 3.42, 13.56):

$$\bar{w} = 1 - sq^2 - t(1 - q)^2$$
$$\Delta q = -(s + t)q(1 - q)(q - \hat{q}), \quad \hat{q} = t/(s + t)$$
$$\varphi(q) = C\bar{w}^{2N}/[q(1 - q)]$$

Assuming that Δq is related to q approximately in this way, empirical values of s, t, and $2N_e$ may be obtained. It was found by trial that $s = 0.86$ and $t = 0.60$ give the mean better than the rather rough direct estimates 0.90 and 0.625, respectively. How well the observed values of Δq are fitted by the theoretical values on this basis is shown in figure 10.4. Agreement is good where observed Δq is based on numbers that are at all adequate. In fitting the observed frequencies for generations 4–10, it was found that $2N_e$ must be treated as approximately 10.7 (instead of 13.03 above or 16). The fit is shown in figure 10.5.

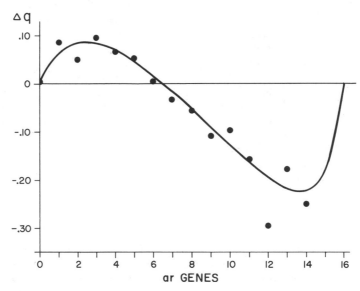

FIG. 10.4. Observed changes in gene frequencies (Δq) (*circles*) in relation to numbers of aristapedia genes in populations of eight flies (*D. melanogaster*), in comparison with those expected, as described in the text. Redrawn from Kerr and Wright (1954*b*, fig. 1); used with permission.

Similar experiments, except that the populations consisted of eight females and eight males, were conducted by Buri (1956) with two brown alleles of *D. melanogaster*. These two alleles, *bw* and *bw*[75] (induced by X-radiation), were carried in a highly inbred strain that also carried scarlet (*st/st*), which made possible accurate discrimination: *bw*[75]/*bw*[75] was bright red-orange, *bw*[75]/*bw* was light orange, and *bw/bw* was white.

Two series, I and II, consisting of 107 and 105 lines, respectively, were carried either to fixation or through the 19th generation. All lines started from heterozygotes ($q = 0.50$) and were maintained by random selection of

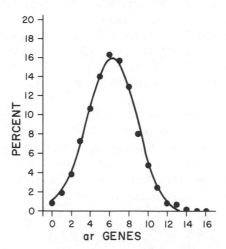

F<small>IG</small>. 10.5. The observed percentages (*circles*) of each type of population accord-ing to the number of aristapedia genes in populations of eight flies. These are compared with the theoretical distribution (*curve*) in an array of populations in which the stable state has been reached with 86% selection against ss^a/ss^a, 60% against ss/ss relative to ss/ss^a, and with $2N_e = 10.7$. Redrawn from Kerr and Wright (1954*b*, fig. 2); used with permission.

eight of each sex to be parents of the next generation. As in the preceding case, the recorded gene frequencies were those of the parental groups. The two series differed in the greater crowding of series I (35-ml vials with medium surface of about 3 sq cm as opposed to 60-ml vials with medium surface of about 12 sq cm).

Preliminary experiments revealed no reliable differences in either viability or net productivity among the genotypes, and this was also the case in the final experiment. There was, however, a complication in the crowded series (I), which makes it desirable to consider first the somewhat simpler results in series II.

There was considerable asymmetry in this with respect to fixation. After 19 generations, bw/bw had become fixed in 13 lines but bw^{75}/bw^{75} in 30 lines. This seems to have been due to a rise in gene frequency of bw^{75} in the first four generations to $q = 0.570$, which was lost later. In the last seven generations the average was 0.497. Estimates of $1/(2N_e)$ were made from the ratio of $q(1 - q)/\sigma^2_{\Delta q}$ for each value of q. These estimates gave the regression of $1/(2N_e) = 0.0427 - 0.0020q$, $SE_b = 0.0009$.

As the slope is of doubtful significance, the value 24.0 for $2N_e$ (from

$1/(2N_e) = 0.0417$) may be taken instead of 32, expected if the contributions of the 16 parents varied only at random.

The steady state distribution should be rectangular except for the usual dip in the subterminal region (vol. 2, pp. 346–50). Unfortunately, the number of generations was hardly enough to expect adequate realization of this situation except in the last three generations. Table 10.8 and figure 10.6 show the class of frequencies in these three generations. The subterminal dip from the rectangular distribution is indicated only approximately because of the considerable reduction in effective size of population. Indeed, the actual rate of fixation in these three generations, $13/204 = 6.37\%$, indicates a value of $2N_e$ of only about 15.7 instead of 24 as indicated by the variances, but that in the preceding three generations indicated 23.9. The numbers are too small for a reliable estimate on this basis.

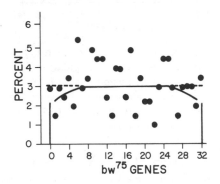

FIG. 10.6. Observed frequencies of bw^{75} in populations of 16 flies (series II) in generations 17–19, in comparison with the nearly rectangular distribution (dips near extremes) expected in the absence of a selective difference between bw^{75} and bw. From data of Buri (1956).

In spite of the great irregularity due to small numbers, this case illustrates fairly well the near-rectangular distributions expected in the absence of selection.

There was also no evidence for systematic selection under the less favorable conditions of series I on the basis of means, but in this case, the estimates of $2N_e$ from $q(1 - q)/\sigma^2_{\Delta q}$ showed a surprising but highly significant regression on gene frequency:

$$1/(2N_e) = \sigma^2_{\Delta q}/q(1 - q) = 0.0451 + 0.0228q; \quad SE_b = 0.0014$$
$$\sigma^2_{\Delta q} = q(1 - q)(1 + 0.506q)/22.18$$

This seems to imply that when there was near-fixation of bw, $2N_e$ approached 22.18, but that with near-fixation of bw^{75}, $2N_e$ approached 14.73.

TABLE 10.8. The distribution of observed numbers of bw^{75} genes in groups of 16 flies (second column) and percentages (third column) for generations 14–19 in series I and generations 17–19 in series II, with calculated percentages (fourth column) as described in the text.

SERIES I

bw^{75}	Obs.	Obs.	Calc.
0	18	4.74	3.01
1	15	3.95	4.07
2	9	2.37	3.98
3	21	5.53	3.89
4	9	2.37	3.81
5	14	3.68	3.72
6	11	2.89	3.64
7	8	2.11	3.57
8	10	2.63	3.49
9	12	3.16	3.42
10	12	3.16	3.35
11	7	1.84	3.28
12	15	3.95	3.22
13	25	6.58	3.16
14	12	3.16	3.10
15	12	3.16	3.04
16	16	4.21	2.98
17	13	3.42	2.92
18	13	3.42	2.87
19	8	2.11	2.82
20	12	3.16	2.77
21	7	1.84	2.72
22	8	2.11	2.67
23	5	1.32	2.62
24	12	3.16	2.58
25	8	2.11	2.53
26	16	4.21	2.49
27	4	1.05	2.45
28	8	2.11	2.41
29	10	2.63	2.37
30	8	2.11	2.33
31	12	3.16	2.29
32	10	2.63	2.45

SERIES II

bw^{75}	Obs.	Obs.	Calc.
0	6	2.94	3.09
1	3	1.47	
2	6	2.94	
3	5	2.45	
4	7	3.43	
5	4	1.96	
6	11	5.39	
7	6	2.94	
8	7	3.43	
9	10	4.90	
10	9	4.41	
11	9	4.41	
12	5	2.45	
13	3	1.47	
14	8	3.92	
15	8	3.92	
16	5	2.45	
17	10	4.90	3.09
18	3	1.47	3.09
19	7	3.43	3.09
20	4	1.96	3.09
21	4	1.96	3.09
22	2	0.98	3.09
23	6	2.94	3.09
24	9	4.41	3.09
25	9	4.41	3.09
26	6	2.94	3.09
27	3	1.47	3.09
28	6	2.94	3.09
29	6	2.94	3.09
30	6	2.54	3.09
31	4	1.96	3.09
32	7	3.43	3.09

SOURCE: Buri 1956.

This would seem to be compatible with the absence of overall selective difference between bw^{75} and bw only if there were more reproductive failure of the former under the conditions of this series but a compensatory superiority in productivity of those that reproduced at all.

Assuming that $\Delta q = 0$, the Fokker-Planck equation for the steady state form of distributions, with decay at rate k, is as follows:

$$(1/2)\frac{d^2}{dq^2}\left(\sigma_{q\Delta}^2\varphi(q)\right) + k\varphi(q) = 0.$$

Let $\varphi(q) = C[1 + C_1q + C_2q^2 + \cdots C_nq^n]$;   and

$$\sigma_{\Delta q}^2 = q(1-q)(1+bq)/2N_0,$$

where N_0 is the effective population number at $q = 0$. On carrying out the differentiation and equating the coefficients of like powers of q,

$$C_1 = (1-b) - 2N_0k$$
$$C_2 = b + (1-b)C_1 - 2N_0kC_1/3$$
$$C_n = bC_{n-2} + (1-b)C_{n-1} - \{4N_0kC_{n-1}/[n(n+1)]\}$$

The system of coefficients may be evaluated by finding by repeated trial the value of $2N_0k$ that will make the series convergent, which requires that the coefficients be kept alternating in sign (vol. 2, p. 390). It has been found that with $b = 0.506$, $2N_0k = 1.2375$ is correct to four decimal places and permits similarly accurate computation of the coefficients up to C_{15}.

$$\varphi_q = C[1 - 0.7435q + 0.4454q^2 - 0.2481q^3 + 0.1335q^4 - 0.0706q^5$$
$$+ 0.0369q^6 - 0.0191q^7 + 0.0099q^8 - 0.0051q^9 + 0.0026q^{10}$$
$$- 0.0013q^{11} + 0.0007q^{12} - 0.0004q^{13} + 0.0002q^{14} - 0.0001q^{15}]$$

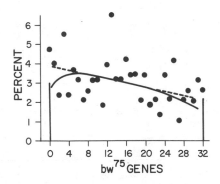

FIG. 10.7. Observed frequencies of bw^{75} in populations of 16 flies (series I) in generations 14–19, in comparison with the theoretical curve (*solid line*) described in the text. From data of Buri (1956).

The terminal frequencies are $\varphi(0) = C$, $\varphi(1) = 0.5410C$, respectively.

For $2N_0 k = 1.2375$ and $2N_0 = 22.18$ from the calculated regression of $1/(2N_e)$ on q, $k = 0.0558$. This value is $0.996 \times \sqrt{[1/(2N_e)]}$ and thus only slightly less than the expected value in the case of a rectangular distribution on the basis of the average class value $[\sigma^2_{\Delta q}/[q(1 - q)]]$.

This distribution is subject to the usual dip near the extremes. Figure 10.7 shows how far the observed frequencies of bw^{75} in generations 14–19 are scattered about the theoretical curve, with the subterminal dip indicated approximately.

Effective Population Numbers

The effective population numbers were less than expected from the number of potential parents in all of these experiments. The observed and calculated values of $2N_e$ and their ratios are as follows:

	Obs.	Calc.	Ratio
$+, f$	11.22	13.53	0.829
$+, B$	9.74	13.53	0.720
ss^a, ss	10.7	16	0.669
$bw, bw^{75}(\text{II})$	24.0	32	0.749
bw (I)	22.2	32	0.693
bw^{75} (I)	14.7	32	0.460

These ratios are in the range of results obtained by Crow and Morton (1955) in extensive studies with *D. melanogaster*. Their values for female parents based on the ratio of $\sigma^2_k/\bar{k}$, where $\bar{k}$ is the number of offspring from an individual, and adjusted to a stationary population ($\bar{k} = 2$), were $N_e/N = 0.71$ from adult progeny and $N_e/N = 0.72$ from eggs. For males they obtained $N_e/N = 0.48$.

They also analyzed previous data on *D. melanogaster* obtained by Alpatov, and obtained estimates of N_e/N ranging between 0.74 and 0.91 under various conditions. From data (Bailey) on the snail *Lymnaea columella*, they obtained adjusted $N_e/N = 0.75$. From data on women in Australia (Powys), England (Pearson), and the United States (Baber and Ross), their estimates ranged from 0.69 to 0.95. These indicated that the effective population number was less than the apparent number because of heterogeneity in reproductive performance beyond that expected from sampling, under widely different conditions, but that the ratio is rarely lower than 0.50 for this reason. There are, of course, other more important reasons by which effective numbers may be much less than apparent numbers, most notably, passage of the population through bottlenecks of small size, illustrated by the experiments at the beginning of this chapter.

Random Drift

The population numbers in the experiments discussed here were so small that they would seem to have no pertinence for conditions in nature. They may, however, be considered as models on exaggerated scales with respect to the inbreeding effect and, in at least two cases of selection, of situations that might occur with respect to pairs of alleles, or of alleles with primary effect on multifactorial quantitative variability.

In any array of such populations with effective size 100 times as great as in the experimental populations and selection coefficients only 1/100 as great, processes would occur only 1/100 as rapidly but the form of the stable state distributions would be nearly the same.

The experiment with forked and the second series with the brown alleles may be considered models of the inbreeding effect on a sex-linked and an autosomal locus, respectively, in almost pure form. The crowded first series with the brown alleles was similar except for the unexpected relation between gene frequency and effective size of population and the apparent compensatory selective differences on two components of productivity. The experiment with Bar gives a model of a case in which one allele tends toward fixation, but not so decisively as to prevent the unfavorable allele from occasionally drifting into fixation against the pressure of selection. Those with aristapedia and spineless illustrate the near-equilibrium from heterozygote advantage, with, however, so much random drift that gene frequency varies almost from one extreme to the other among different populations and at least one of the alleles occasionally drifts into fixation.

The first experiments discussed illustrate how it is possible to have significant sampling drift in rather large populations because of one or more bottlenecks of small size. They also illustrate how selection of the same sort may bring about widely different results in different populations because sampling drift has triggered the crossing of saddles and control by different selective peaks.

CHAPTER 11

Mutation and Selection

The experiments discussed in previous chapters have been based largely on genetic variability already present in the foundation stock. Here we will consider studies of the properties of newly arisen mutations and selection experiments in which current mutation supplies a major portion of the material for selection.

The major kinds of mutation, chromosomal and genic, have been reviewed briefly in volume 1, chapter 3, and in much more detail in all textbooks of genetics; they need not to be gone into here. In this chapter we will be concerned primarily with gene mutations.

In experiments involving mutation, it is customary to use some sort of rough classification by effect. Thus Gustafsson (1947b), in discussing mutations in agricultural plants, found it convenient to classify them as chlorophyl mutants, sterility and lethality mutants, and morphological and physiological mutants with nearly normal viability and fertility. He added the class of chromosomal aberrations in summarizing his results on the frequencies of induced mutations in barley. For each induced morphological mutant, he found about 22 affecting chlorophyll, about 15 that prevented gametogenesis, and about 30 cases of translocation sterility. He estimated that to obtain one mutant with distinctly higher yield, the barley breeder must induce some 700–800 worthless types at the same time.

In *Drosophila* studies, the lethals have been used as the class for which the most reliable counts could be made. They grade, however, into types that occasionally hatch, and 10% viability has been found to be a convenient cutoff point. Mutations with viabilities between 10% and 50% are relatively infrequent, but there is a rapid rise in frequency from 50% to mutations with normal viability, followed by declining frequencies of mutations that appear to be supernormal, at least under the conditions of the test (vol. 1, fig. 6.7b). It is not surprising that there have been widely different judgments on the ratio of detectable detrimentals (D) to lethals (L) (0 to 10% emergence). Table 11.1 gives a condensation of a summary of results of investigations of

TABLE 11.1. Estimates of the ratio of numbers of induced detrimental to lethal mutations in *D. melanogaster.*

	Chromosome	No.	Mutations D/L	Load D/L
Timofeeff-Ressovsky (1935)	I	868	2.1	0.872
Kerkis (1938)	I	277	3.0	. . .
Falk (1955)	III	ca. 56	3.5	0.505
Käfer (1952)	I,II	1,500 each	0.75	0.288
Bonnier and Jonsson (1957)	II	171	0.82	0.215
Greenberg and Crow (1960)	II	465	. . .	0.627
Temin (= Greenberg) (1966)	II	1,083	. . .	0.563
Temin et al. (1969)	{ II	1,855	. . .	0.636
	III	1,855	. . .	0.729

SOURCE: Compiled by Greenberg and Crow (1960) and supplemented by their estimates from the same data of the more reliable ratio of detrimental to lethal load (in homozygous chromosomes). Their estimates of this load ratio from their own data in this and later studies are added.

D. melanogaster, compiled by Greenberg and Crow (1960). Their estimates of the ratio of detrimental to lethal "load" discussed later is also shown.

The data of Timofeeff-Ressovsky (1935) and of Kerkis (1938) were subdivided according to degree of crowding, but the differences were much less than those between observers. Käfer's data were subdivided by dosage, but there was no significant relation.

Mukai (1969), using a very different technique, estimated the ratio of spontaneous occurrence of slightly detrimental to lethal or near-lethal mutations as greater than 20 to 1 and obtained a similar ratio for induced mutations, but as shown later, his method depends on assumptions that probably exaggerate the ratio some threefold. Even so, his results indicate a considerably higher ratio than detectable by the methods used in obtaining the data of table 11.1.

Mutation studies usually include "visibles," of near-normal viability, as well as lethals, near-lethals, and detrimentals. The observed ratios vary greatly, in obvious dependence on the care taken to discriminate and test slight differences from wild-type flies in morphology, eye, or body color. In Muller's (1927) first series of experiments on the induction of mutations in *D. melanogaster* by X-radiation, he reported 86 lethals, 17 semilethals, and 19 visibles. Later estimates have centered around 10% visibles.

Synthetic Lethals

While recessive lethals constitute the most reliable index of mutation rate, their reliability has been somewhat impaired by the discovery by Dobzhansky

(1946) that "synthetic" lethals (ones depending on the presence of more than one gene) are not uncommon. In a study by Dobzhansky and Spassky (1960), 20 normally viable strains of *D. melanogaster* were produced, each of which was homozygous in its second chromosome. They crossed these in all of the possible 190 ways. Females from each cross were mated with males with appropriate dominant markers. Second chromosomes from each of 10 sons were tested for homozygous viability in the usual way. Among the 1,900 recombination chromosomes, 77 were lethal when homozygous (4.0%). These were much more numerous than expected from spontaneous mutation. Moreover, exhaustive tests for localization of the lethal gene in 12 cases were successful in only two. In the other 10, the lethal effect was lost on outcrossing.

Temin et al. (1969) isolated single second and third chromosomes from a wild stock of *D. melanogaster* in each of 1,855 lines. The numbers carrying lethals (viability < 0.10), severe detrimentals (viability 0.10–0.50), and mild detrimentals (viability > 0.50) were found among progeny from mating $Cy/+ Me/+ \times Cy/+ Me/+$, expected to produce $Cy/+ Me/+$, $Cy/+ +/+$, $+/+ Me/+$, and $+/+ +/+$ in the ratio 4:2:2:1 in the absence of differences in viability. Lethals were found in 426 second chromosomes and in 418 third chromosomes (among which 97 segregated together in close agreement with expected 96.0). There were only nine synthetic lethals from second and third chromosomes that showed only mild if any viability loss separately. This is only about 2% of the frequency of either second or third chromosome lethals, a much smaller frequency than suggested by the preceding study.

Such synthetic lethals have been found in other species of *Drosophila*. Magalhães et al. (1964, 1965) interpreted such results in *D. willistoni* as due to interaction of lethals with recessive suppressors which they found to be common. Krimbas (1960) found synthetic sterility of one or both sexes in the chromosomes of *D. willistoni*. There is a case of synthetic sterility in the guinea pig (Wright 1959b) in which two unlinked semidominant genes with fully fertile homozygotes (*sisiDmDm* "silvered," *SiSidmdm* "diminished intensity") give a white anemic double homozygote, *sisidmdm*, which is completely male sterile (vol. 1, p. 77).

Number of Loci

According to Lindsley and Grell (1968), there are about 1,000 loci in intensely studied *D. melanogaster* that have been observed to mutate. Many mutations have been observed at many of these loci, suggesting that the total number of loci is not more than an order of magnitude greater, unless most loci are either much less mutable or give mutations not detectable by the usual techniques.

Intensive studies of short regions in the chromosomes of *D. melanogaster* indicate that the number of loci capable of mutation (including lethal mutation in most cases) corresponds to the number of distinguishable bands in the salivary chromosomes, amounting to at least 5,000 altogether but probably less than 10,000 (Judd and Young 1973).

Estimates of the probable total number of genes can be based on the DNA content of the chromosomes. It was discovered by Boivin, Vandrely, and Vandrely (1948) and confirmed by Mirsky and Ris (1949) and Swift (1950) that the amount of DNA in the haploid sets of each species studied is constant, irrespective of the differentiation of the cells. Table 11.2 shows typical determination of the weight of DNA in units of 10^{-12} gm in haploid sets in diverse organisms (Mirsky and Ris 1951, Ris and Kubai 1970). Multiplication of weight in grams by 6×10^{23} gives the molecular weight, and division of this by 660 (average molecular weight of a nucleotide pair) gives the number of such pairs. Taking 1,000 as a convenient round number for the number of nucleotide pairs per gene (coding protein molecules with about 333 amino acids), the estimated number of genes in millions is about the same as the weight of the haploid set in the units of 10^{-12} gm.

There appears to be room for at least 200,000 genes in the genome of *D. melanogaster*, 200 times the number of loci observed to mutate. If all had essential effects and were capable of lethal mutation with a spontaneous frequency of about 10^{-5} per generation, there would be two or more such mutations per gamete instead of the observed 0.004.

It is unlikely that any of the DNA is incapable of mutation. Thus, it must be supposed that most genes are so nonessential that they are not subject to lethal mutation. One possibility is that most of the DNA is nonsense in which mutational changes have no effect, favorable or unfavorable. Another possibility is that most genes are replicated so that changes in one replication have little effect. As noted (vol. 1, p. 28), Bridges (1935) and Metz (1947) have found cytological evidence of duplication of numerous small blocks in the salivary chromosomes of *Drosophila* and *Sciara*, respectively, and the abundance of pseudoallelic series in *Drosophila* suggests numerous ancient differentiated replications.

There is a difficulty here, however, in that groups of replications would tend to be reduced to a single active component by mutation pressure. Possibly duplication by unequal crossing-over keeps up with mutational decay, a rather slow process. Another possibility is that groups of active replications are maintained by selection for quantitative effects, while protected from drastic effects of inactivation or loss.

TABLE 11.2. Estimated weight of DNA in units of 10^{-12} gm in haploid sets of chromosomes of diverse organisms.

		$\times 10^{-12}$ gm				$\times 10^{-12}$ gm
Viruses	φ × 174	2.6×10^{-6}		Echinodermata	*Lytechinus*	0.9
	Lambda	50×10^{-6}		Tunicata	*Ascidia*	0.16
	T2	208×10^{-6}		Agnatha	*Petromyzon*	2.5
	Fowl pox	382×10^{-6}		Chondrichthyes	*Carcharias longimanus*	3.3
Bacteria	*Mycoplasma*	840×10^{-6}		Osteichthyes	*Acipenser sturio*	1.6
	Escherichia coli	$4,500 \times 10^{-6}$			*Amia*	1.1
Fungi	Yeast	0.02			Teleosts (31 species, 17 families)	0.9
Protophyta	*Euglena gracilis*	3		Amphibia	Lungfish (*Protopterus*)	50
Angiosperms	*Aquilegia alpina*	0.6			*Amphiuma*	84
	Anemone virginiana	10.5			*Necturus*	24
	Tradescantia	58			Toad (*Bufo bufo*)	7.3
Protozoa	*Plasmodium berghei*	0.06		Reptiles	Snapping turtle	2.5
	Trypanosoma evansi	0.2			Alligator	2.5
	Astasia longa	1.5			Black racer snake	1.4
Sponges	Tube sponge	0.06		Birds	Domestic fowl	1.2
Coelenterates	*Cassiopeia*	0.33			Goose	1.5
Nereid		1.45		Mammals	Rat	3.1
Mollusks	Snail (*Tectarius*)	0.7			Calf	3.0
	Squid	4.5			Man	3.2
Crustacea	Crab (*Plagusia*)	1.5				
Insecta	*Drosophila*	0.2				
	Euschistus obscurus	128.0				

SOURCE: Mirsky and Ris 1951 and Ris and Kubai 1970.

Direct studies of DNA in vitro indicate that certain DNA sequences in all higher organisms (bacteria and viruses excluded) are replicated an enormous number of times (Britten and Kohne 1968). Complementary strands of DNA, separated by appropriate treatment in vitro and sheared into lengths of some 400 nucleotides, recognize each other and reassociate under appropriate conditions. In the absence of repeated sequences, the amount of reassociation is proportional to concentration (C_o) and time (t). A bacteriophage with some 2×10^5 nucleotide pairs and probably no repeated sequences follows a sigmoid reassociation curve, relative to $\log C_o t$, reaching 50% reassociation at $C_o t$ 0.3 mole × sec/liter (fig. 11.1). *Escherichia coli*, with

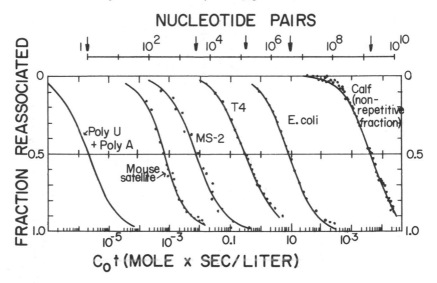

FIG. 11.1. Reassociation of double-stranded nucleic acids from various sources. Redrawn from Britten and Kohne (1968, fig. 2), © 1968 by the American Association for the Advancement of Science; used with permission.

4.5×10^6 nucleotide pairs, requires a value of $C_o t$ some 23 times as great in a parallel sigmoid curve under similar conditions. Calf DNA, with some 3.2×10^9 nucleotide pairs, should show a reassociation curve with $C_o t$ some 700 times that for *E. coli* under the same conditions if composed of similarly nonreplicated sequences. Actually, the form of the reassociation curve was not at all parallel and was obviously composite. The DNA could be separated into fractions by rate of reassociation, one including 40% of the total, reassociated 100,000 times as rapidly as the remaining 60% for which the curve was roughly parallel to that for *E. coli* and at the expected distance for unreplicated units (fig. 11.2). The former is interpreted as consisting of

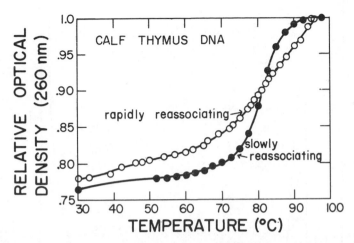

FIG. 11.2. Hyperchromic melting curves for rapidly reassociating and slowly reassociating fractions of calf thymus DNA in 0.08M phosphate buffer. Redrawn from Britten and Kohne (1968, fig. 10), © 1968 by the American Association for the Advancement of Science; used with permission.

some 100,000 replications of the same unit. Similarly Britten and Kohne separated mouse DNA into three components: about 10% (from a chromosome satellite) reassociated as if composed of about a million replications of the same small sequence; about 20% behaved like a heterogeneous mixture of components with 1,000 to 100,000 units; while the remaining 70% behaved as if composed largely of unique sequences but with some replication up to perhaps tenfold (fig. 11.3). They found similar evidence for highly replicated components, making up 30% or more of the total, in all groups of multicellular organisms examined and also protozoa (*Euglena*, dinoflagellates) but very little in yeast and none in any bacteria or viruses.

The occurrence of 30% or more of highly repetitious DNA sequences (each sequence probably at least of the order of 400 nucleotide pairs), does not go far in resolving the discrepancy between estimates of locus number from the number observed to mutate and the amount of DNA, if some 70% of the latter consists of unique sequences. The highly repetitious DNA corresponds perhaps to heterochromatin, if the cytologic evidence that this is highly repetitious is valid. This brings us back to the possibilities indicated above: a vast amount of nonsense sequences, or replications of most sequences, with protection against genetic degeneration by selection for quantitative effects of the replications. If the latter, it must be supposed that the current techniques do not adequately distinguish between uniqueness of most

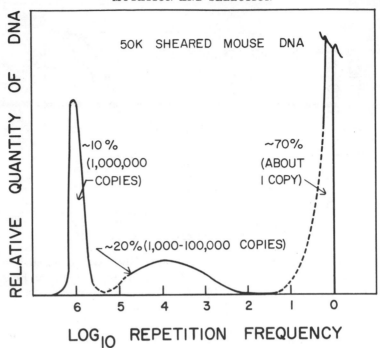

Fig. 11.3. Spectrogram of the frequency of repetition of nucleotide sequences in the DNA of the mouse. The dashed segments of the curve represent regions of considerable uncertainty. Redrawn from Britten and Kohne (1968, fig. 12), © 1968 by the American Association for the Advancement of Science; used with permission.

sequences and small-scale replication. There is a possibility that differentiation of duplications may lead to lack of full recognition in the reassociation process. Britten and Kohne find evidence, indeed, for partial differentiation in many cases in reduced stability of reassociated portions of the DNA shown by a lowering of melting temperatures.

However this may be, it appears that only a small percentage, perhaps no more than 10% of the DNA, consists of genes capable of giving rise to observable major mutations, but there is probably a much larger portion in which mutation may have effects only on quantitative variability, a situation somewhat reminiscent of Mather's hypothesis of oligogenes and polygenes, the latter heterochromatic. Classical genetics has been largely concerned with the former class. Wild species have been expected to be almost homallelic at these loci with respect to wild-type alleles or at least to a group of rather similar wild-type isoalleles.

Population genetics, with its origin in studies of quantitative variability and the effects on it of selection and inbreeding, has been largely concerned with the latter class, with respect to which species are expected to be strongly heterallelic.

Selective Value of Mutations

Mutations with favorable major physiological or morphological effects are expected a priori to be exceedingly rare in natural populations that have been living for a long time under similar conditions, for the obvious reason that if favorable they would long since have been established. The probability for favorable mutation rises under drastic change of conditions, including laboratory conditions and the redefinition of "favorable" as useful to man, in the cases of cultivated plants and domestic animals. Even so, the probability in the latter case is not very great, as noted earlier. In the case of laboratory mutants, practically all of the considerable number of conspicuous mutations (discussed in chap. 9) that have been observed to reach apparent equilibrium at high frequencies in stocks of *Drosophila* ultimately tended toward extinction, indicating that the initial apparent selective advantage of their heterozygotes had been based merely on multifactorial chromosome heterosis, lost as random combination with linked factors is approached.

The situation is different with mutations with minor effects that often appear as merely quantitative with respect to characters that have an intermediate optimum. These may be expected to be almost as likely to be favorable as unfavorable, either in nature or in the laboratory, depending on the rest of the genome.

Degrees of Dominance of Mutations

As pointed out in volume 1, chapter 5, a single dose of a very active gene product is more likely than one with slight activity to exhaust the available amount of its substrate and lead to complete dominance over inactivating mutations of any degree. Conversely, a very slight primary effect on metabolic processes is more likely to be represented in all subsequent metabolic steps by agents that do not greatly deplete their substrates, and thus lead to incomplete dominance of the ultimate observed effect. Some evidence was given in volume 1, chapter 5, that this situation holds, although not without rather frequent, easily explained exceptions. It is desirable to consider in some detail experimental studies of heterozygous effects of mutations that have occurred so recently that there can have been no important selective bias.

Unfortunately the terms *recessive* and *dominant* have usually been used merely to indicate whether a homozygote can be distinguished with more or less reliability from the heterozygotes than its allele. Mendel described statistical differences between heterozygous and homozygous "dominants" in the case of one of his characters of the pea (color of seed coat) and probably made use of this in distinguishing homozygous and heterozygous F_3 progenies where the latter happened to include no recessives in a small test group (Wright in Stern and Sherwood 1966). This was also the situation in the early studies of the genetics of *Drosophila* and of laboratory mammals. There are even now few studies of exact degrees of dominance of large random samples of mutations. It was noted (vol. 1, chap. 5) that such studies of the relatively limited number of known mutations of the guinea pig indicated that where one allele appeared to have a major positive effect this tended to be completely or almost completely dominant, while the secondary "modifiers" tended to have intermediate heterozygotes. Studies of the genetics of multifactorial quantitative variability are most easily interpreted on the basis of some degree of intermediacy of the heterozygotes as most usual (vol. 1, chap. 15).

Overdominance in effects of alleles on characters not concerned directly with fitness seems to be rather uncommon, although a group of striking exceptions were described among the color factors of the guinea pig (vol. 1, chap. 5). The alleles (C, F, P) with primarily positive effects at the loci known to affect the intensity of brown pigmentation in guinea pigs of genotype *Ebb* enter into favorable combinations up to a certain optimum beyond which their combinations dilute the brown color, with the result that in certain combinations Cc^a is slightly more intense than CC, ff is more intense than FF or Ff; and Pp is much more intense than PP (as well as pp), giving overdominance in at least the cases of Cc^a and Pp. The studies of combinations of guinea pig genes illustrated in other ways the relativity of the concept of dominance (vol. 1, chap. 5).

Dominance in the Case of "Recessive" Lethals

The most extensive study of possible heterozygous effects on viability of recently arisen "recessive" lethals, excluding the relatively small class with visible "dominant" effects in the heterozygotes, was that of Stern et al. (1952). They studied 75 spontaneous and X-ray-induced sex-linked mutations of this sort in *D. melanogaster*, all of which had arisen in a particular strain. The results indicated that the heterozygotes of both sorts of mutations had about 3.5% lower viability on the average than the homozygous normals in experiments designed with careful genetic and environmental control.

There was, however, so much variability about this mean that the authors concluded that a few of the mutations were probably heterotic.

Muller and Campbell (Muller 1950) studied 13 autosomal lethal and sublethal mutations, induced by ultraviolet light to avoid deficiencies. They found about 4.5% dominant effect in the heterozygotes.

Controlled tests of random autosomal lethals from a wild population of *D. melanogaster* by Hiraizumi and Crow (1960) showed about a 3% reduction of viability in the heterozygotes, only slightly less than in the newly arisen lethals of the preceding experiments, despite the expected more rapid elimination in nature of the mutations with the more serious heterozygous effects. Their tests were made on a genetic background that was almost wholly of laboratory origin, except for the Y chromosome. We will later go into the controversial question of the selective values of heterozygotes of lethal mutations within wild populations, where the possibilities of selective bias and coadaptation complicate the situation, more than in the case of recent mutations within a laboratory stock.

The most divergent result under laboratory conditions is probably that of Wallace (1965), who determined the heterozygous effect of 17 spontaneous autosomal lethals of *D. melanogaster* in a study involving 800 cultures and some 200,000 flies. He found an average viability loss of 2.1% in the heterozygote, but this was practically all due to one mutation with 31% viability loss in the heterozygote. The average for the other 16 was an insignificant 0.3%, indicating that almost half of the heterozygotes were more or less heterotic and most of the other half very slightly detrimental, on taking account of the variability.

Heterozygous Effects of Random Nonlethal Mutations

Two very extensive studies of random nonlethal mutations have given such unexpected results that they need detailed consideration.

In 1958 Wallace reported on a set of 48 carefully controlled experiments, involving 1,530 cultures and 575,000 flies (*D. melanogaster*). The viability of homozygotes with respect to a quasi-normal second chromosome, of which one homologue had been exposed to 500r of X-radiation, was compared with that of homozygotes in which neither chromosome had been treated. Similar comparisons were made between heterozygotes involving chromosomes of different origin, one the same quasi-normal chromosome as above, rayed or not rayed. The crosses produced other genotypes in which neither chromosome had been rayed in either the treated or control series.

The radiated chromosomes all traced ultimately to a wild-type stock (no. 5 in an experiment discussed later in this chapter) that had been subjected to a cumulative dosage of 250,000r over a period of many generations but that had been carried as a mass culture (no. 18) for about two years without radiation. Four different quasi-normal second chromosomes were isolated by use of crosses with a stock carrying dominant markers, Curly, Lobe, and Plum ($Cy\ L/Pm$) of which Cy and Pm are lethal when homozygous and crossing-over is inhibited by the Curly inversion. These four chromosomes all gave homozygotes in the normal range in viability and fertility. They were introduced by long-continued backcrossing into the $Cy\ L/Pm$ stock. A single $Cy\ L/+$ male was used in each backcross to avoid differentiation of the wild-type chromosome within the line. Homozygous wild type was next extracted from $Cy\ L/+$ parents from such a mating. Half of these were X-rayed (500r), the others were not. Single males $(+)/(+)$ or $+/+$ (representing rayed and nonrayed chromosomes, respectively) were mated with $Cy\ L/Pm$ females and $Pm/(+)$ or $Pm/+$ males were crossed with $Cy\ L/Pm$ females generation after generation. The test vials were from three $Pm/(+)$ males by four $Cy\ L/+$ females controlled by vials from matings of the same sort except that $Pm/+$ males were used. Four distinguishable genotypes are produced in each culture in roughly equal numbers. Precautions were taken to make each test and its control as nearly comparable as possible. Lines carrying each of the four isolated chromosomes were carried on by each of four assistants, who also made the progeny counts, making a block of 16 experiments. The 1958 study dealt with three such sets of sixteen.

Wallace took the frequency of the $Cy\ L/Pm$ segregants as the standard (1.000) in terms of which the frequencies of the other three segregating types were expressed to give relative viabilities. The grand averages of these relative frequencies for the 48 experiments, weighted by the numbers of cultures, are given in table 11.3. There were on the average 12.7 cultures per experiment in the first 32 experiments and nearly twice as many (22.5) in the last 16.

As expected, there is little difference in the case of $Pm/+$, which did not involve a rayed chromosome in either the test or control matings. There is

TABLE 11.3. Relative viabilities of genotypes from matings $Pm/(+) \times Cy\ L/+$ or $Pm/+ \times Cy\ L/+$ in tests of effects of rays on chromosomes.

	$Cy\ L/Pm$ $Cy\ L/Pm$	$Cy\ L/(+)$ $Cy\ L/+$	$Pm/+$ $Pm/+$	$+/(+)$ $+/+$
Test	1.000	1.115	1.137	1.033
Control	1.000	1.094	1.146	1.008
Difference	...	+0.021	−0.009	+0.025

SOURCE: Wallace 1958.

considerable difference, not only in the case of those homozygous except for the induced mutations, $+/(+)$ as opposed to $+/+$, but also in those carrying chromosomes of wholly different origin, $Cy\ L/(+)$ as opposed to $Cy\ L/+$.

The simplest statistical test is that for 48 unweighted paired comparisons (vol. 1, pp. 202–6) (table 11.4).

TABLE 11.4. Tests of differences of genotypes (table 11.3) by method of paired comparisons. The frequencies for $Cy\ L/Pm$ are taken as standard.

Test Control	$Cy\ L/(+)$ $Cy\ L/+$	$Pm/+$ $Pm/+$	$+/(+)$ $+/+$
$\bar{\Delta}$	$+0.0185$	-0.0148	$+0.0206$
$\sigma_{\bar{\Delta}}$	0.00887	0.01003	0.00896
t	$+2.09$	-1.48	$+2.30$
P	0.04	0.15	0.03

Contrary to the usual a priori expectation, the radiation seems to have produced mutations significantly favorable in the otherwise homozygous wild-type flies ($P = 0.03$). This was not wholly unexpected by Wallace, who had stressed the prevalence of heterosis more vigorously than anyone else, but he was surprised by the apparently favorable affect ($P = 0.04$) of the mutations in $Cy\ L/(+)$ in which the controls were presumably so heterozygous that he did not expect a few induced mutations in the wild-type chromosome to add anything to the heterosis. On the basis of later experiments, on which preliminary evidence was already available, he was inclined to interpret this difference in the $Cy\ L/+$ flies as merely an extreme chance deviation and that in the $+/+$ flies as a real effect of heterosis.

Wallace's statistical test did not, however, do full justice to the significance of his results. The differences from the controls, both in the case of $Cy/(+)$ and $+/(+)$, might just as well have been tested relative to $Pm/+$ as standard as to $Cy\ L/Pm$. In either case, chance differences in the standard confuse the differences between test and control. The firmest basis for comparison would seem to be obtained by taking the unweighted average for $Cy\ L/Pm$ and $Pm/+$ as the standard. This can be done by finding this average for each test and control in the 48 experiments and multiplying the published viabilities relative to $Cy\ L/Pm$ by the ratio of $Cy\ L/Pm$ to this average in each case. The results for the 48 paired comparisons are given in table 11.5.

There seems to be no doubt of the reality of the favorable effect of raying on the chromosomal heterozygotes $Cy\ L/(+)$ as well as on those otherwise homozygous $(+)/+$, and the effects are substantially the same (0.0246 vs. 0.0257, respectively). The most significant result of all is, of course, the

TABLE 11.5. Tests of differences between test and control series using the unweighted average of the two unrayed genotypes as standard.

	$0.5\,(Cy\,L/Pm\,+\,Pm/+)$ $0.5\,(Cy\,L/Pm\,+\,Pm/+)$	$Cy\,L(+)$ $Cy\,L/+$	$+/(+)$ $+/+$	$0.5\,[(Cy\,L/(+)\,+\,(+/(+))]$ $0.5\,[(Cy\,L/+\,+\,(+/+))]$
Test	1.000	1.0418	0.9662	1.0040
Control	1.000	1.0172	0.9405	0.9788
$\bar{\Delta}$		$+0.0246$	$+0.0257$	$+0.0252$
$\sigma_{\bar{\Delta}}$		0.00645	0.00613	0.00502
t		3.82	4.20	5.02
P		$\ll 0.001$	$\ll 0.001$	$\ll 0.001$

SOURCE: Calculated from Wallace 1958.

advantage of the rayed chromosomes collectively over the unrayed ones collectively, with $t = 5.0$ and 47 degrees of freedom.

These favorable effects of X rays were so surprising that they required the most careful scrutiny. Wallace made an analysis of variance of the data (with $Cy\,L/Pm$ as standard). This confirmed the significance of the X-ray effect in $+/+$ and probably $Cy\,L/+$. There were highly significant differences among the assistants but no significant interaction between assistant and treatment. The differences among the assistants could have been due to temperature differences in the responses of the genotypes since each assistant used the same shelves throughout and there was a temperature gradient from floor to ceiling in spite of the temperature regulation at 25°C. The possibility of errors in the classification was investigated in a separate experiment, and it was concluded that this could not have affected the differences between test and control to an appreciable extent. In the first 16 experiments the treatment was not coded, giving a possibility for systematic bias, but the differences were close to those in the remaining 32 experiments in which the counts were made blindly. There were also highly significant differences in the responses of the four wild-type chromosomes, but no significant interaction between chromosome and treatment. There were also small but significant differences among the experiments with the same chromosome, but again no significant interaction with the treatment.

A systematic difference between the number of flies per culture due to the treatment might conceivably affect the competitive relation among the genotypes sufficiently to cause a difference between test and control. It was found, however, that there was no appreciable difference in number per culture: an average of 375.9 for the 764 test cultures, 375.8 for the 766 control cultures.

From these and other comparisons, there seems no escape from the conclusion that heterozygosis in chromosomes that had received enough X rays to kill a man had a beneficial effect on these flies, and that this cannot be ascribed wholly to the production of heterozygosis in otherwise homozygous flies because the effect was substantially the same in flies that were heterozygous in chromosomes of different origin.

Wallace conducted six later sets of experiments (1959) with the same chromosomes. These involved 7,753 cultures and $2\frac{3}{4}$ million flies. Three of the new sets were of the same sort as that described, progenies from $Cy\ L/+\ \times\ Pm/(+)$ and from $Cy\ L/+\ \times\ Pm/+$ after the $(+)$ chromosome had been exposed to 500r. In two of the sets, the progeny were from $Pm/+\ \times\ Cy\ L(+)$ and from $Pm/+\ \times\ Cy\ L/+$ so that $Pm/(+)$ and $+/(+)$ carried the rayed chromosome in the progeny. Finally, experiments were made in which two different wild-type chromosomes were brought together, $Cy\ L/+^{1}\ \times\ Pm/(+^{2})$, giving $Cy\ L/Pm$ (used as standard), $Cy\ L/(+^{2})$, $Pm/+^{1}$, and $+^{1}/(+^{2})$, in comparison with $Cy\ L/+^{1}\ \times\ Pm/+^{2}$.

The results of all of these experiments, including those already discussed (designated 2–9 M), are given in table 11.6. The data are not available for paired comparisons within the sets, but presumably the standard error of mean differences would not be very different from those of the first set since the number of cultures was not very different. None of the differences are, however, as great as those in the first set, and the only ones that are

TABLE 11.6. Summary of tests of the effects of heterozygosis in four quasi-normal chromosomes from population no. 18, which had received 500r from X-ray treatment.

Test	$Cy\ L/Pm = 1.000$			$0.5\,[Cy\ L/Pm + (Pm/+)] = 1.000$		
Test	$Cy\ L/(+)$	$Pm/+$	$+/(+)$	$Cy\ L/(+)$	$+/(+)$	No. of Cultures
Control	$Cy\ L/+$	$Pm/+$	$+/+$	$Cy\ L/+$	$+/+$	Test Control
2–9 M	$+0.021$	-0.009	$+0.025$	$+0.024$	$+0.027$	764 766
6–14 F	$+0.015$	$+0.001$	$+0.007$	$+0.014$	$+0.006$	672 676
15–22 F	$+0.005$	$+0.008$	$+0.026$	$+0.001$	$+0.021$	637 636
19–26 M	$+0.008$	-0.007	$+0.010$	$+0.011$	$+0.012$	598 596
	$Cy\ L/Pm = 1.000$			$0.5\,[(Cy\ L/Pm) + (Cy\ L/+)] = 1.000$		
	$Cy\ L/+$	$Pm/(+)$	$+/(+)$	$Pm/(+)$	$+/(+)$	
	$Cy\ L/+$	$Pm/+$	$+/+$	$Pm/+$	$+/+$	
11–18 M	-0.002	-0.002	$+0.007$	-0.001	$+0.008$	637 639
37–43 M	-0.007	$+0.012$	$+0.010$	$+0.015$	$+0.012$	496 499
	$Cy\ L/Pm = 1.000$			$0.5\,[(Cy\ L/Pm) + (Pm/+)] = 1.000$		
	$Cy\ L/(+)$	$Pm/+$	$+^{1}/(+^{2})$	$Cy\ L/(+)$	$+^{1}/(+^{2})$	
	$Cy\ L/+$	$Pm/+$	$+^{1}/+^{2}$	$Cy\ L/+$	$+/+$	
23–34H	-0.003	$+0.007$	$+0.012$	-0.006	$+0.008$	837 839

SOURCE: Wallace 1959.

probably significant at the 0.05 level are the comparison of $+/(+)$ with $+/+$ in 15–22 F, of $Cy\ L/(+)$ vs. $Cy\ L/+$ in 6–14 F, and perhaps of $Pm/(+)$ vs. $Pm/+$ in 37–43 M.

In the case $+/(+)$ vs. $+/+$, all six differences are indeed positive but the average for the later sets is only $+0.012$ and in comparison with $+0.027$ for the first set. Four of the five differences in the case of $Cy\ L/(+)$ vs. $Cy\ L/+$ are positive, but the average for the later sets is only about $+0.008$ in contrast with $+0.024$ for the first set.

The smallness of the differences and the doubtful significance in any of the later sets cannot invalidate statistically the highly significant results of the first set, but it does indicate some unknown difference in conditions. We are left with the conclusion that radiation-induced mutations can affect the viabilities in chromosomal homozygotes $+/(+)$ and chromosomal heterozygotes $Cy\ L/(+)$ under some conditions while there is no convincing evidence that they have any effect in the chromosomal heterozygotes $Pm/(+)$ and $+^1/(+^2)$.

Wallace (1963) reported on another series of experiments, of a basically similar sort but with six wholly different X-rayed chromosomes. Two of these (EX) were from the control stock (no. 3) in a radiation experiment to be described later. Two were from wild flies collected in New Orleans (NO). Two were from wild flies collected in Riverside, California (RC). A seventh stock from wild flies from South Africa (CA) was used in certain crosses in addition to the marker stock $Cy\ L/Pm$.

Four treatments were used: 0r, 250r, 750r, and 2,250r. A series called HOMO was homozygous in one or other of the first six chromosomes except that one chromosome may have been X rayed. A series called INTRA was similar except that the flies were heterozygous for the two chromosomes from the same locality. In a series called INTER, the flies were heterozygous for CA and one of the other six. Altogether 8,189 cultures were produced and $2\frac{1}{2}$ million flies counted.

TABLE 11.7. The relative viabilities of $Cy\ L/+$, $Pm/+$, and $+/+$ flies grouped according to radiation exposure. Matings of type $Pm/(+) \times Cy\ L/+$ or $Pm/+ \times Cy\ L/+$.

Dose	$Cy\ L/(+)$	$Pm/+$	$+/(+)$	Flies per Culture	No. of Cultures
0r	1.154	1.357	1.345	287.8	2,055
250r	1.151	1.358	1.339	288.2	2,050
750r	1.154	1.363	1.343	287.5	2,044
2,250r	1.145	1.354	1.344	285.7	2,040

SOURCE: Reprinted, with permission, from Wallace 1963.

In contrast with some of the preceding results, there were no significant effects of radiation. Grouping the total data according to dosage, the averages were as shown in table 11.7 with $Cy L/Pm$ again the standard.

There were significant differences among the three types of experiments: HOMO, INTRA, and INTER, as shown in table 11.8.

No difference is expected between HOMO and INTRA $Cy L/+$ carrying EX, NO, and RC in equal numbers. INTER $Cy L/+$ may be (and is) different since it carries a different type of chromosome, CA. No differences are expected among any type of $Pm/+$ flies except for possible irregularities in the INTRA culture due to sterility, which reduced the total number of

TABLE 11.8. The relative viabilities of $Cy L/+$, $Pm/+$, and $+/+$ flies grouped according to types of experiment. Matings $Pm/(+) \times Cy L/+$ or $Pm/+ \times Cy L/+$.

	$Cy L/(+)$	$Pm/+$	$+/(+)$	Flies per Culture	No. of Cultures
HOMO	1.137	1.362	1.175	283.4	2,772
INTRA	1.137	1.353	1.387	287.7	2,624
INTER	1.179	1.360	1.475	290.6	2,793

SOURCE: Reprinted, with permission, from Wallace 1963.

cultures. The viabilities of the wild-type flies, $+/+$, increased with increasing heterozygosis as expected.

These experiments as a whole seem to demonstrate that the favorable effect of radiation in genotypes $+/(+)$ and $Cy L/(+)$, in at least some of the earlier experiments where all $+$ chromosomes were derived from a certain stock (with a history of 250,000r accumulated radiation two years earlier), does not apply in general since there was no indication of it in chromosomes from three other wholly unrelated stocks.

Further discussion will be deferred until after consideration of somewhat related experiments by Mukai that also involved counts of millions of flies.

Mukai (1964) started 104 lines, all heterozygous for the same wild-type second chromosome of *D. melanogaster*, by mating an isogenic wild-type strain with a marker stock Cy/Pm of the same genetic background in the other chromosomes, and then mating a single $Pm/+_i$ son with Cy/Pm females followed by 104 separate second backcrosses of $Pm/+_i$ males. These lines were carried on, generation after generation, by the same procedure, $Pm/+_i$ male by Cy/Pm females. The latter were always drawn from the original stock in which spontaneous detrimental mutations would tend to be kept at equilibrium frequencies by natural selection. In contrast to Wallace's experiments, the $+$ chromosomes of the experiments considered first had not been exposed to radiation. Deleterious mutations were initially absent in the chosen + chromosomes, but spontaneous mutations would be expected to accumulate under the enforced heterozygosis.

The second chromosomes were tested from time to time for homozygous viability by making numerous matings of the type $Cy/+_i \times Cy/+_i$ within each line and determining how far the $+_i/+_i$ segregants deviated from the expected frequency (about 33%).

By generation 25, lethals had appeared in 15 of the 101 lines (three having been lost early). The lethal mutation rate was thus about 0.006 ($= (15/101)/25$), which is considerably higher than the usual rate. A few severe detrimentals (frequencies of $+_i/+_i$ less than 20% in the test) also appeared. These lines, like those with lethals, were excluded. After generation 32, however, the accumulation of mild detrimentals brought the viabilities of some lines below 20% and only lethal-bearing lines were excluded. After this generation, duplicate lines were carried to slow down the loss of lines from lethals.

There was a considerable number of lines in generation 32 and a few even up to generation 60, in which the viability index in tests continued to deviate little if any from the expected 33%, indicating that no mutations with appreciable deleterious effect had yet occurred. In each of generations 10, 15, 20, and 25, at least the best five lines were of this sort and were used as contemporary controls. The three lines with highest indexes in generations 52 and 60 collectively were similarly used as the controls for generations 32, 52, and 60.

Some of the statistical results are shown in table 11.19 (Mukai 1964, 1968; Mukai and Yamazaki 1967).

TABLE 11.9. Means and variance components of viability indexes of homozygotes $+_i/+_i$ from matings of type $Cy/+_i \times Cy/+_i$ obtained in the indicated generations as described in the text. The means are compared with those of the best lines (controls).

Generation	10	15	20	25	32	52	60	78
Lines tested	98	97	89	84	80	75	77	44
Flies counted (thousands)	140	240	383	926	428	293	168	...
Viability index (tests)	31.57	27.97	30.85	28.53	28.14	21.32	16.42	8.65
Viability index (controls)	32.94	29.69	32.23	32.99	32.84	33.12	32.41	32.79
Difference (d)	−1.37	−1.72	−1.38	−4.46	−4.70	−11.80	−15.99	−24.14
Genetic variance (σ_d^2)	0.70	2.09	1.85	3.66	5.65	38.95	66.21	37.53
Error variance	10.99	16.90	8.02	6.14	5.78	8.87	6.24	...

SOURCE: Data from Mukai (1964, 1968) and Mukai and Yamazaki (1967).

There was much irregularity in the distribution of viability indexes in the early generations, suggesting that these were affected by differences in conditions, as brought out in chapter 9. Teissier (1942) found that the amount of crowding affected greatly the relative viabilities of $+/+$ and $Cy/+$ segregants. It should be noted also that the genetic variance was small in comparison with the error variance in the early generations.

In Mukai's tests, the distribution of viabilities in generations 10 and 20 were rather compact (ranges only from -5 to $+2$ or $+3$ relative to the mean percentage), while in generations 15 and 25 there was much scattering (-9 to $+5$ in the former, -11 to $+3$ in the latter). In both of these generations the numbers of quasi-normal lines was much less than in the later generation, 32, indicating that unmutated $+/+$ must have been depressed relative to $Cy/+$ in the former. In generation 15, indeed, the best five lines averaged only 29.69%.

An attempt was made to estimate the number of accumulated detrimentals, their average effect on the percentage of segregating homozygotes, and the mutation rate for generations 10 to 25 by formulas proposed by Bateman (1959). These depend on the reasonable assumption that mutations occur at random (Poisson distribution of numbers among lines) and the more questionable assumption that gene effects are additive.

Let n_i be the number of mutations, a_{ix} the effect of a particular mutation, $a_i(= \sum_x^{n_i} a_{ix}/n_i)$ the mean effect, and $d_i(= n_i a_i)$ the total effect, all in line i. Then $\bar{n}(= \sum_i^N n_i/N)$ is the mean number of mutations per line, $\sigma_n^2(= \bar{n})$ is the variance of line numbers under Poisson variability, $\bar{a}$ is the mean effect, σ_a^2 the variance of effects, $\sigma_{\bar{a}}^2(= \sigma_a^2/\bar{n})$ the error variance of $\bar{a}$, and $\bar{d}$ $(= \bar{n}\bar{a})$ the mean and σ_d^2 the variance of line effects.

$$\sigma_d^2 = \sigma_{\bar{n}\bar{a}}^2 = \bar{a}^2\sigma_n^2 + \bar{n}^2\sigma_a^2 = \bar{n}[\bar{a}^2 + \sigma_a^2]$$

The quantities $\bar{d}$ and σ_d^2 are known (table 11.9).

Let $C(= \sigma_a/\bar{a})$ be the coefficient of variability of mutant effects.

$$\sigma_d^2 = \bar{n}\bar{a}^2[1 + C^2] = \bar{d}^2[1 + C^2]/\bar{n}$$

$\bar{n} = d^2[1 + C^2]/\sigma_d^2$ (mean number of mutations per line)

$\bar{a} = \bar{d}/\bar{n}$ (mean effect of mutations)

$u = \bar{n}/g$ (mutation rate [g = number of generations])

The estimation of these quantities depends on the estimation of C.

Mukai (1964) used the regression of $\bar{d}$ and σ_d^2 on number of generations, 0.1261 and 0.1127, respectively, for generations 10, 15, 20, and 25. He took as the minimum estimate of $\bar{n}$ the value for $C^2 = 0$, the case in which all mutations have the same effect, and obtained for generation 25, $\bar{n} = 3.53$, $\bar{a} = 0.894$, and $u = 0.141$.

He took as his upper limit $C^2 = 1$, which yields estimates of $\bar{n}$ and u twice as great as above, $\bar{a}$ half as great. This is the case of a two-point distribution with equal frequencies of $a = 0$ and $a = 2\bar{a}$. Larger values of C^2 are possible;

for example, a negative exponential gives $C^2 = 2$ but mutations with zero effect on viability would be unrecognizable, which indicates that observable C^2 is probably much less than 1.

The above formulas, however, require revision if the condition of additivity of gene effects does not hold. It is quite likely that the accumulation of mutations with individually very slight effects, as long as homeostatic processes are operative, may lead to a rapid falling off of viability as homeostasis breaks down. Mukai found that this was, indeed, the case. He found that relative viability (V) was related to his estimates of the number of accumulated mutations $\bar{n}$ up to generation 60 at his minimum estimated mutation rate, 0.141, by the curve:

$$V = 1 - 0.009813\bar{n} - 0.005550\bar{n}^2$$

Since an interaction effect of this sort invalidates the estimates of mutation rate based on the assumption of additivity, we should express V in terms of generations g by multiplying $\bar{n}$ in the formula by 0.141:

$$V = 1 - 0.001384g - 0.0001103g^2$$

Table 11.10 compares the observed and the calculated values of V.

TABLE 11.10. Observed and calculated relative viabilities (V) by generations.

Generation	10	15	20	25	32	52	60	78
Observed V	0.958	0.942	0.957	0.865	0.857	0.644	0.507	0.264
Calculated V	0.975	0.944	0.928	0.896	0.843	0.629	0.519	0.221

SOURCE: Mukai 1968.

That the formulas for $\bar{n}$ and u are not reliable is shown by considering the estimates they give for later generations (table 11.11).

TABLE 11.11. Estimates of $\bar{n}$ and u for generations 32–78 based on the formulas applied to the regressions of $\bar{d}$ and σ_d^2 for generations 10–25. The estimates for generation 25 were based on the regressions in generations 10–25, as described in the text.

Generation	25	32	52	60	78
$\bar{n}$	3.5	3.9	3.6	3.9	4.2
u	0.141	0.122	0.069	0.064	0.054

There is so much uncertainty in any attempt at transformation of scale in order to bring about additivity that it is desirable to use some other mode of estimation if possible. As Mukai (1964) noted, the chance for a line to remain unchanged is $e^{-\bar{n}}$, where the average number of accumulated mutations within lines is $\bar{n}$, assuming random (Poisson) variability (vol. 1, eq. 8.97). There were six lines among 84 in generation 25 that he considered to be quasi-normal (index range, 32.3 to 34.2; average, 32.99). This yields $\bar{n} = -\log(6/84) = 2.64$ and $u = \bar{n}/25 = 0.106$. He considered that the latter was as close to his estimate 0.141 for mutation rate as could be expected in view of its utilization of only one bit of information. The invalidation of the other method by the interaction effect (not then apparent) seems to leave it, however, as the most unbiased method available.

The estimate of six unmutated lines in generation 25 was, however, decidedly uncertain. There were probably more than six judging from the general depression of viability of $+/+$ relative to $Cy/+$ referred to earlier and from the indications of a larger number in generations 32 and 52. In the former, 10 of the 80 available lines had viability indexes above 31.3%. If these are taken as normal, a mutation rate of 0.065 is implied. There were 6 in 75 lines in generation 52 above 31.6% and 7 in 77 lines in generation 60 above 30.9%. These, if normal, imply mutation rates of 0.049 and 0.040, respectively. Even if only the three lines used as controls for generations 32, 52, and 60 (average 33.1% in generation 52, 32.4% in generation 60) are accepted as normal, the estimated mutation rate is only 0.054. If the mutation rate really were as large as 0.141, there would be an average accumulation of 8.5 mutations per line by generation 60, implying a chance of $e^{-8.5} = 0.00020$ of absence in a given line and of 0.016 of absence in only one of 77 lines. The above consideration points toward a rate of about 0.05 or 0.06 as most probable. This is some eight to ten times that of lethal mutations in the same stock, which, however, seems to have been higher than usual.

Mukai, Chigusa, and Yoshikawa (1965) determined the viability indexes of two kinds of heterozygous wild type: ones with one quasi-normal chromosome $+_o$ and the other random $+_i$, ($+_o/+_i$ from $Cy/+_i \times +_o/+_o$) and ones obtained by random matings among the lines ($+_i/+_j$ from $Cy/+_i \times Cy/+_j$).

The chosen quasi-normal lines were of three sorts: (1) the best line in the generation, represented in what follows by $+_o/+_o$, (2) quasi-normals with respect to the second chromosome from a stock $+_w/+_w$, that traced to the same stock as the preceding but before isolation of the particular second chromosome received by all of the 104 lines, and (3) a stock in which the second chromosomes $+_B/+_B$ were of wholly independent origin. The X and third chromosomes, associated with $+_w/+_w$ and $+_B/+_B$, were, however, all derived from the same isogenic stock as was $+_o/+_o$. The viability indexes and other data for heterozygotes involving these quasi-normal chromosomes are shown in table 11.12.

TABLE 11.12. Comparison of viability indexes (backcross) of mutant heterozygotes with homozygous controls on various genetic backgrounds (interpopulation) $+_B/+_i$, intrapopulation $+_w/+_i$, and homozygous except for accumulated mutations, $+_o/+_i$.

Generation	32	32	32	60	78
Test genotype	$+_B/+_i$	$+_w/+_i$	$+_o/+_i$	$+_o/+_i$	$+_o/+_i$
Matings tested	80	80	80	77	44
Flies (thousands)	446	471	392	375	...
Viability index (test)	54.87	50.27	51.51	52.28	52.41
Viability index (control)	55.99	50.99	49.91	48.62	49.18
Difference	-1.12	-0.72	$+1.60$	$+3.66$	$+3.23$
Genetic variance	0.48	0.22	0.41	1.97	3.61
Error variance	17.24	7.43	5.89	3.96	...

SOURCE: Mukai, Chigusa, and Yosikawa (1965).

Where the genetic background was heterozygous ($+_B/+_i$, $+_w/+_i$), the viability indexes of the heterozygotes were slightly but significantly lower (0.05 level) than their controls. Where, however, there was a homozygous background except for the few accumulated mutations, $+_o/+_i$ was very significantly superior to the controls $+_o/+_o$ in all generations tested (32, 60, 78), with increasing excess up to a certain point, although the homozygotes in the lines were declining markedly as already shown (28.14 in generation 32, 16.42 in generation 60, 8.65 in generation 78).

This result agrees with Wallace's finding of a favorable effect of random mutations in an otherwise homozygous background in his first major series (but not with the absence of effect in his later series). The contrasting result on a heterozygous background agrees to some extent with the general tendency of Wallace's results, although not with all cases.

The existence of a heterotic effect of the $+_i$ chromosome in an otherwise homozygous background $+_o/+_i$ does not in itself prove that it is due to the same mutations responsible for the decline of homozygotes $+_i/+_i$. However, a tabulation of the relative viabilities of $+_o/+_i$ according to the homozygous averages of lines in generations 32, 60, 78, and 85 (Mukai 1968) shows clearly that heterozygous advantage varies inversely with homozygous defect down to about 50% and then declines with further homozygous defect (table 11.13), indicating that the same mutations were involved.

The behavior of random heterozygotes between lines, $+_i/+_j$, was utterly different (Mukai and Yamazaki 1967). In generation 32, matings of type $Cy/+_i \times Cy/+_j$ were made between adjacently numbered lines, 1×2, 2×3, 3×4, and so on. The same sort of matings (RS) were also made in generation 52 but, in addition, matings (RA) were made between all lines,

TABLE 11.13. Viabilities of $+_o/+_i$ relative to the controls, according to the relative viabilities of $+_i/+_i$.

$+_i/+_i$	$+_o/+_i$	Lines	$+_i/+_i$	$+_o/+_i$	Lines
0.90–1.00	1.025	41	0.40–0.49	1.084	17
0.80–0.89	1.035	47	0.30–0.39	1.085	21
0.70–0.79	1.045	23	0.20–0.29	1.076	28
0.60–0.69	1.076	7	0.10–0.19	1.076	25
0.50–0.59	1.092	11	0–0.09	1.062	25

SOURCE: Mukai 1968.

two apart in numbers, 1×3, 2×4, 3×5, and so on. Both series may be considered random with respect to mutations. The controls were the same as for the homozygotes of table 11.10. The results are shown in table 11.14.

These heterozygotes were very significantly inferior to the controls. Those of generation 32 should have been heterozygous in about the same number of loci on the average as the heterozygotes of type $+_o/+_i$ of generation 60, yet they deviate from the controls in opposite directions (relative viability of $+_i/+_j$, 0.900; of $+_o/+_i$, 1.075).

TABLE 11.14. Comparisons of heterozygotes, $+_i/+_j$, between wild-type second chromosomes of lines, both of which had usually but not always accumulated deleterious mutations.

Generation	32RS	52RS	52RA
Line crosses tested	77	69	68
Flies (thousands)	397	292	243
Viability index of tests $(+_i/+_j)$	29.56	24.78	25.88
Viability index of controls $(+_o/+_o)$	32.84	33.12	33.12
Difference	−3.28	−8.34	−7.24
Genetic variance	8.30	37.32	29.77
Error variance	5.85	8.64	10.26

SOURCE: Mukai and Yamazaki 1967.

Heterozygotes $+_i/+_j$ are, nevertheless, heterotic with respect to the homozygotes $+_i/+_i$ and $+_o/+_o$ where these are similar, as shown by the excess of their relative viabilities over the average for the homozygotes (relative viabilities 0.900 vs. 0.857 in generation 32, 0.764 vs. 0.644 in generation 52).

The interpretation is complicated by the fact that a few of the lines had accumulated no mutations, so that heterozygotes involving them would actually be of the type $+_o/+_i$ (heterotic with respect to homozygous normal) or perhaps even homozygous normal themselves. The distributions of viability indexes were indeed strongly bimodal, including in all cases a small group superior on the average to the controls, which Mukai interpreted as having at least one quasi-normal chromosome. The most reliable separation was obtained by tabulating the viabilities of the line crosses against the sums of the viability indexes of the parental homozygotes, as shown for generation 32 in figure 11.4. There is clear separation of two groups in spite of overlap with respect to parental viabilities, with 14 in the superior group and 63 in the inferior. This separation is due to a strong negative correlation (-0.75) in the superior group, similar to the negative relation between $+_o/+_i$ and $+_i/+_i$ of table 11.13, and a strong positive correlation ($+0.67$) in the large group. The same situation was found in both 52RS and 52RA, with 13 in the superior group in the former and 15 in the latter.

Letting M and N represent lines with and without mutations, with frequencies $(1 - q)$ and q, respectively, the distribution of random crossbred progenies can be represented by $(1 - q)^2 M/M + 2q(1 - q)M/N + q^2 N/N$. Solving for q gives 0.095 in generation 32 (7.3 normal lines) and 0.108 in generation 52 (7.7 normal lines). Assuming that there were actually seven such lines in both generations (all in generation 32 having persisted to generation 52 unchanged), the estimated mutation rate is 0.075 from the former, 0.044 from the latter, indicating about 0.060 as most probable, in fair agreement with the estimate of 0.054 from the minimum of three unmutated lines in generation 60.

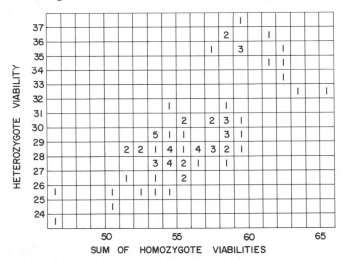

Fig. 11.4. Correlation between viabilities of heterozygotes and sums of corresponding homozygote gene viabilities in generation 32. Redrawn from Mukai and Yamazaki (1967, fig. 6); used with permission.

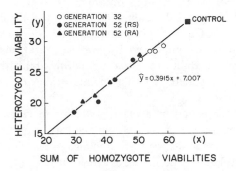

Fɪɢ. 11.5. Relationships between viabilities of homozygotes and heterozygotes carrying chromosomes that both have new mutant genes. Redrawn from Mukai and Yamazaki (1967, fig. 7); used with permission.

The positive correlation between the viabilities of heterozygotes, interpreted as having mutations in both chromosomes, and those of homozygotes, indicates here also that the responsible genes are the same. The regularity of the relation in their case is shown in figure 11.5. The regression equation is

$$\hat{y} = 0.392x + 7.007$$

This indicates that there was about 39% dominance of the deleterious effects of those mutations. Since this is less than 50%, overdominance with respect to the homozygous parental types is implied, although not quite such strong overdominance as when all of the heterozygotes, including those actually of types $+_o/+_i$, are considered.

Mukai, Yoshikawa, and Sano (1966) also made extensive experiments to test whether X-ray-induced mutations have the same sort of effects as accumulations of spontaneous ones. They used the same dosage, 500r, as Wallace. Heterozygotes carrying an X-rayed chromosome $+/(+)$ were compared with isogenic wild type $+/+$. The relative viabilities of the heterozygotes were significantly superior (0.01 level) by $+0.027$ and significantly more flies were reared per line (790 vs. 767). The heterotic effect agrees with Wallace's first series, but this is not true of the greater number reared.

On the other hand, there were no significant differences in relative viability in similar comparisons in three widely different heterozygous backgrounds. In this case, those carrying a rayed chromosome were significantly fewer in numbers per line. Comparisons were made in more than 300 lines in each case, and the numbers of flies reared were much greater (900–1,700) in test and control in each case, in comparison with the experiments on a homozygous background.

The interpretation of the results obtained by Wallace and by Mukai is obviously a matter of great difficulty. The major points are listed below:

1. The viabilities of *Drosophila*, homozygous for certain quasi-normal second chromosomes, are improved by *random* mutations (shown to be of lowered homozygous viability in Mukai's experiments) if these are restricted to one homologue. This holds whether the mutations are induced by X rays or occur spontaneously.

2. There is, on the contrary, a marked decline in viability if both homologues carry such mutations, even though the total amount of heterozygosis is the same. The average degree of dominance of the deleterious effect was about 39% in Mukai's experiments, which, being less than 50%, gives heterosis with respect to the two corresponding homozygotes.

3. In most cases, the occurrence of random mutations in one of two quasi-normal homologues of different origin (heterozygous background) either has no appreciable effect or a slightly deleterious one.

4. In at least one extensively studied case, however, such mutations had almost as a strong a heterotic effect on a heterozygous background as on a homozygous one, $(Cy/(+)$ vs. $Cy/+$ in Wallace's first major experiment).

5. The heterotic effect of random mutations in one member of an otherwise homozygous pair of second chromosomes does not occur in all cases. It appears to be a property of particular chromosomes. In Wallace's experiments, it occurred after X-irradiation from quasi-normal chromosomes from a stock that had been recovering for two years from exposure over a period of generations to 250,000r. It did not occur in similarly extensive experiments with six second chromosomes from diverse natural or untreated mass populations in Wallace's later experiments. It occurred in Mukai's experiments with both spontaneous and X-ray-induced mutations in derivatives of a single normal second chromosome.

6. The heterotic effect of random mutations varied greatly in different experiments conducted by Wallace with the same chromosomes.

The greatest difficulty is in accounting for the qualitative difference in the results according to whether the mutations are restricted to one homologue or are present in both. Nothing in genetics is better attested than the lack of dependence of the effects of mutations on their arrangement in the genome, unless located very close together. The physiologically interacting components of a cistron function better if all of the positive alleles are assembled in one homologue, but such components are separated by only a small fraction of a crossover unit in *Drosophila* and presumably all other higher organisms, although more loosely in bacteria.

Ordinary gene effects are believed to trace to the specific patterns of protein molecules, synthesized in the ribosomes in the cytoplasm under control of molecules of messenger RNA though the mediation of small complementary molecules of transfer RNA that carry amino acids to the site of synthesis. These molecules of RNA, in turn, are synthesized as complements of different DNA molecules in the chromosomes. Only very small portions of the total DNA in the chromosome are involved in synthesizing any single molecule of messenger RNA, an amount consisting of a single cistron. Coupling-repulsion differences in the joint effects of ordinary mutations would thus seem limited to the extent of a cistron and cannot be invoked to account for the differences of this sort, presumably occurring at random throughout a chromosome.

The possibility that these viability mutations may all occur at a single extraordinarily mutable complex locus was ruled out (Mukai and Yamazaki 1967) by the production of numerous recombinant chromosomes from pairs of lines with similarly low homozygous viabilities. These showed enormously increased genetic variance as a group (19.2 vs. 7.1 in one case, 11.6 vs. 1.4 in another) in which homozygous viability indexes were nearly identical at 26%. There was, indeed, recovery of a few apparently normal chromosomes.

It would thus seem necessary to postulate a radically different sort of mutation from those with effects that trace through the transcription-translation process. Moreover, this sort of pretranscription effect must be more important than the latter with respect to viability, either in magnitude or frequency, to outweigh ordinary mutant effects on viability, and it must be such that ones occurring at random permit the alleles assembled in a normal homologue to function more efficiently than if both homologues are normal.

A number of effects on the behavior of chromosomes as wholes or as portions more extensive than compound loci are indeed known. The position effects emanating from heterochromatin, introduced into a euchromatic region of a chromosome, are of this sort. There seems to be a temporary modifying effect on genes over considerable distances, which results in modified transcription and translation so that the normal gene effects are modified in an irregular way, giving rise to mosaics. The effect is not transmitted along the germ line, however, indicating that no irreversible mutation has occurred. The effect rarely spreads over as much as 5% of the length of the chromosome from the introduced heterochromatin (Demerec 1940); other effects on chromosome behavior as a whole are the negative effect of an extra Y chromosome on position effects, elimination of paternal chromosomes in monopolar mitosis in *Sciara* (Metz 1938), or random inactivation of part or all of one of the X chromosomes in somatic cells of female mammals (Lyon 1961; L. B. Russell 1961). These sorts do not seem very promising as models for the action of the viability mutants.

The problem of accounting for the heterotic effect of these mutants is not as formidable a problem as the existence of the coupling-repulsion difference, but nevertheless it is rather difficult. It is possible, if not very plausible, that the activity of the normal chromosome may be more efficient without a competing normal chromosome than with it. More plausible, perhaps, is an interpretation in terms of an advantage in versatility, although it is not clear that random mutations would be likely to give a useful alternative to action of the normal homologue. The usual interpretation of chromosome or genome heterosis (complementarity with respect to favorable gene actions with greater than semidominance) applies to the heterosis of $+_i/+_j$, relative to the two homozygotes $+_i/+_i$ and $+_j/+_j$, but not to that of $+_o/+_i$ to $+_o/+_o$.

The usual dependence of the latter on a homozygous background fits well with the concept of increased versatility, since a heterozygous background would be expected to contribute so much versatility that the effect of a few random mutations would be negligible. The case of the superiority of Wallace's $Cy\ L/(+)$ to $Cy\ L/+$ suggests, however, that the specific properties of the chromosomes are more important than the general nature of the genetic background.

The absence of the heterotic effect in tests of second chromosomes of relatively direct natural origin reinforces this idea. This and the rather low consistency where it does occur (in Wallace's experiments) point to a threshold, which the more fully normal chromosomes are safely above.

These experiments have uncovered extraordinarily interesting phenomena that seem not to be rare anomalies but rather have to do with the most important viability mutations, other than lethal or near-lethal, in at least the second chromosomes of *D. melanogaster*. It is evident, however, that more investigation is necessary before the frequency of occurrence and the interpretation are sufficiently understood to give a firm foundation for interpretations in population genetics.

Course of Evolution from Change in Mutation Rate

We consider next the course of change in gene frequency and mean selective value after an abrupt change in mutation rate, under the usual assumption that the effects of mutations are injurious. It has been concluded (vol. 2, pp. 35–36) that the mean selective value reaches an equilibrium that depends on the mutation rate, independent of the severity of the opposed selection. The course of change, however, depends very much on the severity (Wright 1950b, 1955).

Genotype	f	w
AA	$(1 - q)^2$	1
AA'	$2q(1 - q)$	$1 - s$
$A'A'$	q^2	$1 - s'$

$$\bar{w} = 1 - 2sq - (s' - 2s)q^2$$

$$\Delta q = v(1 - q) - [s + (s' - 2s)q]q(1 - q)$$

If the selection pressure against the heterozygotes, $2sq(1 - q)$, is much greater than the mutation pressure, $v(1 - q)$, q is kept so small that the homozygotes play a negligible role. The term $(1 - q)$ in the selection and mutation pressures may also be ignored in obtaining useful approximations, giving $\Delta q \approx v - sq$.

We assume that equilibrium has been reached at mutation rate v so that $q_0 = \hat{q} = v/s$ and $\bar{w} = 1 - 2v$.

Assume now that the mutation rate becomes kv so that $\Delta q \approx kv - sq$ and gene frequency starts moving toward $q_\infty = kv/s$ and mean selective value toward $\bar{w}_\infty = 1 - 2kv$.

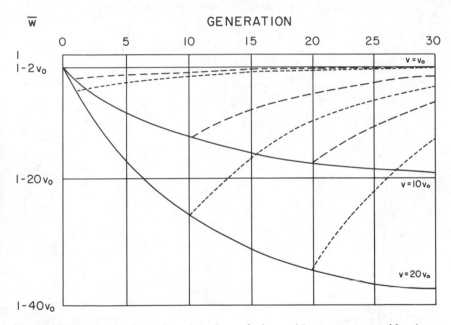

Fig. 11.6. Mean selective values ($\bar{w}$) of populations with respect to semidominant mutations, 10% inferior to type in productivity, in successive generations under 10- or 20-fold increase in mutation rate (*solid lines*), and under return to the normal rate (v_0) after 1, 10, or 20 generations (*broken lines*). Redrawn from Wright (1955, fig. 1).

Then

$$q_1 - q_0 = kv - sq_0 = s(q_\infty - q_0)$$
$$q_\infty - q_1 = (1 - s)(q_\infty - q_0)$$
$$q_2 - q_1 = s(q_\infty - q_1)$$
$$q_\infty - q_2 = (1 - s)(q_\infty - q_1) = (1 - s)^2(q_\infty - q_0).$$

By induction,

$$q_\infty - q_n = (1 - s)^n(q_\infty - q_0) \quad \text{and} \quad \bar{w}_\infty - \bar{w}_n = (1 - s)^n(\bar{w}_\infty - \bar{w}_0).$$

If, after n generations in which the mutation rate is kv, it returns to its original value v, there is return toward the initial equilibrium, q_0, by analogous formulas:

$$q_{n+m} - q_0 = (1 - s)^m(q_n - q_0)$$
$$\bar{w}_{n+m} - \bar{w}_0 = (1 - s)^m(\bar{w}_n - \bar{w}_0)$$

These situations are illustrated in figure 11.6. A mutation that causes a 10% selective disadvantage of the heterozygote has reached equilibrium with the mutation rate, v_0, causing a reduction of mean selective value, $\bar{w}_0 = 1 - 2v_0$. With a 10- or 20-fold increase in mutation rate at generation 0,

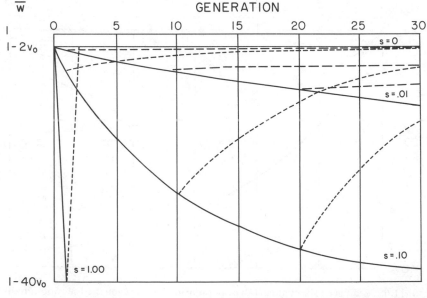

FIG. 11.7. Mean selective values ($\bar{w}$) of populations with respect to semidominant mutations with 1%, 10%, or 100% selective disadvantage, under a 20-fold increase in mutation rate (*solid lines*), and return to the normal rate after 1, 10, or 20 generations (*broken lines*). Redrawn from Wright (1955, fig. 2).

$\bar{w}$ declines toward the new equilibrium at $1 - 20v_0$ or $1 - 40v_0$, respectively (*solid curves*). It takes about seven generations to go halfway, seven more to go half the remaining distance, and so on.

The dotted lines show the course of recovery on returning to mutation rate v at generations 1, 10, or 20. The half period is again about seven generations.

Figure 11.7 compares these courses at three different selective disadvantages, $s = 0.01$ or 0.10 or 1 with mutation rate shifting from v to $20v$ and returning to v at generations 1, 10, or 20.

In the case of a dominant lethal ($s = 1$), the population goes at once all the way to the new equilibrium and fully recovers at once on return of the mutation rate to v since dominant lethals are not transmitted.

In the case of a mutation with only a 1% disadvantage as a heterozygote, the half period of decline is about 69 generations, ten times that for a 10% selective disadvantage. Thus it takes a long time for much apparent damage to occur, but the recovery from that which does occur is equally slow, with half period 69 generations.

The case of a completely recessive mutation is more complicated.

Genotype	f	w	$\bar{w} = 1 - sq^2$
AA	$(1 - q)^2$	1	$\Delta q = v(1 - q)^2 - sq^2 \approx v - sq^2$
Aa	$2q(1 - q)$	1	$\hat{q} = (v/s)^{1/2}$
aa	q^2	$1 - s$	$\bar{w} = 1 - v$

With a k-fold increase in mutation rate, $q_\infty^2 = kv/s$ and $\bar{w}_\infty = 1 - kv$. Putting $\Delta q = dq/dt$,

$$2(kvs)^{1/2}t = \log\left\{\frac{[(kv)^{1/2} + qs^{1/2}][k^{1/2} - 1]}{[(kv)^{1/2} - qs^{1/2}][k^{1/2} + 1]}\right\}$$

$$q_n = \left(\frac{kv}{s}\right)^{1/2}\left\{\frac{(k^{1/2} + 1)\exp[2(kvs)^{1/2}n] - [k^{1/2} - 1]}{(k^{1/2} + 1)\exp[2(kvs)^{1/2}n] + [k^{1/2} - 1]}\right\}$$

$$\bar{w}_n = 1 - sq_n^2$$

As long as n is so small that $\exp[2(kvs)^{1/2}n] \approx 1 + 2(kvs)^{1/2}n$,

$$q_n \approx (v/s)^{1/2}[1 + (k - 1)(vs)^{1/2}n].$$

Letting $\bar{w}_n = 1 - e_n v$,

$$e_n \approx 1 + 2(k - 1)(vs)^{1/2}n, \qquad \bar{w}_n = 1 - e_n v.$$

Figure 11.8 shows the very gradual decline in mean selective value with a 10- or 20-fold increase in the mutation rate and the very strong selection where $v_o s = 10^{-5}$. Even if the recessive does not reproduce at all ($s = 1$) but v is typical (10^{-5}) and k is as much as 20-fold, $\bar{w}$ has declined only about one-third of the way to equilibrium in 30 generations. The curve is sigmoid in this case, and thus declines more rapidly later but more slowly still later.

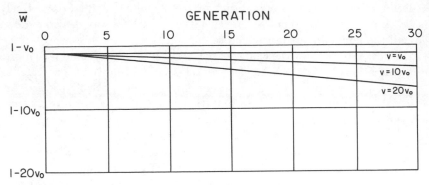

FIG. 11.8. Mean selective values ($\bar{w}$) of populations with respect to recessive mutation, under 10- or 20-fold increase in mutation rate $v_0 s = 10^5$. Redrawn from Wright (1955, fig. 3).

Recovery for m generations after n generations at an increased rate follows the approximate course:

$$2(vs)^{1/2}t = \log \left[\frac{(v^{1/2} + s^{1/2}q)(v^{1/2} - s^{1/2}q_n)}{(v^{1/2} - s^{1/2}q)(v^{1/2} + s^{1/2}q_n)} \right]$$

$$q_{(n+m)} = \sqrt{\frac{v}{s}} \left[\frac{(v^{1/2} + s^{1/2}q_n) \exp [2(vs)^{1/2}m] - (v^{1/2} - s^{1/2}qn)}{(v^{1/2} + s^{1/2}q_n) \exp [2(vs)^{1/2}m] + (v^{1/2} - s^{1/2}qn)} \right]$$

$$e_{(n+m)} = \frac{sq^2_{(n+m)}}{v} \left[\frac{(e^{1/2} + 1) \exp [2(vs)^{1/2}m] + (e^{1/2}n - 1)^2}{(e^{1/2} + 1) \exp [2(vs)^{1/2}n] - (e^{1/2}n - 1)} \right]$$

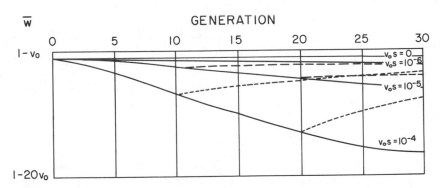

FIG. 11.9. Mean selective values ($\bar{w}$) of populations with respect to recessive mutations for which $v_0 s = 10^{-6}$, 10^{-5}, or 10^{-4}, under 20-fold increase in mutation (*solid lines*) rate and under return to the normal rate after 10 or 20 generations in the last two cases (*broken lines*). Redrawn from Wright (1955, fig. 4).

If m is small enough to permit the approximation $\exp[2(vs)^{1/2}m] \approx 1 + 2(vs)^{1/2}n$,

$$e_{(n+m)} = e_n - 2(e_n - 1)(e_n vs)^{1/2}m; \qquad \bar{w}_{(n+m)} = 1 - e_{(n+m)}v.$$

Figure 11.9 shows the course of decline of $\bar{w}$ for three mutations for which initially $v_o s = 10^{-6}, 10^{-5},$ or 10^{-4}, followed by 20-fold increases in the rates, and by return to the original rates after 10 or 20 generations. The lowest curve applies to such very extreme cases as that of a recessive lethal with a mutation rate of 10^{-4} per generation. The curve for $v_o s = 10^{-6}$ applies to a moderately detrimental type of mutation at a fairly typical mutation rate, for example a 10% disadvantage and a mutation rate of 10^{-5} per generation.

TABLE 11.15. *Top*, Relative incidence of expected detrimental genetic traits for 10 generations with diverse selective disadvantages of the heterozygote ($s = 1/2$ to $1/64$) or, in the last two columns, of recessives for which $vs = 10^{-5}$ or 10^{-6}, under a mutation rate twice that and in which equilibrium had been attained. *Bottom*, The corresponding figures where the doubled rate occurs for only one generation.

Generation	Selective Disadvantage of Heterozygote (s_1)							Recessive Disadvantage (vs_2)	
	$s_1=1$	$s_1=1/2$	$s_1=1/4$	$s_1=1/8$	$s_1=1/16$	$s_1=1/32$	$s_1=1/64$	10^{-5}	10^{-6}
0	1.000	1.000	1.000	1.000	1.000	1.000	1.000	1.000	1.000
1	2.000	1.500	1.250	1.125	1.062	1.031	1.016	1.006	1.002
2	2.000	1.750	1.437	1.234	1.121	1.061	1.031	1.013	1.004
3	2.000	1.875	1.578	1.330	1.176	1.091	1.046	1.019	1.006
4	2.000	1.937	1.684	1.414	1.227	1.119	1.061	1.025	1.008
5	2.000	1.969	1.763	1.487	1.276	1.147	1.076	1.031	1.010
6	2.000	1.984	1.822	1.551	1.321	1.173	1.090	1.038	1.012
7	2.000	1.992	1.866	1.607	1.363	1.199	1.104	1.044	1.014
8	2.000	1.996	1.900	1.656	1.403	1.224	1.118	1.050	1.016
9	2.000	1.998	1.925	1.699	1.441	1.248	1.132	1.056	1.018
10	2.000	1.999	1.944	1.737	1.475	1.272	1.146	1.062	1.020
∞	2.000	2.000	2.000	2.000	2.000	2.000	2.000	2.000	2.000
0	1.000	1.000	1.000	1.000	1.000	1.000	1.000	1.000	1.000
1	2.000	1.500	1.250	1.125	1.062	1.031	1.016	1.006	1.002
2	1.000	1.250	1.187	1.109	1.059	1.030	1.015		
3	1.000	1.125	1.141	1.096	1.055	1.029	1.015		
4	1.000	1.062	1.105	1.084	1.051	1.029	1.015		
5	1.000	1.031	1.079	1.073	1.048	1.028	1.015		
6	1.000	1.016	1.059	1.064	1.045	1.027	1.015		
7	1.000	1.008	1.044	1.056	1.042	1.026	1.014		
8	1.000	1.004	1.033	1.049	1.040	1.025	1.014		
9	1.000	1.002	1.025	1.043	1.037	1.024	1.014		
10	1.000	1.001	1.019	1.038	1.035	1.024	1.014		
∞	1.000	1.000	1.000	1.000	1.000	1.000	1.000	1.000	1.000

Very little appears to have happened even in 30 generations. The recovery in this case (not shown) is, however, very slow.

Table 11.15 (*top*) gives the expected incidence of detrimental genetic traits for ten generations with diverse selective disadvantages of the heterozygote ($s = 1/2$ to $s = 1/64$) or, in the last two columns, of recessives, $vs = 10^{-5}$ or 10^{-6}, under exposure, generation after generation, to a dose of radiation that doubles the original rate of mutation, taking the initial incidence as 1.000.

Table 11.15 (*bottom*) gives the corresponding figures where the doubling dose occurs for only one generation.

The most extensive investigations of the effects of either a single massive exposure or of chronic exposure on the characteristics of populations of animals seem to have been those of Wallace (1950 and later) with *D. melanogaster*. The following populations all traced to the same laboratory stocks, but the first three were started from flies extracted by the use of marked chromosomes in such a way as to insure the initial absence of any lethals or near-lethals. Populations 17 and 18 were derived from flies of generation 125 of population 5, and population 19 was derived from generation 126 of population 6, both of which had received some 250,000r of gamma radiation over these generations. The "large" population had about 10,000 adults in the cages, the "small" less than 1,000 because of differences in nutrition.

Population	Population Size	Initial Exposure	Later Exposure
1	Large	♂7,000r, ♀1,000r	None
2	Large	♂1,000r, ♀1,000r	None
3	Large	None	None
5	Small	None	5.1r/hr = (2,000r/generation)
6	Large	None	5.1r/hr = (2,000r/generation)
7	Large	None	0.9r/hr (300r/generation)
17	Small	⎧None but ⎫	None
18	Large	⎨ 250,000r ⎬	None
19	Large	⎩ ancestral ⎭	None

About 200 males were taken each two weeks and mated with females with marked second chromosomes $Cy\ L/Pm$. A $Cy\ L/+$ male was backcrossed, and male and female $Cy\ L/+$ (with the same + chromosome) were mated to produce about 2 $Cy\ L/+:1\ +/+$ if no lethal or near-lethal mutation was present in the wild-type chromosome, but only $Cy\ L/+$ if a lethal was present.

Population 1 at once acquired 18.3% lethals, and population 2, 4.6% (second generation 8.5%) in contrast with 0.8% in the controls (population 3). The percentages in populations 1 and 2 rapidly declined in the absence of further radiation, probably largely from the selective elimination of semi-

sterile translocations, reaching 10.9% and 4.3%, respectively, in six generations. Thereafter the percentages gradually rose in close parallelism with the
controls, reaching 14.5% in population 1, 9.2% in population 2, and 5.2%
in population 3 in generations 12 and 13, presumably from the accumulation
of spontaneous mutations. The mutation rate was about 0.4%/generation
(Wallace 1950).

Population 2 was dropped but populations 1 and 3 were continued and
populations 5, 6, and 7 were started from generation 20 of population 1 and
carried for more than 120 generations under chronic irradiation (Wallace
1956). Their courses are shown in figure 11.10 with respect to extracted
second chromosome homozygotes.

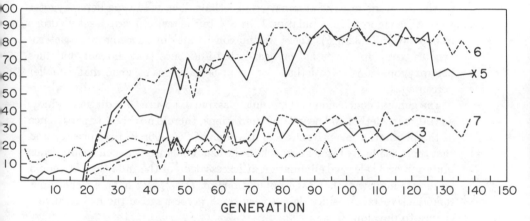

GENERATION

Fig. 11.10. Percentages of lethal-bearing second chromosomes in populations of
D. melanogaster that started with none. Population 1 was large (some 10,000 flies)
with initial exposures of the males to 7,000r and the females to 1,000r, but no
radiation thereafter. Population 3 was a large control population with no exposure to radiation. Populations 5, 6, and 7 were started from population 1 in
generation 20. Population 5 was kept small (about 1,000 flies), while population
6 was large. Both were exposed to 2,000r per generation. Population 7 (large) was
exposed to only 300r per generation. Redrawn from Wallace (1956, fig. 5).

There is little difference in the later courses of population 3 with no
exposure, population 1 with a heavy initial dose but no treatment thereafter,
and population 7 with chronic radiation of only 300r per generation. All came
to equilibrium between new mutation, spontaneous (populations 1 and 3)
or induced (population 7), and selective elimination, at 20% to 40% lethals.
In the last 30 generations, population 7 was a little higher than the controls,
which for 70 generations were somewhat higher than population 1, which
had started with the highest percentage because of its initial massive dose.

Populations 5 and 6, treated with 2,000r per generation, reached equilibrium at much higher percentages of lethals, some 80%. Small or large population size made no consistent difference.

Populations 17 and 18 started from populations 5 and 6 at the high level (about 86%) but with no further exposure, declined in the course of two dozen generations to 56% and 28%, respectively. Population 19, with a slightly lower initial percentage (about 80%), declined very little (77% at the end). The characteristics of the rare homozygotes of random second chromosomes had much less to do with the elimination of lethals than those of the heterozygotes. Tests of the latter, made by bringing together two different second chromosomes with the help of markers, did not, however, throw much light on the different rates of elimination of lethals. The wild-type heterozygotes were most vigorous in population 1 (34.8% in the test crosses), least vigorous in population 5 (32.0%). There was no consistent difference between heterozygotes carrying one lethal or near-lethal and ones that did not, but flies heterozygous for two lethals or near-lethals gave somewhat smaller percentages.

The general conclusion is that while massive exposure to radiation induces a great many lethal and near-lethal mutations, there is no threat to persistence of the population even from accumulations of 250,000r in the ancestry and that there is a fairly rapid approach to the usual balance between spontaneous mutation and selective elimination after cessation of the radiation. It should be noted, however, that cage populations of *Drosophila* have an enormous reproductive excess, since only a very small percentage of the larvae have a chance to develop.

There have been other studies of the effects of acute and chronic radiation on *Drosophila* populations with results consistent with these of Wallace. Thus, Sankaranarayanan (1964) tested the effects of doses of 2,000r, 4,000r, 6,000r, and 7,000r applied to males only. In some cases there was only one acute dose, in others exposures up to 25 generations, followed in all cases by an opportunity for recovery. Viabilities declined for a few generations and then fluctuated about the equilibriums. Recovery to nearly or fully the normal level occurred in three to seven generations.

While not an experimental study, it is appropriate to refer here to a study of Grüneberg et al. (1966) of possible genetic effects on a mammal, of high natural radioactivity in a region of South India. The black rat, *Rattus rattus*, inhabited a rather isolated strip (nearly an island) along the coast of Karala, composed of monazite sand emitting radiation at 7.5 times the rate of inland control areas. The strip was 14 miles long and 150 yards to half a mile wide. A thorough study was made of skeletal and dental characters of 438 rats from eight locations in the strip and from eight locations at short distances inland.

Fertility and morbidity were also studied. No significant differences were found in any characters.

Natural Selection for Reproductive Success in the Laboratory

A truly representative sample of a species can hardly be expected to improve in general fitness by means of natural selection under conditions similar to those to which it has always been exposed in nature. A subnormal sample may, however, improve, and induced mutations may speed up the process. Dobzhansky and Spassky (1947) tested this in lines of *D. pseudoobscura*. They produced three lines homozygous in different second chromosomes and four lines homozygous in different fourth chromosomes by the usual technique of crossing with a strain carrying a dominant marker (lethal when homozygous and associated with an inversion that prevents crossing-over) and then extracting homozygotes of an unmarked chromosome from a single heterozygote. In the segregating generation, all were inferior to the heterozygotes (frequency less than $33\frac{1}{3}\%$) at 25.5°C and all but two at 21°C. Each line was divided into four sublines. Two were maintained as homozygotes and two as heterozygotes, balanced by the marked chromosome. In each case, one was treated with 1,000r X rays every generation, the other was left untreated. All were maintained at 21°C for 25 generations, with severe crowding of the homozygous but not of the heterozygous cultures. By this time the more inferior homozygotes had improved sufficiently to be maintained at 25.5°C and this was done for 25 more generations. Frequent tests

TABLE 11.16. Percentage of homozygous segregants from marked heterozygotes at 25.5°C initially and in generation 50 of lines of *D. pseudoobscura* as described in the text.

LINE	INITIAL	HOMOZYGOUS LINES IN GENERATION 50		BALANCED LINES IN GENERATION 50	
		Untreated	X-rayed	Untreated	X-rayed
PA 748	9.9	29.7	34.4	15.3	14.8
PA 784	20.5	32.6	34.4	18.1	3.1
KA 667	21.6	16.7	38.1	0	0
AA 955	17.6	35.1	23.9	0	0.3
AA 1035	10.2	27.2	17.7	0	0
AA 1105	23.4	29.6	30.4	27.0	0
PA 851	26.4	22.9	26.4	29.6	20.2
Average	19.4	27.7	29.3	12.9	5.5

SOURCE: Data from Dobzhansky and Spassky 1947.

of the relative viability of homozygotes were made in all lines by obtaining segregation for marked heterozygotes and by noting the proportion of unmarked segregants. The results in generation 50 are compared with the initial proportion of such segregants at 25.5°C in table 11.16.

Five of the untreated and six of the X-rayed homozygous lines improved. The average percentages of unmarked segregants rose from 19.4% to 27.7% and 29.3%, respectively. Most of this improvement was no doubt due to selection for coadaptive changes in the unfixed chromosomes rather than of favorable mutations, since the X-rayed lines are only slightly superior as a group to the untreated ones.

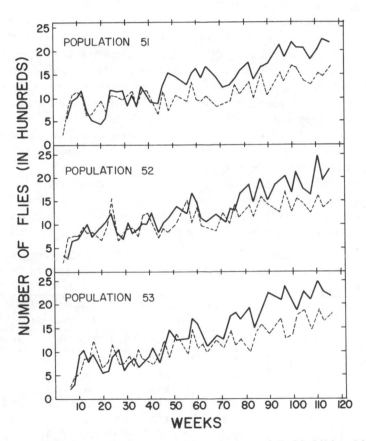

FIG. 11.11. Comparisons of three paired populations of *D. birchii* in which one (*solid line*) in each case was treated with 2,000r (X rays) in each of three generations before the numbers of flies began to be counted, while the other (*dotted line*) was untreated. Redrawn from Ayala (1969, fig. 1); used with permission.

Among the balanced lines, three untreated and five X-rayed lines acquired lethal or near-lethal mutations in the unmarked chromosome, sheltered from selection by the marked one.

Moreover, recognizable defects in some of the lines—slow development, low fertility, and structural abnormalities of homozygotes—were improved in the homozygous lines.

Experiments in which X-irradiation played a more significant role in selection can be cited. Ayala (1966, 1969) compared untreated and X-rayed populations of *D. serrata* and *D. birchii* under conditions of intense competition. The radiation consisted of 2,000r of X-irradiation given the males in each of three generations before observations on population size and weekly productivity began. The course of change was erratic in the case of the *D. serrata* populations but the irradiated ones significantly improved, relative to the untreated. In the *D. birchii* series (fig. 11.11), progress of both was relatively uniform and again was systematically greater in the irradiated lines. In three separate experiments with *D. birchii*, the regression coefficients of population size (in hundreds) on time (in weeks) were 6.9, 8.3, and 6.4 in the control lines and nearly twice as great (11.5, 15.2, and 12.9) in the X-rayed lines. In weekly productivity the regressions were 2.1, 1.4, and 1.4 in the control lines but 3.9, 4.1, and 4.7 in the X-rayed lines. There was no doubt of the significance in all cases. In this case there seems no doubt that favorable mutations increased the material for selection.

While discussing artificial selection for a character with little relation to fitness, it may be appropriate to note that Scossirolli (1954) found that X-irradiation made possible continued successful selection for a number of sternopleural bristles in *D. melanogaster* after a plateau had been reached.

Selection Experiments with Bacteria

The mode of origin of strains of bacteria resistant to a particular virus or poison was for many years an unsettled question. According to one view, resistance is induced by direct action of the virus or poison on cells that do not succumb, analogous to immunity induced in higher organisms by success in overcoming a disease. The alternative was that resistant mutations occur at a definite rate independent of exposure to the agent. Luria and Delbrück (1943) noted that the distribution of numbers of resistant cells in subcultures from the same culture, plated on agar on which virus had been spread, should show merely the sampling variance, that of a Poisson distribution, with variance equal to the mean, whichever hypothesis is true. Tests of small cultures that had had an opportunity to grow since origin from the same culture should, on the other hand, show a greater variance of numbers of resistants than the mean, under the mutation hypothesis, because of the occurrence of groups from clones that had mutated several generations before testing. This would not be the case if resistance arises only at the time of testing with virus. Their experimental work showed very much greater variances than could be accounted for by the induction hypothesis.

Still more direct evidence was obtained by Lederberg and Lederberg (1952) and by Cavalli-Sforza and Lederberg (1956) in experiments in which they were successful in increasing resistance to streptomycin along lineages that had never been exposed, on the basis of the resistance shown by cells derived by replicate plating.

Experiments by Miller and Bohnhoff (1947) illustrate the production both of mutant strains resistant to a drug, and of strains that require the same drug for their growth. Meningococcus was grown on plates containing streptomycin in concentrations ranging from 10 mg/ml to several thousand. Normal colonies developed at the lowest concentrations but not at 40 mg/ml or more. About the same small number of large yellow colonies of a variant type (A) were found in all concentrations up to 2,000 mg/ml. These were resistant, not only in vitro but in vivo, being as virulent for mice as the original culture and producing infection with maximal doses of streptomycin. A second type of mutant culture (B), pearl gray in color, appeared only in media containing at least 40 mg/ml. These grow best at 100–400 mg/ml. Cultures for type B were not virulent to mice not treated with streptomycin but were fatal to mice treated with 500–5,000 mg/ml (fig. 11.12).

Novick and Szilard (1950a,b) devised an apparatus, the "chemostat," for keeping a bacterial suspension growing at a uniform rate indefinitely at an unchanging concentration. The bacteria ($E.\ coli$ in their experiments) were grown in a tube through which a steady stream of nutrient was kept flowing from a storage tank. The nutrient had a high concentration of all required growth factors except for one controlling factor (for example, tryptophan). The rate of inflow (and outflow) was adjusted to maintain the concentration of bacteria at a desired level. Mutational resistance to bacteriophage T5 was studied with tryptophan as the controlling factor. The number of resistants present per milliliter could easily be determined in samples and was found to rise linearly as expected under the pressure of recurrent neutral mutation $dq/dt = v(1 - q) - sq(1 - q)$, $\hat{q} = v/s$. Mutation to T4 resistance also behaved in this way. The occurrence of any mutation more favorable than the original type, growing under a shortage of tryptophan, should rapidly displace not only the original type, but all previously neutral mutations, such as T4 resistance, which had reached equilibrium. The concentrations of both were, in fact, typically observed to drop sharply at some time. In the case of T5 resistance, linear increase started after a delay of 32 generations along a line parallel to the previous one, indicating that it was still neutral in the modified

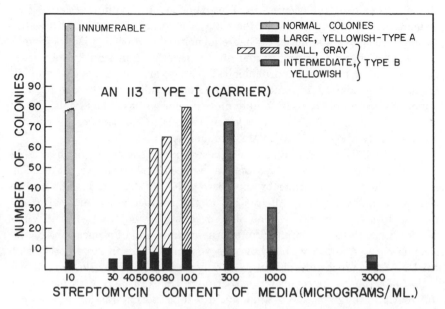

FIG. 11.12. Numbers of colonies of *Meningococcus* of indicated sorts that appeared on media with different concentrations of streptomycin. Redrawn from Miller and Bohnhoff (1947, fig. 1); used with permission.

genome. Mutation to T4 resistance arrived at a considerably lower level, indicating that it was even more strongly selected against in the new background. In the course of an experiment, run for 450 generations, such evolutionary steps occurred on the average at intervals of about 50 generations.

The device made possible unusually accurate determination of mutation rates. The effects of mutagenic agents (theophylline, caffeine, theobromine, paraxanthine, 8-azoguanine) (Novick and Szilard 1951) and of antimutagens (guanosine, adenosine, inosine) (Novick and Szilard 1952) could be evaluated. An unexpected result was that mutation rate for resistance to T5 was constant per unit of time rather than per generation over a threefold range of generation time whether the limiting factor was tryptophan, arginine, phosphorus, or nitrogen (Novick and Szilard 1952). Where lactate was the controlling factor, T5 resistance reached an equilibrium indicating slower growth than the original type.

Summary

The largest class of easily definable mutations in the higher animals that have been studied most intensively, and hence the class best adapted for comparing mutation rates, has been that of lethal mutations, in spite of occasional complications from the occurrence of lethal effects due to interaction of two or more genes. In higher plants, the class of chlorophyll mutations possibly occupies this place. Estimates of the frequencies of mutations with less severe detrimental effects on viability and ones with effects on fertility, not associated in either case with morphological change, are much less easily made because of gradation into normal, but from studies in *Drosophila* such mutations are probably considerably more numerous than the lethals. Easily recognizable morphological and color mutations in *Drosophila* are usually observed only about 10% to 20% as frequently as lethals.

Estimates of the total number of loci, based on recurrence rate of observable mutations in such a form as *Drosophila* and ones based on the weight of DNA in the haploid nucleus, differ by some two orders of magnitude. The former seem to include only a small fraction of the total number, even after removing from consideration some 30% or more of the DNA that in vitro studies indicate to be composed of enormous numbers of replications of the same few genes. It is possible, however, that mutations in large portions of the DNA have no phenotypic effects or only slight effects. Another possibility is that most genes are replicated two or more times (usually with functional differentiation), giving protection against inactivation (major mutations) while permitting much quantitative variability. Such compound genes may be maintained as such by selection relating to the latter. If so, most of the material for evolution consists of mutations of this latter type, rather than the conspicuous mutations on which our knowledge of genetics largely depends.

The degree of dominance, including possible overdominance of mutations, is important in the dynamics of evolutionary change at loci. Studies of newly arisen lethal mutations in *Drosophila* indicate that the apparently "recessive" lethals typically have a slight dominant effect in heterozygotes, giving about 3% or 4% disadvantage relative to homozygous normals. There is considerable variability, however, ranging from strongly deleterious to slightly favorable heterozygous effect.

Some very extensive studies of the heterozygous effect of moderately deleterious mutations in *Drosophila* have given such unexpected results that rather detailed discussion seemed desirable. Two independent studies (Wallace and Mukai) have indicated that the presence of random mutations (spontaneous or induced) in one homologue in an otherwise homozygous pair of chromosomes gives a net favorable (heterotic) effect, but that the presence of such mutations in both homologues gives a decided reduction in heterozygous viability, a difference between the effects of the same genes in coupling and repulsion that is very difficult to interpret. Interpretation is not made easier by differences in the behavior of different chromosomes, of the same chromosome in different sets of experiments, and of highly significant heterotic effects of random mutations on a heterozygous genetic background in one extensive series.

While mutations with major morphological or color effects tend to be recessive, those with minor modifying or merely quantitative effects tend to have intermediate heterozygotes according to most of the available evidence.

The course of change of gene frequency following an abrupt change in the mutation rate of genes with more or less dominant, or with completely recessive, deleterious effects are presented mathematically. The changes in mean selective value, even after many generations at a greatly changed rate, tend to be rather slight, especially in the case of recessive mutations, but lead ultimately to a new equilibrium that depends wholly on the mutation rate (Haldane's principle).

Experimental studies of the effects on *Drosophila* populations of either exposure to a single massive dose, or to chronic exposure to irradiation for many generations, agree with the theoretical courses. The number of lethal mutations carried by the population may increase greatly until a new equilibrium is reached (under chronic irradiation). There is no threat to the persistence of the population from genetic effects of irradiation, at least in *Drosophila* populations, which have an enormous surplus reproductive capacity. There is rapid reduction in the frequencies of the accumulated lethals after cessation of irradiation. Data from a mammal population (rats) that have lived for an indefinitely long time on isolated beaches of monazite sand in South India, exposed to more than seven times the amount of normal radiation, have shown no detectable mutant effects. Mutation pressure by itself is usually not a very effective evolutionary process.

Experiments in which *Drosophila* have been exposed to chronic irradiation indicate, however, that natural selection for increased viability under the experimental conditions is significantly furthered by the increased supply of mutations.

Experiments have shown that bacteria acquire resistance to injurious substances by natural selection of resistant mutations rather than by direct induction. Other experiments have demonstrated selection not only of strains resistant to a drug but of others that require it for their growth.

Unusually precise determinations of the rate of induction of mutations by diverse mutagenic agents have been made by use of a device, the chemostat, that maintains a uniform concentration of *E. coli* in a medium in which a substance, necessary for growth, is kept at a constant concentration as the limiting factor. It has been possible to follow the course of evolutionary adaptation to given conditions with remarkable precision in the chemostat. Mutation rate has been found to be uniform per unit of time rather than per generation.

Finally, it may be appropriate to recall from the discussion on pages 233–34 that selection for quickness of duplication of RNA extracted from a bacteriophage and duplication in vitro has resulted in the smallest known self-duplicating entity, a molecule of molecular weight 1.7×10^{-5}.

CHAPTER 12

Theories of Evolution: Mutation and Mass Selection

The Material for Evolution: Darwin and de Vries

Darwin's theory had to be formulated without any knowledge of the laws of heredity. He viewed the material for evolution as consisting either of the omnipresent quantitative variations or the rare "sports" of which the short-legged Ancon sheep was an example. In his judgment, the former were much the more important.

De Vries, on the contrary, concluded that significant steps in evolution depend on abruptly occurring "mutations," affecting all parts of the organism and giving rise to new species at single steps. While he was one of the re-discoverers of Mendelian heredity in 1900, he attributed no significance to the Mendelian differences in the origin of species. Under his theory, significant transformation of character and speciation are the same process.

He did, however, recognize the cumulative effect of selective survival among species in guiding the course of evolution of the higher categories. Thus he wrote in 1906:

Notwithstanding all these apparently unsurmountable difficulties, Darwin discovered the great principle which rules the evolution of organisms. It is the principle of natural selection.—It is the sieve which keeps evolution on the main line, killing all or nearly all that try to go in the other direction. By this means, natural selection is the one directing cause of the broad lines of evolution.

Chromosomal Mutations

The species forming mutations that de Vries believed that he had observed in *Oenothera Lamarckiana* (vol. 1, chap. 1) were found to be chromosome aberrations that involved the number of chromosomes, or the number or arrangement of blocks in them. There is an enormous literature (Stebbins 1950; White 1954) on the differences usually found in the karyotypes of related species. These prove beyond question that chromosome aberrations are often and, indeed, usually, involved in the isolation of a new species.

The principal types of aberrations—polyploidy and aneuploidy, duplication, deletion, transposition, translocation, and inversion—and the extent to which each is likely to contribute to the transformation of characters and to genetic isolation have been discussed briefly in volume 1, chapter 3. The polyploids and polysomics give the most validity to de Vries' theory. Minor duplications and deletions have undoubtedly played considerable roles in phenotypic change but can be treated largely as if gene mutations. The balanced rearrangements cause relatively slight phenotypic change but have undoubtedly contributed to genetic isolation.

Hybridization

Hybridization of species as a source of material for evolution should be touched on here. Hybrids between species in which the chromosomes have become so different that pairing in meiosis fails more or less completely have undoubtedly been of major importance in giving rise to new polyploid (amphidiploid) species that differ drastically from both parent species and are genetically isolated from the first by the production of sterile triploids in backcrosses.

Segregation from more or less fertile hybrids has undoubtedly often provided material for the operation of selection leading to new phenotypes (Lotsy 1908). Of special importance in this connection has been the "introgression" of genetic materials from one species into another by repeated backcrossing (Anderson 1949). Introgression here supplements gene mutation, with the advantage that the blocks of introduced genetic material have already proved their value in another context.

Theories of Abrupt Origin of Higher Categories

Several post-Mendelian evolutionists have gone beyond de Vries in the magnitude of the mutational steps they invoke. Willis (1922, 1940), Goldschmidt (1940), and Schindewolf (1950) maintained not only that species arise by single steps but that genera arise only from generic mutations, families only from family mutations, and so on.

There was no real recognition of selection in Willis' statement that "mutations may, but by no means necessarily must, have some functional advantage attached." He held that "chromosome alterations are probably largely responsible for the mutations that go on." He emphasized the "downward" direction of evolution:

Evolution goes in what one may call the downward direction from family to variety, not in the upward direction required by the theory of natural selection.

In line with this, he maintained in his "age and area" theory that the extent of territory occupied by a taxon is generally in proportion to the time since it arose by mutation.

He found support for his views in the so-called hollow curve of frequencies of genera according to their numbers of species, typically a rectangular hyperbola, $y = c/x$, which Yule (1924) showed to be that expected under his theory.

Willis had evidently not seriously considered what really is expected under natural selection with respect to the apparent "upward" or "downward" direction of evolution. His age and area theory is only a half-truth since it ignores the effect of the large-scale extinction of species and higher categories, indicated by the geologic record. Neither he nor Yule realized that his hollow curve is expected, if extinctions are in balance with the splitting off of new species, irrespective of the mechanism of origin of the latter (Wright 1941a).

Goldschmidt made an absolute distinction between "microevolution" and "macroevolution":

Microevolution within the species proceeds by the accumulation of micro-mutations and occupations of available ecological niches by the preadapted mutants. Microevolution, especially geographic variation, adapts the species to the different conditions existing in the available range of distribution. Micro-evolution does not lead beyond the confines of the species and the typical products of microevolution, the geographic races, are not incipient species. There is no such category as incipient species.

He contrasted this with his concept of "macroevolution":

Species and the higher macroevolutionary steps are completely new genetic systems. The genetical process which is involved consists of a repatterning of the chromosomes which results in a new genetic system. The theory of the gene and of the accumulation of micromutants by selection has to be ruled out of this picture.

Goldschmidt's "systemic mutation" implies action of the chromosomes as wholes on the developmental process. He seems to have been thinking in terms of development by a sort of spatial preformation rather than by a temporal epigenetic complex of chain reactions. His systemic mutations must act prior to the transcription of molecules of mRNA and the translation of this information into that of protein molecules.

There was only the faintest possible recognition of natural selection in his characterizations of the macromutants as "hopeful monsters." It is evident (Wright 1941b) that Goldschmidt had never grasped the creative effect of natural selection, operating generation after generation on small accidental genetic differences and thereby building up large adaptive changes that would be inconceivably improbable as single steps. In rejecting any such trial and error process at the level of small random changes, he was forced to accept a determinative role of catastrophic events of which the adaptive consequences can only be characterized as miraculous.

Willis recognized this dilemma and boldly grasped it:

> The process of evolution appears not to be a matter of natural selection of chance variations of adaptational value. Rather it is working upon some definite law that we do not comprehend. The law probably began its operation with the commencement of life and it is carrying this on according to some definite plan.

> Evolution is no longer a matter of chance but of law. It has no need of any support from natural selection.... It thus comes into line with other sciences which have a mathematical basis.

He made no suggestion as to the mathematical law for a succession of miracles. We are carried back to Lamarck's ladder of life, Nägeli's mysterious perfecting principle, and Osborn's aristogenesis.

Schindewolf (1950), from consideration of the evolutionary steps that seemed to him to be implied by the data of paleontology, concluded that new taxa differed so radically from their precursors in basic morphology that their origins could only be accounted for by abrupt mutational steps giving reproductive isolation from the first. Simpson (1953, 1961), also writing from the standpoint of a paleontologist, found no such necessity. The same general arguments against the mutation concepts of Willis and Goldschmidt apply also to Schindewolf's concept.

Major Gene Mutations

Most of the early geneticists who speculated on evolution were naturally much influenced by de Vries' mutation theory, but most of them soon shifted to the Mendelian "unit characters" as the basis for phenotypical transformation. It was generally believed, following de Vries, that quantitative variability is of a wholly different sort and that selection based on it can have no permanent effects.

Morgan became the leading spokesman for evolution on the basis of major gene mutations. He summed up his views in 1932 in the following statement:

> If we had the complete ancestry of any animal or plant living today, we should expect to find a series of forms, differing at each step by a single mutant change in one or another of the genes, and each a better adapted or differently adapted form from the preceding.

Morgan was clearly concerned more with the material utilized in evolution than with the dynamics of the selection process that is implied in the last phrase. He evidently thought of the species as homallelic or nearly so with respect to all loci at any given time, with progress based on the occasional occurrence and fixation of a favorable mutation.

This concept of evolution is referred to by some population geneticists as the classical theory. It was undoubtedly that which was accepted by most of those who approached the subject from the standpoint of classical genetics, especially that of *Drosophila*. Muller (1929) should be referred to for his clear exposition of the role of natural selection in this connection. Among those who took up consideration of evolutionary dynamics in the 1920s, Haldane came the closest to this viewpoint. We will return to his views later.

Quantitative Variability

The most active center for experimental population genetics in America in the first two or three decades after the rediscovery of Mendelian heredity was the Bussey Institution of Harvard University where W. E. Castle and E. M. East were both firm advocates of Darwinian evolution by the natural selection of quantitative variability. Their hypotheses on the genetic nature of this were, however, widely different.

Castle (1912b) from the first challenged the prevailing doctrine that selection of quantitative variability could not bring about any permanent advance over existing genotypes. He developed the idea that the segregating "unit characters" are subject to continuous variability and that this occurs especially by reciprocal "contamination" in heterozygotes, an idea somewhat similar to the more modern concept of conversion.

That there could be more than pairs of Mendelian alternatives had been clearly demonstrated by Cuénot (1904) in mice, and Castle (1905) showed that in rabbits complete albinism and the Himalayan variety, white with dark ears, nose, and feet, behaved not only as simple recessives of full color but as alleles of each other.

Guinea pigs, like other rodents of the family Caviidae, lack little toe, big toe, and thumb. Starting in 1901 with a male with a rudimentary little toe, Castle produced, by gradual selection, a strain in which the little toe was always perfectly developed (as noted in chap. 7). Similarly, starting from guinea pigs with a few scattered white hairs, he gradually produced a strain that was silvered over its whole body.

His most extensive selection experiment was on the black and white pattern of hooded rats (in which I assisted him from 1912 to 1915) (Castle and Wright 1916). This has been described in some detail in chapter 7. As noted there, he succeeded in carrying both plus and minus lines far beyond the original limits of variability. In the plus strain, an abrupt advance, far greater than usual, was shown to depend on a mutation in the hooding locus of a single male, with an effect intermediate between that in the plus strain and in a strain of self-colored wild rats. Thus it was plausible to suppose that the success of the selection in both directions had been based on smaller mutations at this locus.

On the other hand, the results of crosses between the extremes could only be interpreted on his hypothesis by assuming an enormous amount of "contamination." Castle and Phillips (1914) recognized the alternative possibility that there had been selection of multiple independent modifiers according to the theory of quantitative variability for which their colleague, East, had been getting experimental evidence (vol. 1, chap. 15). They felt, however, that more crucial evidence was needed before abandoning their interpretation of the results with the rats. Such evidence was supplied later, as noted in chapter 7, by the nearly complete convergence of the plus and minus rats, extracted after three backcrosses to the same wild strain (Castle 1919).

The reconciliation of Mendelian heredity with the correlations between relatives in quantitatively varying characters, found by Galton and Pearson before the rediscovery of Mendelian heredity, has been noted in volume 2, chapter 15. Pearson himself (1904, 1909) continued to have strong reservations, but it was made sufficiently clear by Yule (1902, 1906) and especially by Weinberg (1909, 1910) that there was no inconsistency between the experimental and the statistical results.

My own thesis project, under Professor Castle's supervision, was "an intensive study of the inheritance of color and other coat characters in guinea pigs with special reference to graded variation" (Wright 1916). Several continuous series were studied to find out how far they depended on independent genetic factors, major and minor, how far on multiple alleles, stable or otherwise, and how far on nongenetic variability. Hair direction (smooth to full rough), the agouti pattern (dark to light-bellied agouti), eumelanic pigmentation (white to black), and phaeomelanic pigmentation (white to intense "red") were studied. Beneath the continuous gradations were different complex situations (vol. 1, chap. 5). There were interacting major Mendelian differences, involving multiple alleles as well as independent loci, and incomplete dominance in some cases of which the degree depended on modifiers introduced from specific strains. The continuity in all cases depended on a hierarchy of modifiers, ranging from ones that could easily be isolated to multifactorial sets so confounded by nongenetic variability as to be practically unanalyzable.

As discussed in earlier chapters of this volume, inbreeding and selection experiments with a great variety of organisms have indicated that situations of this sort are typical, at least of laboratory stocks. In some cases, as in experiments with *Drosophila*, these were not far removed from wild populations.

It has seemed best to discuss the theories of evolution before taking up the interpretation of genetic variability observed in nature, but it may be stated at once that wild species exhibit the same ingredients as those demonstrated in laboratory stocks.

Many genetic mechanisms that permit the persistence of strong heterallelism in population have been presented in volume 2. Selective advantage of rarity and selective advantage of diversity may maintain high frequencies of two or more alleles without any genetic load. Mutation pressure against adverse selection in general maintains only weak heterallelism, but if selection is directed toward an intermediate optimum, the heterallelism is moderately strong. Under many conditions, net heterozygous advantage over the homozygote maintains strong heterallelism, but in this case at the expense of a considerable genetic load, a concept to be discussed later.

The astronomically great numbers of recombinant genotypes from even a limited number of segregating loci provide the possibility for an enormous number of different interaction systems. These are closer to the characters on which selection acts than are the individual genes, but they are broken up too rapidly under random mating, unless linkage is almost complete, for effective selection among them in a homogeneous population. Mass selection in such a population is practically restricted to the net differences in effect of alleles. A discussion of the conditions under which there may be selection among interaction systems is deferred to the next chapter.

Transformation by Mutation Pressure

From the molecular standpoint, practically every mutation at a locus is expected to be unique. For mutation pressure to be of appreciable direct significance in guiding evolution, large portions of the array of mutations at the same locus, and also mutations at different loci, must have some tendency toward a common direction of effect. The only property likely to be shared by many is the inactivation of developmental processes. Muller (1918) pointed out that enforced heterozygosis, generation after generation, protects inactivating mutations and deletions from adverse selection as long as the normal allele in the homologous chromosome is dominant over inactivation or loss. He was concerned in his 1918 study with mutations in chromosome regions kept heterozygous by balanced lethals. The accumulation of deleterious genes by this means has since become a standard procedure in *Drosophila* genetics. Muller suggested that the largely inactive character of the Y chromosome in *Drosophila* may be due to its being carried normally in heterozygotes, XY. Other examples have been given by Haldane (1933) in a comprehensive discussion of the topic of this section.

Evolutionary advance depends not only on new adaptive transformations of characters but on getting rid of characters that are no longer useful. It has been suggested again and again that the degeneration of organs that have become useless is due to the predominantly inactivating character of mutations. It is merely necessary to compare the number of mutations that reduce wing size or venation, size or pigmentation of eyes, number and size of bristles, and so on in *Drosophila* with the number with the opposite effect, to indicate this possibility. This tendency of mutations to cause morphological degeneration presumably traces to the strong probability that random change in the composition of the protein molecules, controlled by random changes in the DNA, will bring about inactivation of their properties as enzymes.

Thus there can be little doubt that mutation pressure is continually tending to bring about degenerative changes in all organs. The only question with respect to causation, where such changes have actually occurred, is whether there may not be selective processes that bring about much more rapid degeneration than can be accounted for by mutation pressure by itself, in view of the low rate of mutation.

There are two wholly different sorts of selection to consider (Wright 1929c, 1939, 1964a). There is direct selection against useless parts as encumbrances or causes of energy waste. Selection pressure of this sort falls off as degeneration proceeds. It is to be expected, however, that genes that have been involved in the development of the part in question, as long as it is useful, will tend to be displaced by alleles that are more favorable with respect to their pleiotropic effects on related parts that are still useful. Such alleles are expected to be less effective on the average than the previous type allele in maintaining the now useless part. According to this view, the rate of degeneration will tend to be proportional to the amount of reorganization of the related useful parts. If there is conservatism in these other respects, the useless part may be maintained at least as a vestige for a very long time by the persistence of the useful pleiotropic effects of the genes in spite of mutation pressure and adverse selection against the character as an encumbrance and cause of wasted energy. If, on the other hand, related parts undergo radical reorganization, the useless part should degenerate rapidly and completely.

An illustration is given by the loss of the thumb, big toe, and little toe in the family Caviidae, referred to earlier (cf. "Digit Number," vol. 1, p. 463). Presumably this occurred millions of years ago in common ancestors of this large family. Yet, as noted, vestiges of the little toe recur occasionally in guinea pigs and a true breeding strain was produced by Castle in a few years from such a beginning. It was interesting that a normal-seeming plantar pad, absent in unselected guinea pigs, was fully restored by the selection for the little toe. Crosses with normal strains indicated that several independent loci were involved. That development of the little toe was ordinarily just below the threshold was also indicated by the restorative effects of certain environmental conditions. That development of the whole ancestral pentadactyl foot was not very far below the threshold was indicated by its restoration by a combination of the effects of a mutation Px (homozygote monstrous and lethal) and the modifiers for the little toe. It seemed clear that the genes "under which the ancestral pentadactyl foot developed were so deeply involved in the development of other parts of the foot and of the organism as a whole, that most of the genetic systems concerned with the thumbs, big toes, and little toes are still present" (vol. 1, p. 103). The loss in the course of evolution could not have been due merely to mutation pressure affecting useless parts, but must have involved a complex balancing of selection pressures.

Thus mutation pressure has played a major role in the inactivation of parts of the genome that are no longer useful or that are protected from adverse selection by enforced heterozygosis. Mutation pressure has also been directly involved in the degeneration of useless organs, but much more in this respect as a source of material for both direct and indirect effects of natural selection.

Transformation by Mass Selection

This and following sections consider selection based on differences among individuals in characters that affect viability or productivity, directly or indirectly. Selection based on the characters of close relatives, which can be translated into selection among individuals by means of the genetic correlations, is also included here. In evolution in general, familial selection is important, not only for traits restricted to one sex but for traits that involve sacrifice of individual welfare for the benefit of the progeny or even more remote relatives (Hamilton 1963, 1964). In this chapter, however, we are concerned with general aspects of the evolutionary process rather than with the evolution of particular traits.

Evolution by selective substitution of a major gene mutation for a type gene can be expected to lead only to rather coarse adaptation, while such substitutions of mutations with minor effects by selection of quantitative variability is necessary for fine adjustment. There are important differences in the probable frequency of favorable mutations. The mathematical theories, however, are essentially the same.

Some very important distinctions between evolution under conditions that remain practically unchanged over very long periods and the response to persistently changed conditions will also be discussed, but again the basic theory is the same.

There is an important distinction between a really novel favorable mutation and mere recurrences of a given mutation at an appreciable rate, as material for mass selection. It is not enough that a novel favorable mutation appears to give a new direction to evolution. Selection has very little to do with its fate as long as it is rare, because of the dependence of selection pressure on its frequency ($\Delta q = sq(1 - q)$ where s is the momentary selection coefficient). There is a strong probability that it will be lost by accidents of sampling. As discussed in volume 2, chapter 13, the chance of fixation of a mutation with heterozygous advantage s is approximately only $2s$ (Haldane 1927), more accurately, if semidominant, $2s/(1 - e^{4Ns})$ (Fisher 1930; Wright 1931, by different methods). In the case of a completely recessive favorable mutation, the chance was found to be of the order $\sqrt{(s/N)}$ by Haldane (1927) and, more accurately, to be $1.1\sqrt{(s/2N)}$ (Wright 1942), and still more accurately, $\sqrt{(2s/\pi N)} = 1.128\sqrt{(s/2N)}$ (Kimura 1957).

Once a favorable mutation is so firmly established at a moderate absolute frequency (which may be a very small percentage frequency) that the chance of accidental loss is negligible, it tends to advance toward its equilibrium frequency in a practically deterministic fashion according to whatever formula for Δq is appropriate (vol. 2, chap. 3–5). The course of change under simple conditions (dominant, semidominant, and recessive, little or much complication from environmental variability, differences among favorable genes in initial frequency, in magnitudes of effect, and in degrees of dominance) has been discussed in chapter 6 of this volume.

While not the first to examine this question (vol. 2, p. 29), Haldane dealt with it more comprehensively than anyone else (especially in a series of ten papers from 1924–34 reviewed in 1932). It is significant of his emphasis on gene substitution as the basic evolutionary process that his formulas were in terms of the changes (Δu) in the ratio $u = q/(1 - q)$ of a favorable mutant gene to its type allele, instead of in terms of the changes Δq in gene frequency itself. The latter is more appropriate in dealing with multiple minor factors and especially with the shifting balance theory, discussed later, in which there is no definite type allele; a symmetrical treatment of all alleles is preferable to Haldane's formulation.

Rate of Evolution and Genetic Variability

It is evident from the paleontological record that there has been the utmost diversity in the rates of evolution of different organisms. Some have hardly changed for hundreds of millions of years and others have changed markedly in less than 100,000 years. It will be convenient to consider here some of the possible limiting factors under guidance by mass selection without implying that the latter has been the only or the most important mode.

The most obvious limiting factor is lack of adequate material for selection to act on. The relation between the amount of genetic variability and the rate of evolution was the subject of a theorem presented by R. A. Fisher (1930, 1941), which he designated "the fundamental theorem of natural selection."

Fisher came into population genetics from a mathematical rather than a biological background. In his first paper in the field (1918), he somewhat extended earlier studies of the statistical consequences of multifactorial Mendelian heredity with respect to the correlation between relatives under various conditions. In his 1930 book, he emphasized the nearly even chance that a very slight mutational change would be favorable in contrast with the overwhelming probability that a major mutation would be deleterious. Thus in contrast to Haldane, he leaned strongly toward minor mutational effects as the more important material for natural selection. According to Fisher's theorem, "The rate of increase in fitness of any organism at any time is equal to its genetic variance in fitness at that time."

By "genetic variance" he meant the additive component. It should be noted that "fitness" is used in two senses; first as a property of the population that has a rate of increase, and second as a property of individuals that has a genetic variance. The former will be shown to agree with what we have called the fitness function $F(w/\bar{w})$ (where this exists), rather than with the mean selective value, in randomly mating populations.

Fisher defined two properties of each pair of alleles that affect any character: the average excess, a, of the portion of the population with one allele over that with the other (apportioning heterozygotes equally), and the average effect, α, produced in the population, as genetically constituted, by the substitution of one for the other. He showed that the additive contribution of such a pair to the variance is given by $a\alpha q(1 - q)$, where q is the frequency of one allele. On applying these concepts to "fitness," defined for individuals as their net productivity, he showed that the rate of increase of fitness in the population under given conditions is $\sum\alpha(dq/dt)$, and that this is equal to the additive genetic variance $\sum a\alpha q(1 - q)$. Later he generalized this to allow for multiple alleles.

Since I have found Fisher's demonstration rather difficult to follow, I will give one that seems simpler (Wright 1956). This is partly, no doubt, because I assume random mating, which is a little more restricted than Fisher's postulates. As shown by Turner (1970), however, in a penetrating review, Fisher's extension applied to such a special case of assortative mating as to add little.

Random combination is also assumed. As shown in volume 2, chapter 4, this assumption does not in general hold exactly but it holds approximately if the interaction component of the selection coefficient is of lower order than the recombination coefficient. Since the theory depends on this assumption, it is always merely an approximation even for interacting genes in different chromosomes (Moran 1964; Kimura 1965). No restriction is put on the number of loci or of alleles at each locus.

Let $A_i A_j B_k B_l \cdots$ represent any genotype with selective value W_T and frequency f_T.

As shown in volume 2, chapter 15, the variance of any character (including selective value) can be analyzed into additive components from all of the genes and nonadditive components due to dominance and interaction effects. The additive components consist of selective values for each gene, W_{Li} (L for locus, i for allele). It will be convenient to collect all other components into a single independent residual term, R. The total genotypic selective value, W_T, is thus

$$W_T = W_{Ai} + W_{Aj} + W_{Bk} + W_{Bl} + R = G_T + R.$$

The gene frequencies will be represented in the two-locus case by q_{Ai}, q_{Aj}, q_{Bk}, q_{Bl}, and, generally by q_{Li}. The genotypic frequency under the assumption of random combination is then as follows:

$$f_T = \Pi q_{Li}$$

$$W_{Ai} = \sum \sum \sum (W_T q_{Aj} q_{Bk} q_{Bl})$$

$$\overline{W} = \sum W_T f_T = \sum \sum \sum \sum (W_T q_{Ai} q_{Aj} q_{Bk} q_{Bl})$$

$$\frac{\partial \overline{W}}{\partial t} = \sum \left(W_T \frac{\partial f_T}{\partial t} \right) + \sum \left(f_T \frac{\partial \overline{W}}{\partial t} \right)$$

$$\sum \left(W_T \frac{\partial f_T}{\partial t} \right) = 2 \sum \left(W_{Li} \frac{\partial q_{Li}}{\partial t} \right) \qquad \text{Two alleles at each locus}$$

$$\frac{\partial \overline{W}_T}{\partial t} - \sum \left(f_T \frac{\partial W_T}{\partial t} \right) = 2 \sum \left[(W_{Li} - \overline{W}) \frac{\partial q_{Li}}{\partial t} \right] \qquad \text{Since } \overline{W} \frac{\partial q_{Li}}{\partial t} = 0$$

$$\frac{\partial q_{Li}}{\partial t} = q_{Li}(W_{Li} - \overline{W})$$

Thus $\dfrac{\partial \overline{W}}{\partial t} - \dfrac{\overline{\partial W_T}}{\partial t} = 2 \sum [q_{Li}(W_{Li} - \overline{W})^2] = \sigma_G^2.$

The analogous formula for discrete generations is:

$$(\Delta \overline{W} - \overline{\Delta W})/\overline{W} = 2 \sum [(W_{Li} - \overline{W})^2 q_{Li}]/\overline{W} = \sigma_{(G/\overline{w})}^2$$

Assume that the W's and f's are functions of time only because gene frequencies vary with time; then

$$\frac{\partial \overline{W}}{\partial t} = \frac{\partial \overline{W}}{\partial q_{Li}} \frac{\partial q_{Li}}{\partial t}$$

$$\sum f_T \frac{\partial W_T}{\partial t} = \sum \left(f_T \frac{\partial W_T}{\partial q_{Li}} \frac{\partial q_{Li}}{\partial t} \right)$$

If population fitness, $F(W)$ is defined so that

$$\frac{\partial F(W)}{\partial t} = \sum \left[W_T \frac{\partial f_T}{\partial t} \right] = \frac{\partial \overline{W}}{\partial t} - \sum \left(f_T \frac{\partial W_T}{\partial t} \right)$$

$$\frac{\partial F(W)}{\partial t} = \sigma_G^2,$$

the analogous formula for discrete generations is

$$\Delta F(W/\overline{W}) = \sigma_{G/\overline{w}}^2$$

The last two formulas give Fisher's fundamental theorem in terms of continuous and discrete generations, respectively.

Individual selective value, W, is the same as the individual fitness that has a variance in Fisher's formula, $W/\overline{W}$ being that relative to the mean, $\overline{W}$. The mean selective value, $\overline{W}$, is often referred to as the population fitness of Fisher's formula, the rate of change of which equals the above variance. But the theorem holds for this concept of fitness only if $\overline{\Delta W} = 0$ and thus generally breaks down in the presence of frequency-dependent selection. It is the quantity $F(W/\overline{W})$ that changes in accordance with the additive genetic variance and that had best be called the population fitness (as it has been throughout this treatise). Unfortunately $F(W/\overline{W})$ does not exist in most cases in which $W/\overline{W}$ involves more than a single pair of alleles (vol. 2, p. 121). In all cases, however, in which random mating and random combination are assumed, the expression $(\Delta \overline{W} - \overline{\Delta W})$ plays the role of increase of population fitness in the theorem in being equal to the additive genetic variance.

If mating is not at random and there is nonrandom combination for this reason, a correction term can be found. Assume that the genotypic selective values, W_T, are fitted as closely as possible to the sum of the correlated contributions, α_{Li}, from each gene. There is, as before, a residual term, R, due to dominance and interaction. An expression can be obtained for the additive genetic variance by path analysis (Wright 1956). The subscripts of W_{Li} and α_{Li} may be omitted without confusion. Summations are by locus and allele. It will be convenient to number the equations:

(1) $\quad W_T = \sum \sum \alpha + R = G + R$

(2) $\quad \overline{W} = \overline{G} = 2 \sum \sum \alpha q \qquad\qquad\qquad \sum R = 0$

(3) $\quad \dfrac{\partial \overline{W}}{\partial t} = 2 \sum \sum \left(\alpha \dfrac{\partial q}{\partial t} + 2 \sum \sum \left(q \dfrac{\partial \alpha}{\partial t} \right)\right.$

(4) $\quad \Delta \overline{W} = 2 \sum \sum (\alpha \, \Delta q) + 2 \sum \sum (q \, \Delta \alpha) \qquad$ Discrete analogue of 3

(5) $\quad c_{W\alpha} = c_{G\alpha} = 1 \qquad$ Partial regression

(6) $\quad b_{W\alpha} = b_{G\alpha} = \left\{ \sum \sum [(W - \overline{W})(\alpha - \bar{\alpha})q] \right\} / \sigma_a^2 \quad$ Total regression

(7) $\quad p_{G\alpha} = c_{G\alpha} \sigma_\alpha / \sigma_G = \sigma_\alpha / \sigma_G \qquad$ Path coefficient

(8) $\quad r_{G\sigma} = b_{G\alpha} p_{G\alpha} = b_{G\alpha} \sigma_\alpha / \sigma_G \qquad$ Correlation

(9) $\quad 2 \sum \sum (p_{G\alpha} r_{G\alpha}) = 1 \qquad$ Total determination

(10) $\quad 2 \sum \sum (p_{G\alpha} \sigma_G)(r_{G\alpha} \sigma_G) = \sigma_G^2 \qquad$ From 9

(11) $\quad 2 \sum \sum (b_{G\alpha} \sigma_\alpha^2) \qquad$ From 10, 7, 8

(12) $\quad \Delta q = q(W - \overline{W})/\overline{W} \qquad$ Volume 1, equation 3.14

(13) $\quad b_{W\alpha} \sigma_\alpha^2 = \overline{W} \sum \sum [\Delta q(\alpha - \bar{\sigma})] \qquad$ From 6 and 12

(14) $\quad b_{W\alpha} \sigma_\alpha^2 = \overline{W} \sum \sum (\alpha \, \Delta q) \qquad$ Since $\bar{\alpha} \sum \Delta q = 0$

(15) $\quad \Delta \overline{W} = 2 \sum \sum (b_{W\alpha} \sigma_\alpha^2)/\overline{W} + 2 \sum \sum (q \, \Delta \alpha) \qquad$ From 4 and 14

(16) $\quad \left[\Delta \overline{W} - 2 \sum \sum (q \, \Delta \alpha) \right]/\overline{W} = \alpha_{(G/\overline{W})}^2 \qquad$ From 11 and 15

(17) $\quad 2 \sum \sum (q \, \Delta \alpha) = \overline{\Delta W} - \overline{\Delta R} \qquad$ From 1

(18) $\quad [\Delta \overline{W} - \overline{\Delta W} + \overline{\Delta R}]/\overline{W} = \sigma_{(G/\overline{W})}^2 \qquad$ From 16 and 17

(19) $\quad \overline{\Delta R} = \sum (f_T \, \Delta R) = -\sum (R \, \Delta f_T) \qquad$ Since $\sum (Rf_T) = \overline{R} = 0$

(20) $\quad \Delta F(W/\overline{W}) + \overline{\Delta R}/\overline{W} = \sigma_{(G/\overline{W})}^2 \qquad$ From 18

Thus Fisher's theorem does not hold under nonrandom mating, even with population fitness, $F(W/\overline{W})$, defined as before, unless $\overline{\Delta R} = 0$.

Kimura (1958) has analyzed the residual term (using the continuous model) into contributions from dominance and interaction. It is to be noted, however, that there are no appreciable contributions of these sorts if there is random mating and the nonadditive components of selection are more than an order of magnitude less than the recombination coefficients.

There have been many discussions of Fisher's theorem that present its limitations from diverse viewpoints, among them Li (1955a,b), Crow and Kimura (1956), Kempthorne (1957), Moran (1962), and Turner (1970).

One important deduction can be made at once. With a given total genetic variance, the rate of evolution under the individual selection, postulated here, is reduced proportionately by the percentages by which nonadditive effects contribute to this total. Such selection operates only on the net differential effects of the genes, even though adaptation depends more directly on the interaction effects. There can be no selection among interaction systems unless the component genes are so closely linked that the complex behaves largely like a set of multiple alleles.

A particular interaction system is, of course, built up in the course of time since one allele of a locus, with modifying effects on genes or interaction systems already established, is more favorable than another only if it contributes favorably to these. This building up of interaction systems in tandem is a very different matter from selection among interaction systems. Only the latter, discussed in the next chapter, makes use of the enormous amplification of the field of variability brought about by recombination.

Effect of Deviations from Random Combination

As noted, Fisher's theorem holds strictly only under the assumption of random combination of loci. It applies in equilibrium populations with respect to genes with wholly independent effects, in spite of linkage. There is not, indeed, immediate restoration of random combination after a disturbance. As shown earlier (vol. 2, chap. 2), the deviation from random combination falls off per generation by the average recombination percentage in the gametes. This rate is, however, so rapid in relation to the usual rate of evolution that such temporary deviations from random combination can have no appreciable effect on the rate of evolution unless linkage is very close.

Even if there is interaction, but the interactive selection is of lower order than the amount of recombination, the persistent deviation from random combination that results is usually unimportant (vol. 2, chap. 5) and Fisher's theorem holds approximately. In the reverse case, however, the persistent abundance of the favorable combinations and the rarity of the unfavorable ones makes for selection of the interaction system. This is undoubtedly very important at the cistron level. It is also important where larger blocks or even whole chromosomes are maintained largely intact by differences in gene arrangement. This will be discussed later, especially in connection with the polymorphic inversion patterns characteristic of many *Drosophila* species.

Fisher's theorem gives a useful means of comparing probable rates of evolution of homogeneous species under diverse conditions such as whether they have been living for a long time under essentially the same conditions or have been subjected to drastically changed conditions.

Under long persistence of the same conditions and thus continued decrease of σ_G^2, a species may be expected to become either homallelic or polymorphic because at equilibrium with respect to all loci with major affects. In neither case is there a contribution to additive variance and hence there is no evolution. Quantitative variability on the basis of multiple segregating minor factors will, no doubt, continue but there will be no additive component of variability because selection will be directed toward an optimum at the mean (vol. 2, chap. 4). Evolutionary change will depend on the occurrence of novel favorable mutations that may be expected to be exceedingly rare events. They will occur occasionally, however, in view of the nearly infinite possibilities of change at the molecular level, and the establishment of one would provide a basis for others that had been much more improbable before. At best, however, mass selection after long-continued unchanging conditions should be exceedingly slow.

Under drastically changed conditions, many more or less unfavorable genes that have been kept at low to moderate frequencies by adverse selection or one of the other mechanisms for equilibrium, may become more favorable than the current type allele. An enormous amount of additive genetic variance is created and it causes relatively rapid genetic change.

If conditions keep changing at a tempo with which natural selection can keep up, as is undoubtedly the case with a great many species, the tendency toward decrease in additive variance will be balanced by new contributions due to the changing environment, and evolutionary change may continue indefinitely at a relatively rapid rate. It is, however, rather like movement on a treadmill. The species merely holds its own as acquirement of each new rough adaptation accompanies the undoing of an old one.

The Effect of Natural Selection on Linkage Patterns

The great variety in the numbers of chromosomes of organisms and in the amounts of crossing-over within chromosomes indicates that there is no one pattern that is always favored in evolution. Different chromosomes usually segregate independently but there are such cases as in certain of the *Oenothera* species in which groups of chromosomes segregate together. The chance that two random loci are linked at all is small, some 5% or less, in placental mammals; even if in the same chromosome the chance that they are strongly linked is small, judging from the high recombination percentages in both oogenesis and spermatogenesis in the mouse. On the other hand, the chance for linkage is about one-third in *D. melanogaster*, and for genes in the same chromosome the average recombination coefficient is much less than 0.25 because of absence of crossing-over in males. In viruses and bacteria, linkage is rather loose even within groups of cooperating loci (cistrons). In higher organisms, what were considered single loci in earlier studies have often been found to be complexes of very tightly linked elements.

It has long been known that linkage can be profoundly modified by chromosome rearrangements, and that short inversions, duplications, etc., produce considerable changes in chromosome regions. Detlefson and Roberts (1921) brought about considerable changes in recombination percentages of genes in *D. melanogaster* by artificial selection. There have been many more recent studies, reviewed by Bodmer and Parsons (1962), of modification of linkage.

A very large number of independently segregating elements is probably unfavorable because of the irregularities in meiosis that it would tend to bring about. Otherwise the linkage pattern should make no difference in a homallelic species except where association in the same cistron leads, through transcription and translation, to association of cooperating entities in the same gene product. In a heterallelic population, however, it improves fitness if favorably interacting ones, AB and $A'B'$, segregate together in excess over the unfavorably interacting ones, AB' and $A'B$. On the other hand, random assortment of loci tends to be favorable from the standpoint of versatility of the species in encountering a heterogeneous and changing environment. The actual situation indicates that natural selection tends toward an optimal balance for each form under the constraint of what has happened previously. To try to account for the great differences among specific forms is at present, however, a highly speculative matter.

Fisher (1930) surmised that in such a case as indicated above for the A and B loci, selection would automatically tend to bring about closer linkage. Kimura (1956) showed mathematically that if one locus is maintained in balanced polymorphism by heterozygote superiority, the second locus will remain polymorphic if linkage is sufficiently close and that any mutation (inversion, for example) that increases linkage will tend to be established by selection. Nei (1967) investigated different mathematical models (structural changes, independent modifier loci) and concluded that natural selection would always tend to decrease recombination of interacting loci. He found that stable polymorphisms are more favorable for establishing close linkage in such cases than unstable and transient polymorphisms.

Turner (1967), using various mathematical models, found that the direction of change for the linkage of two loci that interact so as to produce a deviation from random combination ("gamete excess" is his terminology) always tends toward closer linkage. The question asked in the title of his paper, "Why does the genotype not congeal," was answered by considering the situation with three or more loci. Here fitness will sometimes decrease when linkage becomes tighter. There is a balancing among the effects of multiple loci that tends toward maintenance of an optimum pattern of linkage.

It should be noted that the formation of cistrons can occur by differentiation of gene duplications (Wright 1959b) without affecting other loci, and that this is the most plausible mechanism where the components carry through successive steps in a metabolic transformation. Formation of gene complexes by reduction of linkage seems more plausible where there is a favorable cooperation of physiologically unrelated phenotypes.

Limitations on the Rate of Evolution in Homogeneous Populations

It is desirable to consider the limitations on the rate of evolution in homogeneous populations from other viewpoints than its equality to the amount of additive genetic variance.

After the additive genetic variance has been exhausted by selection under long continuance of the same conditions, the limiting factor becomes, as noted, the rate of occurrence of novel favorable mutations. The view usually taken on first consideration of the subject is, indeed, that mutation rate is always the limiting factor. This, however, is far from being the case if conditions are continually changing, as they usually are, at least at intervals of tens of thousands of years. As frequently noted, most loci are expected to be at least weakly heterallelic in a typical species and a great many to be strongly so. Change of conditions increases the additive genetic variance by reversing the order of the selective values of alleles and insures further evolution without any relation to mutation rate. This can occur again and again, affecting selection at other loci as well as at those previously affected.

Apportionment of Selective Intensities

The necessary apportionment of selective intensities among characters, if numbers are to be maintained, is a principle that has always been painfully familiar to breeders of livestock. Wriedt (1925, 1930) wrote vigorously on the folly of breeders, engaged in improvement in useful respects, of paying attention to fancy points, however highly regarded in the show ring.

In a wild species, the number of individuals in a species that occupies a niche in nature tend to be regulated by density-dependent factors. The offspring that are best adapted in the more important respects will fill up the niche, irrespective of shortcomings in minor respects that would have been selected against in the absence of differences in the former (Wright 1931):

The greater the number of unfixed genes in a population, the smaller must be the average effectiveness of selection for each one of them. The more intense the selection in one respect, the less it can be in the others. The selection coefficient for a given gene is thus in general a function of the entire system of gene frequencies.

This negative relation between selective intensities applies most directly to characters and thus to loci affecting different characters (different interaction systems). It cannot be expected to apply strongly to different loci in any case, however, unless the variabilities are largely genetic. Haldane (1932) dealt with this in clonal selection. The application to quantitative variability in diploid populations has been discussed in chapter 6 of this volume (see also Wright 1969a).

Generations per Gene Substitution

The limitations on the rate of evolution can best be described in terms of the average number of generations per gene substitution (G). Since conditions may be such that there is no appreciable evolution over periods of hundreds of millions of years, the question of most interest is how small the average number of generations per gene substitution may be under various kinds of favorable conditions. The apportionment of selection intensities among characters and genes enters into this, but not always to an important extent.

It will be convenient to consider first a case that is very simple in principle although impossible concretely. This is the case of an indefinitely large array of haploid clones in which a rare allele is present at each of a great many loci and the combinations are random. This implies an impossibly large number. If each rare gene has an initial frequency of 10^{-3}, the frequency of the combination of 100 of these is 10^{-300}. If the genes are not combined at random, selection of each modifies the frequencies of the others. To avoid such modification, it will be assumed that there is total selection of the rare genes in order of importance. It will also be assumed that each individual is selected on the basis of only one gene. It will further be assumed that the population size is maintained by selecting as parents a proportion p that is the reciprocal of the reproductive capacity, excluding nonselective losses.

In table 12.1A, all initial rare gene frequencies are $q_0 = 0.001$ and there is a tenfold effective reproductive capacity; thus $p = 0.10$. The first selection of

10% includes all clones with rare gene a_1, all with rare gene a_2, and so on to gene a_{100}. The frequencies of these rise from 0.001 to 0.01, leaving the others still at 0.001 (under random combination). A second selection of the best 10% as parents includes all with a_1 to a_{10}, raising their frequencies to 0.1, leaving those selected in only the first generation at 0.01 and those not selected at all at 0.001. The next selection of the best 10% fixes a_1, leaving 9 at 0.1, 90 at 0.01, and the others at 0.001. The other rare genes may now be fixed by selections of the same sort, one generation each for the nine at 0.1, a generation to bring the next ten to 0.1, ten generations to fix these in succession, and so on (table 12.1).

It appears that while three generations are required to fix the first gene, the average number of generations per gene substitution is 1.11. The average would not be appreciably altered if the initial frequency had been smaller; 1.111 if $q_0 = 10^{-4}$.

More generally, if the percentage selected as parents of the next generation, in order to maintain population size, is p and the initial gene frequencies are all q_0, the frequencies of p/q_0 rare genes shift from q_0 to q_0/p in the first generation, those of p/q_0 rare genes shift from q_0/p to q_0/p^2 in the second generation. The average number of generations per gene substitution is about $1/(1 - p)$, irrespective of q_0.

The case of haploids with twofold effective reproductive capacity and $p = 0.50$, $q_0 = 0.001$ is shown in table 12.1B. The average number of generations per gene substitution comes out 1.998 using a fractional estimate for the last generation in which selection of 50% would remove some of the desired genes. This again is approximately $1/(1 - p)$.

The situation is wholly different if there is incomplete simultaneous selection of rare genes, since random combination is destroyed. Given clones A_1A_2, A_1a_2, a_1A_2, and a_1a_2 in random combination, the last with frequency q_0^2, simultaneous selection of a_1 and a_2 clones equally but incompletely soon eliminates A_1A_2, with a_1a_2 still very rare. For several generations such selection leaves A_1a_2 and a_1A_2 both close to 50% each before a_1a_2 rises to moderately large frequencies. It takes 3.75 generations per gene substitution instead of about 2 if $p = 0.50$. If there were no a_1a_2 clones at all at first, this combination could, of course, never appear (except by mutation, which is not taken into account here).

It might seem that the minimum number of generations per gene substitution is 1, but this is only under the assumption that each individual is selected at only a single locus. Suppose that the initial gene frequencies of several genes are 0.1 and that these are combined at random so that the frequency of a_1a_2 is 0.01, of $a_1a_2a_3$ is 0.001, and so on. If there is 100-fold reproductive capacity and $p = 0.01$, a_1a_2 can be fixed in the first generation, implying that incomplete dominance. Numbers after selection and contributions to cost ($q + \Delta q$) are given in parentheses. The last contribution to cost is the fraction of the usual amount of culling ($1 - p$), necessary for fixation of A. G is the average number of generations per gene.

(A) HAPLOID ($q_0 = 0.001, p = 10\%$)

Generation	Per 1,000		q	$q + \Delta q$
	A'	A		
0	1	999 (99)	0.001	(0.010)
1	10	990 (90)	0.010	(0.100)
2	100	900 (0)	0.100	(1.000)
3	1,000	0	1.000	...
4				
	900/900 = 1.000			$G = 1.110$

(B) HAPLOID ($q_0 = 0.001, p = 50\%$)

Generation	Per 1,000		q	$q + \Delta q$
	A'	A		
0	1	999 (499)	0.001	(0.002)
1	2	998 (498)	0.002	(0.004)
2	4	996 (496)	0.004	(0.008)
3	8	992 (492)	0.008	(0.016)
4	16	984 (484)	0.016	(0.032)
5	32	968 (468)	0.032	(0.064)
6	64	936 (436)	0.064	(0.128)
7	128	872 (372)	0.128	(0.256)
8	256	744 (244)	0.256	(0.512)
9	512	488 (0)	0.512	(0.976)
10	1,000	...	1.000	...
11				
	488/500 = 0.976			$G = 1.998$

(C) DIPLOID, SEMIDOMINANT ($q_0 = 0.001, p = 10\%$)

Per 10,000			q	$q + \Delta q$
$A'A'$	AA'	AA		
0	20	9,980 (980)	0.001	(0.010)
1	198	9,801 (801)	0.010	(0.100)
100	1,800 (900)	8,100 (0)	0.100	(0.550)
3,025	4,950 (0)	2,025 (0)	0.550	(0.775)
10,000	0	0	1.000	
6,975/9,000 = 0.775				$G = 1.435$

(D) DIPLOID, SEMIDOMINANT ($q_0 = 0.001, p = 50\%$)

Per 1,000,000			q	$q + \Delta q$
$A'A'$	AA'	AA		
0	2,000	998,000 (498,000)	0.001	(0.002)
4	3,992	996,004 (456,004)	0.002	(0.004)
16	7,968	992,016 (492,016)	0.004	(0.008)
64	15,872	984,064 (484,064)	0.008	(0.016)
256	31,488	968,256 (468,256)	0.016	(0.032)
1,024	61,952	937,024 (437,024)	0.032	(0.064)
4,096	119,808	876,096 (376,096)	0.064	(0.128)
16,384	223,232	760,384 (260,384)	0.128	(0.256)
65,536	380,928	553,536 (53,536)	0.256	(0.512)
262,144	499,712 (237,856)	238,144 (0)	0.512	(0.762)
580,644	362,712 (0)	56,644 (0)	0.762	(0.839)
1,000,000			1.000	
419,356/500,000 = 0.839				$G = 2.623$

$G = 0.5$ with no change in population size; if the reproductive capacity is only tenfold but $p = 0.01$, a_1a_2 can again be fixed in the first generation, $G = 0.5$, but population size is reduced to only 10% of its previous value. This sort of discrepancy between reproductive capacity and intensity of selection obviously cannot continue long without leading to extinction.

A more realistic process of total selection of rare genes is possible with diploids since random combination is approached in each generation. It is supposed in table 12.1C that a great many rare semidominant genes (all $q_0 = 0.001$) have simultaneously become favorable. With tenfold reproductive capacity, selection of the best 10% of the individuals as parents has the same effect as in the array of haploids, as long as only type homozygotes are culled. Efficiency of selection is reduced when elimination of heterozygotes becomes necessary. The average number of generations per gene substitution is 1.435 instead of 1.110, an increase of 29%.

With twofold reproductive capacity and 50% selection of the best individuals (table 12.1D), $G = 2.623$ instead of 1.998, in haploids, about 31% greater.

In the case of a complete recessive, selection is much less efficient than with partial dominance, so that fixation proceeds more slowly. Initial equilibrium frequencies tend to be higher however. If $q_0 = 0.01$ and $p = 50\%$, with twofold reproductive capacity, $G = 5.2$ instead of 2.6.

While the calculations in the cases of diploids were based on selection of genes in order, the continual restoration of approximate random combination should make the value of G independent of order.

Truncation Selection of Quantitative Variability

We here consider truncation selection of the additive effects of multiple strongly heterallelic loci on the same character instead of the separate selection of weakly heterallelic loci as above. This differs markedly in that individuals tend to be selected because of carrying multiple favorable genes. Only the case of multiple equivalent semidominant genes will be considered here, except for brief references to cases already considered in chapter 6. It will be assumed for simplicity that all have the initial frequency $q_0 = 0.10$ and reach 0.90. The average number of generations per gene substitution will be referred to as G. Using the symbolism of chapter 6 otherwise,

Genotype	f	Effect	
AA	$(1-q)^2$	0	$M = 2n\alpha q$
AA'	$2q(1-q)$	α	$\sigma_T^2 = 2nq(1-q)\alpha^2 + \sigma_E^2$
$A'A'$	q^2	2α	$\Delta q = (z/p)q(1-q)(\alpha/\sigma_T)$

The simplest case is that of complete determination by heredity:

$$(\sigma_E^2 = 0), \qquad \alpha/\sigma_T = 1/\sqrt{[2nq(1-q)]},$$

$$\Delta q = k\sqrt{[q(1-q)]} \quad \text{where } k = (z/p)/\sqrt{(2n)}.$$

Letting t be the number of generations to pass from $q = 0.50$ to a specified value,

$$t = (1/k) \sin^{-1}(2q - 1).$$

With semidominance, it requires $2t$ generations to pass from q_0 to $1 - q_0$, giving $2t = \pi/k = 4.443\sqrt{(n)}/(z/p)$ for $q_0 = 0$, $2t = 2.623\sqrt{(n)}/(z/p)$ for $q_0 = 0.10$. The number of generations per gene substitutions is $G = 2t/n$. Values of G, where q changes from 0.10 to 0.90, are given in table 12.2 for various values of p (and hence z/p) and n, with $k^2 = 1$ (left).

TABLE 12.2. Number of generations per gene substitution, G, where $q_0 = 0.10$ to 0.90 with different proportions, p, selected as parents and 1, 10, or 100 equivalent semidominant genes in the cases of $h^2 = 1$ and h^2 very small, with $\alpha/\sigma_E^2 = 0.02$. The gene effect, α, is the same irrespective of numbers.

p	z/p	$h^2 = 1$			s	$h^2 \approx 0$		
		$n = 1$	$n = 10$	$n = 100$		$n = 1$	$n = 10$	$n = 100$
0.90	0.195	13.45	4.25	1.35	0.0039	1,127	113	11.1
0.50	0.798	3.29	1.04	0.33	0.0160	275	28	2.7
0.10	1.755	1.50	0.47	0.15	0.0351	125	13	1.2
0.01	2.665	0.98	0.31	0.10	0.0533	82	8	0.8

It is more likely for heritability to be very low ($h^2 \approx 0$) in nature, with respect to quantitative variability of a character related to fitness, than high. It is accordingly desirable to consider the limiting case in which the total variance differs only negligibly from σ_E^2 even with 100 loci, at all of which the favorable allele contributes the amount α.

In this case, $\Delta q = sq(1-q)$ where $s = (z/p)(\alpha/\sigma_E)$.

$$t = (1/s) \log[(1 - q_0)/q_E]$$

If $\alpha/\sigma_2 = 0.02$, $G = (2 \log 9)/sn = 4.394/sh$.

Values of G, with $h^2 \approx 0$, are given in table 12.2 (right) for the same values of p and n as for $h^2 = 1$.

In the case of complete determination by heredity ($h^2 = 1$), the average number of generations per gene substitution ($q = 0.10$ to 0.90) is always rather low and may be less than 1.0 under conditions that are not extreme ($G = 0.47$ for $p = 0.10$, $n = 10$). This is in spite of interference among the selection processes of loci that cause the value of G to fall off only as $1/\sqrt{n}$.

Where h^2 is very small ($h^2 \approx 0$), the value of G ($q_0 = 0.10$ to 0.90) tends to be much larger, but since it falls off as n, in the practical absence of interference among the selection processes of loci (assumed to be always the same), G varies as $1/n$ and falls to fairly low values with large n ($G = 1.2$ for $p = 0.10$, $n = 100$).

If it is assumed that heritability is constant, irrespective of the number of segregating loci, α varies as $1/\sqrt{n}$. In this case G also varies as $1/\sqrt{n}$ as under complete determination by heredity, but for a different reason.

Several less regular cases of truncation selection were considered in chapter 6. In all of these z/p was taken as 1, which implies selection of $p = 38\%$ as parents.

In figure 6.3, there were five classes of semidominant favorable genes that differed only in their initial frequencies: ten at 0.02, two at 0.10, one at 0.50, two at 0.90, and ten at 0.98. The 25 genes reached fixation in 14 generations, $G = 0.56$, but the 12 that started at 0.10 or 0.02 averaged $G = 1.17$ in spite of competition in the early generations from the genes with higher initial frequencies.

The case of five semidominant favorable genes with different effects, α, 0.8α, 0.6α, 0.4α, and 0.2α, but all with initial frequency 0.10, was shown in figure 6.4. Apportionment of selective intensities is indicated by the long delay in the fixation of the less important genes. All five, however, reached fixation in 18 generations, $G = 3.6$.

A more extreme case of this sort was shown in figure 6.5, in which there was one major gene (effect α) and ten minor ones (effects 0.1α each). All started from $q_0 = 0.10$. The 11 genes were fixed in 13 generations, $G = 1.18$.

A case of five favorable genes with different degrees of dominance: recessiveness, semidominance, dominance, two kinds of overdominance, all with the same contribution at equilibrium, was shown in figure 6.7. All except the dominant ($q = 0.95$) practically reached fixation or equilibrium by generation 26, after starting from $q = 0.10$. The value of G at generation 26 was 5.2.

These cases again show that only a moderate number of generations are used per gene substitution in attaining near-equilibrium where there is complete determination by heredity and truncation selection. As noted, nongenetic variability increases the number for near-fixation and much more for complete fixation.

The Cost of Natural Selection

We come now to Haldane's (1957) theory of the "cost of natural selection." This applied to cases in which a change in conditions has caused the type gene to become less favorable than an allele, assumed to be present at frequency q_0. His formula was for the total losses of former type genes as represented in individuals before substitution by the new type gene is complete. The population was assumed to maintain a constant size, ($\overline{W} = 1$). For a semidominant favorable gene with low heritability and slight selective advantage,

Genotype	f	w	W	
AA	$(1-q)^2$	$1-s$	$(1-s)/\bar{w}$	$\bar{w} = 1 - s(1-q), \overline{W}=1$
AA'	$2q(1-q)$	$1-0.5s$	$(1-0.5s)/\bar{w}$	$\dfrac{\delta q}{\delta t} = sq(1-q)/2\bar{w}$
$A'A'$	q^2	1	$1/\bar{w}$	$\delta t = 2\bar{w}\,\delta q/[sq(1-q)]$

$$C = \int_0^\infty (1-\bar{w})\,dt = 2 \int_0^1 \left[\left(\frac{1-s}{q}\right) + s\right] dq = 2s(1-q_0) - 2(1-s)\log_e q_0$$

For very small s, $C = -2 \log_e q_0 = -4.6 \log_{10} q_0$. Note that this is independent of s.

A formula can readily be derived for any degree of dominance, h, of the old type allele (Crow 1970):

$$C = -\frac{1}{1-h}\left[\log_e q_0 + h \log_e \frac{h}{1-h-(1-2h)q_0}\right].$$

If heritability is low, there is, as noted, little competition among loci with respect to selection so that multiple equivalent genes are fixed independently. The total cost is then the sum of the costs for the separate loci and the average is the same as for one of them. Table 12.3 lists some values of C (one locus) for various degrees of dominance h of the old type allele and initial gene frequency q_0 of the favorable allele. The results for haploids are the same as for a completely recessive type gene.

TABLE 12.3. Values of the cost (C) of fixation of a locus for various degrees of dominance (h) and various initial gene frequencies q_0.

h \ q_0	10^{-1}	10^{-2}	10^{-3}	10^{-4}
0	2.3	4.6	6.9	9.2
0.5	4.6	9.2	13.8	18.4
0.9	8.5	21.9	49.2	72.2
1.0	11	104	1000.0	10008.0

The values for complete dominance of the old type gene becomes enormous at low initial frequencies of the favorable allele. With semidominance or dominance of the favorable allele, the value of C is small to moderate.

The cost for substitution of a semidominant gene, q_0 to $1 - q_0$, is $2 \log_e (1 - q_0)/q_0$; that for the number of generations $2t$ is $- (2/s) \log_e (1 - q_0)/q_0$. Thus $2t = C/s$. This is also the average cost per locus, $\bar{C}$, if there are multiple equivalent loci, but the number of generations per gene substitutions is $G = 2t/n = \bar{C}/sn$ and $\bar{C} = snG$. For selection at a single locus ($n = 1$), the cost tends to be much less than the number of loci per gene substitution, but for multiple loci and total cost, $C = n\bar{C}$, the reverse tends to be true.

Haldane took $q_0 = 10^{-4}$ and costs of 10 to 100 as typical for genes maintained by balance between recurrent mutation and adverse selection before the change in conditions that made them favorable ($q_0 = v/s$ if appreciable dominance). He conjectured that 30 would be a fair average, and taking $s = 0.10$ as fairly typical, he suggested that 300 ($= G$) would be typical of the number of generations for fixation of the gene in question. He noted that this is so large that evolution is necessarily an extremely slow process.

This estimate has seemed to some authors to be incompatible with the probable rates of evolution in certain cases, as judged by paleontological evidence. Thus, Dodson (1962) held that human evolution must have been much more rapid. The ratio of amino acid substitution in proteins has been held by Kimura (1968) and King and Jukes (1969) to have been much too rapid for selective fixation if Haldane's estimate is correct. They hold that fixation by random drift, which entails no cost, is indicated. This question will be discussed in volume 4. It may be noted here, however, that Haldane made no allowance for the possibility of simultaneous selection at multiple loci, under which his estimate should be divided by n to give the number of generations used up per gene substitution.

Haldane's term *the cost of natural selection* and his definition in terms of losses of the type genes over the whole period of fixation relative to a selective value of 1 for the favorable allele, seem to suggest that these losses are a direct cost to the species. This, however, was not his meaning. The species is always better off and suffers fewer losses because of the selection (of the type considered here, $F(W/\bar{W}) = \bar{W}$) than if there had been no selection. If the species actually suffers a decline in population size following the postulated deterioration of the environment, the loss must be attributed to the latter, not to the selection process, which partially alleviates it. Brues (1964) made numerical comparison of the "cost" of evolving by selection and the greater cost of not evolving.

There have been several criticisms of Haldane's failure to take explicit account of changes in population size occurring under changed conditions (Feller 1967; Moran 1967). Haldane assumed that the population fluctuated about a constant size, irrespective of environmental action and selection. His formula depended on the relative selective values and it is not apparent that other assumptions would affect it seriously.

Others have criticized Haldane's assumption of constancy of the relative selection values of genotypes (Sved 1968; Maynard Smith 1968a; O'Donald 1969). In particular, these authors have held that one or another model of truncation selection is more realistic and would greatly reduce the number of generations devoted to each gene substitution. Smith proposed a model in which more than 25,000 genes could be selected for simultaneously. This model was criticized as unrealistic by O'Donald, who proposed a different truncation model under which the number selected for simultaneously was still very large. The critical parameter, G, the average number of generations per gene substitution, was not, however, calculated in these cases.

The source of the dissatisfaction with Haldane's "cost" seems to be in its composite character. The maximum rate of evolution depends (inversely) on G, which is only one of the factors of cost. The maintenance of population size requires that the proportion, p, of the progeny selected for parentage be as large as the reciprocal of the reproduction capacity, offspring per parent, after allowing for accidental losses. This is an inverse function of the selection intensity (z/p). The closely related selection coefficient, s, is another factor of cost. The number of genes that can be selected simultaneously is still another factor in cost per locus, $\bar{C} = snG$. Cost as a parameter has the attractive feature that it is practically independent of s (if the latter is small) because of the inverse relation between s and G. It is, however, unfortunately easy to slip into interpretations in terms of only one of its factors.

Endurance of Environmental Deterioration

Felsenstein (1971) has proposed a different concept of "cost," the proportional reproductive surplus, d, necessary for persistence of population size in the face of deleterious environmental changes. He restricted his discussion to haploid populations. He assumed that deleterious environmental changes occur at intervals of k generations. At each such change, a type gene becomes disadvantageous, $W_A = 1 - s$, while its allele (initial frequency q_0) becomes favorable enough to maintain population size when fixed ($W_{A'} = 1$). Viability effects are assumed to be multiplicative. He arrived at a formula for the relations among d, k, and q_0 under which population size changes from N' to N. As with Haldane's cost, the selection coefficient, s, cancels out.

$$N = N'(1 + d)^k q_0$$

The relations for persistence of population size thus are as follows:

$$(1 + d)^k q_0 = 1$$
$$k \log_e (1 + d) = -\log_e q_0$$

Haldane's cost (for small s) is thus $k \log_e (1 + d)$ in Felsenstein's model, and Felsenstein's cost is $d = (1 - q_0^{1/k})/q^{1/k}$.

Following are portions of Felsenstein's tables. Table 12.4A gives the reproductive surplus necessary for maintenance of population size with various intervals, k, between environmental change and various initial gene frequencies, q_0. Table 13.4B gives the minimum interval, k, between the environmental changes that the species can endure with a given reproductive capacity $(1 + d)$ and values of q_0.

TABLE 12.4. A gives the reproduction surplus, d, necessary for maintenance of population size under various intervals, k, between environmental changes and various initial gene frequencies, q_0. B gives the minimum intervals, k, that the species can endure with given $1 + d$ and q_0.

A	k			B	1 + d	
q_0	1	10	100	1.1	2	10
10^{-1}	$d = 9$	0.259	0.023	$k = 24.2$	3.3	1
10^{-2}	99	0.584	0.047	48.3	6.6	2
10^{-3}	999	0.995	0.071	72.5	10.0	3
10^{-4}	9,999	1.512	0.096	96.6	13.3	4
10^{-5}	99,999	2.162	0.122	120.8	16.6	5

It is to be noted that Felsenstein's $1 + d$ is equivalent to $1/p$ as used here, so that $d = (1 - p)/p$. It seems surprising at first that the strength of selection makes no difference in the reproductive surplus needed to maintain population size with given k and q_0, but strong selection uses up the surplus rapidly, weak selection slowly.

Felsenstein's model assumes that the selective value of the substituting allele rises from whatever it was before the environmental change to an absolute $W_A = 1$, while the old allele declines to $W_{A'} = 1 - s$; this is in contrast with the assumption that $1:1 - s$ are merely the relative values and that the mean of the absolute values merely fluctuates about $\overline{W} = 1$ because of density regulation. This assumption restricts severely the amount of environmental deterioration that the population can endure.

Evolution Based on Correlations with Relatives

The process of selection based on the grades of relatives does not differ essentially from mass selection of individuals, except that the regression of offspring on selected parents is multiplied by the regression of the parent on the graded relative or relatives.

The remarkable evolution of different instincts manifested only in the sterile worker castes of different species of termites or of social hymenoptera was necessarily based on the genetic correlation between the reproductive castes and the workers, as brought out by Darwin (1859), since the caste system clearly long antedated the divergence of families, genera, and species.

It has often been suggested that the evolution of "altruistic" traits, especially those in which individuals benefit from self-sacrifice of parents or siblings, may be due to natural selection based on genetic correlations with the self-sacrificing individuals (Hamilton 1963, 1964; Wynne-Edwards 1962, 1963).

This process has often been called "group selection." It differs from the form of "intergroup selection" (Wright 1931) involved in the shifting balance process. Because of the confusion, the latter has usually been referred to here by the more explicit term "interdeme selection." The latter also can bring about selection of altruistic traits, but with emphasis on benefit to the whole population rather than merely the family. The conditions under which interdeme selection can outweigh adverse individual selection are somewhat restricted (Haldane 1932; Wright 1945; Eshel 1972). There is much less restriction if the benefit from self-sacrifice applies more to genotypically similar individuals than to all individuals (Boorman 1974). This subject will be discussed in a later chapter.

Summary

In this chapter, we have been concerned with three main questions: (1) the kinds of variability that provide material for evolutionary change, (2) the processes by which such change may take place in a large homogeneous population, and (3) the limitations on the rate of evolution in such a population. The kinds of variability were reviewed only briefly. The question whether evolution makes most use of the omnipresent quantitative variability or the rare "sports" was considered by Darwin, who favored the former. De Vries favored the latter, holding that new species can arise only from mutations of species rank.

The mutations actually observed by de Vries were found to be chromosome aberrations. Polyploidy, following wide hybridization, has, indeed, proved to be a common mode of origin of new species of plants, giving reproductive isolation and wide difference in character from the parent species. Translocation and other balanced rearrangements have undoubtedly also been very important in contributing to the isolation of species, but have had little to do with the building up of their phenotypic differences. In the latter respect, the unbalanced minor aberrations have undoubtedly played an important role in the long run but can be grouped with the gene mutations as following essentially the same principles of transmission. Hybridization without tetraploidy has supplemented these primary sources of genetic variability, especially in the form of Anderson's introgression.

The theories of Willis, Goldschmidt, Schindewolf, and others who have held that not only species but each of the higher categories can have arisen only from mutations of appropriate magnitude are not scientifically acceptable explanations of adaptive evolution. They invoke a succession of miracles.

Most geneticists, led by Morgan, early turned from de Vries' species mutations to the major gene mutations as the principal basis for phenotypic evolution. Others, however, notably Castle and East, returned essentially to quantitative variability as the primary material for selection. Castle's original interpretation of this, as due to continual variability of major genes, gave way (partly from crucial experiments that he made himself) to interpretations in terms of multiple independent minor genic differences, demonstrated experimentally especially by East and his associates. The latter view had meanwhile been shown to be fully in harmony with the statistical study of correlations between relatives.

The general conclusion with respect to the genetic basis for phenotypic evolution came to be that there is, in general, strong heterallelism with respect to alleles with minor differences in effect, at a great many loci, because of various sorts of balancing of evolutionary pressures; occasionally strong heterallelism with respect to alleles with major differences in effect giving conspicuous polymorphisms; and at least weak heterallelism with respect to both major and minor effects at all loci in any large species. In all of these cases there are multiple alleles, at least at the molecular level, because of the enormous number of possible one-step mutations in genes and, in many cases phenotypically distinguishable ones. Selection studies especially have demonstrated the omnipresence of genetic variability.

Recombination enormously amplifies the amount of variability giving rise to numerous different interaction systems, but these are in general broken up too rapidly in a panmictic population to provide material for mass selection other than by net gene effects.

Mutation pressure is too scattered in effect to contribute by itself to evolutionary trends, except in the case of effects based on inactivations or actual loss. Such mutations are, however, undoubtedly important in pruning the genome. They also undoubtedly tend to bring about the degeneration of organs that have become useless in the course of evolution but, because of the slowness of mutation rates, it is probable that two kinds of selection are more important in the latter respect: (1) selection toward reduction of useless organs as encumbrances and energy wastes and (2) selection of alleles at the loci involved in the development of the organ in question according to pleiotropic effects on other characters. Useless organs are rapidly lost if related parts are undergoing reorganization, but they may be retained a long time as vestiges if related parts are conservative.

Most new favorable mutations are lost by accidents of sampling, the chance of establishment of one with a dominant effect being only $2s$ where s is its selective advantage. Once established at a safe frequency, a mutation tends to advance toward an equilibrium frequency or to fixation in a deterministic fashion, based on its net selective advantage. This is the process implied by Morgan and worked out mathematically, especially by Haldane. The overall rate of advance of fitness, $F(W/\overline{W})$, in a panmictic population is equal to its additive genetic variance with respect to selective value (W) according to a theorem presented by Fisher. This holds to a good approximation for pairs of genes for which the amount of interactive selection is of lower order than that of recombination. A particular interaction system is built up since mutations are favorably selected only if they are in harmony with the inter- action system already established. With a given total variance, the rate of evolution is cut down by just the proportion that interaction variance con- tributes to the total. The effects of deviation from random combination in permitting some selection among interaction systems is discussed briefly.

In a panmictic population that has lived for a very long time under essen- tially the same conditions, the additive genetic variance becomes reduced to a minimum maintained by the rate of occurrence of novel favorable mutations, and evolution thus becomes extremely slow. This seems to be the only condition under which a mutation rate is the limiting factor for evolution.

Under changed conditions, the changes in the system of selective values at heterallelic loci contribute to the additive genetic variance. Evolutionary readaptation may proceed relatively rapidly irrespective of mutation rate.

If conditions keep changing at a tempo with which evolution can keep pace, relatively rapid evolution may continue indefinitely, irrespective of mutation rate, but it is somewhat like movement on a treadmill as different interaction systems are built up and broken down, one after the other, within the virtually infinite array of possible gene combinations.

The conditions under which natural selection may change the linkage pattern are discussed briefly. The apportionment of selective intensities among characters, and ultimately genes, puts some limitation on the rate of evolution where heritabilities are high, but not much where low.

The most useful parameter in estimating the maximum rate of evolution in terms of selective intensities and initial gene frequencies is the average number of generations per gene substitution (or near-substitution), G. Where numerous genes with fully heritable effects come under selection simultaneously and are all so rare that individuals may be assumed to be selected on the basis of single genes, the value of G is about $1/(1 - p)$ until fixation is approached, where p is the proportion selected as parents (after allowing for nonselective losses). This tends to be increased by about 30% in the later generations because of the imperfect selection of the favorable gene in heterozygotes. It tends to be reduced, however, because of possible selection of individuals for more than one locus as the favored alleles approach fixation.

Quantitative variability tends to be determined by numerous strongly heterallelic loci. If there is strong heritability, G may be very low, less than 1 under reasonable conditions, in spite of competitive apportionment of selective intensities among the loci. If, however, heritability is low, G tends to be higher, unless the number of roughly equivalent loci is very great or p is very small.

The cost of natural selection per locus as defined by Haldane is a composite, $C = snG$, of various concepts that bear on the possible rate of evolution. One factor is G, the generations per gene substitution, which defines the rate under specified condition as noted above. The number of generations required for fixation of each particular gene varies inversely with its selective advantage, s, which is also a factor of cost, with the consequence that C is practically independent of s if small. Another factor of the mean cost per locus is the number, n, of loci under simultaneous selection. While C is an attractive parameter because of its near-independence of s when the latter is small, its interpretation depends on knowledge of its factors.

Finally, the reproductive capacity of a species is an important limiting factor on its rate of evolution. Environmental deterioration is the principal determinant of evolutionary change in a homogeneous population, but the reproductive capacity of the species sets a limit to the tempo of environmental change the species can endure.

The situation where the species is subdivided into partially isolated, more or less differentiated demes is very different from that in the essentially panmictic populations considered in this chapter. This subject and the very important role of ecologic opportunity in evolutionary rates will be dealt with in the next chapter.

CHAPTER 13

Shifting Balance Theories of Evolution

There has been much meaningless discussion of evolutionary dynamics since Darwin proposed his theory of natural selection, because of failure to grasp his basic concept. The essence of his theory was that evolution comes about from the interaction of a random and a directed process. The random process that he invoked was the occurrence of small *random* variations, which he supposed provided the raw material for natural selection, a process *directed* by the necessity for adaptation to the environment, and one that builds up, step by step, changes that would be inconceivably improbable at a single step. Willis, Goldschmidt, Schindewolf, and even de Vries did not fully grasp this point and preferred a miraculous single step to a guided succession of little steps.

The failure to recognize any third alternative to 100% chance and 100% determinism is reflected in such statements as that progress by natural selection has a probability comparable to that of the successful typing of the Bible by a monkey, punching keys at random, and that there must accordingly be some unknown natural law, comparable to those postulated for stellar evolution. Natural selection is more like the children's game of 20 questions, in which it is possible to arrive at the correct one out of about a million objects by a succession of 20 questions each answered merely by yes or no, than it is like a succession either of wholly random events or wholly determinate ones (Wright 1967*b*).

There is frequent misunderstanding of what the evolutionist means by *random* in this connection. It is not a mathematical term here. There is no implication that each "random" variation does not have an adequate physical cause. All that is meant is that the variations are not correlated with the course subsequently taken by evolution under the guidance of selection. The nature of the variations is, of course, severely limited by the accumulated results of past evolution. The latter gives a basis for departure of small new "random" variations at each step.

The Darwinian process discussed in the preceding chapter consisted of mutation as the random genetic process and natural selection among individuals (or families) as the guiding principle. In this chapter we are concerned with random processes other than mutations. Those with a specifiable variance contribute to a "random drift." Others consist of major unique events, which may affect the course of organic evolution much as historical accidents affect human history. Either random drift or major unique events are significant only as they lead to new positions of equilibrium in the frequencies of interacting sets of genes.

Random Drift

The word *drift* has come to be used frequently in population genetics but not always in the same sense. I am probably responsible for introducing it, though with no intention of making it a technical term. In a comment on a paper by R. A. Fisher (1928a), I (Wright 1929a) wrote:

> In consequence of these facts there will be a gradual drift of the heterozygote toward wild type.

In this case, *drift* referred to the results of a directed process, natural selection. It was used in a different sense in the following (Wright 1929b):

> Being random, such variations largely neutralize each other, but there is a second order drift which cannot be ignored.

Here *drift* referred to the cumulative effect of random processes, the accidents of sampling in gametogenesis. I continued to use it in such expressions as *drifting at random* in this connection, and it was taken up by others as a technical term in this second sense. I avoided using it in this way until much later (Wright 1956), and then emphasized the importance of always specifying whether "random drift" or "steady drift" was intended, and in the former case of specifying whether it arose from accidents of sampling or from fluctuations in the systematic pressures. The convenient term *sampling drift* has been suggested by Cain and Currey (1963) for the sort due to accidents of sampling in gametogenesis.

The components of change in gene frequency in the case of a local population, surrounded by others, may be written as follows in the simplest case, assuming that all coefficients are so small that order of effect is unimportant:

$$\Delta q = v(1 - q) - uq - m(q - Q) + sq(1 - q),$$

(vol. 2, eq. 3.1 and 3.46),

where v and u are, as usual, the rates of mutation to and from the gene in question, respectively; m is the effective amount of immigration, representative of the species as a whole (gene frequency Q); and s is the momentary selective advantage of the gene over its alleles.

The variance of such changes is, in this simple case,

$$\sigma_{\Delta q}^2 = (1/2N)q(1 - q) + \sigma_m^2(q - Q)^2 + m^2\sigma_Q^2 + \sigma_s^2 q^2(1 - q)^2$$

(vol. 2, eq. 13.45, 13.63, 13.64 and 13.65).

The first term is the sampling variance, N being the effective size of population; the next two terms are those due to fluctuations in the amount and nature of the immigration; and the last term is that due to fluctuations in the selection coefficient.

These variances do not constitute the random drift with respect to the gene in question. This is the cumulative result of the systematic pressure, Δq, toward the equilibrium value, $\hat{q}$ (derived from $\Delta q = 0$) and the variances. This result is the stochastic distribution with given frequencies at all other loci:

$$\varphi(q) = (C/\sigma_{\Delta q}^2) \exp [2(\Delta q/\sigma_{\Delta q}^2) \, dq, \qquad \text{where} \quad \int_0^1 \varphi(q) \, dq = 1$$

(vol. 2, eq. 13.42).

Illustrations of distributions involving different components have been given in volume 2, chapter 13.

The random drift for an array of loci refers to the multidimensional stochastic distribution for an array including all alleles at each locus. If there are multiple alternative interaction systems, the stochastic distribution has multiple "peaks," "saddles," and "pits." The set of gene frequencies tends to fluctuate about a particular peak for long periods of time before crossing a shallow two-locus saddle to come under control of a usually higher peak. The random drift at a given time had best be understood to be restricted to the portion of the total distribution within which all Δq's are directed toward the current equilibrium point for all loci.

If all genotypic selective values are constant, the component of the systematic pressure due to selection is approximately

$$\Delta q = q(1 - q)(\partial \overline{W}/\partial q)/2\overline{W} \qquad \text{(vol. 2, eq. 3.24, 4.6).}$$

The corresponding distribution, ignoring mutation pressure, is

$$\varphi(q_1, q_2 \cdots q_n) = C\overline{W}^{2N}\prod[q_i^{2NmQ_i - 1}] \qquad \text{(vol. 2, eq. 14.11).}$$

The product term has a factor for each allele at each locus. The selection term $\overline{W}^{2N}$ shows how the multiple peaks in the "surface" of mean selective values, $\overline{W}$, are reflected in enormously exaggerated form in the stochastic distribution.

More general formulas must be used if there is frequency dependence of the genotypic selective values. Representing genotype frequencies by f and, assuming that interactive selection is of lower order than recombination so that deviations from random combination are unimportant,

$$\Delta q = 0.5q(1 - q) \sum [(W_i/\overline{W}) \, \partial f_i/\partial q] \qquad \text{(vol. 2, eq. 4.1).}$$

Fitness, $F(W_i/\overline{W})$, has been defined as $\int \sum (W/\overline{W})(\partial f/\partial q)\, dq$ if this exists, giving

$$\Delta q = 0.5q(1 - q)\, \partial F(W_i/\overline{W})/\partial q$$

$$\varphi(q_1, q_2 \cdots q_n) = e^{2NF(W_i/\overline{W})} \prod_i [q_i{}^{4NmQi-1}] \qquad \text{(vol. 2, p. 395)}.$$

In this case, there is a "surface" of fitness values that may be very different from the "surface" of mean selective values of the same array of loci. The systematic pressures are here directed toward a set of gene frequencies characterized by a fitness peak.

If the fitness function does not exist, there still may be many different selective "goals" toward one of which the current systematic pressures are directed. It will be convenient for the sake of visualization, however, to refer to fitness surfaces and control by fitness peaks, even though the mathematical situation may sometimes be more complicated.

The additional Darwinian processes considered in this chapter are ones in which random processes other than mutation either lead to fixation or, if not, may carry the set of gene frequencies across a saddle to come under control of a usually higher peak in the surface of the fitness function, which may or may not be the same as a peak in the surface of mean selective values, or more generally under control of a different selective goal. Random processes here include major unique events as well as random drift. The guiding principle, if present, may be either mass selection among individuals (or families) or among local populations by asymmetric diffusion.

Conspectus of Theories of Species Transformation

It will be convenient at this point to try to list possible theories under the general postulate of multiple fitness peaks. The pre-Mendelian stages of evolution, by which the eukaryotic cell with perfected processes of mitosis and meiosis was arrived at, are here excluded, as are the modes of origin of new species, and of new higher taxa. The following processes were listed in 1929 with a brief evaluation of probable importance, and were reevaluated after further consideration (Wright 1929c, 1932b) (fig. 13.1). Process A was not in this list but had previously been discussed and largely rejected (Wright 1929a).

A. *Mutation Pressure.* The minor importance of this process in the sloughing off of useless characters in comparison with natural selection against such characters as encumbrances, or for alleles with favorable pleiotropic effects on related characters, has been discussed in the previous chapter.

B. *Mass Selection under Constant Conditions.* "In too large a freely interbreeding population, there is great variability but such a close approach of all gene frequencies to equilibrium that there is no evolution under static conditions" (Wright 1929c).

This statement clearly was too strong. As noted later, "With an unlimited chain of possible gene transformations, new favorable ones should arise from time to time, and gradually displace the hitherto more favorable genes, but with the most extreme slowness even in terms of geologic time" (Wright 1931). This process has been discussed in chapter 12.

C. *Mass Selection under Changing Conditions.* "Changed conditions cause a usually slight and reversible shift of gene frequencies to new equilibrium points" (Wright 1929c).

On further consideration, it was recognized that the genetic changes would actually be largely irreversible, both because of the indefinitely great number of possible alleles at loci and because of the building up and tearing down of successive interaction systems, guided by the succession of environmental changes.

"Here we undoubtedly have an important evolutionary process and one that has been generally recognized" (Wright 1932b). This is the process stressed as most important of those considered in chapter 12.

D. *Pure Sampling Drift.* The differentiation among closely inbred lines of plants and animals, discussed in chapters 2 and 3, is the result of almost wholly random fixation, although even in such lines sufficiently deleterious genes cannot be fixed. The typical reduction in vigor in such lines is also a consequence of this process, resulting from the frequently deleterious nature of the genes that become fixed.

With respect to natural populations, Gulick (1872, 1905) was the first to attribute importance to the accumulation of statistical variations of heredity in bringing about the divergence of isolated portions of species. His conclusions were based on extensive collections of land snails (Achatinellidae) in the mountain valleys of Oahu in the Hawaiian Islands. He found that each valley, often each grove of trees, had its own characteristic type, differing from others in "nonutilitarian" respects. He attributed the divergence in part to accidental differences in the original colonists, but mainly to the separate accumulations of random variations, leading to different selective interactions of the heredities with the environment. He insisted on the absence of significant environmental differences among the localities. His studies came too early for any Mendelian interpretations, but his conclusions were qualitatively similar to those from the Mendelian interpretation of the effects of inbreeding.

Quantitatively, however, the situation in the large populations of mountain valleys is very different from that in inbred lines. I (Wright 1929c) took a very negative view in 1929 of the importance of pure sampling drift in natural populations, except for ones reduced to the verge of extinction: "In too small a population, there is little variation, little effect of selection, and thus a static condition, modified occasionally by chance fixation of a new mutation, leading to degeneration and extinction."

This evaluation was essentially repeated in many later papers. Nevertheless, the hypothesis that random fixation is a significant alternative to natural selection in phenotypic evolution has frequently been attributed to me (Fisher and Ford 1947, 1950; cf. Wright 1948, 1951*b*). I did ascribe an important role to random drift (not fixation) as an adjunct to natural selection, as will be discussed later, but this is a very different matter.

Assume that the alleles at a locus are grouped in a favorable class with selective advantages over an unfavorable class; that there is semidominance; and that u is the mutation rate from each class to the other. The frequency of the favorable class changes according to the following formula:

$$\Delta q = u(1 - 2q) + sq(1 - q), \quad \text{with} \quad \hat{q} = [(s - 2u) + \sqrt{(s^2 + 4u^2)}]/2s$$

or about $[0.5 + s/8u]$ if $2u \gg s$, $\sqrt{(0.5)}$ if $s = 2u$, and $1 - (u/s)(1 - u/s)$ if $s \gg 2u$.

$$\varphi(q) = Ce^{4Nsq}[q(1 - q)]^{4Nu - 1}$$

If $4Ns > 4Nu > 1$, there is near-fixation of the favorable class but largely random fixation of the alleles included within it. This is process B as far as the class as a whole is concerned, but random drift with respect to members with the same value of s.

If both $4Ns$ and $4Nu$ are less than 1, there is a U-shaped distribution of class frequencies with minor deviations from symmetry according to the term e^{4Nsq}. There is largely random fixation of all genes (process D).

If $4Ns$ is considerably greater than 1, but $4Nu$ less than 1, there is again a U-shaped distribution of class frequencies but strong bias toward higher frequencies of the favorable class.

If $4Nu$ is considerably greater but $4Ns$ less than 1, there is strong heterallelism.

These cases are illustrated in volume 2 in the first column of figure 13.6 for different values of $4Nu$ and s/u. These figures show the transition from U-shaped distributions with $4Nu = 0.1$, to ones with restricted random drift ($4Nu = 1$, $4Ns = 10$; $4Nu = 10$) and the shift from control largely by the equilibrium between the opposed mutation pressures (slightly above $\hat{q} = 0.5$) to control by the balance between favorable selection and opposed mutation (nearly $1 - u/s$). It is to be noted that the random drift can be considerable, extending over more than half of the range with appreciable frequencies even if both $4Nu$ and $4Ns$ are as large as 10. The second column in this figure shows similar distributions in the case of complete dominance of the favorable class.

While the distinction between processes B and D was originally made on the basis of size of population (N in relation to $1/4u$), it is evident that the relation of selective intensity to mutation rate and hence to size of population is also important. The same population, of course, may show evolutionary changes of types B and D at different loci.

While the rate of fixation of a particular neutral mutation varies inversely with the size of the population, being $1/2N$ (vol. 2, p. 347), the rate of fixation of mutations as a class, purely by chance, is independent of size of populations, because the number of genes subject to mutation varies directly with population size. This rate is thus simply the rate of recurrence v (Wright 1941c, vol. 2, p. 390).

Recently, studies of amino acid replacements in protein molecules, in comparisons among organisms of the most diverse sorts, have led authors who were well aware of the implausibility of pure sampling drift as a cause of phenotypic evolution (except in extremely small populations) to the conclusion that it has played a major role in these cases in which complete neutrality at many sites is plausible (Kimura 1968; King and Jukes 1969). This molecular evolution is a very different matter, however, from phenotypic evolution at the level of the organism, which we are here concerned with. It will be discussed in volume 4.

With respect to Gulick's observations, it seems probable that the populations of land snails that he studied were much too large to owe their divergence to any important extent to pure sampling drift. Gulick himself suggested an interaction with selection, which leads us to process E.

E. *Shifting Balance in Homogeneous Populations.* "With intermediate size of population, there is a continual shifting of gene frequencies and consequent alteration of all selection coefficients, leading to irreversible and largely fortuitous but not degenerative changes, even under static conditions. The absolute rate, however, is slow, being limited by mutation pressure" (Wright 1929c).

On further consideration, I took an even less favorable view: "The rate of progress is extremely slow since change of gene frequency is of the order of the reciprocal of the effective population size and this reciprocal must be of the order of the mutation rate in order to meet the conditions for this case" (Wright 1932b).

This process depends on the occasional crossing of a saddle in a multidimensional surface of fitness values leading from control by one peak to control by a higher one. The stochastic distribution for a particular gene in the absence of frequency dependence and under specified condition in the rest of the genome and the environment is

$$\varphi(q) = C\overline{W}^{2N}q^{4Nv-1}(1-q)^{4Nu-1} \quad \text{where} \quad \int_0^1 \varphi(q)\, dq = 1.$$

Here $\overline{W}^{2N}$ represents a multidimension surface of mean selective values relative to the frequencies of genes at all loci in an interaction system. There must be numerous peaks, saddles, and pits in this surface to provide the basis for a continuing Darwinian process in which the random drift in the entire set provides the random element and mass selection of individuals or families, following the accidental crossing of a two-locus saddle, carries the set a step farther in adaptation.

The conditions are intermediate between those for processes B and D. The factor $4Nu$ cannot be much greater than 1 in order that there may be adequate random drift, and cannot be so much less than 1 as to lead to near-fixation. The term $\overline{W}^{2N}$ can be written approximately $e^{2(\overline{W}-1)N}$. Writing $\bar{s} = \overline{W} - 1$ for the mean selective advantage of the gene in question, at gene frequency q (with given frequencies at all other loci), it may be seen that $2\bar{s}N$ also cannot differ much from 1 if fixation is to be avoided ($2\bar{s}N \approx 1 \approx 4\overline{N}u$). Thus, if $u = 10^{-5}$, process E can occur in populations of the order of 25,000. If u is actually larger, as suggested from observations on the persistence of clones and supposedly isogenic lines (chap. 5), process E can occur only in populations of correspondingly smaller size.

There must be rather extensive random drift to given an appreciable chance for the crossing of any particular two-locus saddle. On the other hand, if a large number of loci are involved in interaction systems, there are an enormous number of two-locus saddles that may be crossed.

The most favorable case for a large number of selective peaks is, as often noted, that of a quantitatively varying character affected by many loci with more or less equivalent effects, and the optimum not far from the mean. For continual progress, the optima must differ somewhat in mean selective value because of slight pleiotropic effects. There can be many evolutionary steps without any novel mutation, and progress can continue indefinitely without changes in condition from the array of new sets provided by each novel mutation. Progress is thus not sharply limited by the rate of mutation, contrary to the early statement. Nevertheless it would be a very slow process, even in populations within the narrow size range in which it can occur at all. Very rarely, the crossing of a shallow saddle may bring the set of gene frequencies under control of a peak characterized by selective values several orders of magnitude higher than previously and lead to a rapid advance.

Random drift may be due to fluctuations in the selection coefficients (Wright 1931, 1940, 1948), as well as to accidents of sampling, in which case the process is favored by large population size. The most significant sort is probably that due to fluctuations in the positions of the optima of the quantitatively varying characters (Wright 1935b). Following each sufficient shift in an optimum, the set of gene frequencies moves toward one or another of the many equilibrium sets at the new optimum, under mass selection. Fluctuations back and forth between optima should shuffle the set toward control by the highest selective peak for each character more effectively than the sampling drift previously considered. Since only mass selection is involved, this process might be looked on in a sense as coming under process C, with environmental fluctuations rather than uniquely changing environments. On the other hand, the multiple interaction systems with more or less similar selective peaks at a given moment provide a large random element, lacking where process C consists in a succession of directional selections based on mutations and polymorphisms.

A shift to control of variability by a new selective peak may also come about from a unique event that, having no variance, cannot be represented by a stochastic distribution. Thus a large population may go through a bottleneck of extremely small size, during which large random changes occur in the gene frequencies at most heterallelic loci and the population emerges with these under control of a new selective peak, which in general gives less adaptation than before. This effect, in the case in which a colony is started by a few stray members of a species, is one to which Gulick attracted importance. This has been called the "founder principle" by Mayr (1942). It is an extreme sampling effect, but not so extreme as where there is a succession of such bottlenecks in a region in which colonies are continually being started, becoming extinct, and being restarted by stray individuals from another such colony (Wright 1940). This process is especially likely to lead to favorable peak shifts, contrary to the preceding process.

A unique environmental change may also cause a shift to control by a new selective peak. There may be a succession of unpredictable unique environmental changes at sufficient intervals to permit some selective adjustment at each. This, however, is identical with process C in cases in which mass selection at each step tends to be directed toward an intermediate optimum instead of toward an extreme.

F. *Shifting Balance in a Structured Species.* "Finally, in a large but subdivided population, there is a continually shifting differentiation among the local races, even under static conditions, which through intergroup selection brings about indefinitely continuing, irreversible, adaptive and much more rapid evolution of the species as a whole" (Wright 1929c).

This was the evolutionary process advocated as much the most important in the 1929 and 1931 papers. Since 1932, I have treated both processes C and F as of major importance and the combination of subdivision into demes (F), associated with changes in conditions (C), as most conducive to rapid evolution.

The effects of processes A to F were illustrated in the 1932 paper by diagrams, shown here as figure 13.1. In these, a small portion of the surface of selective values was given token representation by contours.

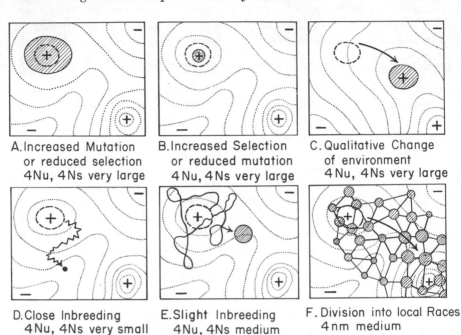

A. Increased Mutation
or reduced selection
4Nu, 4Ns very large

B. Increased Selection
or reduced mutation
4Nu, 4Ns very large

C. Qualitative Change
of environment
4Nu, 4Ns very large

D. Close Inbreeding
4Nu, 4Ns very small

E. Slight Inbreeding
4Nu, 4Ns medium

F. Division into local Races
4nm medium

FIG. 13.1. Hypothetical multidimensional field of gene combinations (represented by two dimensions), with fitness contours. Field initially occupied by a population indicated by heavy broken contour. Field occupied later indicated by crosshatched area (multiple subpopulations in F). Courses indicated in C, D, E, and F by arrows. Effective population numbers: N (total), n (local); coefficients: u (mutation), s (selection), m (immigration). Redrawn from Wright (1932b, fig. 4).

Misunderstandings

Many authors have bracketed the theories proposed by Haldane, Fisher, and myself around 1930 as essentially the same and as merely a mathematical formulation of the "classical" theory of which Morgan was the leading proponent. The most extensive recent critique of these theories along this line was that of Mayr (1959) (cf. Wright 1960c; Haldane 1964), Mayr's viewpoint is indicated by the following quotation:

> In order to permit mathematical treatment, numerous simplifying assumptions had to be made such as that of an absolute selective value of a given gene....

This period was one of gross oversimplification. Evolutionary change was essentially presented as an input or output of genes, as the adding of certain beans to a bean bag and the withdrawal of others.

The last sentence applies to the "classical" theory. Simple cases were expressed in mathematical form by Castle (1903) and Norton (in Punnett 1915), the implications of whose tables were developed by Chetverikov (1926) and much elaborated by Haldane (1924 and later). This needed doing, and Haldane (1964) has given an adequate "defense of bean bag genetics."

There was, however, much more to Haldane's theory, which, contrary to the first sentence in the above quotation from Mayr's paper, included mathematical studies of factor interaction and frequency dependence. The term *bean bag genetics* hardly applies at all to Fisher's "fundamental theorem," which was concerned with the relation between rate of evolution and the amount of additive genetic variance, which he interpreted as due largely to the minor factors of quantitative variability. Fisher also did not ignore factor interaction in showing that the rate of evolution in a homogeneous population with a given genetic variance is reduced proportionately by the amount of the latter contributed by nonadditive interactions. Finally, Mayr was obviously completely unaware of the nature of my shifting balance theory under which the significant steps in evolution are shifts from control by one interaction system to control by a superior one, rather than addition or removal of single genes.

As noted earlier, the hypothesis that random drift is a significant alternative to natural selection in the phenotypic evolution of species has very frequently been erroneously attributed to me.

Most of the misrepresentations of this sort were probably merely repetitions of earlier ones, mistakenly supposed to be based on study of the original papers.

Because of these widespread misrepresentations, it is desirable to go rather explicitly into the special premises of the shifting balance theory, as applied to subdivided populations (Wright 1970).

1. *A Very Large Amount of Heterallelism with Respect to Minor Factors.* The latter are in the main the multiple factors responsible for the quantitative variability of characters. This is a necessary premise since the effectiveness of process F depends on there being widely ranging stochastic distributions at a great many loci, and this occurs only if the selective differences at these loci are very small. Utilization of major mutations in evolution is by no means excluded, but these are usually so strongly selected against at first that establishment ordinarily requires an adaptive process among minor modifiers.

2. *Pleiotropy in the Effects of Most Allelic Differences.* The usual existence

of a complex network of metabolic and intercellular developmental processes between the primary reactions of gene products and the character differences that are the objects of selection insures such pleiotropy. This is a necessary premise for evolutionary progress under process F in the important class of cases of quantitative variability with intermediate optima.

3. *Multiple Fitness Peaks in the Field of Genotypic Frequencies.* The occurrence of joint reactions in the above networks implies extensive interactions in effects of loci on characters. In many cases, all characters contribute to a single definable character, fitness, which has in general numerous potential peak values, relative to the multidimensional field of gene frequencies, each corresponding to a different harmonious interaction system. If there is no definable fitness function, there may be numerous discrete goals of selection relative to this field, determining alternative evolutionary paths. These contrast with the single best genotype that would necessarily be present if each gene substitution were favorable or unfavorable in itself. Moreover, even if there is additivity with respect to a quantitatively varying character, the usual intermediacy of the optimum grade insures the potentiality for a great many selective peaks and pleiotropy insures that these be at different levels.

4. *Multiple Partially Isolated Demes.* Most species include many randomly breeding local populations (demes) that are sufficiently isolated, if only by distance and slowness of diffusion, to permit differentiation of their sets of gene frequencies, but that are not so isolated as to prevent the gradual spreading of favorable gene complexes throughout the species from their centers of origin. There may be great differences in gene frequencies without conspicuous phenotypic differences.

Phases in the Shifting Balance Process of Evolution

Change of gene frequency has been treated in this treatise as the elementary evolutionary process since it permits reduction of the effects of all factors to a common basis.

Where, however, conditions are favorable for the shifting balance process there are significant evolutionary steps that consist in the shifting of control over changes in the gene frequencies from one fitness peak to a higher one. In principle the process may continue indefinitely, step by step, with little or no novel mutation and without necessarily involving any environmental differences among localities or any systematic environmental changes in time. In the first statement of the three phases involved in each step, no such differences or changes are assumed. The first two phases apply to process E as well as to F.

1. *Phase of Random Drift.* In each deme, the set of gene frequencies drifts at random in a multidimensional stochastic distribution about the equilibrium set characteristic of a particular fitness peak or goal. The set of equilibrium values is the resultant of three sorts of pressures on the gene frequencies: those due to recurrent mutation, to recurrent immigration from other demes, and to selection. The fluctuations in the gene frequencies responsible for the stochastic distribution (or random drift) may be due to accidents of sampling or to fluctuations in the coefficients measuring the various pressures.

2. *Phase of Mass Selection.* From time to time, the set of gene frequencies drifts far enough to cross one of the innumerable two-factor saddles in the surface of fitness values, in one of the demes. There ensues a period of relatively rapid change in this deme, dominated by selection among individuals (or families) until the set approaches the equilibrium associated with the newly controlling fitness peak, about which it now drifts at random and thus returns to the first phase, but in general at a higher level.

3. *Phase of Interdeme Selection.* A deme in which the set of gene frequencies comes under control of a fitness peak superior to those controlling the sets at neighboring demes tends to produce a greater surplus population and, by excess dispersion, systematically shifts the position of equilibrium of these toward its own position until the same saddle is crossed in them, and they all move autonomously to control by the same fitness peak. This process tends to spread through the species in concentric circles. Two such circles, spreading from different centers, may overlap and give rise to a new center that combines the two different favorable interaction systems and becomes a still more active population source. The virtually infinite field of interaction systems may be explored in this way with only a small number of novel mutations, as alleles which had been rare come in time to displace the previously more abundant ones.

A Model with Six Homallelic Selective Peaks

The interplay of random drift and selection within a deme is illustrated in a simple model in figures 13.2–13.5 (Wright 1964*b*), In figure 13.2 the grade of a character is assumed to be determined by four equivalent, semidominant genes (A, B, C, and D) with the optimum grade at the midvalue. It is further assumed that genes A and B have small favorable semidominant pleiotropic effects. Taking account of these, there are six selective peaks, all homallelic, at three different levels as shown in figure 13.3 The lowest is homallelic in $abCD$ ($\overline{W} = 1.000$), there are four at an intermediate level ($aBCd$, $AbCd$,

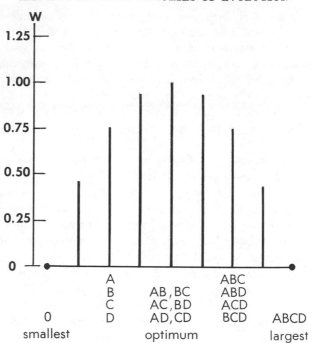

FIG. 13.2. Contributions of the combinations of four pairs of alleles, based on their additive effects on a quantitatively varying character that has an intermediate optimum. The 16 kinds of homozygotes are designated by the plus factors, *A*, *B*, *C*, *D*, which they carry (*AB* representing *AABBccdd*). Redrawn from Wright (1964*b*, fig. 8).

aBcD, and *AbcD*, all with $\overline{W} = 1.125$), and one at the highest level (*ABcd*, $\overline{W} = 1.250$). These are all separated by at least two gene replacements.

Any set of frequencies at these loci can be represented by a point in a four-dimensional space, one dimension for the frequencies at each locus. Figure 13.4 represents two surfaces in this space. One connects the lowest peak, homallelic *abCD* ($\overline{W} = 1.000$), with one of the four intermediate peaks, homallelic *aBCd* ($\overline{W} = 1.125$). There is a saddle at $q_B = 1/3$, $q_d = 1/6$ with $\overline{W} = 0.990$. The other surface connects this intermediate peak with the highest one, homallelic *ABcd*, through a saddle at $q_A = 1/3$, $q_c = 1/6$, with $\overline{W} = 1.115$. These calculations are based on the selective values in figure 13.2 and 13.3 and equation 4.75 in volume 2.

The directions in which sets of gene frequencies tend to move under selection are indicated by arrows. It is assumed that mutation pressure (at first) and immigration pressure in general, prevent any from being completely

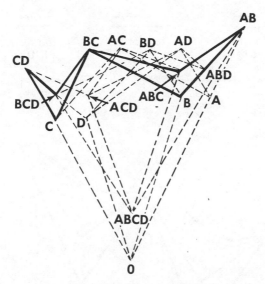

FIG. 13.3. Total selective values of the 16 genotypes of figure 13.2 on the basis of the contributions indicated there, supplemented by equal semidominant pleiotropic effects of genes A and B. Redrawn from Wright (1964b, fig. 9).

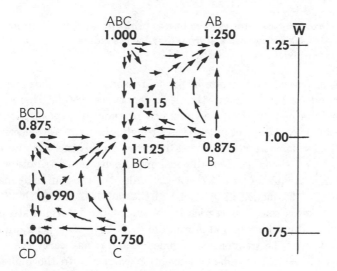

FIG. 13.4. Two of the surfaces in the four-dimensional space of figure 13.3, that connecting the lowest peak, $CD(= ccddCCDD)$, to one of the four intermediate ones, $BC(= aaBBCCdd)$, and that connecting the latter to the highest peak $(AB = AABBccdd)$. Trajectories are indicated by arrows. The selective values at peaks, pits, and saddles are given. Redrawn from Wright (1964b, fig. 10).

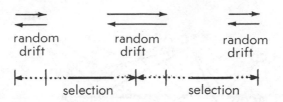

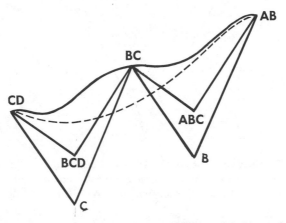

Fig. 13.5. Profile of trajectories from pits to peaks (*straight lines*), from two-locus saddles to peaks (*solid curves*), and from four-locus saddle to peaks (*broken curve*). Redrawn from Wright (1964*b*, fig. 11).

fixed. It is also assumed that, for historical reasons, the population is initially characterized by the lowest of the peak genotypes, *abCD*. There is opportunity for improvement from the assumed presence of genes *A*, *B*, *c*, and *d* at low frequencies, but a set must pass through a lower mean selective value than 1.000 by some random process in order to advance. If the effective population size is small, a four-dimensional stochastic distribution extends from the high frequency at *abCD*. Occasionally the random drift may carry the set of gene frequencies across one of the saddles, each with $\overline{W}$ at 0.99, after which mass selection carries it to one of the intermediate peaks at $\overline{W} = 1.125$. After a long period of random drift about this peak, an extreme joint deviation of the frequencies of genes *A* and *c* may carry the set across the saddle ($\overline{W} = 1.115$) and mass selection carries it to the highest peak, homallelic *ABcd*, $\overline{W} = 1.250$.

Figure 13.5 shows in profile the selective values along this path from lowest to highest peak, and illustrates the alternation of the long periods of random drift (phase 1) and of rapid mass selection (phase 2). As shown, the two-locus

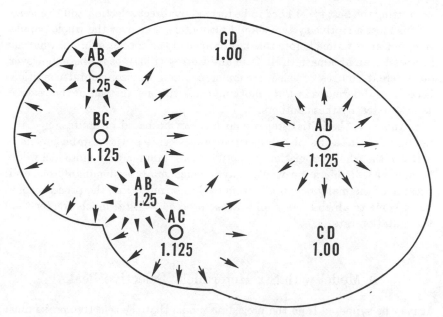

FIG. 13.6. Diagram of a region characterized initially by the lowest peak, *CD*, in which intermediate peaks *BC*, *AC*, and *AD* (but not *BD*) have been arrived at locally and the highest peak, *AB*, has been arrived at from *BC*, and from overlap of *BC* and *AC*. Redrawn from Wright (1964*b*, fig. 12).

saddles are considerably less depressed than the four-locus saddle ($\overline{W} =$ 0.973).

These figures, restricted to one deme, do not represent the third phase of process F. This is done in figure 13.6, representing an area occupied by the species, assumed to include a great many demes, all of type *abCD* initially. At first there is only a pattern of trivial local differences due to random drifting of the gene frequencies in the demes separately. These are indicated as crossing two-factor saddles, leading to control by one or another of the intermediate selective peaks (*aBCd*, *AbcD*, and *AbCd* in fig. 13.5). Excess population growth and dispersion at these demes tends to pull the sets of gene frequencies of neighboring ones toward the same levels until the same saddles are crossed and progress becomes autonomous.

The influence of the first of these (*aBCd*) is represented as having spread the furthest and as having given rise to a deme that has reached the highest level (*ABcd*) and has spread concentrically a short distance.

A different process is represented where the spreading circles of influence of the demes that reached the peaks *aBCd* and *AbcD* have come to overlap, permitting the best type *ABcd* to be formed by direct selection and to spread.

This most adaptive system should ultimately take over the whole population, but at any time before this has happened the species exhibits what has been called an area pattern by Cain and Currey (1963), based here, however, on different courses of selection under common environmental conditions because of differences in the genotypic constitutions, instead of the reverse, as postulated by these authors.

In this case the initial step would have to be based on mutation. Later immigration takes its place until the process ceases with establishment of *ABcd* in the whole population. In actual cases, innumerably more loci would be involved, immigration would play a much more predominant role, and, because of the many new peaks from each new mutation, the process would never come to an end but would rather proceed at rates much greater than the mutation rate.

A Model with Six Heterallelic Selective Peaks

It may be supposed from the preceding model that the selective peaks must be homallelic. We will consider next a model in which the six selective peaks are all heterallelic (Wright 1965a). This again is a case in which selective values fall off as the square of the deviation from an optimum, but there are only three loci, equivalent, with semidominance, and optimum with three plus factors, apart from pleiotropic effects (table 13.1).

TABLE 13.1. Basic selective values of genotypes according to number of equivalent, semidominant plus factors with optimum 3, apart from indicated pleiotropic effects.

Plus Factors	Basic Selective Value	Pleiotropic Effect
6	$1 - 9s = 0.91$	A 0.005
5	$1 - 4s = 0.96$	B 0.003
4	$1 - s = 0.99$	C 0.002
3	$1 = 1.00$	
2	$1 - s = 0.99$	
1	$1 - 4s = 0.96$	
0	$1 - 9s = 0.91$	

SOURCE: Reprinted, with permission, from Wright 1965a.

Genotype $AABBCC$, with basic selective value 0.910, has selective value 0.930 on allowing for pleiotropic effects.

It is assumed that the ratio of s to c (here amount of recombination) is so small that random combination may be assumed (if $c = 0.5$, $s/c = 0.02$). Letting p, q, and r be the frequencies of genes A, B, and C, respectively, $\overline{W} = 1 - 0.01\{2p(1-p) + 2q(1-q) + 2r(1-r) + 2[(p+q+r) - 3]^2 + 0.010p + 0.006q + 0.004r\}$.

TABLE 13.2. Coordinates p, q, and r, and mean selective values, $\overline{W}$, of pits $(-)$, peaks $(+)$, and saddles (corner, S_0; two-factor, S_2, and three-factor, S_3) in the interaction system of table 13.1.

	Genotype			p	q	r	$\overline{W}$
$\cdots$	aa	bb	cc	0	0	0	0.910
$\cdots$	AA	BB	CC	1	1	1	0.930
S_0	aa	bb	CC	0	0	1	0.994
S_0	aa	BB	cc	0	1	0	0.996
S_0	AA	bb	cc	1	0	0	1.000
S_0	aa	BB	CC	0	1	1	1.000
S_0	AA	bb	CC	1	0	1	1.004
S_0	AA	BB	cc	1	1	0	1.006
S_3	A, a	B, b	C, c	0.450	0.550	0.600	0.99510
$+$	aa	B, b	CC	0	0.650	1	1.00245
S_2	aa	B, b	C, c	0	0.850	0.900	1.00185
$+$	aa	BB	C, c	0	1	0.600	1.00320
S_2	A, a	BB	C, c	0.150	1	0.300	1.00185
$+$	A, a	BB	cc	0.750	1	0	1.00725
S_2	A, a	B, b	cc	0.850	0.950	0	1.00710
$+$	AA	B, b	cc	1	0.650	0	1.00845
S_2	AA	B, b	C, c	1	0.183	0.233	1.00518
$+$	AA	bb	C, c	1	0	0.600	1.00720
S_2	A, a	bb	C, c	0.817	0	0.967	1.00518
$+$	A, a	bb	CC	0.750	0	1	1.00525
S_2	A, a	B, b	CC	0.183	0.283	1	1.00043

SOURCE: Reprinted, with permission, from Wright 1965a.

The trajectories are derived from the approximate equation $q_i = 0.5q_i(1 - q_i)\, \partial\overline{W}/\partial q_i$. The equilibrium points are those at which all Δq's are zero. Table 13.2 gives the coordinates p, q, and r and the values of $\overline{W}$ at the various selective pits $(-)$, saddles (S_0, S_2, S_3), and peaks $(+)$. Those for the two-factor saddles (S_2) are put between the adjacent peaks. S_3 is the triple saddle. The locations of these points are shown in figure 13.7.

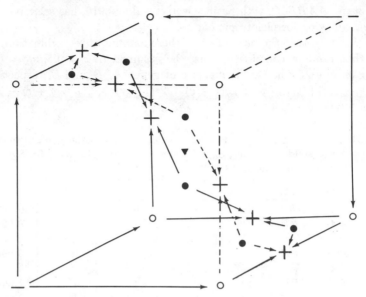

Fig. 13.7. Coordinate system (*solid lines in foreground, broken lines in background*) for three pairs of alleles, showing locations of peaks (+), pits (−), corner saddles (*open circles*), two-locus saddles (*solid circles*), and the three-locus saddle (*triangle*), of the interaction system of tables 13.1 and 13.2. Redrawn from Wright (1965*a*, fig. 3); used with permission.

With linkage there would be no changes in the values for the pits (−), the other corners (S_0), and the peaks (+), but the saddles between the peaks (S_2) and the triple saddle (S_3) would be shallower.

Figure 13.8 shows how a species that had arrived at a system controlled by the lowest peak (*aaBbCC*, $\overline{W} = 1.00245$) might progress by either of two routes toward the highest peak (*AABbcc*, $\overline{W} = 1.00845$). It is assumed that there are a great many partially isolated demes that become differentiated by random drift. There is much more chance that the set of frequencies will cross the shallow saddle (−0.00060) in the route indicated by the solid line than across the deeper saddle (−0.00202), indicated by the broken line. If, however, a deme happens to be successful along this second route, the ensuing progress by selection will be much greater and is much more likely to continue for another step (to *AAbbCc*) than is the first route (which is likely to stop at *aaBBCc*). If both sets are attained in different demes and each spreads concentrically until they are in contact, inbreeding provides the material for direct selection toward the highest peak (*AABbcc*, $\overline{W} = 1.00845$) without the necessity for any further random differentiation.

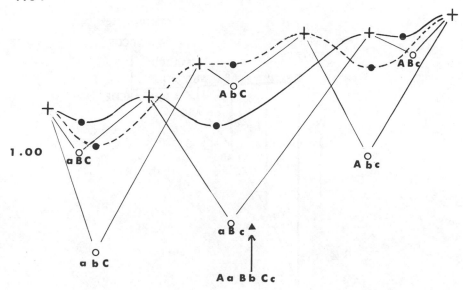

W

1.01

1.00

0.99

a B C

a b C

A b C

a B c

A a Bb Cc

A b c

A B c

FIG. 13.8. Profile of trajectories connecting pits, saddles, and peaks of the system in figure 13.7. Redrawn from Wright (1965a, fig. 4); used with permission.

A Model for the Fixation of a Major Mutation

Mutations with major effects are almost certain to have seriously deleterious net effects and thus to be selected against so strongly that they are kept at very low frequencies by the balance between the pressures of selection and recurrent mutation. The effect may, however, give the basis for a major advance if deleterious side effects can be removed by a suitable combination of modifiers (Wright 1941b, 1963a,b). There is a possibility that random drift of the frequencies of these modifiers occurring independently in many demes may, in some deme at some time, lead to a combination that neutralizes the deleterious effects of the major mutation, present at a low frequency. The latter will then rise in frequency and give the deme in which this occurs such an advantage that the whole complex, major mutation and array of modifiers, gradually spreads through the whole species by interdeme selection. This situation is illustrated in figure 13.9.

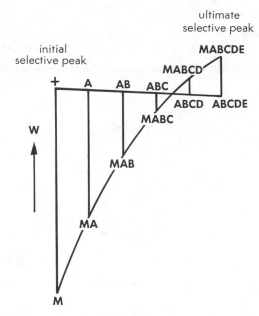

FIG. 13.9. Combinations of an initially deleterious mutation, M, with modifiers, A, B, C, D, and E, which in cases $MABCD$ and $MABCDE$ are superior to any combination that lacks the mutant. Redrawn from Wright (1964b, fig. 13).

Effectiveness of Process F

In process E the store of genetic variability depends primarily on the balance between adverse selection and recurrent mutation. As noted, the selection pressure cannot be much greater than the mutation pressure to permit progress by this process. A store may be maintained by the various mechanisms for polymorphisms: selective advantage of rarity or diversity, heterotic alleles, and so forth. Again, the selective pressures that maintain these must be weak in order to permit the amount of random drift required for process E. In either of these cases, the genes must be involved in interaction systems with multiple fitness peaks. These relations and the narrow range of population sizes ($N \approx 1/4u$) that permit shifting balance are so severe that this process can be of little or no importance of itself.

In process F, on the other hand, the store of genetic variability is kept large by the favoring of different alleles in different places (either because of differences in interaction systems or environments) and local stores are also kept large by the balancing of local selection pressures by immigration from neighboring demes.

The process resulting from balancing of sampling drift by natural selection within each deme is similar in principle to that in the whole population in process E, but it would be enormously more rapid. This is because immigration pressure, coefficients mQ and $m(1 - Q)$, take the place of v and u, respectively, in the equation but tend to be several orders of magnitude larger on the average and may be of any amount. The condition that $N \approx 1/4mQ$ implies that the effective size of demes should be several orders of magnitude less that of the whole population in process E and the conditions $s \approx 1/2N$ implies only moderately small, instead of almost impossibly small, selection coefficients for the most rapid evolution by this process. Again the absence of any narrow restriction on the value of m and, hence on N and s, is favorable. Evolution under process F should proceed thousands of times as rapidly in each deme as under process E and, since it may proceed more or less independently in thousands of demes, the chance of making a step from control of one fitness peak to control by another is millions of times as great as under the most favorable assumptions for process E.

The most serious criticisms of the theory have been that local populations are in general not small enough to give a basis for appreciable sampling drift, or sufficiently isolated for enough differentiation of their genetic systems to give an adequate basis for interpopulation selection (Simpson 1953; Mayr 1963). There is theoretically a uniform stochastic distribution of gene frequencies from $q = 0$ to $q = 1$ if $Nm = 0.5$ (vol. 2, p. 363) and one with about half this range if $Nm = 5$. Taken literally, the value $Nm = 0.5$ means only one immigrant every other generation and $Nm = 5$ means five immigrants per generation into the deme in question. These figures seem impossibly small, but it must be recalled that both N and m refer to effective values that, for reasons discussed in volume 2, chapter 8, are both usually much smaller than census figures. It has been shown that there may be a large amount of random differentiation of neighborhoods in a two-dimensional continuum and much more in a one-dimensional one. Theoretically there is no difficulty here.

The question whether there actually is differentiation among demes because of differences in condition of selection (m and s of the same order) or because of acquirement of different interaction systems under sampling drift and local mass selection in spite of identical environments will be discussed in volume 4.

The two kinds of differentiation are not easy to distinguish and are apt to be combined in nature. Sampling drift, if accepted at all, is apt to be discussed as a factor to be considered only in accounting for trivial differences in gene frequencies among localities. Thus Cain and Currey (1963), on finding striking difference among extensive populations of the snail, *Cepaea nemoralis*, for which there was no obvious environmental explanation (for example, protective coloration), stated:

These area effects are too homogeneous over large areas containing very large numbers of individuals to be due to sampling drift or any other random process acting at present.

They concluded that the cause must be some subtle climatic or other environmental difference. This may well be correct, and such differences should certainly be thoroughly looked for, but this conclusion cannot be established by elimination. The color factors in this case may conceivably enter into interaction systems that have become differentiated by the process in which random drift acts as a trigger to bring about passage from one interaction system to another, in the absence of any environmental differences.

The illustrations given above were in terms of sampling drift, but it must not be overlooked that this is not the only sort of random drift. Fluctuations in the position of an optimum, taking place over a large area, may, for example, result in the chance acquirement of different interaction systems in different places with consequent "area effects," even though there are no contemporary environmental differences at any time. It is not difficult to establish the existence of widespread differentiation among local populations of large species, but it is usually very difficult to make a trustworthy evaluation of the roles of differences in environment and differences arising from the interplay of random processes and selection.

The spreading of interaction systems throughout species by differential population growth and diffusion among demes in most cases can only be inferred, but there is historical evidence at least of differential migration in the case of mankind.

In these illustrations, it has been assumed that the intensity of genotype selection is constant so that fitness, $F(W/\overline{W})$, is the same as mean selective value, $\overline{W}$. Illustrations could be given in which this is not the case, so that the peaks and saddles concerned in phases 1 and 2 are those of the surface of fitness values. With respect to phase 3, however, the relative rates of population growth, measured by mean absolute selective value, $\overline{W}$, and capacity to migrate successfully, determine the influence of local populations on the course of evolution of the species as a whole. Where $F(W/\overline{W})$, $\overline{W}$, and migratory capacities differ, the course of evolution of the species as a whole is determined by some sort of a compromise.

Evolution of Clonal Mixtures

Where there is exclusive uniparental reproduction, the only random source of genetic variability is mutation. There is no amplification by recombination.

This phase of the evolutionary process is thus not nearly as rich as under biparental reproduction.

On the other hand, with a given amount of genetic variability, mass selection is enormously more effective among clones than in a panmictic population because it is wholly according to the genotype as a whole in the former, merely according to the net effects of the separate genes (with only minor qualifications from linkage) in the latter.

Mass selection of clones in a region exposed throughout to the same conditions should soon reduce the array to a single clone. Further evolution must wait on mutation, and there is no chance, except from mutations, to readapt to a change in environment. On the other hand, different clones are expected to be arrived at in different environments, preventing reduction of any widely ranging species to a single clone.

If there is a means of widespread dispersal, such a species may adapt itself with maximum rapidity to any local opportunity.

With respect to evolution of the species as an enormous array of similar clones, mutation and differential selection according to local conditions, both balanced by widespread diffusion, provide the random principle, while selection according to general adaptiveness provides the guiding principle.

An early comparison of the relative advantages and disadvantages of uniparental and biparental reproduction was made by Muller (1932b). More recent comparisons have been made by Crow and Kimura (1965a, 1969) and by Maynard Smith (1968b), with different emphases.

The capacities for rapid local adaptation and for evolution of the species as a whole are greatly enhanced if prevailing uniparental reproduction is interrupted occasionally by crossing and recombination. This combines the advantage of selection according to the genotype as a whole, characteristic of uniparental reproduction, and that of a large and continually renewed field of variability, characteristic of biparental reproduction, and thus may be looked on as an extreme form of process F (Wright 1931). The differences in the genotypes favored in different localities prevents reduction to a single clone by selection.

It may appear that this should be a more effective mechanism than fine-scaled subdivision of a large diploid species. Judging from the results, this has clearly not been the case. Biparental reproduction is the prevailing mode throughout the groups that have evolved the most, such as the vertebrates, the insects, and the angiosperms.

Uniparental reproduction with occasional crossing is a violent process in which the balance between very rapid genotypic selection, and recombination, on which progress depends, is continually tending to be destroyed. It is a process that lends itself better to a succession of readaptations to violent changes in local conditions of the treadmill sort than to the orderly evolution of the species as a whole.

Ecologic Opportunity

We have been concerned in this and the preceding chapter with ways in which evolutionary processes may bring about the enormous changes and the extraordinarily detailed adaptations that have occurred, probably in much less than a billion years of eukaryotic evolution. There is, however, no necessity that there be continuing change in all cases. Some forms appear hardly to have changed at all since the early Cambrian some half billion years ago; others have progressed rather steadily along relatively narrow lines; and still others have evolved with relatively explosive speed and often have branched in many directions to give rise to new, higher categories (bradytely, horotely, and tachytely of Simpson [1944]).

The limiting factor is, undoubtedly, ecologic opportunity. No matter how favorable the population structure, if the ecologic pressure of other species is such as to restrict the one in question to a narrow niche in which the only opportunity is for gradual perfection of the adaptations it has already developed, evolution can be very slow.

If, however, the species is presented with a major ecologic opportunity, such as may arise if the species is among the few survivors of a devastating change of conditions to which it happens to be somewhat preadapted, or if it is the first of its general kind to reach unoccupied territory, or if a slowly perfected succession of special adaptations happens at a certain stage to open up an extensive new way of life, attainment of a favorable population structure permits exploration of the opportunity at a very rapid rate in terms of geologic time (Wright 1942, 1949a,b).

Conclusions on Theories of Evolution

The first conclusion is that there are many valid theoretical modes of evolution. Evolution under exclusive uniparental reproduction or under such reproduction predominantly, but with occasional crossbreeding, or under nearly exclusive biparental reproduction are processes with very different consequences. There are also very different processes within the last category. There are diverse kinds of genetic material and diverse kinds of selection.

The first evolutionary hypothesis associated with modern genetics was de Vries' mutation theory, supported later by Willis, Goldschmidt, and Schindewolf. This postulated a kind of mutation that produces at the same time reproductive isolation and major phenotypic change. This hypothesis has been found to have only limited validity. The best examples are the many amphidiploid species of plants. Other chromosome mutations either give reproductive isolation with little phenotypic change (translocations) or ones that have no isolating effect and behave essentially like gene mutations (duplications and deletions) but that add or subtract from the amount of available genetic material. Inversions form a special class with important evolutionary possibilities but no isolating effect. There seem to be few if any gene mutations that may be considered as species-forming by themselves.

Discussion of the relative evolutionary importance of conspicuous variants and of ordinary quantitative variability goes back to Darwin, who favored the latter. This discussion has continued in terms of the major, demonstrably Mendelian gene mutations as opposed to quantitative variability now interpreted as polygenic Mendelian for statistical reasons. Closely associated are alternatives with respect to amounts of heterallelism, little or much, respectively.

The evolutionary views of Castle and East were based on the universal occurrence of quantitative variability, supposed by Castle to be due to multiple alleles until a crucial experiment led to his acceptance of the multiple factor theory for which East had obtained abundant evidence. Under this view, most loci are heterallelic.

Those working with *Drosophila*, led by Morgan, adopted what has been called the classical theory under which the loci in wild species are supposed to be homallelic with respect to "wild-type" genes, except for rare deleterious mutations of no evolutionary significance, and much rarer favorable ones, that occasionally replace previous wild-type genes by mass selection. The mathematical theory for this process under a great diversity of conditions has been due principally to Haldane, who, however, did not restrict himself to it.

As Castle's assistant, I was much impressed by the success of his selection experiments, as well as by East's results. Shortly thereafter, as a staff member of the Animal Husbandry Division of the U.S. Bureau of Animal Industry, I attempted to formulate the roles of inbreeding and artificial selection in the improvement of breeds of livestock (Wright 1922c). Later I applied my conclusions to evolution in nature (Wright 1929c, 1931, 1932b). Among the conclusions of the 1931 paper was the statement

Evolution as a process of cumulative change, depends on a proper balance of the conditions which at each level of organization—gene, chromosome, cell, individual, local race—make for genetic homogeneity or genetic heterogeneity of the species.

This was followed by a brief review of the expected consequences of various postulates, ending in the statement

Finally in a large population, divided and subdivided into partially isolated local races of small size, there is a continually shifting differentiation among the latter (intensified by local differences in selection, but occurring under uniform and static conditions) which inevitably brings about an indefinitely continuing, irreversible, adaptive and much more rapid evolution of the species.

The final sentence of the paper affirmed that "conditions in nature are often such as to bring about the state of poise among opposing tendencies on which the indefinitely continuing evolutionary process depends."

This paper was obviously opposed to the classical theory, not in support of it (the "bean bag theory") as stated by Mayr (1959). It was a theory of evolution as a continually shifting balance. It is distinguished from later "balance" theories in being based primarily on the universal heterallelism of quantitative variability instead of the more restricted inversion polymorphisms of Dobzhansky or the sporadic, conspicuous polymorphisms of Ford, and especially in depending on shifting states of balance among all factors instead of merely on states of balance of chromosome arrangements or single loci.

The primary cleavage adopted in this volume has been between modes of evolution possible in large, essentially panmictic species (chap. 12) and those occurring either in small to moderately large populations or ones of any size but so structured as to be far from panmictic (chap. 13).

In large panmictic species, the material for selection is largely restricted to the net effects of alleles. Combinations are broken up too rapidly to permit effective selection among interaction systems (unless the loci are so closely linked that alleles at different loci behave almost as if alleles of each other). Natural selection is restricted to differences among individuals, or groups of close relatives.

The progress of mass selection under a constant set of conditions is limited by the extremely rare occurrence of novel favorable mutations. Under change of conditions, at a tempo with which mass selection can keep step, there is, on the other hand, rapid, indefinitely continuing evolutionary change, but change largely of the treadmill type in which continual readaptation is accompanied by the breaking down at previous adaptations.

The fixation of a gene by mass selection tends, however, to be followed by a succession of favorable modifiers that build up a favorable interaction system.

Turning to the processes discussed in this chapter, no evolutionary significance was attributed in the 1931 paper to the extensive sampling drift (inbreeding effect) expected in species with very small populations, other than probable extinction.

In a homogeneous species of such a size that there is little likelihood of fixation of appreciably deleterious genes, there is theoretically a zone of balance between the tendency toward wide stochastic deviations at numerous loci and selection of favorable combinations. This constitutes an evolutionary process in which the steps consist of crossing of "saddles" toward higher peaks in the "surfaces" of selective values. The conditions are, however, so restricted that this process was rejected (Wright 1932b) as one of no appreciable significance.

The situation is much the most favorable for evolution where there can be selection among interaction systems. This can occur, within widely ranging species, divided into small populations, sufficiently isolated to permit wide stochastic deviations in numerous loci but not so isolated as to prevent excess diffusion from those centers that happen to have acquired the most adaptive interaction systems because of local stochastic crossings of saddles, followed by local mass selection.

It should be noted that where there is frequency dependence, the function of the selective values and genotype frequencies that determines the courses of differentiation of the separate local populations may differ considerably from that which controls the amounts of diffusion from them. In spite of this complication, the general result is a succession of steps in which inferior interaction systems are replaced by superior ones among those possible in the current field of gene frequency combinations of the species.

While this concept of evolutionary progress was developed with quantitative variability as the basis for selection, the process can play an important role in the fixation of recurrent major mutations that would be advantageous except for the usual deleterious side effects. The local establishment of an array of modifiers that suppress these side effects followed by selective diffusion of the whole complex, major mutation and its modifiers, may establish this complex in the species as a whole, something which could not occur by mass selection in a panmictic population.

The multidimensional stochastic distribution of the array of gene frequencies that constitutes random drift depends not only on accidents of sampling but also on fluctuations in the selection and diffusion coefficients. Unique historical accidents also contribute to the local crossings of "saddles" from such causes as discussed. It is noted, however, that sampling drift has special importance because it occurs automatically at all heterallelic loci in which the leading alleles differ only slightly in momentary selective value. There are likely to be many thousands of such loci in contrast with relatively small numbers subject to other aspects of random drift.

As shown by Dobzhansky, polymorphic chromosome arrangements give a basis for the simultaneous building up of different interaction systems that give adaptation to different conditions among those encountered by a species in a heterogeneous environment. Discussion of this topic is deferred to volume 4.

Evolution according to the genotype as a whole occurs necessarily in a population with exclusive or predominant uniparental reproduction. If there is exclusive uniparental reproduction, the field of variability is, however, rapidly exhausted by elimination of all but the best clone, limiting further evolution to the rate of occurrence of favorable mutations. If, however, there is a possibility for occasional crossing of clones with diverse genotypes, a wide field of variability is maintained. This process is very effective in producing clones well adapted to any transient environmental situation. It is less effective for general progress in adaptation to prevailing conditions than the shifting balance process in a finely structured population with biparental reproduction.

I have continued to use the general term "shifting balance" when referring to the most favorable complex of processes because of the impossibility of giving a full specification concisely. The expression, "evolution from random local differentiation and selective diffusion," gives the main features but does not specify the nature of the local process. Such expressions as "interaction system shift" or the more figurative "selective peak shift" describe the results fairly concisely but not the process itself.

The consequences where situations vary greatly among localities or in time are fairly obvious. If local population numbers are sufficiently small in only a portion of the species range, favorable interaction systems may be established in this portion but cannot spread farther until the situation becomes favorable elsewhere. If environmental conditions differ significantly in different parts of the range, correspondingly different interaction systems tend to be established. The resulting pattern is difficult to distinguish from one arrived at wholly by local mass selection, but the interaction systems should be more adaptive than mere gene substitutions. Periods in which population structure is favorable for selective peak shifts may alternate with unfavorable periods. Such evolution as occurs in the former will be arrested in the latter, except as novel favorable mutations or environmental changes bring about changes by mass selection.

The conditions most favorable for one phase of the shifting balance process need not, however, be the most favorable for the others. The most favorable population structure for the first phase (sampling drift-saddle crossing) was described as follows (Wright 1940): "The populations in certain regions are liable to frequent extinction with reestablishment by rare migrants."

A structure consisting of larger, more stable local populations is more favorable for the second (local mass selection) and third (selective diffusion) phases. Human evolution, in the relatively brief historical period, has, for example, undoubtedly proceeded by the second and especially the third phase enormously more than by the first phase. In general, a mixture of population structures, both in time and space, would seem more favorable than any consistent pattern.

The most favorable conditions for the first phase happen to be precisely those for a high probability of the fixation of a translocation, semisterile as a heterozygote (Wright 1941c).

The chance of fixation in a population of plants with exclusive sexual reproduction is of the order 10^{-3} if the effective population number is 10. It is of the order 2×10^{-6} in groups of 20 individuals and of the order 3×10^{-14} in groups of 50 individuals.

The chances were given as somewhat better in animal populations of a given size, but were still very slight. The most favorable case in animals is, however, the occurrence of numerous colonies that become extinct from time to time to be reestablished by single fertilized females.

It is necessary to conclude from the extreme rarity of translocations within species, but great frequency of such differences between closely related ones, that a large proportion of species have their origins in the partial reproductive isolation provided by the fixation of a translocation in a very small colony.

Multiple selective peak shifts are especially likely to have been established in the portion of the range of the parent species within which a translocation becomes fixed and also during the early history of the new species while still broken up into small transient colonies, since the fertilized females that start new colonies are likely to come from the more successful ones.

This situation gives rise to a curious similarity of the course of evolution from the shifting balance process to that supposed to occur under de Vries' mutation theory. In both cases there would be an association between the "chromosome repatternings" of Goldschmidt that provide the basis for reproductive isolation, and for the major phenotypic steps. Goldschmidt supposed that these were two aspects of the same event, while under the view taken here, they merely tend to be associated because both are favored by the same local situation. Goldschmidt also considered that there is a "bridgeless gap" between the minor phenotype differences within species and those that distinguish species. There is something of a gap between a mere change of gene frequency or even a gene substitution and a selective peak shift. In this case some such shifts would, indeed, be expected to arise also after the new species has become large, but even so, the occurrence of many more during the period of origin would give something of a gap.

CHAPTER 14

Genetic Load and Genetic Variability

Various aspects of genetic variability have been discussed in volumes 1 and 2 and in earlier chapters of this volume. We are here concerned with certain concepts that have been widely discussed as the "genetic loads" of populations. The most obvious is the burden of extreme deviants that segregate out in populations, generation after generation, and that may be described in terms of percentages. The term *genetic load* was coined, however, for the special case of detrimental mutations that persist for limited periods before extinction by selection, during which each is considered to be responsible on the average for a unit amount of damage, defined as the product of its selective disadvantage and the generations of persistence. Of more general application is the concept of genetic load as the percentage deviation of the population mean of selective values from the best genotype. Closely related is the deviation of the population mean from a selective peak, which may be at a set of gene frequencies, instead of being the best single genotype. This leads to comparison of the various ways by which populations may maintain stores of genetic variability and the relation to the genetic load.

Genetic Load in Man

H. J. Muller, after discovering the mutagenic effect of X rays in 1927, and noting the almost invariably detrimental effect of the resulting mutations, became deeply concerned over the effects of the increasing amount of radiation to which mankind is being exposed.

In a report of a Committee of the National Academy of Sciences in 1956, it was estimated that the 4 or 5 rads per 30 years from natural sources is supplemented by some 3 rads from diagnostic and therapeutic uses of X rays in the United States, and by small amounts, about 0.1 rads, from the fallout from testing atomic weapons. It was noted that much larger amounts are to be expected when atomic power plants come to supply a substantial portion of the energy needs of the country. It should be noted, however, that according to more recent estimates the input from natural sources averages only 3 rads per 30 years.

In publicizing this danger, Muller needed a measure of the damage to human populations from given increases in radiation. For this purpose, he introduced the concept of the "genetic load" of abnormal heredity, imposed by such radiation (Muller 1950). He drew on an almost forgotten paper by Danforth (1923) in developing the thesis that, with negligibly rare exceptions, every mutation contributes substantially the same amount of damage in the long run, irrespective of the severity of its effect on individuals. The more drastic ones affect only a few individuals, but the less drastic ones affect many before being eliminated. The elimination of each mutation he referred to as a "genetic death," whether due to immediate lethality, shortening of life span, or reduction of fecundity.

Danforth had made the first attempt to estimate the rate of mutation of genes in man, on the hypothesis of a balance between this rate and that of elimination by adverse selection. He suggested that the rate of recurrence (v in the symbolism used here) of a mutation with heterozygous effects could be estimated from the ratio of the equilibrium frequency ($\hat{q}$) to the average number of generations ($\bar{n}$) of persistence of individual mutations in the population, which he used as an inverse measure of the selective disadvantage. This method was not followed up by him or by others, probably because of the practical impossibility of making valid estimates of average persistence from recorded family histories. These are usually too short and too unreliable with respect to manifestation of traits, especially in the earlier generations. There is, moreover, the likelihood that the trait may skip generations because of irregular penetrance (as in polydactyly and syndactyly, the two traits that he himself considered).

Muller based his thesis essentially on the constancy of the product $hs\bar{n}$, implied by Danforth, where s is the selective disadvantage of the homozygous mutation and h is the degree of dominance (1 if complete). As discussed in chapter 11, most so-called recessives, including recessive lethals, have some dominant effect. In complete recessives, moreover, the quantity $s\bar{n}$ is constant if $\bar{n}$ is interpreted as the number of manifestations of the homozygous gene before extinction.

A related principle had been put forth by Haldane (1937) (vol. 2, pp. 35–36) and was discussed by Muller in his 1950 paper. According to this principle, recurrent mutations with detrimental effects as heterozygotes reduce the mean selective value at equilibrium by approximately twice their rates of recurrence ($\overline{W} \approx 1 - 2v$) unless h is very nearly zero, irrespective of the severity of the selection. With complete recessiveness, Haldane showed that there is half as much reduction ($\overline{W} = 1 - v$).

There is an important difference between the Danforth-Muller concept of a reciprocal relation between severity and average persistence and Haldane's principle of the independence of the effect of a recurrent mutation on the mean selective value of the population and its severity. The mean ($\bar{w}$) of the relative selective values assigned genotypes in a population is a parameter that is primarily useful in connection with estimates of the rates of change of gene frequencies. Its value is always relative to an arbitrary choice of a standard of reference and thus is meaningless as a property of the population, except insofar as a standard can be chosen that gives it meaning. In the case of the absolute mean selective value ($\bar{W}$) of the population, a standard is implied such that $\bar{W} - 1$ is the rate of increase of its effective size ($\bar{W} - 1 = (N - N')/N'$ where N' is the effective size of the preceding generation [vol. 2, pp. 30–31]). Haldane's principle affirms that $\bar{W}$ in a population in which there is a balance between the effect of recurrent mutation and adverse selection tends to be smaller by a certain amount, usually $2v$, than it would have been if there had been no recurrence of the kind of mutation in question. If interpreted as damage, it is merely in this sense. Muller's principle, while assuming that there is such an equilibrium, relates directly to the total damage to individuals from each individual mutation before its extinction.

Because of the indicated near-constancy of damage wrought by each individual mutation in the long run, Muller maintained that the total damage to a human population from increased exposure to radiation (total number of genetic deaths) is the same as the additional number of induced mutations. He maintained that this could be estimated from the linear relation between dosage and rate of induction in experimental animals, after making allowance for the relative numbers of genes.

This view was widely accepted. Thus in the 1956 report of the Committee of the National Academy of Sciences, referred to above, estimates by six members, including Muller, of the total number of genetic deaths expected from exposure of the whole population of the United States to an additional dose of 10 rads, averaged 5,000,000.

This is an alarming figure, even though its impact would be distributed over some 40 or 50 generations. There are two questions, however, that should be considered. Does Muller's principle of equal damage really apply to all mutations or merely to a small class? Where it does apply in its algebraic sense; how far does it imply equality in concrete damage in a population?

On the first of these questions, Muller himself made certain exclusions in his calculations. He, of course, excluded all mutations with favorable heterozygous effects, even though deleterious when homozygous. These reach equilibriums and tend to persist indefinitely, with effects that, on the average, are beneficial. He considered such mutations to be very rare, although responsible for an excessive amount of variability when present because of their relatively high equilibrium frequencies. For practical reasons, he also excluded mutations with less than 10% detrimental effect as homozygotes and took 5% as a typical degree of dominance at this cutoff point from his evidence from *Drosophila*, and hence $hs = 0.005$. He estimated that detrimental nonlethal mutations down to this level are about five times as numerous as lethals, again from *Drosophila* evidence.

Muller postulated a "type" allele at each locus and supposed that virtually all mutations, apart from a few heterotic ones, are deleterious, with the implication that the ideal situation in man would be homozygosis in nearly all type genes and hence a population in which all individuals of the same sex are almost as much alike as identical twins. This viewpoint overlooks the positive value (within limits) of genetic variability, in giving versatility to a population in dealing with a varied and ever-changing environment, and in mankind in particular, in giving a basis for a far-reaching division of labor. The principle of equal damage from all mutations obviously does not apply here at all.

This consideration, however, does not necessarily affect Muller's actual calculations very drastically because of the lower limit he set for deleteriousness in making them. It does, however, introduce a considerable element of uncertainty.

Muller considered possible complications from factor interaction ("synergism"). He properly concluded that this would not have an important effect on the calulations where the equilibrium frequencies of the mutations are as low as expected where based on the balance between recurrent mutation and much stronger adverse selection.

This brings us to the second question, the relation between the count of genetic deaths and the concrete damage to persons and to the population as a whole. These are difficult to evaluate, but it seems clear that the equality indicated by the algebraic formulation is far from true.

The damage to society can be estimated rather more objectively than personal illness, pain, and frustration by classifying human phenotypes according to the ratio of social contributions to social cost, the latter in terms of cost of rearing, education, and maintaining standards of living (Wright 1960a).

The largest classes of phenotypes (fig. 14.1) are probably (1) those in which there is an approximate balance between rather modest costs and equally modest contributions. These grade into classes (2) in which much greater costs are balanced by much greater contributions. There are also those (3) who cost little and contribute much and those (4) who cost much but contribute

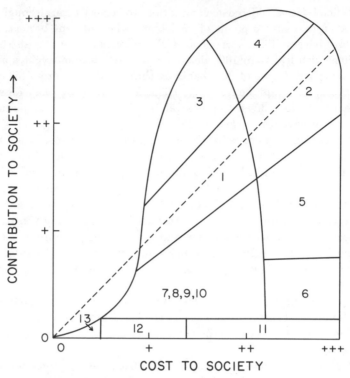

Fɪɢ. 14.1. Diagram indicating the costs and contributions to society of various phenotypic classes (numbered as in text). Redrawn from Wright (1960a, fig. 1).

much more. Mutations that leave persons in any of these classes cannot be considered seriously detrimental to society, although some of them may cause considerable personal distress. There is social damage from those (5) who cost significantly more than they contribute at whatever level, not because of any genetic mental or physical disability but because of their reliance on inherited wealth. There is much more social damage from those (6) who contribute negatively because of antisocial activities, and here genotype probably plays more role, though how much is a controversial question. This brings us to several classes in which cost is in general much greater than contributions for reasons in which genotype very frequently (although not always) plays an important role: (7) subnormal physical constitution and health, (8) low mentality but not complete dependence, (9) physical breakdown relatively early in adult life, (10) mental breakdown after maturity, especially from one of the major psychoses. Perhaps the most damaging class socially with an unquestionably large genetic component is that (11) in which there is complete physical or mental incapacity throughout a lifetime of more or less normal length. The class (12) of those who also are completely incapacitated but who die early is less costly to society. Finally the class (13) of deaths at or before birth is responsible for relatively little social cost but probably includes the largest portion of mutations above Muller's cutoff point.

We have not included differences in fecundity in the criteria for classes. There is relatively little social damage from the class of mutations that reduce fecundity or even cause sterility in otherwise normal carriers in a population in balance with its resources, and there is some advantage to a society threatened by overpopulation.

If we remove from the genetic load a great many mutations that contribute usefully at least as heterozygotes to the versatility of the population and remove lethals that act at or before birth as far less damaging than those left, the total will be very much reduced. Nevertheless, enough are left to fully justify Muller's campaign to reduce exposure to radiation as much as is practicable in view of important advantages in its use.

Generalized Load

Dobzhansky (1955) suggested that another sort of load, which he called the "balanced load," might be more important in natural population than the "mutation" load. He was referring to segregation of the homozygotes in populations in which two or more alleles are kept at equilibrium frequencies by the superiority of the heterozygotes.

Load is here considered from a different point of view from that which Muller had in mind. For Dobzhansky, the load carried by the population consisted of the percentage of deviant individuals that appeared in each generation because of genetic segregation. For Muller it consisted of the total effects of detrimental mutations, adding up in each case to a single genetic death by the time of its extinction by selection. A mutation that increases the average fitness of the population because of its heterozygous effect does not contribute to the load at all in Muller's sense, although contributing generation after generation indefinitely to the load in Dobzhansky's sense, because of the segregation of deleterious homozygotes.

Morton, Crow, and Muller (1956) preferred the term *segregation load* to *balanced load* for the type considered by Dobzhansky, because the mutation load also involves a state of *balance*. There is still some ambiguity, however, since both forms of load reflect *segregation* of inferior genotypes. Segregation from states of balance is also involved in other types of load proposed later.

The term *heterotic load* will be used here for the type that depends on heterozygous superiority. It may be objected that the homozygotes, not the heterozygotes, constitute the load, but at least it refers unambiguously to the load associated with heterotic loci.

Morton, Crow, and Muller (1956) proposed a definition of the genetic load designed to include both of these types. This was extended later by Crow to various other types. In a later paper (1970) Crow gave three definitions, the broadest of which was as follows:

The genetic load is the fraction in which the population mean differs from a reference genotype. The trait measured is usually fitness or some component thereof and the reference genotype is usually the maximum among the actually or theoretically available types.

Thus he had in mind a family of parameters rather than one narrowly defined one. Care must obviously always be taken in applying the concept to specify the precise sense in which it is being used.

Using w_R for the relative selective value of the reference genotype, $L = (w_R - \bar{w})/w_R$.

The value of L can be obtained at once for many cases from the selective values and their means (vol. 2), where the reference genotype is that with maximum selective value, and the selective values of others are given in relation to it as the standard. The load in this case is merely $1 - \bar{w}$.

If the application is to size of population, it is convenient, as noted earlier, to use absolute selective values such that $\overline{W} = N/N'$, where N and N' are the effective sizes of population in the same and preceding generations, respectively, under specified conditions with respect to environment and the rest of the genome (vol. 2, p. 30):

$$L = (w_R - \overline{W})/w_R.$$

In the case in which the reference genotype is that with maximum selective value, $w_R = w_m$ and the population maintains a constant size ($\overline{W} = 1$),

$$L = (w_m - 1)/w_m.$$

If, in this case, selection is wholly by differential mortality, L is the fraction of the offspring generation (after allowing for accidental deaths) that is lost selectively. Parents produce an average of w_m offspring per parent, excluding those lost accidentally, and this is reduced to an average of one per parent by selective mortality.

If, however, selection is wholly by differential fecundity, the average number of offspring per parent, excluding those lost accidentally, is one. Parents of some genotypes have fewer offspring, others have more, and w_m is the maximum.

A third extreme possibility is that a constant population size is imposed by the carrying capacity of the environmental niche occupied by the species. There may be a limited number of territories, and only those individuals that successfully hold a territory reproduce. The number of offspring per parent may be the same for all genotypes and may vary greatly from generation to generation according to conditions, provided that the total numbers of females equals or exceeds the number of territories. Selection is exhibited by the ratio of the number of territories occupied by a genotype to its initial frequency.

In general, the operation of selection is more complex than in these simple cases. There seems to be no general biological meaning of load as defined above other than the percentage difference of the mean selective value of the population in question from that of some reference genotype.

Multiple Loci

The foregoing formulas for load apply to sets of total genotypes as well as to single loci. It will be well to consider the relation between total load, L_T, and that at one of the loci, L_i.

If multiple loci contribute independently to the total selective value,

$$\bar{w}_T = \prod w_i.$$

Assuming that the reference genotypes at each locus are those of maximum selective value and that the w_i's are relative to these so that $\bar{w}_i = 1 - L_i$,

$$\bar{w}_T = \prod (1 - L_i) \approx e^{-\Sigma L_i} \quad \text{if the } L\text{'s are small, and}$$

$$L_T \approx 1 - e^{-\Sigma L_i} \approx \sum L_i.$$

These formulas must be used with caution since interactions with respect to selective value are usually present. If many loci are involved and loads per locus are not very small, the genotype that combines the reference genotypes of all loci is in general so rare theoretically that neither it nor anything approaching it exists in a finite population. In this case, L_T becomes practically meaningless, even if there are no interaction effects.

Deviation of a Population from a Selective Peak

There are special cases that come under the definition of generalized load that are of significance for the shifting balance theory of evolution. These are ones in which the "reference genotype" is the genotype, or genotypic array, of a selective peak of some sort and the "population mean" refers to that of any population involving the loci in question, not necessarily one at equilibrium. These, however, do not measure loads *within* populations but measure the extent to which the selective value or fitness of a given population falls short of the peak that controls its evolution at the time. The symbol $D(\overline{W})$ will be used here for the deviation from the mean selective value $\hat{W}$ of the peak population and $D[(F(W/\overline{W})]$ will be used similarly for the deviation with respect to the peak fitness, $\hat{F}(W/\overline{W})$:

$$D(\overline{W}) = (\hat{W} - \overline{W})/\hat{W}$$

$$D[F(W/\overline{W})] = [\hat{F}(W/\overline{W}) - F(W/\overline{W})]/\hat{F}(W/\overline{W}).$$

Analogous formulas may be used for the difference between a given peak and a reference one.

In cases in which the selective values of genotypes are not independent of the gene frequencies, $F(W/\overline{W})$ exists for all pairs of alleles but only in restricted cases if there are multiple alleles or loci. There are, however, selective goals in these cases toward which the population moves within the system of gene frequencies, along paths that can be determined, but there is no single metric (vol. 2, pp. 120–22).

There is hardly enough advantage in using $D(\overline{W})$ and $D[F(W/\overline{W})]$ over merely specifying the values of $\overline{W}$ and $F(W/\overline{W})$ at the peak and elsewhere, as done throughout volume 2, to justify the addition of new symbols for general use. They will, however, be convenient in this chapter in connection with values of L.

Mutation Load

In the case of a haploid population, it may be found at once that the load $1 - \bar{w}$, for a pair of alleles is the mutation rate v. In a diploid population with relative selective values $1:1 - hs:1 - s$ for genotypes AA, AA', and $A'A'$, respectively, and equilibrium frequencies, the load is, as noted earlier, approximately $2v$ unless h is close to zero or greater than 1, while for a completely recessive mutation ($h = 0$) it is v, both in accordance with Haldane's principle (vol. 2, p. 35).

The load is so small in these cases that there may be contributions from a great many loci with substantially independent selection, even though interactions would occur if combinations were produced. Thus even with 10,000 loci and independent contributions of the order of 10^{-5}, the total load would be only about 10%.

In this case, the wild-type genotype is the selective peak. The deviation of a population, whether at equilibrium or not, is the same as the load. Thus $D(\overline{W}) = L$ in the case of mutation load.

The situation under inbreeding is of considerable interest (Crow 1970). The genotype frequencies and mean selective values of the randomly bred (w_R) and inbred (w_I) components of $\bar{w}$ are as follows (vol. 2, p. 244):

Geno-type	F	w_R	w_I
AA	$(1-F)(1-q)^2 + F(1-q)$	1	1
AA'	$2(1-F)q(1-q)$	$1-hs$	
$A'A'$	$(1-F)q^2 + Fq$	$1-s$	$1-s$

$$\bar{w}_R = 1 - sq[2h + (1-2h)q]$$
$$\bar{w}_I = 1 - sq$$
$$\bar{w} = (1-F)\bar{w}_R + F\bar{w}_I$$

$$L = 1 - \bar{w} = sq\{2h + (1-2h)F + (1-2h)(1-F)q\}$$

It makes a great deal of difference whether the inbreeding is initiated in a randomly bred population that has reached equilibrium with mutation [$\hat{q} = \sqrt{(v/s)}$ if $h = 0$, $\hat{q} = v/hs$ if $h \gg 0$], and is carried out so rapidly that there is no appreciable change in gene frequency (possible if s is small), or the new equilibrium is attained.

If equilibrium is reached under inbreeding (vol. 2, eq. 10.3 rearranged, with hs substituted for s and s for t)

$$\Delta q = v(1 - q) - sq(1 - q)[h + (1 - h)F + (1 - 2h)(1 - F)q].$$

The equation for determination of $\hat{q}$ is thus

$$(1 - 2h)(1 - F)\hat{q}^2 + [h + (1 - h)F]\hat{q} - v/s = 0$$

$$\hat{q} \approx v/\{s[h + (1 - h)F]\} \qquad \text{(if } h \text{ and } F \text{ are not both close to 0)}.$$

In the case of fixation in each line ($F = 1$), $\hat{q} = v/s$ and the load at equilibrium is $\hat{L} = v$, as in the case of clones.

With incomplete fixation and exact semidominance ($h = 0.5$), $\hat{q} = 2v/[s(1 + F)]$, $L = sq$, and $\hat{L} = 2v/(1 + F)$. With complete dominance ($h = 1$), $q \approx v/s$ and $L = v(2 - F)$, which does not differ very much from the values for semidominance. Intermediate degrees of dominance give intermediate results. In the case of a completely recessive mutation, $\hat{L} = v$ irrespective of F. There is a transition to $\hat{L} = 2v/(1 + F)$ with values of h between 0 and 0.5. The approximate loads with unchanged q from random mating and with attainment of equilibrium under inbreeding are summarized in table 14.1.

At first, it seems surprising that gradual inbreeding tends to reduce the equilibrium load (unless there is complete dominance). This is because of the reduction of the equilibrium frequency (unless $h = 1$).

It is possible for the equilibrium load to exceed $2v$, but this occurs to an appreciable extent only if s approaches v in smallness and the heterozygotes are inferior to both homozygotes. Thus if $h = 1.5$ and $s = 4v$, the equilibrium load is $2.5v$ (Kimura 1961).

TABLE 14.1. The approximate load with various degrees of dominance (h) and inbreeding (F) in cases of no change from q for $F = 0$ and of equilibrium $\hat{q}$ for given F.

	L with No Change from $\hat{q}$ for $F = 0$				L with $\hat{q}$ under Given F		
	$h = 0$	0.5	1		$h = 0$	0.5	1
$F = 0$	v	$2v$	$2v$	$F = 0$	v	$2v$	$2v$
$0 \ll F < 1$	$F\sqrt{(sv)}$	$2v$	$v(2 - F)$	$0 \ll F < 1$	v	$2v/(1 + F)$	$v(2 - F)$
$F = 1$	$\sqrt{(sv)}$	$2v$	v	$F = 1$	v	v	v

In sex-linked mutations, the load is $1.5v$ (Haldane 1937) or $0.5(v_M + 2v_F)$ if the mutation rates of males, v_M, and of females, v_F, differ (Kimura 1961).

Mutation Load in Small Populations

In small populations, the equilibrium gene frequency declines (vol. 2, pp. 364–66) but the distribution spreads out so much that it extends beyond the equilibrium value for large populations. Because of this, the genetic load rises (Kimura, Maruyama, and Crow 1963).

Using v for the rate of origin of gene A' and u for reverse mutation, the distribution and load are

$$\varphi(q) = C\overline{W}^{2N}q^{4Nv-1}(1 - q)^{4Nu-1}$$
$$L = sq[2h + (1 - 2h)q].$$

The mean contribution of the locus to the mutation load is

$$\bar{L} = \int_0^1 L\varphi(q)\, dq \Big/ \int_0^1 \varphi(q)\, dq.$$

The paper by Kimura, Maruyama, and Crow (1963) must be consulted for the methods used for approximate integration. It is shown that for small $2Ns$ ($2Ns < 5$),

$$L \approx s/[1 + (u/v)e^{2Ns}].$$

Figure 14.2 shows the relation of mutational load to effective population size for various values of s and h, with $v = 10^{-5}$ and $u = 10^{-6}$.

It may be seen that in a sufficiently small population, the load approaches s, irrespective of degree of dominance, and thus is generally much larger than the load in large populations ($2v$ to v). The latter values are approached in populations as small as 100 if $s = 0.1$, but only in populations of more than 5,000 if s is as small as 0.001. There is thus a wide range of population sizes in which the load is many times as great under weak selection as under strong selection. For example, if $N = 900$ and $h = 0$, the load is approximately 50 times as large for $s = 0.001$ as for $s = 0.01$.

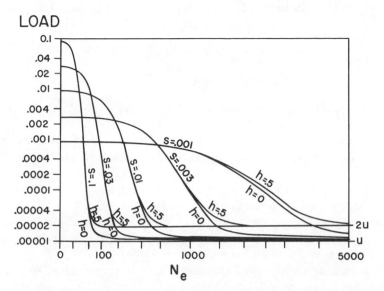

LOAD

N_e

FIG. 14.2. Relations between mutational load (ordinate) and effective population size (abscissa) for various values of s, with $v = 10^{-5}$, $u = 10^{-6}$. Reprinted, by permission, from Kimura, Maruyama, and Crow (1963, fig. 1).

A consequence is that a population large enough to have a small load, $2v$, if panmictic, has a much larger load if subdivided into sufficiently small demes.

It is important to note that in this case, as in general, where mutation pressure is balanced by adverse selection, the selective peak is at homozygous wild type and the load measures the percentage by which the population falls below the selective peak $L = D(\overline{W})$.

Immigration Load

The balance between immigration into a partially isolated population, and local selection against the introduced genes, mathematically resembles the balance between recurrent mutations and adverse selection because of the linearity of both immigration and mutation pressure (vol. 2, pp. 27–28, 36–38).

Letting m be amount of replacement by immigrants, Q the frequency of an introduced gene, and s the local selection against the latter,

$$\Delta q = (mQ + v)(1 - q) - [m(1 - Q) + u]q - sq(1 - q).$$

Thus mQ can be substituted for v and $m(1 - Q)$ for u in formulas

$$\hat{q} \approx mQ/s \quad \text{if } m \ll s$$

$$\hat{q} \approx 1 - \sqrt{(1 - Q)} \quad \text{if } m = s, \text{ and}$$

$$\hat{q} \approx Q\left[1 - \frac{s}{m}(1 - Q)\right], \quad \text{if } m \gg s.$$

In an equilibrium population, $\overline{W} = 1 - 2s\hat{q} = 1 - 2mQ$ if $m \ll s$; $L = 2mQ$.

At the selective peak, $\overline{W} = 1$. Thus immigration pulls down the mean selective value of the local population by the amount of the load $D(\overline{W}) = 2mQ$.

If m is much greater than s, the local population tends to be swamped with respect to the gene in question, with $L = D(\overline{W}) = 2sQ$.

The effects of selection with other than semidominance can be deduced from the formulas in volume 2 (p. 37). In the case of introduction of a recessive gene with $m \ll s$, $L = D(\overline{W}) = mQ$.

The load from multiple loci affecting the same character is likely to be complicated by interaction effects much more than in the comparable case of recurrent mutation because of the usually much higher frequencies of the introduced genes.

The Heterotic Load

We consider the heterotic load here from the standpoint of the formula $L = (w_m - \bar{w})/w_m$ where the genotype of reference is the heterozygote. The two-allele case is as follows (vol. 2, eq. 3.32–3.42):

Genotype	f	w	
AA	$(1 - q)^2$	$1 - t$	$\bar{w} = 1 - sq^2 - t(1 - q)^2$
AA'	$2q(1 - q)$	1	$\hat{q} = t/(s + t), 1 - \hat{q} = s/(s + t)$
$A'A'$	q^2	$1 - s$	$L = 1 - \bar{w} = st/(s + t)$ at equilibrium

Mutation pressures are generally so much less than the selection pressures that they need not be taken into account (see, however, vol. 2, eq. 3.43).

In this case no diploid population can exhibit the maximum selective value. This is approached, however, as the number of alleles increases. If there are n alleles and all heterozygotes have the same value, $w_{ij} = 1$, while all homozygotes are lower, $w_{ii} = 1 - s_{ii}$:

$$\bar{w} = 1 - \sum (s_{ii}q_i^2) \qquad \text{(vol. 2, eq. 3.39)}$$

$$L = 1 / \sum (1/s_{ii}) = \tilde{s}_{ii}/n$$

$$\text{(at equilibrium, Crow [1970])}$$

Here $\bar{s}_{ii}$ is the harmonic mean of the selective disadvantage of the homozygotes.

The selective peak in this case is at the array of equilibrium frequencies of the alleles at the locus, disregarding slight displacement of the latter due to recurrent mutations. Thus $D(\overline{W}) \approx 0$. The difference from the load, L, contrasts with the identity in the cases of mutation and immigration loads, in which the equilibrium frequency is carried away from the selective peak by nonselective pressures.

The heterotic load at a locus is generally much greater than the mutational load, where it is present at all. The total from many loci, under the assumption that selection acts independently on them, $L_T = 1 - \exp\left(-\sum L_i\right)$ is thus likely to be very nearly 1. It has often been suggested that the reproductive surplus of species would be so depleted by the presence of many heterotic loci that the species would be unable to maintain its numbers. This has provoked much discussion, since it had been held by some, notably Dobzhansky, Lerner, and Wallace, that a very large proportion of all loci are heterotic.

As pointed out by Sved, Reed, and Bodmer (1967) and by Milkman (1967), the load relative to the best genotype, where many loci are heterotic, may give a grossly exaggerated impression of the strain on the reproductive capacity. This is because the best genotype is practically never present, or at all closely approached, in a finite population.

Consider the case of equal selective disadvantage of both of two homozygotes, $s = t = 0.10$, $L = 0.05$ at each locus. With 100 loci, selected independently, $L_T = 1 - e^{-5} = 0.9993$, a load so close to 1 that it would seem to be inevitably fatal to the species. With gene frequency array $(0.5A + 0.5A')$, the distribution of heterozygotes (Het) and homozygotes (Hom) would be $(0.5\,\text{Het} + 0.5\,\text{Hom})^{100}$ for 100 loci, with mean 50 Het, standard deviations 5 Het, and thus would range from about 35 to 65 heterozygous loci. The genotype with heterozygosis at all loci is so far outside this range that neither it nor anything approaching it ever occurs in the population. It is absurd to suppose that all genotypes produce as many offspring as would be produced by this 100-fold heterozygote, if it existed, but that these are reduced to infinitesimal numbers by the cumulative selective coefficients where there is homozygosis at 35 or more loci. Thus the load relative to the 100-fold heterozygote is meaningless. It would be more appropriate to take as the reference genotype one that is above the average by only two or three standard deviations.

This is on the assumption that selection is independent at the loci. The actual selection process in such a population would probably involve much correlation among loci. It might well correspond roughly to the culling of all individuals with selective values below a certain degree of heterozygosis. If this is at 50% heterozygosis in the above case, 50% are culled and 50% are left to replenish the population. If at 55% heterozygosis, 16% are left as parents. If population size is regulated by environmental factors, the culling point may vary enormously from generation to generation but always be such as to leave about the same number of parents.

Lerner (1954) accepted the occurrence of heterosis at a great many loci as the main basis of genetic variability and the primary factor in maintaining what he called "genetic homeostasis," the tendency of populations to maintain a genetic composition leading to an optimum balance, capable of adjusting rapidly to changing conditions. He suggested that this does not cause excessive variability because of a ceiling beyond which additional heterozygosis contributes little or nothing to fitness.

Sved, Reed, and Bodmer (1967) and King (1967) discussed the mathematics of this concept at some length. Under this there is a wide range of genotypes with substantially the same optimum fitness. This grade can then be used in the formula as w_m and avoid the absurdity of a reference genotype that is enormously superior in principle to any of those that actually exist. The load for the total genotype is not, of course, the sum of contributions from single loci when segregating by themselves.

If selection relates primarily to fecundity (a character on which heterosis usually has major effects), this may be supposed to be adjusted so as to maintain the species at a roughly constant size at average levels of heterosis. The fecundity of the nonexistent genotype with 100% heterosis may imply an enormously large w_m, and hence load, but there is no problem of accounting for the persistence of the species. In this case also, however, it is probable that there is a ceiling beyond which additional heterotic loci are without effect. This again makes it impossible to attribute unit loads to loci.

It is more likely that the heterotic effect per locus would fall off gradually than abruptly at a certain ceiling. Assume that it falls off inversely with the standard deviation, σ_T, of the number of heterotic loci, $\sigma_T = \sqrt{[np(1 - p)]}$, where p is the frequency of heterozygosis:

Genotype	f	w	
AA	$(1 - q)^2$	$1 - (t/\sigma_T)$	$\bar{w} = 1 - \dfrac{s}{\sigma_T} q^2 - \dfrac{t}{\sigma_T}(1 - q)^2$
AA'	$2q(1 - q)$	1	$\hat{q} = t/(s + t), \quad 1 - \hat{q} = s/(s + t)$
$A'A'$	q^2	$1 - (s/\sigma_T)$	$L_i = st/[\sigma_T(s + t)]$
			$= st/\{\sqrt{[np(1 - p)]}(s + t)\}$
			$L_T = \sum_{i}^{n} L_i = st[n/p(1 - p)]^{1/2}/(s + t)$

If only one locus were segregating and $s = t, q = 0.5, \sigma_T = 0.5$, and $L = s$. If 10,000 equivalent loci were segregating, $\sigma_T = 50$, the total load is $100s$, only 100 times that with one segregating.

Relative Amounts of Mutational and Heterotic Loads

Some light is thrown on the frequency of heterotic loci, relative to ones in which recurrent mutation is balanced by selection, by a method proposed by Morton, Crow, and Muller (1956). This was based on determination of the inbreeding depression found where the inbreeding is so rapid that gene frequencies may be assumed to remain unchanged.

In the case of a pair of heterotic alleles, continued inbreeding without change of gene frequency leads to the array of lines $[qAA - (1 - q)A'A']$. Here the inbred load, L_I, is $t + (s - t)\hat{q} = 2st/(s + t)$, in comparison with $L_R = st/(s + t)$ under random mating. Thus $L_I/L_R = 2$ with two alleles.

With n alleles and equivalent heterozygotes the ratio similarly is $L_I/L_R = n$. Heterozygotes that are less favorable than the best are, however, held at low frequencies or eliminated. Thus there are not likely to be many alleles sufficiently equivalent to give a large ratio of L_I/L_R.

This ratio for heterotic loci is to be compared with that for mutational loads. In the case of complete recessives, this is indicated by table 14.1 to be $L_I/L_R = \sqrt{(s/v)}$, which would practically always be much larger than that for heterotic loci. For a recessive lethal, $s = 1$ and with $v = 10^{-5}$, $L_I/L_R = 316$. If $s = 0.10$, $L_I/L_R = 100$. If there is some dominant effect, h, the ratio as shown earlier is approximately $1/(2h)$. Thus if $h = 0.03$, $L_I/L_R = 33$, which is still very large compared with typical values for heterotic loci. With exact semidominance, however, there is no inbreeding depression and $L_I/L_R = 1$.

Morton, Crow, and Muller (1956) proposed to estimate L_I by extrapolation from inbreeding effects on such characters as mortality percentages or percentage incidence of abnormalities observed in the progeny of cousin marriages of various degrees ($F = 1/16$ for first cousins, $F = 1/32$ for first cousins once removed, and $F = 1/64$ for second cousins). This is permissible because the effects of genes are expected to be additive on an appropriate scale. Thus, on fitting the observed percentages to $Y = A + BF$ (or to $-\log Y = A + BF$ if a logarithmic scale is used), the ratio B/A would give an estimate of L_I/L_R if the variability is wholly genetic.

The variability of multifactorial characters of this sort never is wholly genetic, however, so that in principle the value of A should be corrected so as to reduce it to the additive genetic effect. No correction is required for B.

Thus the ratio of B to the genetic component of A for such characters is probably always much greater than observed B/A.

Even without correction, observed B/A tends to be so large in adequate human data as to indicate that the mutational load is much more important than the heterotic load with respect to inbreeding depression.

It is not necessarily implied that the mutational load is the larger within a randomly bred population. The gene frequencies of deleterious mutations are kept so low ($\hat{q} = v/hs$ if any dominance) under random mating that the segregation of a great many causes little load, while the gene frequency of the more deleterious allele maintained by overdominance is typically rather large. The total load in this case tends, however, to be much less than the sum of the loads for single heterotic loci, as just indicated. Even so, the heterotic load for the same number of loci would be much the greater.

Similarly a much larger portion of the variability in a randomly breeding population is likely to be due to a few heterotic loci than to a much larger number of those with mutation-selection balance.

Persistent Strong Heterallelism

Probably the main reason for the assumption that species carry an enormous number of heterotic loci is that this seemed the simplest way of accounting for the persistent genetic variability of species. The main objection has been the intolerable genetic load this seemed to entail. As shown above, there are a number of plausible considerations that reduce this apparent load. Nevertheless, a considerable load seems inevitable. It is desirable to review briefly here alternative ways of maintaining strong heterallelism, some of which entail little or no load.

As often noted earlier, natural selection of quantitative variability must usually be directed toward an optimum grade. It is both mathematically convenient and biologically plausible to assume that selective value falls off as the square of the deviation from the optimum (vol. 2, pp. 105–19). In spite of the fact that each plus gene is more favorable than its allele in combinations sufficiently below the optimum, less favorable otherwise, selection pressure by itself tends ultimately to fix all, or all but one, of the genes of some combination that is at or near one of the many selective peaks (Wright 1935b; vol. 2, pp. 72–93, 99–105).

Using V for the value on the quantitative scale and V_0 for the optimum on this scale, as in volume 2,

$$w = 1 - s(V - V_0)^2 = 1 - s[\sigma_V^2 + (\bar{V} - V_0)^2].$$

Taking α_i as the heterozygous effect of a semidominant plus gene,

$$\bar{V} = 2 \sum q_i \alpha_i$$

$$\sigma_V^2 = 2 \sum q_i(1 - q_i)\alpha_i^2$$

$$\Delta q_i = v(1 - q_i) - uq_i - sq_i(1 - q_i)[(1 - 2q_i)\alpha_i^2 + 2\alpha_i(\bar{V} - V_0)]$$

Assuming that the mean is at the optimum $\bar{V} = V_0$,

$$\hat{q} = (1/4)\{1 - \sqrt{[1 - (8v/(s\alpha^2)]}\} \approx v/s\alpha^2$$

or

$$(1/4)\{3 + \sqrt{[1 - (8u/s\alpha^2)]}\} \approx 1 - u/s\alpha^2$$

The net selection coefficient $s\alpha^2$ for a gene is expected to be very small if there are many genes and each has only a slight effect, especially if there is much environmental variation. If $u = v = 10^{-5}$ and $s\alpha^2 = 0.001$, $\hat{q} \approx 0.01$ or 0.99. If $s\alpha^2 = 10^{-4}$, $\hat{q} \approx 0.10$ or 0.90. Considerable heterallelism is thus likely to be maintained by recurrent mutation.

From another standpoint, assume that there are an even number of loci with equivalent effects. The extreme deviations from the midpoint (with $n/2$ plus genes) are $\pm n\alpha$. Assuming that there is lethality at these extremes,

$$w = 1 - sn^2\alpha^2 = 0$$

$$s\alpha^2 = 1/n^2$$

$$\hat{q} = (1/4)[1 - \sqrt{(1 - 8n^2v)}] \approx n^2v$$

or

$$\hat{q} = (1/4)[3 + \sqrt{(1 - 8n^2u)}] \approx 1 - n^2u$$

Again, assume that $u = v = 10^{-5}$. If $n = 10$, $\hat{q} = 0.001$ or 0.999; if $n = 31$, $\hat{q} = 0.01$ or 0.99; if $n = 67$, $\hat{q} = 0.05$ or 0.95.

$$L = D(\bar{W}) = 2\hat{q}(1 - \hat{q})/n \approx 2nv$$

The mutational load, which is also the percentage decline from the selective peak, is approximately the same as with semidominant definitely deleterious mutations at the same number of loci. The equilibrium frequencies in the present case, however, rise with increase in the number of loci. They require rather large numbers of equivalent loci, however, to give heterallelism at the 0.01 or the 0.05 levels.

The extremes may, however, be far from lethality. If selection is only $1/k$ times as strong ($s\alpha^2 = 1/kn^2$), it only requires $67/\sqrt{k}$ loci for q to be about 0.05 or 0.95. Thus weak selection directed toward the optimum of a quantitatively varying character, dependent on many loci, can maintain moderate heterallelism at all of the loci involved.

A balance between immigration and local selection in each of numerous local populations that are sufficiently isolated to maintain different adaptation to some extent, either to the same or different conditions, not only insures strong heterallelism in the species as a whole, but also, because of the immigration, in each local population. Moreover, such subdivision of the species will be very effective in maintaining heterallelism at many loci if there are many quantitatively varying characters in which selection is directed toward an optimum. This holds if it is the same optimum everywhere for the same reason that mutation pressure maintains more heterallelism in this case than in that of unequivocally deleterious genes. It holds, a fortiori, if the optimum varies from place to place. Under this condition especially, different selective peaks tend to be in control in different places (Wright 1935*b*, 1956).

Another way in which strong heterallelism may be maintained at many loci is by selective advantage of rarity and disadvantage of overabundance. A number of models have been discussed in volume 2, chapter 5. As noted there, this sort of selection is to be expected where a genetically homogeneous population has a variety of niches open to it, and selective values within these niches are related negatively to population density. As noted in chapter 9 of the present volume, this sort of selection has also been found extensively in connection with selective mating, especially with respect to discrimination by females among genotypes of males.

As a simple example, we will take the special case dealt with in volume 2 (pp. 136–38), in which there is linear dependence of the selective values of homozygotes on the gene frequencies with parameters chosen to avoid any complication from heterosis.

Geno-type	f	w	
AA	q^2	$1 + s_1 - (s_1 + s_2)q$	$\bar{w} = 1 - [s_1 - (s_1 + s_2)q](1-2q)$
AA'	$2q(1 - q)$	1	$\sum w \dfrac{\partial f}{\partial q} = 2[s_1 - (s_1 + s_2)q]/\bar{w}$
$A'A'$	$(1 - q)^2$	$1 - s_1 + (s_1 + s_2)q$	$\hat{q} = s_1/(s_1 + s_2)$

$$F(w/\bar{w}) = 1 + 2s_1q - (s_1 + s_2)q^2 \text{ if } s_1 \text{ and } s_2 \text{ are small}$$

The peak mean selective value is at $q = (3s_1 + s_2)/4(s_1 + s_2)$ at which $\bar{w} = 1 + (s_1 - s_2)^2/8(s_1 + s_2)$. This, however, is not at the equilibrium frequency $\hat{q} = s_1/(s_1 + s_2)$ at which $\bar{w}$, and in fact all w's, have the same selective value 1.0. Thus $D(\bar{w}) = (s_1 - s_2)^2/8(s_1 + s_2)$ at equilibrium. There is no load at equilibrium; however, $L = 1 - \bar{w} = 0$ in marked contrast with the load in a population in heterotic equilibrium. The peak fitness is, of course, at the equilibrium gene frequency, $D[F(w/\bar{w})] = 0$.

The load is also zero in the equilibrium population under a number of other cases of rarity advantage discussed in volume 2, in which the fitness peak is at an intermediate gene frequency. These include the following sets. In all of them, the population is at the fitness peak at equilibrium so that $D[F(w/\bar{w})] = 0$.

Set	f	w	$\bar{w}$
5.1	$A-$ $1 - q^2$	$1 - s_1(1 - q)^2$	$\bar{w} = 1 - s_1(1 - q^2)^2 - s_2q^4$
			$\hat{q} = [s_1/(s_1 + s_2)]^{1/2}$
	aa q^2	$1 - s_2q^2$	$w_{A-} = w_{aa} = \bar{w} = 1 - [st/$
			$(s + t)]$ at $\hat{q}$
5.10	$A-$ $1 - q^2$	$1 + [s_1/(1 - q^2)]$	$\bar{w} = 1 + s_1 + s_2$
			$\hat{q}_i = [s_2/(s_1 + s_2)]^{1/2}$
	aa q^2	$1 + [s_2/q^2]$	$w_{A-} = w_{aa} = \bar{w} = 1 + $
			$s_1 + s_2$ at $\hat{q}$
5.52	A_1 q_1	$1 - s_1q_1$	$\bar{w} = 1 - \sum s_iq_i^2$
			$\hat{q}_i = \sum s_iq_i^2/s_i$
	A_2 q_2	$1 - s_2q_2$	$w_{A_1} = w_{A_2} = w_{A_3} = \bar{w} = 1 - $
			$\sum s_iq_i^2$ (at equilibrium)
	A_3 q_3	$1 - s_3q_3$	
5.58	A_1 q_1	$1 + (s_1/q_1)$	$\bar{w} = 1 + \sum s_i,$ $\hat{q}_i = s_i/\sum s_i$
	A_2 q_2	$1 + (s_2/q_2)$	$w_{A_1} = w_{A_2} = w_{A_3} = \bar{w} = 1 + $
			$\sum s_i$ at equilibrium
	A_3 q_3	$1 + (s_3/q_3)$	
5.65	AB pq	$1 + s - 2pq$	$\bar{w} = 1 + spq - 2sp^2q^2$
			$pq = 0.5$ at equilibrium
	Ab $p(1 - q)$	1	All w's $= \bar{w} = 1$ along equilib-
			rium ridge
	aB $(1 - q)q$	1	
	ab $(1 - p)(1 - q)$	1	

As shown in volume 2 (p. 121), cases in which the fitness function exists in the presence of multiple alleles or multiple loci are probably exceptional. There still may be one or more sets of intermediate gene frequencies toward which the population tends to move. The movement may be along a spiral course, as in the three-allele case illustrated in volume 2 (figure 5.12a). Again $L = 0$.

5.85 $(s = t)$	f	w
A_1	q_1	$1 - s(q_1 + q_2)$ $\quad \bar{w} = 1 - (1/2)s - (1/2)s \sum q^2$
A_2	q_2	$1 - s(q_2 + q_3)$ $\quad \hat{q}_1 = \hat{q}_2 = \hat{q}_3 = 1/3$
A_3	q_2	$1 - s(q_3 + q_1)$ $\quad w_{A_1} = w_{A_2} = w_{A_3} = \bar{w} = 1 - (2/3)s$

In the two-locus case in volume 2 (eq. 5.91), the selective values of the gametes are functions of the frequencies of the complementary gametes. In the symmetrical case, $s_{11} = s_{10} = s_{01} = s_{00}$, $L = 0$.

5.91	f	w
AB	pq	$1 + s(1 - p)(1 - q)$ $\quad \bar{w} = 1 + (\sum s)\, p(1 - p)q(1 - q)$
Ab	$p(1 - q)$	$1 + s(1 - p)q$ $\quad \hat{q}_{AB} = \hat{q}_{Ab} = \hat{q}_{aB} = \hat{q}_{ab} = 1/2$
aB	$(1 - p)q$	$1 + sp(1 - q)$ $\quad w_{AB} = w_{Ab} = w_{aB} = w_{ab} = w =$
		$\qquad\qquad\qquad 1 + (s/4)$
ab	$(1 - p)(1 - q)$	$1 + spq$

Most species must adapt to heterogeneous environments. The capacity to produce a variety of genotypes is an advantage. Thus there is probably almost always some selection for diversity itself. Following is a simple model (not treated in vol. 2):

Genotype	f	w	
A_1A_1	$(1 - q)^2$	1	$\bar{w} = 1 + 2sq(1 - q)[a - 2q \times (1-q)]$
A_1A_2	$2q(1 - q)$	$1 + s[a - 2q(1 - q)]$	$\sum w \dfrac{\partial f}{\partial q} = 2s(1 - 2q)[(a - 2q(1 - q)]$
A_2A_2	q^2	1	$\Delta q = sq(1 - q)(1 - 2q) \times [a - 2q(1 - q)]$
With small s,			$F(w/\bar{w}) = 1 + 2sq(1 - q) \times [a - q(1 - q)]$

$$\hat{q} = (1/2)[1 \pm \sqrt{(1 - 2a)}] \text{ (stable), or } 1/2 \text{ (unstable).}$$

Figure 14.3 shows the values of Δq, $\bar{w}$, and $F(w/\bar{w})$ for $a = 3/8$. There is stable equilibrium if $q = 1/4$ or $3/4$, but $\bar{w}$ is maximum at $(1/2)[1 \pm \sqrt{(5/8)}] = 0.105$ or 0.895. Since $a = 2q(1 - q) = 0$, $w_{A_1A_1} = w_{A_1A_2} = w_{A_2A_2} = \bar{w} = 1$. Here again, $L = 0$. The fitness at equilibrium is $1 + (9/128)s$. Since this is the fitness peak, there is no deviation at equilibrium from the peak values, $D[F(w/\bar{w})] = 0$.

It should be added that under changing environmental conditions, gene frequencies lag behind the changes in the equilibrium frequencies with the

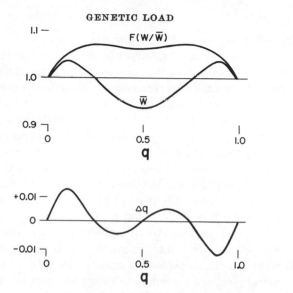

Fig. 14.3. Relations of mean selective value $\overline{W}$, fitness $F(W/\overline{W})$, and of Δq to gene frequency, q, under the model of selective advantage of diversity described in the text.

consequence that some genetic load is to be expected from alleles subject to frequency-dependent selection, in spite of theoretical absence of load at equilibrium.

Summary

This chapter has been concerned with various aspects of genetic variability, centering in the concept of genetic "load."

The most obvious way of dealing with the burden imposed by the occurrence of extreme deviants, generation after generation, in a population in genetic equilibrium, is to determine the percentage of each sort. These include a mutational load due to segregation of recurrent deleterious mutations, a heterotic load due to segregation of homozygotes where equilibrium is maintained by heterozygous advantage, a threshold load due to failure of developmental homeostasis below a threshold in the combination effects of nongenetic and genetic factors, and other sorts.

The term *genetic load* was, however, first proposed by H. J. Muller exclusively for the mutational load and this in a different sense from that above: the total damage from an array of individual mutations, each over the period from origin to extinction by selection. The unit load, one "genetic death," is the product of the selective disadvantage and the number of generations of persistence. The total load from n mutations is thus n genetic deaths. The advantageous heterotic mutations must be excluded as well as the minor mutations that contribute advantageously to versatility in encountering a heterogeneous environment. There is the further qualification that Muller's unit is unrealistic in treating selective disadvantage as equivalent to either personal or social damage per generation in the human case with which he was concerned. A lethal mutation that ends life shortly after conception is enormously less damaging than one that results in a personally distressing incapacity, burdensome to society, through a long life.

A third concept of load includes not only a mutational load but also a heterotic one and many others. The measure here is the percentage by which the mean selective value of the population falls short of that of an appropriate reference genotype, usually that at the maximum.

The mutation load per locus in an equilibrium population is very small, typically twice the mutation rate, in contrast with one "genetic death" per individual mutation. The total load is again practically the sum of the loads from the separate loci since combinations are too rare to cause much complication from correlated selection. The mutational load is, however, considerably greater in small populations subject to much sampling drift.

Individual heterotic loci, on the other hand, typically cause much larger loads than the preceding, but the total load from many is probably always much less than the sum, because of correlated selection involved in interaction, often including threshold effects.

The relative importance of the mutational and heterotic loads has been much discussed. Part of the disagreement has been due to confusion between the actual numbers of the two kinds of loci, their contributions to variability (to which a few heterotic loci contribute much more than many with mutation-selection balance), and their contributions to inbreeding depression (in which the zygote frequencies of the deleterious mutations are much increased). A method of evaluating the last of these indicates greater importance of the "mutational loci" for inbreeding depression; probably greater for numbers of loci, but less for contributions to variability.

There is an immigrational load in partially isolated demes, due to the balance between the introduction of locally deleterious genes (advantageous elsewhere) and locally adverse selection. This must, in general, be much larger for loci to which it applies at all than the mutational load and may be expected to be correspondingly less additive.

The reference genotype (or genotypic array) may be that of a selective peak either in the surface of mean selective values or in that of fitness, where these differ and the latter exists at all. The formula for load (last sense) here measures the deviation of the contour on which the population lies, from the peak in question. Two contours are of interest in the theories of evolution discussed in chapters 12 and 13: that relative to the peak of fitness values, for mass selection within homogeneous populations (species or demes), and that relative to the peak of mean selective values, for selection among demes. In cases in which the peak is a single genotype and the equilibrium depends on balancing of selection by some other pressure, the load (third sense) and deviation from the peak are necessarily the same. In the heterotic case, on the other hand, the load (relative to the maximum genotypes) tends to be large and the deviation from the heterallelic selective peak (which is the same for both mean selective values and fitness if selection values are independent of gene frequencies) is zero (apart from complications from recurrent mutation or from sampling drift). Where equilibrium depends wholly on an advantage of rarity, or on selection for diversity, there is no load in the third sense in an equilibrium population and no deviation of such a population from the fitness peak. The latter deviates in one direction or the other from the peak of mean selective value.

Among the thousands or tens of thousands of loci of a higher organism, there is reason to believe that a large proportion are strongly heterallelic ($\hat{q}$ less than 0.95 for the leading allele). In only a small proportion of these can this heterallelism be maintained by strongly manifested heterozygote advantage. Similarly, only a small proportion can be due to the balancing of a strong local selective advantage of one allele, capable of maintaining it against strong diffusion from neighboring regions in which this allele is not the most favorable. A population that occupies a definite niche tends to maintain itself, irrespective of unfavorable traits, if all individuals have one or more such traits, thereby greatly reducing all but the strongest selection pressures.

A larger proportion of strongly heterallelic loci may be due to strong rarity advantage, especially if conditions are uniform and constant, because of the absence of load at equilibrium.

Nevertheless most strong heterallelism must be maintained by balancing of only very weak selection pressures, or balancing of weak selection by weak diffusion or by mutation. In the last case, the selection must be less than 20 times the mutation rate ($v/s > 0.05$) but, as discussed in chapter 5, the rate of change of isogenic lines suggests that total mutation rates are probably much greater than the usually assumed value, about 10^{-5} per generation based on easily detectable mutations. Selection coefficients of the order of 10^{-3} per generation or more may be compatible with maintenance of strong heterallelism by balancing mutation.

The loci characterized by weak heterallelism are presumably ones in which recurrent mutation is overbalanced by stronger selective advantage of the leading allele over all others than implied above.

CHAPTER 15

The Evolution of Dominance

Mendel (1866) described one member of each of his seven pairs of contrasting traits of the pea as dominant, which he defined as applying to those "traits which pass into hybrid association entirely or almost entirely unchanged" (translation of Stern and Sherwood 1966). Thus, he recognized that dominance is not always complete. Nevertheless, after the rediscovery of the principles of heredity, dominance was sometimes treated as if it were as essential a principle as segregation.

This ceased when a considerable number of cases were found in which the heterozygotes were on the average about halfway between the homozygotes and usually highly variable, although segregation was strictly Mendelian. Goldschmidt (1911 and later) emphasized the ready modifiability of such heterozygotes by either genetic or environmental factors, and made this one of the cornerstones of his version of the theory that genes act by controlling the rates of processes. Morgan (1919), who rarely agreed with Goldschmidt, agreed on this. He described a number of cases in *Drosophila melanogaster* in which intermediate heterozygotes were profoundly modified by environmental conditions, and he also described an experiment in which he shifted the incidence of normality of the Notch mutation in heterozygous females (a sex-linked deficiency, lethal in males and homozygous females) from occasional to 50%, by selection. Prior to this, Bridges (1913) had observed that red eye color becomes incompletely dominant over white (sex-linked, normally recessive) in the simultaneous presence of sex-linked vermillion and third-chromosomal pink. Lancefield (1918) discovered a third-chromosomal modifier that causes intermediacy of sex-linked forked, ordinarily completely recessive. Timofeeff-Ressovsky (1927) produced, by selection, strains of *D. funebris* in which wild type was incompletely dominant over two, usually fully recessive, venation abnormalities.

Wilson (1908) showed from a compilation of herdbook records of Shorthorn cattle that roan is probably the heterozygote of the other common colors, red and red-eared white. There were, however, a rather large number of exceptions, which he accounted for by the wide range of variability of the roan color. There were attempts by others to explain the exceptions by postulating two or more loci, but consideration of the genotypic frequencies expected under the various hypotheses proved conclusively that Wilson's one-locus hypotheses, complicated by minor modifiers, was correct (Wright 1917b). A later analysis of data from the English, American, and Canadian herdbooks is presented in chapter 16. There seems no doubt that heterozygotes may range from self through dark roan and light roan almost, if not quite, to red-eared white, in accordance with the genetic background.

It is, indeed, probable that selection has made white with dark ears completely dominant or completely recessive in other breeds. Whether this is true or not, the following quotation makes clear my position in 1917 on the modifiability of dominance. The quality of the color, black, dun, or red, depends on a different locus.

The famous Chillingham cattle are white with dark points. They occasionally throw black or dun colored calves. Here white is clearly dominant. A very similar white with dark points crops out from time to time in Pembroke and in Highland cattle although selected against. Here white is clearly recessive. Thus, we have whites of very similar appearance which are respectively recessive, dominant, or neither in Pembroke, Park cattle, and Shorthorns. Is it necessary to suppose that three independent factors are involved? The simplest explanation seems to be that there is a continuous series of physiological conditions between a very strong self tendency through roan to a very strong white tendency. Two hypotheses are at present equally possible. There may be one Mendelian pair which stands out in importance of effect, but which, nevertheless cooperates with lesser factors which determine the general level in the series; or there may be a series of allelomorphs.

Dosage-Response Curves

The first theory of the nature of dominance was that of Bateson in (1909):

All observations point to a conclusion of great importance, namely that a dominant character is the condition due to the presence of a definite factor, while the corresponding recessive owes its condition to the absence of the same character.

This view became untenable with respect to the presence or absence of the genes themselves with the discovery of qualitatively diverse multiple alleles and reverse mutation, but (contrary to Ford [1930]) this in no way invalidates it in terms of the degree of activity of the gene product. The hypothesis is that there are diminishing returns as activity is increased, until a ceiling is reached, because of limitation of some other factor in the reaction. This accords with the usual situation with respect to the effects of multiple alleles on a character:

increasing dominance of higher over lower members of the series up to the highest, which is apt to be completely dominant over all others (vol. 1, chap. 5).

The application of the general thesis that degree of dominance is a function of the relation between gene dosage and the physiological response of the reaction of its product with those of other genes and with the developmental and environmental conditions requires investigation of the effects of substitutions at the locus in question under a great variety of conditions.

The simplest cases are those in which these effects are independent of all other loci. This is at least approached in the case of antigenic effects, with respect to which some very extensive multiple allelic series exhibit codominance, indicating that antigenic specificity is very directly related to the specificity differences of the primary gene products. More surprising are the sets of alternative color patterns in a considerable number of animals in which codominance takes the form of dominance of the pigmentation in any regions determined by one allele over absence of pigment in the same region in the pattern determined by another allele and vice versa, indicating that the specificity of each allelic product is manifested by response to equally specific local developmental conditions (vol. 1, chap. 5).

There are, indeed, cases in which lower alleles in what seems a graded series of allelic effects are completely dominant over all still lower ones. These cases can be accounted for by postulating that the specificity of the primary gene products is manifested in the range of substrates with which it reacts, instead of in the degree of activity. The series A, A^r, a in the guinea pig, determining an extensive yellow band, a narrow yellow band, and no yellow band, respectively, in otherwise eumelanic hair, seems to be of this sort.

Where the allelic effects suggest merely grades of activity, the observed response is not, in general, that of the primary gene product. There is likely to be a chain of processes intervening between the reaction of the latter with its substrate and the observed response. Successive reactions in the chains may be affected by different loci. The study of such systems has revealed a number of different ways in which the dosage-response curve for one locus may be modified, with consequent changes in degrees of dominance. The coat colors of guinea pigs have been studied extensively from this point of view. The more important data have been given in volume 1, chapter 5.

Dominance among Genes Affecting Coat Color in Guinea Pigs

Castle (1912a) produced, by selection, a strain of guinea pigs of exceptionally dilute colors (very light sepia, very light yellow), and albinos ($c^a c^a$). He produced another strain, selected for extreme intensity of black. This also carried albinism. It was shown (Wright 1915) that the former carried an allele, c^d, of the albino gene, c^a, the latter carried gene C. On crossing the extreme dilute-selection stock with the albinos of the intense stock, the sepias, $E\ c^d c^a$, and yellows, $ee\ c^d c^a$, were greatly intensified. There was also intensification of extracted homozygotes $E\ c^d c^d$ and $ee\ c^d c^d$ but to a lesser extent than the heterozygote, so that the degree of dominance of gene c^d was increased (Wright 1916).

Castle's dilute-selection strain was the source of a modifier, diminished, dm, demonstrated many years later (Wright 1959b), and it probably also carried another modifier of intensity, silvering, si (known to be present in Castle's colony), judging from a novel color variety, white with pale spots and reduced eye color (1912a), but not seen again until the combination $sisi\ dmdm$ was produced and described in the 1959 paper. The considerable variation in the degrees of dominance of c^d over c^a, according to the presence of E or ee, and the number of representatives of the semidominant modifiers Si and Dm, may be seen from the table on page 77 in volume 1. The range of values of the coefficient of dominance, h, is from 0.56 to 0.66 in the sepias and from 0.23 to 0.51 in the yellows. The values below 0.50 were interpreted as due to a threshold for yellow, indicated by other lines of evidence.

The determination of the effects on phaeomelanic (ee) and eumelanic pigmentation by the combination of 15 compounds of the c-series of alleles (C, c^k, c^d, c^r, c^a) with E, e; P, p; F, f; and B, b revealed differences of the degrees of dominance within the c-series in most cases. In the case of genes c^d and c^a, there was a range from complete dominance of c^d (eye color with P, later pelages with $EbbP$) through diverse grades of intermediacy to complete dominance of albinism (with $eeff$) (Wright 1927). The same tortoiseshell animal, $e^p e^p c^d c^a ffP$, showed complete dominance of c^d in the eye color, intermediacy within sepia areas, and dominance of c^a within yellow areas.

The pattern of interaction indicated a certain sequence of reactions

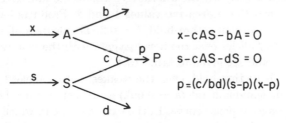

FIG. 15.1. Theoretical relation of the rate (p) of production of a gene product (P) to activity of gene product (x), rate (s) of production of the substrate (S), and rate constants (b, c, d).

illustrated in volume 1 (fig. 5.11). Figure 15.1 (=5.5A, vol. 1) represents the reaction between gene product (A) and substrate (S), applicable to homozygotes and to heterozygotes with an inactive allele. In the case considered, x is the activity of the gene product, s the rate of supply of substrate, c the rate constant for the reaction (product P), b the rate constant for diversion of the activity of the gene product, and d the rate constant from diversion of the substrate. These lead, during a steady state, to the following equations:

$$p = k(x - p)(s - p), \qquad k = c/bd$$
$$p = 0.5\{(x + s + 1/k) - [(x + s + 1/k)^2 - 4sx]^{1/2}\}$$

The dosage-response curve is a hyperbola rising from zero for the homozygous inactive allele toward the asymptote, s. There may, however, be destruction up to a certain level, giving a threshold t:

$$p' = p - t$$

The amount of substrate, S, may be affected by differences at other loci, as may be the constants b and d because of competitive reactions. The product of the first reaction is the dosage for the second. If complete dominance is imposed at the first step by limitation of s, AA and Aa have the same dosage at the second step.

Figure 15.2 and table 15.1 are intended to illustrate different ways in which dominance may be increased at the first step: (1) There may be substitution of a higher allele with a greater value of x, (2) lowering of the amount of substrate, S, (3) reduced competition with another reaction of the gene product, reducing b, or (4) reduced competition with another reaction of the substrate, reducing d. The last two increase the parameter $k = c/bd$. Finally, (5) there may be reduction of a threshold, t. In figure 15.2 the alleles A_4, A_3, A_2, A_1, and a are assigned the values 8, 4, 2, 1, and 0, respectively. Substrate rate s is assigned the alternative values 0.5 and 1. The rate constant k is assigned the alternative values 1 and 2. Products have been calculated with and without a threshold, $t = 0.2$. The degrees of dominance of each of the higher alleles over the heterozygote with the inactive allele have been calculated and are given in table 15.1.

Where there is no threshold, even the weakest active allele A_1 is considerably more than semidominant ($h = 0.652$) as expected from the curvilinear form of the dosage-response curve. In the extreme case in which the reaction is of the bottleneck type, $p = s$ or $p = x$, whichever is smaller, $h = 0.50$. Lowering the ceiling by 50% raises the degree of dominance only from 0.652 to 0.698. Doubling the rate constant k causes about the same change, 0.652

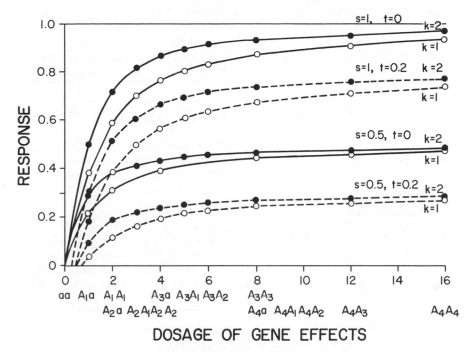

FIG. 15.2. Hypothetical ratio of genotypic products (ordinate) to dosages of multiple alleles a, A_1, A_2, A_3, and A_4 with different rates ($s = 0.5, 1$) of production at substrate, different values of k ($k = 2$, *solid circles*; $k = 1$, *open circles*), and either no threshold ($t = 0$, *solid lines*) or a threshold ($t = 0.2$, *broken lines*).

to 0.695, while substituting an allele A_2 with twice as much activity raises h from 0.652 to 0.767. If there is a threshold at $p = 0.2$, the degree of dominance is reduced to 0.472, and if the ceiling is also lowered 50%, dominance is almost reversed (to 0.169) and could be completely reversed with slight changes in the coefficients. On the other hand, doubling the rate constant k and putting the threshold at $t = 0.2$ has relatively little effect ($h' = 0.600$). Starting from a higher dosage $A_3 = 4$, $h = 0.871$, lowering the ceiling 50% increases dominance only to 0.887; doubling the rate constant increases it to 0.924. There is relatively little effect of introducing a threshold ($h' = 0.833$).

The most important conclusion is that there is a strong bias toward a high degree of dominance of the most active allele at a locus with major effects, over lower alleles since the reaction of its product is especially likely to be limited by the amount of substrate. Even if it is not seriously limited at the first step, it is likely, for the same reason, to be limited at the next or a later step, and limitation at any one step is enough.

TABLE 15.1. Degrees of dominance, h, of alleles with active products, A_1 to A_4, over complete inactivation, a, under diverse conditions. Effect, p, of twofold difference in rate constant ($k = 1$ or 2), of twofold differences in ceiling ($S = 0.5$ or 1.0), of threshold, $t = 0.2$, $p' = (p - t)$, $h' =$ corresponding degree of dominance, and of differences in the primary products of five alleles, $x = 0, 1, 2, 4, 8$.

	x	S	$k = 1$				$k = 2$			
			p'	h'	p	h	p'	h'	p	h
aa	0	1	0		0		0		0	
A_1a	1	1	0.182	0.472	0.382	0.652	0.300	0.600	0.500	0.695
A_1A_1, A_2a	2	1	0.386	0.684	0.586	0.767	0.519	0.784	0.719	0.833
A_2A_1, A_3a	4	1	0.564	0.833	0.764	0.871	0.663	0.903	0.863	0.924
A_3A_3, A_4a	8	1	0.677	0.918	0.877	0.935	0.734	0.956	0.934	0.964
A_4A_4	16	1	0.738		0.938		0.768		0.968	
aa	0	0.5	0		0		0		0	
A_1a	1	0.5	0.019	0.169	0.219	0.698	0.093	0.510	0.293	0.767
A_1A_1, A_2a	2	0.5	0.114	0.595	0.314	0.802	0.182	0.763	0.382	0.871
A_2A_2, A_3a	4	0.5	0.192	0.796	0.392	0.887	0.238	0.887	0.438	0.935
A_3A_3, A_4a	8	0.5	0.242	0.896	0.442	0.940	0.269	0.945	0.469	0.968
A_4A_4	16	0.5	0.220		0.470		0.284		0.484	

If, on the other hand, a gene has only a minor modifying effect on the character, its primary product may well be limited only slightly by the amount of substrate available, and there is less likelihood of serious limitation at subsequent steps than in the preceding case. Thus, intermediacy of the heterozygote is likely to be characteristic of minor modifying factors as well as of lower alleles of loci at which the most active allele is completely or nearly completely dominant.

Dominance in Other Interaction Systems

Some other interaction systems in which degrees of dominance at loci are modified by substitutions at others have been given in volume 1, chapter 5. In the case of the pattern of epidermal ridges, manifested by irregularities in hair direction in the guinea pig, the gene R, essential for the common series of rosette patterns, ranging from reversal of hair direction on the toes (R-MM) to a pattern of rosettes over all parts of the coat (R-mm), is completely dominant in all combinations, including crosses with smooth-furred wild species ($rrMM$). Its major modifier, M, ranges, however, from nearly complete dominance to less than semidominance according to the genetic background of minor unanalyzed modifiers (Wright 1916, 1949c).

The gene St was found to be completely dominant when introduced into otherwise smooth-furred strains (rr), in which it was responsible merely for a strong rosette in the forehead. In the presence of R, however, there was considerable reciprocal inhibition of the two rosette patterns (Wright 1950a). In association with R-mm, $StSt$ has a stronger inhibitory effect than $Stst$, showing that excess secondary product carries through the reaction of the gene product in spite of apparent complete dominance of St over st in all other known combinations, and manifests itself in this excess inhibitory effect. This is analogous to the cases of the color factors C and P, which seem to be completely dominant over their alleles in most combinations but in browns (E-bb) reveal a difference between CCP- and Cc^aP-, and between C-PP and C-Pp, respectively. There is the difference that the excess C and P products revealed in browns have inhibitory effects on pigmentation, giving apparent overdominance of Cc^a and Pp.

Changing dominance was also exhibited by gene Re (eye rosette) involved in a third sort of rosette pattern, according to whether associated with R or rr, and manifested in this case in amount of penetrance.

A study of interactions among genes affecting flower color in the functionally tetraploid *Dahlia variabilis* by Lawrence and Scott-Moncrieff (1935) is especially instructive because of the opportunity to study dosage effects in up to four representatives of the gene. The number of representatives is indicated by the superscript in the following formulas. As may be seen in volume 1 (fig. 5.12), there are rising responses to increase in dosage of gene A in the presence of y^4b^4, ranging from none with a^4 to considerable cyanine with A^3a and A^4. In the case of B in the presence of $y^4i^4a^4$, the response rises from zero with b^4 to heavy cyanine with Bb^3 and to heavy pelargonin with B^2b^2 to B^4. The response of gene I in the presence of $y^4a^4b^4$ is from zero with i^4 or Ii^3 to presence of ivory apigenin with I^2i^2 to I^4, an example of a threshold. The response to Y in the presence of a^4b^4 merely exhibits dominance with one dose, no yellow flavone with y^4, full amount with Yy^3 to Y^4, but in the presence of A shows an effect on anthocyanin pigment (cyanine with y^4, pelargonin with Yy^3, followed by progressive inhibition of this in Y^2y^2 to Y^4). These are only a portion of the array of dominance modifications shown as aspects of a complex pattern of interactions in this case.

Threshold Dominance

Even if dominance is not imposed by a limiting factor at some reaction in the metabolic chain leading to the observed character, it may be imposed by a threshold or ceiling in the character itself. The extent of a spotting pattern can only range between 0 and 100%, but there may be combinations of genes favorable to one or the other extreme beyond ones that can bring these about. The situation is similar with respect to the range between absence and complete development of an organ. In one genetic background the hetero-zygote at a locus, as well as the unfavorable homozygote, many give 0, while in a different genetic background the same heterozygote and the other homozygote may give 100%. In this sort of extreme case, dominance has no essential relation to metabolic activity of the gene product.

I first observed the modification of dominance by selection as Dr. W. E. Castle's assistant (1912–15) in his selection experiment with hooded rats described in chapter 7.

The selections in each direction were made exclusively in the hooded rats, *ss*. The amount of white in the patterns was shifted from roughly 50% to 90% in the plus series and to about 10% in the minus series. In crosses with a strain of wild rats (all self-colored, genotype *SS*) the former yielded hetero-zygotes that were almost or quite self-colored while the latter yielded hooded intermediates. Thus selection, restricted to the hooded homozygotes, carried with it a correlated modification of heterozygotes that markedly altered the degree of dominance.

Castle seems to have taken this correlated effect so much for granted that he refers to it only casually in his accounts of the progress of selection in the homozygotes. I found, however (Wright 1923c, 1928), that the situation was essentially the same in the guinea pig except for the much more extensive nongenetic variability in the latter.

There were wide differences in the degrees of dominance from crosses of various piebald inbred strains, *ss*, with median amounts of white ranging from 15% to 87%, with two unrelated self-colored strains, *SS*. Experiments were later designed to make it possible to compare *SS*, *Ss*, and *ss* on each of three different genetic backgrounds (Wright and Chase 1936). The results are shown in part in volume 1 (table 5.2). With 7/8 of strain *D*, all of *SS* and 96% of *Ss* were self-colored (4% *Ss* had mere traces of white), while all *ss* showed white but not very much. With 15/16 of strain 13, 50% of *SS* were self-colored and the other 50% were slightly beyond threshold for traces of white, while all of *Ss* showed some white and all of *ss* showed a great deal. Crosses of strains *A* (*SS*) with *B* (*ss*) gave intermediate results.

The occurrence of the little toe in the guinea pig has a lower threshold at which a rudiment develops and a ceiling of perfect development (vol. 1, fig. 5.14, 5.15). Inbred strains ranged from 100% with no little toe (three-toed) through ones with various percentages to one (*D*) produced by Castle by selection, with 100% perfect little toes (four-toed). The differences were found to depend on multiple factors (equivalent to four equal loci with semidomi-nance, except as affected by threshold and ceiling). Crosses of four-toed strain *D* with three-toed strain no. 2 showed complete dominance of the three-toed condition. Crosses of *D* with a strain (no. 35) with 69% three-toed, 31%

with usually rudimentary little toes, gave near-dominance of the four-toed condition. Thus the direction of dominance was almost reversed according to the location of one of the parents on the scale of digit tendencies. All of the other results were in harmony with the threshold interpretation of dominance (Wright 1934*d*).

In meristic characters (vol. 1, chap. 6) there is a succession of thresholds, one for each increase in the number of units. The intervals between them need not be uniform in terms of gene dosages. It is, indeed, usual to find more or less homeostatic regulation at a particular number (canalization of Waddington [1942]). This is illustrated in volume 1 (fig. 6.9 and 11.8) for the number of tentaculocysts in the jellyfish, ephyra (standardized at 8) and the number of ray florets in a strain of *Chrysanthemum segatum* (standardized at 13). The number of scutellar bristles in wild *D. melanogaster* is standardized

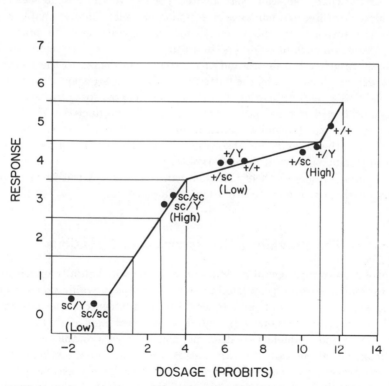

FIG. 15.3. Relation of number of scutellar bristles (ordinate) to dosages due to genotypes *sc/Y*, *sc/sc*, +/*sc*, +/*Y*, and +/+ after low or high selection. The dosage range between the threshold and ceiling for four bristles is much greater than for other numbers. From data of Rendel (1959); used with permission.

at 4 but was increased to 14 by selection (Payne 1918*a,b*) starting from rare exceptions with five bristles. The sex-linked mutation scute typically has two bristles but varies from none to three. Rendel (1959) produced a low-selection line in which *sc/sc* averaged 0.25, *sc/*+ averaged 3.99, and +/+, 4.00, and thus practically complete dominance of +. He established normalized ranges for the bristle intervals by use of probits (5 + *pri p*, vol. 1, p. 170). These ranges were substantially the same for all stocks, low, intermediate, and high. The total range from the threshold for one bristle to that for seven was about 14 probit units, but that encompassing four bristles was about half of this, leaving only seven probit units for the five other intervals. Figure 15.3 shows the relation between dosage (probits) and response (average bristle number) according to Rendel's calculations.

It appears, however, that there would be a rather high degree of dominance even if the character scale were related linearly to the dosage scale. The considerable degree of dominance of + under plus selection presumably represents the response in the terminal metabolic reaction. The practically complete dominance of wild type in an unselected or low selected strain is thus the result of the imposition of a rather high degree in the metabolic reactions, supplemented by the flattening of the dosage-response curve in the interval that gives four bristles. The absorption of extra dosage up to a certain point in this region is comparable to that of *CPF* product (vol. 1, p. 76) in brown guinea pigs beyond a certain level, except that in the latter case the effect, after the capacity for absorption of excess is saturated, is inhibitory instead of moving toward increase (vol. 1, p. 76).

The dosage-response theory of dominance was developed in the earlier studies without reference to evolutionary implication.

The Prevailing Recessiveness of Mutations

An evolution toward complete dominance at loci with initially intermediate heterozygotes has been postulated to account for the prevailing recessiveness of mutations, first by East and Jones (1919) and later by Fisher (1928*a*).

The problem is a real one, although it is somewhat exaggerated by natural selection in the frequencies within populations. Unfavorable, appreciably dominant mutations cannot escape natural selection and so are kept at very low frequencies, $\hat{q} = v/hs$. Unfavorable recessives largely escape selection, and, if equally recurrent (rate v), are kept at much higher frequencies $\hat{q} = \sqrt{(v/s)}$ (vol. 2, chap. 2). We are concerned here, however, only with the frequencies of different degrees of dominance of mutations before alteration by selection.

It is difficult to get statistics wholly unbiased by selection, but there seems no doubt that the majority of nonlethal mutations, with effect sufficiently conspicuous to be recognizable without special tests, are recessive or nearly so.

Fisher (1928a) supported this contention by compiling a table of the nonlethal mutations of *D. melanogaster* from the data of Morgan, Bridges, and Sturtevant (1925) (as well as by less extensive data from other organisms):

	Recessive	Intermediate	Completely Dominant	Total
Autosomal	130	9	0	139
Sex-linked	78	4	0	82

These referred to *Drosophila* mutants that had accumulated before Muller (1927) discovered induction by X-radiation. There are certainly various sorts of possible bias that may distort the actual frequencies somewhat, but the frequencies observed since in arrays of induced mutations indicate a similar enormous excess of recessives in spite of the more direct recognition of those with appreciable dominance.

A more serious qualification for the present purpose is in the way in which the terms *dominant* and *recessive* have been used by geneticists, including Mendel (1866). It is stated in the paper (p. 70) from which the data were taken that

It is to be remembered that the distinction between a dominant and a recessive factor is largely an arbitrary one, depending to a great extent upon the line of demarkation in the F₂ groups of the three classes of *AA*, *Aa*, and *aa*.

Emerson, Beadle, and Fraser (1935) tabulated the mutants of maize known at that time, including multiple alleles. There were 296 recessives and 54 dominants (not all completely dominant). These included mutants extracted from populations, with bias favoring recessives, and observed mutations, favoring dominants.

As mentioned earlier (vol. 1, chap. 5), I have paid special attention to measuring the grades of dominance in sets of alleles of the guinea pig from 1916 on. The wild type was completely dominant over lower alleles at only five loci (with ten lower alleles) among 13 loci in which the differences were recognizable on a wild-type background. There were six loci at which the heterozygotes were typically intermediate and two at which wild type seemed completely recessive. The heterozygotes between nontype alleles were intermediate in seven of eight cases. The heterozygotes between alleles with effects not recognizable on a wild-type background were intermediate in all of the three loci.

The mutant genes were indeed, with one exception, ones that had accumulated over many years of domestication and they are hardly an unbiased lot. The data illustrate, nevertheless, that the dominance of wild type is far from invariable.

Evolutionary Theories of Dominance

According to East and Jones (1919, pp. 179–80),

One may be led to inquire why it is that most of the experimentally observed mutations are recessive and less favorable to the best development of the organism. We do not know, but we may hazard a guess. The repeated appearance and disappearance of certain mutations is merely a type of variability which has probably been a constant feature of the organism for a long period and has been subjected to natural selection in the same way as any other character. In other words, may not the tendency to produce dominant unfavorable variation have been reduced to the minimum by natural selection? Conversely a tendency to produce unfavorable recessive mutation has been tolerated because the latter are protected in hybrid combinations by their dominant favorable allelomorphs.

The selection pressure on modifiers of particular kinds of recurrent mutation would obviously be exceedingly small. A more plausible theory was advanced by Fisher (1928a) and has attracted much attention since then.

Fisher (1928a) started from the paradox that mutations are usually recessive to the wild-type genes but in the course of evolution the latter are replaced by mutations that in turn become dominant type genes. He wrote (1928a, p. 116),

This difficulty will lose its force if it appears that there is a tendency always at work in nature which modifies the response of the organism to each mutant gene in such a way that the wild type tends to *become* dominant.

His hypothesis was that this acquirement of dominance is due to an enormously long continued selection of specific modifiers of the rare heterozygotes, a process that ultimately causes the latter to resemble wild type.

With respect to the physiology of dominance, Fisher (1928a, p. 123) considered the "pristine" character of heterozygotes to be intermediacy. Fisher (1930, pp. 65–66) stated:

A priori it would be reasonable to suppose that at the first appearance of a mutation, the reaction of the heterozygote would be controlled equally by the chemical activity of the two homologous genes, and that this would generally, though not necessarily in every individual case, lead to a heterozygote somatically intermediate between the two homozygotes.

He showed, by consideration of the contributions made to descendants, that the selective intensity of his heterozygote modifiers would be of the order of the mutation rate of the type gene, if it brought about dominance of the latter by itself. He took this rate to be about 10^{-6} per generation. Modifiers with a minor effect on dominance would be subject to a correspondingly weaker selection.

He believed that the theory had very important evolutionary implications, in addition to accounting for the prevailing recessiveness of mutations (Fisher 1928b, p. 574)

Perhaps the most illuminating sidelight of all which it throws upon the evolutionary theory arises from the fact that it reveals an effect of natural selection which has nothing to do with adaptation of the species to the conditions of its environment.... It has now been shown that the same agency as a minute byproduct of its activity must also tend to modify dominance, and if the recessiveness of each several mutation be referred to this cause, the vast number of reactions which must have been so modified, give a measure of its efficiency which might have startled even a Weismann.

Critique of Fisher's Theory

The paradox from which Fisher started loses impressiveness on taking account of the extremely small proportion of mutations that ever become type genes. The latter is not a random sample. Nearly all of the conspicuous observed mutations are probably ones with partially or completely inactivated products, with little chance of ever being seized upon and established as type genes by natural selection. Most of those seized upon are probably ones with only slight effects, for cogent reasons on which Fisher (1930) himself put much stress. These may be expected to have intermediate heterozygotes initially and may remain so after establishment. The species may continue in a state in which most frequently occurring mutations are recessives, as observed, while gene substitution involves little if any change in dominance (Wright 1929a).

There is no doubt about the minuteness of the selection pressure implied by Fisher's theory. His conclusions that this is at most of the order of the mutation rate of the type gene was verified by a wholly different method (Wright 1929a, 1934a; also vol. 2, pp. 69–71). Since then, O'Donald (1967b) has obtained a similarly minute selective intensity from the standpoint of the additive genetic variance of Fisher's "fundamental theorem."

My primary criticism of Fisher's theory was that practically all of his dominance modifiers might be expected to have some effect on homozygous wild type that would take precedence over their minute effects on the heterozygotes, because of the much greater abundance of the former. His selection pressure would, in short, fall below the pleiotropic threshold (vol. 2, pp. 69–71).

Fisher argued (1934) that the pleiotropic effects on the modifiers, some favorable, others unfavorable, would balance, leaving the minute effect as modifiers of the heterozygote to prevail. The net pressure on each modifier must, however, be considered separately. Three cases may be distinguished. First, an effect tending to increase the fitness of the organism would have tended to bring about near-fixation of the modifier from the first, in which case its frequency could not rise appreciably. It would already be exerting practically as much influence tending to cause dominance of wild type as possible. Second, an effect tending to reduce fitness would hold the frequency of the modifier at a low equilibrium value so strongly that a counterpressure as a dominance modifier of the heterozygote of lower order of magnitude would be of no significance. Third, an effect as a modifier of homozygous type of the same sort as that of the heterozygote would give a much greater selection pressure toward its fixation and take precedence as long as the homozygote was below the ceiling for this sort of change. If the type gene were not initially completely dominant, such a modifier would tend to make it more so, assuming that the correlated effect on the heterozygote was much greater, as expected of a typical dosage-response curve, rising toward a horizontal asymptote. This is precisely the sort of dominance modifiers implied by the dosage-response theory of dominance, and has nothing to do with Fisher's specific modifiers of the heterozygote.

The smallness of the selection pressure resulting from natural selection of modifiers of the rare heterozygotes also runs into difficulty from being only of the order of the mutation pressure even in the most favorable case, because it is then no more effective than the mutation rates in determining its change in gene frequency (assuming that these are of the same order as the mutation rate of the primary gene) (Wright 1929a). It also runs into difficulty because of being overwhelmed by random drift unless the effective population size is of higher order than its reciprocal (Wright 1929b). The probability distribution for the modifier, assuming semidominance, is as follows (vol. 2, eq. 13.30 with $t = 0$ and $m = 0$):

$$\varphi(q) = Ce^{4Nsq}q^{4Nv-1}(1 - q)^{4Nu-1}$$

$$\text{If } 4Ns = 4Nv = 4Nu = 1, \qquad \varphi(q) = Ce^{q}.$$

With no selection, the distribution is rectangular. The frequencies of all possible gene frequencies of the modifiers are uniform as the latter drifts at random over the entire range, 0 to 1. Under the conditions assumed above, this distribution is tilted toward higher frequencies in the neighborhood of fixation, but at fixation ($q = 1$), it is only 1.7 times as frequent as when $q = 0$. With larger effective populations, there is, indeed, more drift toward fixation but with ones of lower order the effect of selection is wholly negligible. The actual change in gene frequency from accidents of sampling has a standard deviation of $\sqrt{[q(1 - q)/2N]}$ that is enormously larger than the selective pressure if $s = 1/4N$ (Wright 1929b). Crosby (1963) has stressed both of these difficulties as a result of experience with simulation experiments.

The Dosage-Response Theory of Dominance

Thus there are a number of reasons why the minute selection pressure implied by Fisher's theory would be ineffective. The problem of the prevailing recessiveness of observed mutations remains, and requires an alternative explanation. As noted, Fisher held that there was no possible alternative.

The alternative that I stressed (1929a) was that under the dosage-response theory, the type gene at a locus with major effects would usually be as dominant from the first over the most common type of mutation (inactivation), as observation indicates, and thus would require no evolutionary process. Loci with minor effects, even though considered to be much more abundant in the genome as a whole, were excluded because of not providing many of the type alleles at which mutations are observed. As noted above, I considered it probable that they would show only semidominance if they could be observed. Various sorts of modifiers of dominance were listed, but I did not think then that selection of favorable ones in the cases in which there was not complete or nearly complete dominance of the new wild-type gene from this first could carry the grade of the heterozygote very much beyond semidominance before selection of the modifiers of the homozygote, the process referred to in the preceding section, would cease.

The basic difference between Fisher's and my viewpoint lies in our concepts of the physiology of dominance. In a reference to a paper of mine (1925) on the c-series of alleles in guinea pigs (in the backgrounds $EPFB$ and $eePFB$), he suggested that the mutations c^k, c^d, c^r, and c^a from type C have been recurring through the whole history of mammals (and he might well have extended this to that of vertebrates, since recessive albinism occurs in all classes). According to his theory, gene C is dominant over each of the others because of selection of specific modifiers of Cc^k, Cc^d, Cc^r, and Cc^a, respectively, while the six heterozygotes, c^kc^d, c^kc^r, c^kc^a, c^dc^r, c^dc^a, and c^rc^a, are intermediate between the corresponding homozygotes (as I had shown them to be) because they are rare to the second order, giving inadequate time for selection of favorable modifiers. This contrasts with the interpretations of the grades of all 15 genotypes on the basis of dosage-response curves with dominance of C, the most active, due to limitations of substrate, probably in the primary reaction. Fisher's concept that competition between the gene effects in each zygote results in a pristine tendency toward an intermediate heterozygote, subject to change by specific modifiers, is very different.

The questions whether the most common mutations are ones with more or less inactivated products, as I assumed, or ones with competing products, as postulated by Fisher, may be considered here on the basis of later evidence (discussed in part in Wright [1934a]).

All mutations that are not deleterious are, no doubt, chemically different; and this applies, except for redundancy of the genetic code, to their protein products. Competition is thus possible, but mere reduction or loss of activity of the gene product seems a priori likely to be more significant. No certain conclusion is possible a priori, however.

There is abundant evidence that recessive mutations characteristically exhibit pseudodominance in heterozygotes with known chromosomal deletions. More significant is the observation that the accumulation of more than two representatives of recessives in the genome results in phenotypes that tend to approach wild type, if they change at all, rather than give a more extreme defect, indicating that they are like partially inactivated type genes, rather than active competitors (Stern 1929; Muller 1932a). Moreover, where there is clear evidence of qualitatively different effects, as in antigens and certain allelic series of color patterns, the heterozygotes tend to exhibit codominance rather than competition.

In the case of the c-series of alleles of the guinea pig, there is strong reason to believe that these determine qualitatively different tyrosinases because of the lack of parallelism in their effects on phaeomelanin and eumelanin. Tests of tyrosinase and dopa-oxidase activities, described in volume 1, chapter 5, indicate, however, that these alleles are directly concerned with degrees of activity in these two respects considered separately, with C most active in both and c^a wholly inactive (at birth). Genes E, P, and F also seem clearly concerned with tyrosinase activity, in contrast with their alleles e, p, and f.

The numerous investigations in biochemical genetics initiated by Beadle and Tatum in 1941 have enormously increased the evidence that the dominant allele is usually concerned with carrying through a particular metabolic reaction (often by means of a demonstrable enzyme), while the recessive allele is associated merely with absence of the reaction.

It should be added that major type genes often turn out to be mixtures of isoalleles with only slightly different effects, but all dominant or nearly so over recurrent mutations with no effect.

Other Criticisms

Haldane (1930) made similar criticisms of Fisher's theory on the basis of a somewhat similar physiological interpretation of gene action, but he emphasized that essential genes with inadequate effects under some conditions tend to be replaced by more active alleles, which are more completely dominant over inactivation.

Plunkett (1932a,b) reported on a tendency at low temperatures for wild-type *Drosophila* to exhibit defects similar to those of many mutations and suggested that natural selection of modifiers of homozygous wild type that give a greater factor of safety against this tendency would be expected to bring about a still greater improvement of the heterozygote as a by-product. This is the same sort of selective modification of homozygotes implied by the form of the dosage-response curve, but it would be more effective in increasing dominance than I had supposed. I had envisioned only slight quantitative variability in homozygotes with responses close to the ceiling. Under Plunkett's suggestion the average response may still be considerably below the ceiling when the norm is at the ceiling, permitting selection to raise the response of the heterozygote while giving a factor of safety to the homozygotes.

Muller (1932a) made a similar suggestion. His terms *amorph*, *hypomorph*, and *hypermorph*, according to degrees of activity relative to type, and *neomorph* and *antimorph*, for qualitatively different sorts of activity, fit in with the dosage-response theory. I reviewed the subject (1934a) in a paper devoted primarily to developing this theory further, and noted my acceptance of the supplementary views of Haldane, Plunkett, and Muller.

East (1935) reviewed the subject and in the main agreed with my interpretation. He concluded:

(a) That most recessive genes are genetic isomers that are defective physiologically and are easy to detect on this account, (b) that such mutations are not material that is useful in evolutionary processes, and (c) that non-destructive mutations occur with a high degree of frequency, that they are not recessive or dominant to wild type but are active pattern formers just as are active wild type genes, and are difficult to detect on this account.

Haldane came back to the subject in 1939. From mathematical study of the effectiveness of Fisher's process in populations that are mainly self-fertilized, he concluded that the intensity of selection must be less than in outbred species, but since dominance is often more common in inbred than outbred species, Fisher's theory needed modification on this account as well as others.

It may be added here that the theory of the limitation of the rate of evolution developed by Haldane (1957) (chap. 12) makes the fixation of the host of modifiers that would have "startled even a Weismann," in the earlier quotation from Fisher, difficult to accommodate in geologic time.

Selection for Dominance during Substitution

Another possible alternative to Fisher's hypothesis is that dominance developed while the new type gene was being substituted for the old one. This substitution could have been due either to a change in the environmental conditions, as in the well-known replacement of light-colored moths by melanic variants in heavily industrialized regions during the last century and a half, or it may go back to remote ancestors of the modern species, as suggested by Fisher for the type allele of albinism. In either case, selection would operate over a period in which heterozygotes were relatively abundant. This avoids the extreme slightness of the selection of modifiers that was the basis for the criticism of Fisher's hypothesis. In my first paper on the subject (Wright 1929a), I agreed with the possibility of modification of dominance by selection of modifiers of the heterozygote where the latter were abundant:

If for any reason, the proportion of heterozygous mutants reaches the same order as that of type, selection of modifiers approaches the order of direct selection in its effects and might well become of evolutionary importance.

In the present case, however, even this sort of selection is not very effective since the selection of modifiers cannot go very far before the new type gene has approached fixation and heterozygotes are again rare. Haldane (1956) demonstrated this in an investigation of the case of the melanic moths referred to above. In a computer study by Crosby of the most favorable case (that of a semidominant modifier starting from a frequency of 0.50, and producing dominance by itself when homozygous), selection raised its frequency only to 0.67 and 0.63 in two trials. Mayo (1966) reached similar results in similar experiments with one, two, or ten modifiers.

The pattern of selective values for a primary pair of alleles, A and a, and a semidominant modifier, M, used in the discussion of the pleiotropic threshold in volume 2 (p. 69) is repeated here, except that r replaces $1 - p$ as the frequency of new type genes.

Modifier	f	aa $(1 - r)^2$	Aa $2r(1 - r)$	AA r^2
MM	q^2	$1 - s$	$1 - hs(1 - x)$	1
Mm	$2q(1 - q)$	$1 - s$	$1 - hs(1 - 0.5x)$	1
m	$(1 - q)^2$	$1 - s$	$1 - hs$	1

$$\bar{w} = 1 - s[1 - 2(1 - h)r + (1 - 2h)r^2 - 2hxqr(1 - r)]$$

(vol. 2, eq. 4.18)

$$\Delta r = sr(1 - r)[1 - h + hxq - (1 - 2h + 2hxq)r]$$

mutation term negligible

$$\Delta q = shxr(1 - r)q(1 - q)$$

$$\Delta q/\Delta r = hxq(1 - q)/[1 - h + hxq - (1 - 2h + 2hxq)r]$$

It will be noted that the ratio of the change in frequency of the modifier to that of the new type gene is independent of the selective advantage of the latter over its predecessor (in agreement with the findings of the above authors). The ratio is roughly proportional to the degree of dominance, h, and to the modifying action of M if both of these are small. Assume, however, that there is initially semidominance of A ($h = 0.5$) and that MM brings complete dominance ($x = 1$):

$$\bar{w} = 1 - s(1 - r)(1 - qr)$$

$$\Delta r = 0.5sr(1 - r)(1 + q - 2qr)$$

$$\Delta q = 0.5sr(1 - r)q(1 - q)$$

$$\Delta q/\Delta r = q(1 - q)/(1 + q - 2qr)$$

The ratio rises from about $q(1 - q)/(1 + q)$ initially ($r = 0$) to $q(1 - q)$, which is 0.25 at most, when the type gene is half fixed ($r = 0.5$) and approaches q as the type gene approaches fixation. It is evident that selection of a modifier of the heterozygote such as M can never carry A more than about one-fourth of the way toward complete dominance during the period in which the frequency of A reaches equilibrium with mutation. Further increase in dominance can occur by modifiers of AA with correlative effect on Aa (Plunkett-Muller), or by replacing A by a more active allele (Haldane), or, if a change in environment favors somewhat lower activity, by direct selection of a modifier that lowers the upper limit of the dosage-response curve.

Finally, it may be added that none of these processes are significant in cases in which the new type differs little in activity from its predecessor and is almost or completely dominant over inactivating mutations from the first. They apply significantly only to major neomorphs in Muller's terminology.

Selection for Overdominance

There have been a number of papers on the possible evolution of overdominance by specific modifiers of the heterozygote, either during establishment of a new mutant gene (Parsons and Bodmer 1961; Bodmer 1963) or during long-continued selection against disadvantageous mutations by extension of

Fisher's theory (O'Donald 1967*b*). These encounter the same difficulties as with mere dominance.

The simplest process is probably the occurrence of a mutation inferior to type in one effect, superior in a pleiotropic effect, both sufficiently developed in the heterozygote to give the latter a net selective advantage over both homozygotes. Human sickle-cell anemia is such a mutation in malarial regions.

Where there is an intermediate optimum with respect to quantitative variability, all loci tend toward fixation but the heterozygote of the last heterallelic one may be closer to the optimum than either homozygote. It is doubtful, however, that such fixation would go far because of balancing of the slight adverse selection pressures by recurrent mutation.

Finally, a gene that is disadvantageous in one respect may acquire an independent favorable effect by mutation, with sufficient dominance of both favorable effects to give the heterozygote a net advantage. This differs from the first only in whether the mutation occurs in the type allele or in any allele carried at low frequency.

Artificial Selection

Under artificial conditions, it is possible to select intermediate heterozygotes directly according to whether they resemble one or the other homozygote. Since there always seems to be an abundance of genetic variability in stocks in the absence of long-continued inbreeding, the effective intensity of the selection is limited only by the amount of uncontrollable nongenetic variability. The process of modifying the dominance of intermediate heterozygotes is thus expected to be as successful as artificial selection in general.

A number of early experiments in which this proved to be the case have been cited. It is unnecessary to go into the details of successful experiments of this sort made more recently by Ford (1940), with the moth *Abraxas grossulariata*, and by Fisher and Holt (1944), with the length of the tail of a short-tailed mutant variety of the mouse. These have no bearing on Fisher's theory for the prevailing recessiveness of deleterious mutations beyond giving additional evidence of the genetic variability of intermediate heterozygotes, for which abundant evidence had long been available.

Later Discussions

Fisher's latest statement on his theory, apart from that in the reprinting of *The Genetical Theory of Natural Selection* (1930) in 1958, was in 1949. Referring specifically to populations in which disadvantageous mutations are maintained at low frequencies by recurrent mutation, he wrote:

Natural selection will constantly favor modifying factors, tending to render each type of heterozygote more normal or in other words to render the disadvantageous mutant more recessive; and that such recessive genes, though each rare compared with its normal allelomorph, exist in great numbers in different parts of the germ plasm. This theory, although strenuously combated by Prof. Sewall Wright, appears now to be generally acceptable. Since it was put forward, moreover, its only questionable premise, that the reaction of the heterozygote is readily modified by the selection of modifying factors, has been decisively verified by E. B. Ford with the magpie moth, *Abraxas grossulariata*, and by Fisher and Holt, with the common house mouse, *Mus musculus*.

What he calls the only questionable premise was, as just noted, common knowledge, before he proposed his theory, to those who had experimented with intermediate heterozygotes. I certainly did not doubt it in view of the evidence in my early papers already cited. The verification of the modifiability of intermediate heterozygotes by direct selection was, as also just noted, wholly irrelevant to his thesis. The really questionable premises were that selection pressures of the order of the mutation rate are at all effective, and that the prevailing near-dominance of wild-type genes over their mutations cannot be accounted for in other more plausible ways.

Crosby (1963) strongly criticized Fisher's theory and held (correctly) that earlier criticisms had understated the case in making unduly favorable assumptions. In simulation experiments, he found that the expected minute selective changes in the modifier frequencies were swamped by enormously greater random fluctuations and "half of his experiments which ran for about 100 generations ended with less dominance than they began with." He also noted that recurrent mutation of the modifiers themselves might take precedence over the selection pressures. As noted earlier, I had made both of these criticisms but Crosby stressed them more strongly. Also, as already noted, Crosby found in simulation experiments that selection of modifiers of the heterozygotes of new favorable mutations could not go far before the latter had approached fixation.

Crosby disagreed with me in rejecting the interpretation of degree of dominance by the dosage-response relations in the successions of gene products leading to the observed characters. He argued for determination by "control of competition among gene replicas for sites on microsomal particles." He does not show how this can account for the observed complicated patterns of factor interaction or how it can be reconciled with codominance, both of which can be interpreted by the dosage-response theory.

A paper by Sheppard and Ford (1966) was devoted primarily to criticism of Crosby's conclusions but included those that Haldane and I had reached. The seven statements in which I was mentioned were as follows, with numbering for convenience of reference.

1. Crosby's criticisms fall into two parts. Firstly, he maintains as did Wright [1929a,b] and Haldane [1930] that the selective advantage of genes modifying dominance, being of the same order of magnitude as the mutation rate, is too small to have any evolutionary effect.

2. Secondly, he criticizes, as did Wright [1934a], the basic assumption that a new mutation, when it first appears, produces a phenotype somewhat intermediate between those of the two homozygotes.

3. Even if such modifiers are present, it can still be claimed as Crosby does that these frequencies will be governed not by their effects or dominance but by the selective value of the other effects. Wright [1934a] also made this claim. However, it has already been partly answered by Fisher [1934].

4. Crosby, like Wright [1934a] before him, questions the assumption that dominance or recessiveness is evolved and is not an intrinsic property of a particular allelomorph.

5. The fact that dominance is not an attribute of an allelomorph but of a character is fatal to Wright's [1934a] theory that dominance would be expected from what is known about the activity of enzymes.

6. There is abundant experimental evidence to show that dominance is not an attribute of an allelomorph but of a character and that dominance, at least in some polymorphic situations, is dependent on the presence of a particular gene complex and has been evolved. No purely physiological explanation of the nature of dominance, such as that of Wright [1934a] or Crosby [1963], will explain these observations.

7. Crosby's views, like those of Sewall Wright, are based on the complete fallacy that dominance is a property of genes, whereas it is a property of characters. This invalidates the interpretations, alternative to selection, of both authors.

The first of these statements is misleading in implying that it was the smallness, as such, of the selective intensity in Fisher's theory, to which I objected. This is similar to statements of others (O'Donald 1967a,b, 1968; Parsons and Bodmer 1961; and many others, who have merely touched on the subject) that my objection was that the process was too slow. Fisher (1929), at the time of his reply to my first paper, recognized what my point really was:

The real difficulty, however, felt by Professor Wright, is not so simple as that the selective action upon dominance modifiers is so small that there has not been time for it to have had an appreciable effect. He makes no mention of the time probably available for these changes in the course of evolution and agrees with me as to the probability of demonstrating such selection experimentally. Where he really differs from me is in my assumption that a small selective intensity of say 1/50,000 the magnitude of a larger one will produce the same effect in

50,000 times the time. He suggests that the gene ratio of the modifying factors will either be held in stable equilibrium by more powerful forces so that a minute selective intensity will merely shift to a minute extent this position of equilibrium and produce no progressive effect—as if the complex of gene ratios were a gel rather than a sol—or be irresistibly increased or decreased by selective agencies so powerful that a minute additional selective intensity can only delay or accelerate the extinction of the less favored gene without ever being able to determine which gene shall be extinguished—much as wind blowing along a railroad will not exert any affect in accumulating rolling stock at the leeward terminal. This is a criticism genuinely aimed at my theory. I do make the assumption and the assumption really may be doubted.

The second statement by Sheppard and Ford also does not adequately express my position. I did not question that a new type mutation when it first appears may, if a neomorph, give an intermediate heterozygote. I held, however, that there is a strong bias in favor of dominance of most new type genes over their common mutations, assuming that most of these are in the direction of inactivation: hypomorphs or amorphs.

The third statement by Sheppard and Ford is correct, but their statement that it had "already been partly answered by Fisher [1934]" is not, for reasons given earlier.

Their remaining statements all have in common the assertion that I consider dominance a property or attribute of a gene. This is, of course, directly contrary to my thesis in all papers bearing on the subject from 1916 on, including those to which they refer.

The most recent review at the time of writing is that of Sved and Mayo (1970). They started with Fisher's theory of 1928. They evidently attempted to deal fairly with the objections but were also evidently almost as unfamiliar with earlier studies of dominance and its modifiability as was Fisher. They wrote:

The theory of evolution of dominance was put forward in 1928 in the absence of any direct evidence that dominance was normally subject to modification. In fact, although the evolutionary aspect constitutes the most novel aspect of the theory, this was one of the first statements that it is not necessarily a property of the gene in question but rather a property of the whole genome.

Fisher's paper was, indeed, the first to attract attention to an evolutionary interpretation, probably because the general acceptance by geneticists of his premise that dominance is modifiable made it seem plausible. The authors' next sentence largely refuted the preceding ones:

Before this, Wright [1927] had given examples in the guinea pig where dominance at a locus was affected by the genotype at other loci.

The point of my 1927 paper was not, however, to give new examples of such a well-known principle as the modifiability of dominance, but rather to explore systematically the pattern of changes in effects including dominance, for all pairs in a set of five alleles at the c-locus, in all possible combinations with four other loci. The complicated but orderly pattern seemed to give some insight into dominance as an aspect of factor interaction.

Later in the paper, the authors wrote:

It is necessary at this point to emphasize the distinction between dealing with the phenomenon of dominance at the level of the gene and at the level of the genotype or phenotype. Explanations such as Wright's and Haldane's are primarily at the level of the gene but, as emphasized by Ford [1964] and Sheppard and Ford [1966], Fisher's theory is directed at the level of the phenotype.

Here again the attitude toward dominance attributed to me seems to be the precise opposite of that which I had taken in all papers from 1916 on, in which dominance had been treated as an aspect of the whole interaction system including environmental factors, and including its relation to necessary thresholds and ceilings, imposed by the nature of the character. Perhaps the objection has been to interpretations in terms of activities of gene products, but any physiological interpretations of dominance (including Fisher's) must necessarily be in these terms. The contrast actually seems to be between dealing with actually observable modifying effects of gene substitutions on dominance at other loci, and assignment of abstract properties, such as specific modification of heterozygotes, without effects, correlated or otherwise, on homozygotes, a type of modifier not ordinarily found.

The only definite conclusion arrived at by Sved and Mayo was that all mathematical arguments agree that the selective pressure favoring modifiers of the heterozygote are of the order of the mutation rate. They recognized that selection of homozygous wild type for a factor of safety against environmental effects, as proposed by Plunkett and Muller, would be more effective but they do not reach any conclusion as to whether Fisher's scheme would operate at all.

Selection among Demes

So far discussion has been on the effects of natural selection in essentially homogeneous populations. Since dominance is an aspect of factor interaction, it seems likely that its evolution along the lines proposed by Plunkett and Muller would be favored by subdivision of the population into partially isolated demes. The differentiation among demes in the frequencies of weakly selected modifying genes in an interaction system gives the basis for intergroup selection of those systems that have the most favorable general properties, including factors of safety for all nearly dominant genes.

Summary

Dominance is not a character of the ordinary sort that evolves by changes in the frequencies of a limited number of genes, but rather a relation between genotype and phenotype that applies separately to all loci.

The cases of codominance of antigenic properties and of some alternative color patterns indicate that the two genes in a diploid zygote produce separate products that act as independent physiologic agents in development except as they come to be restrained by common limiting factors.

There are two polar types of limitation. One consists of an inadequate supply of substrate for which there is competition between the gene products or between products of later reactions. At the other extreme are limits in the possible variation of the phenotype, such as those at 0 or 100% color in a piebald pattern, or those with complete absence or perfect development of an organ. In general the degree of dominance as determined by the system of metabolic processes is distorted, not only by phenotypic thresholds, but also by nonlinearity in phenotypic expression at intervening grades.

In many multiple allelic series with graded effects on a character, the dominance relations indicate that the gene products, although chemically different, compete for action on the same limited substrate and may be treated as differing merely quantitatively in activity in this respect. There may be pleiotropic effects in which the relative grades of activity are different. In some series, the dominance relations are more complicated. There may be differences in the range of substrates that contribute to the same phenotype.

Whatever, the nature of the limiting factors, the dominance relations for a pair of alleles, and for multiple alleles that can be treated as differing quantitatively in their effects on the character under consideration, may be represented by an array of dosage-response curves, one for each combination of genes at other loci and developmental and environmental conditions. Each observed curve of this sort may in principle be considered the resultant of a succession of dosage-response curves, corresponding to the successive reactions between primary effect and the observed phenotype. The response for each reaction provides the dosage for the next.

The degree of dominance indicated by the total dosage-response curve is, in consequence, the resultant of various sorts of interactions. The degree may be increased by a reduction of the amount of a substrate at any step; by reduction of a competing reaction, either of a product of the gene in question or of its substrate; or by a lowering of a threshold due to destructive factors.

Where there is a phenotypic threshold or ceiling, any factor that shifts the position of the heterozygote in the multifactorial dosage-response curve to below the former or above the latter causes complete dominance of one or the other allele, irrespective of activity.

An essential gene, as judged by the severity of the defect caused by inactivating mutations, is more likely to have its effect restricted by a limiting factor at some stage between the reaction of its primary product and the observed phenotype than a gene with merely a minor modifying effect. The former is thus more likely to be strongly dominant over less active alleles than the latter.

It may be assumed that there is a particular degree of activity at each major locus in an interaction system that is optimal under given conditions. The system will be more stable under varying conditions if each of these degrees of activity is determined by an allele with excess activity restrained by a similarly stabilized limiting factor, and thus with response close to the asymptote of the dosage-response curve, than if determined by an allele with just the right response in the rising portion of the curve. Because of occasional extreme environmental disturbances, a considerable excess is advantageous. This is likely to be so great that the correlated response of the rare heterozygote is also brought fairly close to the asymptote, thus giving a high degree of dominance.

The shifting fine adjustments to changing conditions are brought about most readily by changes in the frequencies of largely semidominant modifiers. These would cause only minor changes in the nearly complete dominance of the major factors over the recurrent inactivating mutations.

Two or more active alleles may be maintained at high frequencies by frequency dependent selection with no consistent dominance in relation to each other, or by an advantageous heterotic effect with overdominance. All of these tend to be strongly dominant over their recurrent inactivating mutations.

From time to time substitutions occur at the major loci. If in the direction of increased activity, the new type gene is initially more strongly dominant over inactivating mutations than was its predecessor. If in the opposite direction, it will be less strongly dominant at first but there is likely to be a readjustment among the modifiers that restores the former degree. Substitution at any one major locus may be expected to change more or less the dosage-response curves and the degree of dominance at other loci, but again only until selective readjustments have occurred among the modifiers.

Under this view, evolution of the interaction system may go on indefinitely, largely by changes in the frequencies of modifiers that remain semidominant on the average but occasionally by substitution at major loci that shift the degrees of dominance of these over their recurrent mutations only slightly and temporarily.

At rare intervals, a major neomorphic mutation becomes established by selection. There is a good chance, from the nature of dosage-response curves, that it will be strongly dominant over inactivating mutations from the first, but if merely semidominant, selection of modifiers will give it an adequate margin of safety and thus typically strong dominance.

Discussion of the evolution of dominance has centered around a very different theory proposed by R. A. Fisher (1928a). His basic assumption was that all heterozygotes tend to be intermediate unless specific modifiers of their grades are assembled by selection. He held that a newly established type gene is typically only semidominant over its recurrent mutations until specific modifiers of the rare heterozygote shift the grade of the latter toward identity with that of the homozygote, thereby making the new type gene dominant. He showed that the pressure of natural selection on the frequencies of such modifiers was of the order of the mutation pressure or less, and held that this proved the efficiency in the long run of even the slightest favorable selection.

The theory encounters the difficulty that the postulated selection pressure would be negligible compared to that from even very slight effects on the much more abundant homozygotes, and would be overwhelmed by effects of mutation pressure and random drift, unless the effective population is very large. His contention that there is no alternative was in part due to an exaggerated view of the prevalence of complete dominance, but primarily to a physiological interpretation that ignores the likelihood of a strong bias toward dominance of major essential genes over the most common sort of mutation. Also ignored was the expected adjustment of the degrees of dominance of all major genes in an interaction system toward stabilization by selection of modifiers of the abundant homozygotes.

Another alternative, that selection of specific modifiers of the heterozygote occurs largely in the period before the new type gene has reached its equilibrium frequency and while the heterozygotes are abundant, can only account for an increase of dominance that goes about one-fourth of the way from semidominance to complete dominance.

There is no difficulty in accounting for cases of overdominance. There is also no difficulty in accounting for the ready modification of dominance by direct artificial selection of heterozygotes.

The general conclusion is that the prevailing dominance of major genes over their recurrent mutations reflects an optimal adjustment of the dosage-response curves of all such loci in the same interaction system to varying conditions. Of primary importance in this adjustment is a strong bias toward dominance of the favorable alleles at all major loci in the system. This is supplemented by direct selection of minor modifiers (themselves expected to be largely semidominant) for factors of safety at all loci against extreme environmental effects. Adjustment is facilitated if there is differentiation among partially isolated local populations, permitting selection among these of genetic systems as wholes, by means of differential population growth and diffusion.

The bias toward dominance of genes with major favorable effects accounts adequately for the usual inbreeding depression discussed in chapters 2 and 3.

CHAPTER 16

Breeds of Livestock

The breeds of livestock developed by a process that was rather more like the evolution of natural species than were selection experiments in the laboratory. The numbers of individuals per generation were much larger than in the latter, and selection, insofar as deliberate, was based largely on the overall merit of the animals (in the eyes of the breeder), rather than on the truncation of a frequency distribution according to a single metric character. While the rate of change from the wild ancestors was much greater than in nature, it was much slower than in the laboratory. Moreover, natural selection probably played a more important role through most of the process than did deliberate artificial selection. This probably applied to adaptations to life under domestication and to the maintenance of vigorous growth and fecundity. Finally, selection among breeds, and among herds within breeds as sources of sires, have been of very great importance in addition to mass selection among individuals on the basis of their own merit or that of parents, siblings, or offspring. It will be well to follow the example of Darwin in considering aspects of livestock breeding before discussing variability in wild species.

Population Studies of the Color of Shorthorn Cattle

Breeds of livestock have usually been selected for uniformity of color, but in some cases the occurrence of two or more colors has been favored. The first cases of Mendelian inheritance in livestock were established from studies of the occurrence of rare recessive segregants (for example, of red calves in black breeds of cattle) and from herdbook tabulations of the parentage of registered animals where more than one color was recognized. The latter could be studied from both the standpoint of the genetics of individuals and that of populations.

Wilson (1908) showed from the former standpoint that most of the data on the common colors of Shorthorn cattle—red, roan, and white with red ears, in data compiled by Barrington and Pearson (1906)—could be accounted for on the hypothesis that roan is the heterozygote of red and white. There were, however, a considerable number of exceptions. Wentworth (1913b) found exceptions that were clearly not herdbook errors and proposed a two-locus hypothesis that largely eliminated these. An analysis of Wentworth's data showed, however, that his hypothesis was wholly untenable from the population standpoint and that Wilson's one-locus hypothesis must be correct, except for phenotypic overlaps, after allowing for herdbook errors (Wright 1917b).

Wentworth's data were not wholly random samples of the breed. Tabulations (unpublished) were made later from the British, American, and Canadian herdbooks of 1921, 1,000 males and 1,000 females from each. These were analyzed from the standpoint of Wilson's hypothesis (A) and two two-locus hypotheses, Wentworth's (B) and a threshold hypothesis (C), which reduced the number of exceptions still more. Reds with white spots were grouped with reds.

Phenotype	Wilson (A)	Wentworth (B)	Threshold (C)
Red	RR	$R\text{-}roro$	$R_1R_1R_2R_2$; $R_1R_1R_2r_2$; $R_1r_1R_2R_2$
Roan	Rr	$R\text{-}Ro\text{-}$	$R_1R_1r_2r_2$; $R_1r_1R_2r_2$; $r_1r_1R_2R_2$
White	rr	$rr\text{-}$	$R_1r_1r_2r_2$; $r_1r_1R_2r_2$; $r_1r_1r_2r_2$

Table 16.1 gives the colors of the parents of the 6,000 calves. There were 225 exceptions to Wilson's hypothesis after checking for errors of tabulation. Table 7.8, volume 1, inadvertently gave the number of matings of each type from the original tabulation, but the difference is not great. In contrast with the 12 possible kinds of exceptions, only three kinds are possible under hypothesis B and only two kinds under hypothesis C. The actual number of exceptions were 88 and 5, respectively, but the apparent improvement over hypothesis A must be discounted because of the greater elasticities of B and C.

An important question is whether there was differential registration of male and female calves in relation to color. There seems to have been some in the British herdbook (probability 0.01 from χ^2), largely from discrimination against white males, but none was indicated in the other cases or in the combined data (probability 0.62). There appears to be no serious error in combining the data for the sexes.

Table 16.2 shows the frequencies of the three colors in the separate tabulations from the three herdbooks. The slight difference between the British and American samples presumably reflects the discrimination against

TABLE 16.1. Parents of 6,000 Shorthorns registered in 1921 according to color (1,000 males and 1,000 females from the British, American, and Canadian herdbooks). The total numbers of matings (observed, o) are compared with those expected (c) on the hypothesis of random mating by color.

PARENTS		OFFSPRING			TOTAL			
Dam	Sire	Red	Roan	White	o	c	$o - c$	$\dfrac{(o - c)^3}{c}$
Red	Red	1,140	(86)	(5)	1,231	1,229	$+ 2$	0.00
Red	Roan	843	854	(24)	1,721	1,750	$- 29$	0.48
Roan	Red	442	437	(22)	901	911	$- 10$	0.11
Red	White	(45)	263	(7)	315	288	$+ 27$	2.53
White	Red	(13)	106	(6)	125	117	$+ 8$	0.55
Roan	Roan	359	681	281	1,321	1,298	$+ 23$	0.41
Roan	White	(12)	110	78	200	213	$- 13$	0.79
White	Roan	(3)	89	81	173	167	$+ 6$	0.21
White	White	(0)	(2)	11	13	27	$- 14$	7.26
Total		2,857	2,628	515	6,000	6,000	0	12.34
%		47.62	43.80	8.58	100.0			

white males in the former. The Canadian sample was clearly a little darker than the American (in 1921).

The frequencies of red, roan, and white in the sample of 2,000 registered calves (O) from each of the herdbooks, British, American, and Canadian, are shown in table 16.2, together with those of their sires (S) and dams (D) with midparent (P) and three differences. There were fewer reds and more roans among the sires than among the dams in all three herdbooks, but most strikingly in the American. The white sires were greatly in excess in the American sample, slightly in excess in the Canadian, but in deficit in the British. The differences between the sexes (S − D) are highly significant ($\chi^2 = 75$ in the British sample, 393 in the American, and 26 in the Canadian). It can be assumed that there was little or no selection of dams and thus much selection of sires. The British sample (D) was the lightest and the American sample the darkest in the parental generation, but the American calves (O) of 1921 were the lightest because of the very strong selection of sires. It is to be noted that there is little difference between the calves and the parental average (O − P), especially with respect to the estimated frequency of the white gene (0.5 roan + white).

An important question for population analysis is the closeness of approach to random mating. The deviations in the three samples were all significant by the χ^2 test, but not consistent. Assigning values of 0, 1, and 2 to the three colors, there was a correlation of -0.117 in the British data, $+0.068$ in the

TABLE 16.2. Frequencies of red, roan, and white in tabulations from three herdbooks, and estimates of the frequency of gene r, from $q = (0.5$ roan and white).

1921 HERDBOOK	COLOR	FREQUENCIES				DIFFERENCES		
		Dam (D)	Sire (S)	Midparent (P)	Offspring (O)	S − D	O − P	O − D
British	Red	0.485	0.360	0.423	0.454	−0.125	+0.031	−0.031
	Roan	0.458	0.595	0.526	0.472	+0.137	−0.054	+0.014
	White	0.057	0.045	0.051	0.074	−0.012	+0.023	+0.017
	(1/2)Ro + W	0.286	0.343	0.314	0.310	+0.057	−0.004	+0.024
American	Red	0.601	0.300	0.451	0.453	−0.301	+0.002	−0.148
	Roan	0.349	0.544	0.447	0.443	+0.195	−0.004	+0.094
	White	0.050	0.156	0.102	0.104	+0.106	+0.002	+0.054
	(1/2)Ro + W	0.224	0.427	0.326	0.326	+0.203	0	+0.102
Canadian	Red	0.547	0.468	0.508	0.521	−0.079	+0.013	−0.026
	Roan	0.404	0.469	0.436	0.400	+0.065	−0.036	−0.004
	White	0.049	0.063	0.056	0.079	+0.014	+0.023	+0.030
	(1/2)Ro + W	0.251	0.298	0.274	0.278	+0.047	+0.004	+0.027

American, but only -0.005 in the Canadian, all with standard errors of 0.022. In a combination of all of the data, the correlation was -0.022 ± 0.013 and thus not significant. There is significance by the total χ^2 test (12.34, $P = 0.016$), but more than half of this (7.26) is from the smallest class, white × white. There can be no serious error in treating the 6,000 matings as random.

With 8.8% white, 53.6% roan among the total sires, the overall frequencies of the white gene r was 0.356 in sires under the one-locus hypothesis. Similarly, the frequency of r in dams was 0.254. The expected frequency in the offspring is given by the product of the two frequency arrays $(0.644R + 0.356r)$ $(0.746R \times 0.254r)$, giving the expectation shown in table 16.3. The Hardy-Weinberg frequencies from $(0.695R + 0.305r)^2$ are given for comparison.

TABLE 16.3. Observed and expected frequencies under the one-locus hypothesis (A).

	Red	Roan	White
Observed	0.476	0.438	0.086
Expected	0.481	0.429	0.090
Hardy-Weinberg	0.483	0.424	0.093

There is good agreement between observation and expectation from this standpoint ($P = 0.10$). The slightness of the effect of even rather strong differences between the sires and dams is shown in the Hardy-Weinberg frequencies. It may be added that the Canadian and American frequencies agreed well by themselves ($\chi^2 = 3.2$, 0.5, respectively, $df = 2$), but the English disagreed ($\chi^2 = 19.2$, $P < 0.001$) because of the discrimination against registering white calves and the rather marked disassortative mating.

The fairly good agreement of the monofactorial interpretation does not, of course, prove it to be correct. A similar test of the other two hypotheses, however, definitely rules them out, as well as any other hypothesis involving more than one major pair of alleles. As indicated above, no serious error can be made by ignoring the different frequencies of the sires and dams and the departures from random mating.

Under hypothesis B, the proportion of whites (0.0858) is r^2, giving $r = 0.293$, and the proportion of reds among nonwhites (0.521) is ro^2, giving $ro = 0.722$. The frequencies of the genotypes under random combination is given by terms of $(0.707R + 0.293r)^2(0.278Ro + 0.722ro)^2$. The proportions within each phenotype are shown in table 16.4.

TABLE 16.4. Genotypic frequencies within each phenotype under hypothesis B.

Red	Roan	White
RR roro 0.547	*RR RoRo* 0.088	*rr RoRo* 0.077
Rr roro 0.453	*RR Roro* 0.459	*rr Roro* 0.402
	Rr RoRo 0.073	*rr roro* 0.521
	Rr Roro 0.380	
1.000	1.000	1.000

The expectation from each type of mating can be estimated approximately by assigning these relative genotypic frequencies to the parents.

Under hypothesis C, the gene frequency array is approximately

$$[(1 - x)R_1 + xr_1]^2[(1 - y)R_2 + yr_2]^2.$$

Red $(1 - x)^2(1 - y)^2 + 2(1 - x)^2y(1 - y) + 2x(1 - x)(1 - y)^2 = 0.476$

Roan $(1 - x)^2y^2 + 4x(1 - x)y(1 - y) + x^2(1 - y)^2 \qquad = 0.438$

White $x^2y^2 + 2x^2y(1 - y) + 2x(1 - x)y^2 \qquad = 0.086$

These may be solved by iteration, giving the array $(0.902R_1 + 0.098r_1)^2$ $(0.337R_2 + 0.0663r_2)^2$. The genotypic frequencies within each phenotype are given in table 16.5.

The expected proportions from the various matings, grouping reciprocals, are compared with the observed proportions in table 16.6. The most crucial matings are those of red × white and roan × roan. Neither hypothesis B

TABLE 16.5. Genotypic frequencies within each phenotype under hypothesis C.

Red	Roan	White
$R_1R_1R_2R_2$ 0.194	$R_1R_1r_2r_2$ 0.817	$R_1r_1r_2r_2$ 0.901
$R_1r_1R_2R_2$ 0.764	$R_1r_1R_2r_2$ 0.180	$r_1r_1R_2r_2$ 0.050
$R_1R_1R_2r_2$ 0.042	$r_1r_1R_2R_2$ 0.003	$r_1r_1r_2r_2$ 0.049
1.000	1.000	1.000

nor C comes at all close to accounting for the high proportion of roans from red × white or for the relatively low proportion from roan × roan. There seems to be no other hypothesis involving more than one major pair of alleles capable of accounting for the opposite deviations from the observed frequencies. Hypothesis A, on the other hand, fits all cases roughly, and the deviations can fairly easily be accounted for by segregation of minor factors, causing overlapping of designations, and occasionally by errors of registration.

TABLE 16.6. Comparison of observed frequencies of various colors of 6,000 Shorthorn calves from various types of mating with those expected under hypotheses A (Wilson 1908), B (Wentworth 1913b), and C (threshold).

Parents	Red × Red (1,231)			Red × Roan (2,622)			Red × White (440)		
Offspring	Red	Roan	White	Red	Roan	White	Red	Roan	White
Observed	92.6	7.0	0.4	49.0	49.2	1.8	13.2	83.9	2.9
Expected A	100.0	0	0	50.0	50.0	0	0	100.0	0
Expected B	94.9	0	5.1	39.8	55.1	5.1	55.8	21.5	22.7
Expected C	83.8	16.2	0	58.8	39.4	1.8	28.4	50.5	21.1

Parents	Roan × Roan (1,321)			Roan × White (373)			White × White (13)		
Offspring	Red	Roan	White	Red	Roan	White	Red	Roan	White
Observed	27.2	51.5	21.3	4.0	53.4	42.6	0	15.4	84.6
Expected A	25.0	50.0	25.0	0	50.0	50.0	0	0	100.0
Expected B	16.7	78.2	5.1	23.4	53.9	22.7	0	0	100.0
Expected C	8.4	83.2	8.4	2.1	45.6	52.2	0	22.6	77.4

Animals genetically dark roan (Rr) have undoubtedly sometimes been called red and, conversely, animals genetically red (RR) but with white spotting have sometimes been called roan. Similarly, animals that are genetically very light roan (Rr), especially if strongly white-spotted, are likely to be called white (Wentworth), and some animals of genotype rr may show enough red beyond that normally present (ears and feet) to be called light roan. Roan certainly varies, almost from self-red to white, and it is probable that there is actual overlap.

Table 16.7 shows estimates of the amount of overlap required to approximate closely the observed results from all types of matings. These were obtained by assigning trial values to the frequencies of reds and whites of genotype Rr, all others being derived from the marginal totals (table 16.3). The array of gene frequencies from each color could thus be obtained and used to estimate the genotypes and, from these, the color frequencies expected from each of the six types of mating. The observed difference from a 6×3 table differs from that expected (after grouping the very small classes, white from red $\times$ red and red from white $\times$ white [probably errors] with the roans in each case) with $\chi^2 = 12.2$, which with 8 degrees of freedom has probabilities between 0.10 and 0.20 and thus fits reasonably well. There would probably, however, be somewhat closer agreement between phenotype and genotype, but for the occasional errors of registration.

TABLE 16.7. Estimated relative phenotypic frequencies for each genotype under the unifactorial hypothesis.

	Red	Roan	White	Total
RR	46.1	2.0	0	48.1
Rr	1.5	40.2	1.2	42.9
rr	0	1.6	7.4	9.0
Total	47.6	43.8	8.6	100.0

The three colors have been maintained in the Shorthorn breed by a preference of many breeders for roan and in spite of selection against white, which is in part natural, the white cows having an excess tendency to a particular type of sterility, "white heifer disease" (Rendel 1952).

The frequencies of the colors have varied greatly according to selection. The year 1921 happened to be a time when selection was moving toward the lighter colors, especially in America, where the increase in the frequency of the "white" gene r had the extraordinarily high value $\Delta q = 0.102$, in comparison with 0.028 in Canada and 0.024 in England.

Inbreeding in Livestock

The cattle of northeastern England, after crossing extensively with imported Dutch cattle from the 16th to the early 18th centuries, acquired a high reputation for the production of both beef and milk. The Shorthorn breed, now among the most numerous breeds of cattle in the world (United States, Argentina, and Canada, as well as the British Isles), traces to an extraordinary extent to two herds of such cattle, those of Charles and Robert Colling, who began farming in 1782. Pedigree records have been published since 1822 when the *Coates's Shorthorn Herd Book* was commenced, but this volume carried the record of the more important animals for about half a century and thus into the founding period of the modern breed (Wallace 1923).

The inbreeding coefficients in the foundation period can be calculated from the formula $F = \sum [(1/2)^n(1 + F_A)]$, in which n is the number of animals in a path connecting the uniting gametes and F_A is the inbreeding coefficient of the common ancestor of the path (vol. 2, chap. 7). The pedigrees of the most important of the founding bulls, Favourite (252), born in 1793, and his most famous son, Comet (115), are given in figure 16.1 with minor omissions.

Terms from remote, omitted ancestors raise the coefficients for Favourite to 0.192 and for Comet to 0.471.

Complete pedigree studies were made of the 64 cows of the famous Duchess strain of Thomas Bates (Wright 1923*b*), of which Darwin (1868) stated:

For thirteen years, he bred most closely in-and-in, and during the next seventeen years, although he had the most exalted notion of his own stock, he thrice infused fresh blood into his herd. It is said that he did this not to improve the form of his animals, but on account of their lessened fertility.

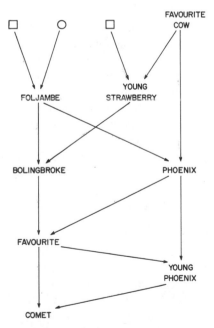

FIG. 16.1. Pedigree of the Shorthorn bulls, Favourite (252) and Comet (115).

	A	n	$(1/2)^n(1 + F_A)$	
Favourite (252)	Foljambe	3	0.125×1	$= 0.125$
	Favourite Cow	4	0.063×1	$= 0.063$
			$F =$	$\overline{0.188}$
Comet (115)	Favourite	2	0.250×1.19	$= 0.298$
	Phoenix	3	0.125×1	$= 0.125$
	Foljambe	5	0.031×1	$= 0.031$
	Favourite Cow	6	0.016×1	$= 0.016$
			$F =$	$\overline{0.470}$

Duchess 1st (born 1808) was a daughter of Comet and five generations in the straight female line from a foundation cow of the herdbook. Duchess 59th and 62nd were eight generations later than Duchess 1st in the straight female line. Complete pedigrees of either would include $2^{14} - 1$ entries (16,383) and it would hardly seem practicable to find all of the independent ancestral connections in order to calculate F. Actually, the labor of calculation was not great after tabulating the straight female lines of all "Duchesses" after Duchess 1st, all their sires, and the sires and dams of the sires. In the course of the work, a list was made of the correlations between gametes of important animals, which could then be used whenever two of them appeared on opposite sides of these condensed pedigrees. This left few cases

TABLE 16.8. Summary of coefficients of inbreeding and relationship for the 64 Duchesses, including Duchess 1st bred by Charles Colling, and the eight generations bred by Thomas Bates.

						RELATIONSHIP (%)		
GENERATION	No.	INBREEDING (%)				To Favourite (252)		
FROM	OF	Indi-			Between	Indi-		
DUCHESS 1ST	Cows	vidual	Sire	Dam	Parents	vidual	Sire	Dam
0 (Duchess 1st)	1	40.8	47.1	31.5	58.7	76.3	80.5	72.7
1	4	35.6	21.6	40.8	54.0	68.6	62.5	76.3
2	10	47.1	43.4	37.1	67.0	69.2	72.1	69.4
3	14	42.4	38.0	43.9	60.2	67.9	67.8	68.7
4	7	38.3	38.6	43.7	54.7	66.8	64.9	67.3
5	9	36.7	41.9	42.5	51.8	67.3	64.7	67.4
6	10	43.4	35.5	33.9	64.4	62.0	61.7	66.2
7	7	36.4	33.2	34.9	54.3	56.7	56.7	57.9
8	2	42.8	32.1	41.7	62.6	57.4	56.7	60.5
Total	64	40.9	37.5	39.6	59.2	65.6	65.1	67.2

SOURCE: Reprinted, with permission, from Wright 1923b (table 3).

that needed to be taken up separately. A systematic method for such analysis has been given by Cruden (1949).

Table 16.8 shows the coefficients of inbreeding and relationship (as percentages) for Duchess 1st and the eight generations of later Duchesses. The extraordinarily high inbreeding and the strong correlations with Favourite, maintained by Bates from 1812 to 1849, are brought out. The inbreeding coefficients are shown in time in figure 16.2.

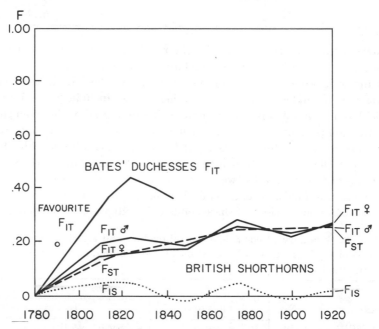

FIG. 16.2. Inbreeding coefficients of British Shorthorn cattle from foundation period 1780 to 1920. Inbreeding of bulls and cows relative to the foundation stock, F_{IT}, of the breed as a whole, and of the Bates' Duchesses are in solid lines. Those for hypothetical offspring of randomly mated animals, F_{ST}, are in a broken line. Those for individuals relative to the contemporary breed, F_{IS}, are in a dotted line. The inbreeding coefficient of the foundation bull, Favourite, is indicated by a circle. Reprinted, by permission, from Wright (1965b).

The breed expanded so greatly in numbers that little further inbreeding would be expected after the foundation period were it not that the bulls that were used either came from a very limited number of herds, of which that of Bates was one of the most important, in the first half of the 19th century, or were the sons or grandsons of such bulls. The consequences of this pattern may be determined by finding the average coefficient of inbreeding F of the breed as a whole and related statistics.

Average Inbreeding of Whole Breeds

The preferred method used (Wright and McPhee 1925) was to take a random sample of the individuals registered in the *Coates's Herd Book* in a given year and trace a single *random* line back of each parent. Coin tossing was used to obtain a random succession of sires and dams. Such paired lines either do or do not show any common entry. No attention needs to be paid to the generation in which a tie occurs, since remoteness is exactly compensated for by increase in the number of ties of which the one observed is representative. Let n_s and n_d be the number of generations back to the common ancestor in the two lines. The average value of F from m such two-line pedigrees with k observed ties is

$$\bar{F} = \frac{1}{m} \sum^{k} [2^{n_s + n_d}(1/2)^{n_s + n_d + 1}(1 + F_A)] = \frac{k}{2m}(1 + \bar{F}_A).$$

The standard error of the proportion of ties, $p = k/m$, is $\sqrt{[p(1 - p)/m]}$. That for F is given approximately by multiplying by the ratio F/p. The estimate of $\bar{F}$ from two-line pedigrees of Bates' Duchesses was 0.422 ± 0.016, in comparison with 0.409 from determination from the 64 full pedigrees.

The Inbreeding Coefficient for Sex Linkage

The foregoing discussion relates to the inbreeding coefficient for autosomal loci. The situation is more complicated for sex-linked loci, in which it applies only to females. Only ties between random lines back of sire and dam, which do not involve successive males, contribute at all (vol. 2, chap. 7).

Possible Contributory Lines Back of Sire by Generations								
	0	1	2	3	4	5	6	7	n_s
Males	1	0	1	1	2	3	5	8	$f(n_s - 1)$
Females		1	1	2	3	5	8	13	$f(n_s)$
Total	1	1	2	3	5	8	13	21	$f(n_s + 1)$
Possible Contributory Lines Back of Dam by Generations								
	0	1	2	3	4	5	6	7	n_d
Males		1	1	2	3	5	8	13	$f(n_d)$
Females	1	1	2	3	5	8	13	21	$f(n_d + 1)$
Total	1	2	3	5	8	13	21	34	$f(n_d + 2)$

Of the 2^{n_s} lines tracing to ancestors of the n_s ancetral generations back of the sire, only $f(n_s + 1)$ involves no successive males, where n_s is the n_sth Fibonacci number, starting from $f(1) = 1, f(2) = 1$, and $f(n) = f(n - 1) + f(n - 2)$.

In the case of the dam, $f(n_d + 2)$ of the lines in the n_dth ancestral generation involve no successive males.

The contribution of a tie between the paternal and maternal lines in a complete pedigree of a female is $(1/2)^{n_f}(1 + F_A)$, if no two males are in succession, where n_f is the number of females, including the dam in the path connecting the uniting gametes. Each tie in a two-line pedigree is representative of $f(n_s + 1)f(n_d + 2)$ possible ties. The estimated value of F for sex-linkage from m two-line pedigrees is

$$F = \frac{1}{m} \sum^k [f(n_s + 1)f(n_d + 2)(1 + F_A)]/2^{n_f}$$

The Autosomal Inbreeding in Shorthorns

Estimates were made for British Shorthorns from random two-line (or four-line) pedigrees for various years from the foundation period (McPhee and Wright 1925) (table 16.9).

The values of F for 1810, representing the foundation period, were 19.1% for bulls and 14.3% for cows. They rose to 25.4% and 26.7%, respectively, by 1920. The greatest contributions were in the first 30 years or so, when the numbers were small, but there was a substantial increase later.

It is conceivable that these figures might be due merely to a high frequency of consanguine matings and that the cessation of such matings would at once reduce F to zero. This can be tested by associating the one-line pedigrees of sires with those of dams at random and finding the number of such pairs with ties. Treating the actual Shorthorn breed of the time as a subpopulation, S, of all possible populations, T, that might have been derived similarly from the foundation stock, the coefficient is of the type F_{ST} (vol. 2, chap. 12), while the estimate from the actual matings is F_{IT}. The amount of inbreeding relative to the array of contemporary gametes in the breed is $F_{IS} = (F_{IT} - F_{ST})/(1 - F_{ST})$. F_{IS} is thus a measure of the amount of contemporary consanguine mating, while F_{ST} measures the cumulative inbreeding of the breed as a whole. F_{IS} is negative if F_{ST} is greater than F_{IT} because of prevailing crossbreeding of distinct strains within the breed.

Table 16.9 shows these statistics for British Shorthorns at six dates after the foundation period, about 1780. In 1920, coefficients are given not only

TABLE 16.9. Coefficients of inbreeding and relationship (percentages) in the Shorthorn breed in Great Britain, including those for the 100 Dairy Shorthorns with the highest recorded milk yields in 1920, and 100 with average milk yields. The last row refers to the 20 leading Shorthorn prize-winning sires at the Chicago International Livestock Exposition 1918–22 on assigning appropriate weights to each as sires, grandsires, and great-grandsires of first-, second-, and third-prize winners.

Date	F_{IT} (%)			F_{ST} (%)	F_{IS} (%)	Relationship (%)		
	♂	♀	Average			Favourite	Champion of England	Breed
1780	0	0	0	...	...	...	...	...
1810	19.1 ± 1.5	14.3 + 1.3	16.6 ± 1.0	12.8 ± 2.3	+4.4	44.3	26.3	22.0
1825	21.9 ± 2.1	16.1 ± 3.7	19.9 ± 1.7	16.1 ± 2.2	+4.5	51.3	29.9	26.8
1850	18.4 ± 1.9	16.9 ± 3.7	18.0 ± 1.7	20.0 ± 2.8	−2.5	50.1	26.1	33.9
1875	25.9 ± 2.1	29.0 ± 3.8	27.4 ± 1.8	24.1 ± 3.1	+4.3	57.6	32.9	37.8
1900	23.2 ± 2.1	22.6 ± 3.8	22.9 ± 1.7	24.1 ± 3.0	−1.6	52.1	39.2	39.3
1920	25.4 ± 4.0	26.7 ± 4.1	26.0 ± 2.8	24.9 ± 3.1	+1.5	55.2	45.5	39.5
Dairy (high)	...	26.9 ± 2.8	...		+0.7	55.7	41.2	38.0
Dairy (average)	...	28.0 ± 2.7	...		+5.3	56.6	42.7	37.6
Sires IPW	25.0	...	...	24.8	+0.3	55.0	47.0	33.0

SOURCE: Coefficients for 1780–1920 were from McPhee and Wright (1925) (F_{IT} from table 1, relationships from table 3) (F_{IT} from table 1, F_{ST} and F_{IS} are recalculated from slightly different coefficients in the above paper. Those for the prize-winning sires were also given therein from calculation based on Malin's (1923) data. The coefficients for the Dairy Shorthorns are from McPhee and Wright (1926).

for random samples of males and females from the breed as a whole, but for the 100 Dairy Shorthorns with the highest milk records in that year and 100 random ones (records initiated in 1905) (McPhee and Wright 1926). Coefficients are also given for the 20 leading prize-winning sires in the Chicago International Livestock Exposition of 1918–22, based on a tabulation by Malin (1923) in which bulls were given weights as sires (32), grandsires (16), and great-grandsires (8) of first-prize winners; three-fourths and three-eighths, respectively, for second- and third-prize winners. The relationships to the bulls Favourite, born 1793, and Champion of England (17526), born 1859, and between random animals of the breed as a whole are shown.

The high values of F_{ST} show that there was increasing inbreeding of the breed as a whole, and not merely contemporary consanguine mating (F_{IS}). This implies that the effective population numbers were enormously smaller than the apparent numbers. Three principal ways have been listed (vol. 2, chap. 8) in which this may occur: (1) an extreme sex ratio, (2) differences in productivity beyond expectation under random sampling, and (3) passage through bottlenecks of small size.

The effect of sex ratio is given by the formula $N_e = 4N_m N_f/(N_m + N_f)$, in which N_m and N_f are the numbers of mature males and females, respectively. This approaches $4N_m$ if there are many more females than males. This obviously applies here, as in livestock breeding in general. The effect of differences in productivity is given by $N_e = 4N/(2 + \sigma_k^2)$ in a static population, where σ_k^2 is the variance of the number of offspring that reach maturity. This is also a very important factor in livestock breeding in which sires of generally recognized merit produce many more sons, grandsons, and great-grandsons that are selected to become sires than the average. Finally the effective number is the harmonic mean of the varying numbers over a series of generations and thus is dominated by the bottlenecks of small size. This, of course, played an important role in the foundation period during which the coefficient F_{ST} rose to 12.8%. The other two principles must have kept N_e rather small, at least until 1875, by which time F_{ST} had risen to 24.1%. There was not much change thereafter (24.9% in 1920). The theoretical rate of increase per generation is $1/(2N_e)$, giving $N_e = 1/(2\Delta F_{ST})$. In this period, 1810–75, about 12 generations, $\Delta F_{ST} = 0.113/12 = 0.0094$, giving $N_e = 53$. The effective number of sires during this period was thus only about 13 per generation. It is about the same for the whole period 1780–1920.

The early history of the breed was dominated by bulls produced by the Colling brothers. The exceedingly intense inbreeding to Charles Collings' bull Favourite has already been noted. The breed of 1920 had a closer relationship to Favourite than of offspring to parent in a randomly breeding population.

In the period up to 1850, bulls came predominantly from such herds as those of Thomas Bates and the Booths, based on the Colling brothers' herds, and the registered bulls were considerably more inbred than the cows. Up to 1825, there was much consanguine mating ($F_{IS} = 4.4\%$ in 1810, 4.5% in 1825), indicating the building up of inbred lines illustrated by the Bates' Duchesses. The negative value $F_{IS} = -2.5\%$ in 1850 is in marked contrast and indicates a period in which outcrossing of strains within the breed tended to overbalance consanguine mating. The exceedingly inbred Bates strain had previously been the most favored, but came to be recognized in this period as inclined toward sterility.

The value of F_{IS} of 4.3% in 1875 indicates renewed building up of distinct strains. The most influential herd came to be that of Amos Cruickshank of Sittyton, near Aberdeen. His most famous bull was Champion of England (17526, born 1859). Cruickshank produced a decidedly different and blockier type, with broader chest, wider back, and deeper ribs than those produced earlier. Bulls of these two types are illustrated in figures 16.3 and 16.4. "Scotch" Shorthorns made over the breed to a considerable extent. While Champion of England was strongly related to Favourite (45%), his relationship to the breed of 1850 (26.1%) was less than the average (33.9%), but this had risen to 32.9% by 1875, to 39.2% in 1900, and to 45.5% by 1920, indicating the spread of his influence. The low value of F_{IS} in 1900 (-1.5%) is probably a reflection of this process.

The enthusiasm for Scotch Shorthorns reached America somewhat later, but during the first decade of the 20th century it brought about a similar transformation. This centered in such imported bulls as Whitehall Sultan (163573). He and his son Avondale were the leading sires of prize winners at the International Livestock Exposition in the period 1918–22, studied by Malin. Nine others of the group of 20 were descendants of Whitehall Sultan. As shown in the bottom line of table 16.9, this group was not differentiated to any significant extent from the British breed of 1920.

Whitehall Sultan was a white bull, and his sons, grandsons, and great-grandsons, who largely made over the American Shorthorn, were much lighter colored than the average. The leading 20 sires of prize winners in 1918–22 included only 2 reds, 11 roans, and 7 whites, and thus a gene frequency of r of 0.625, far above the breed average, 0.36. The extraordinary lightening of the color of the American Shorthorns in 1921 was probably more a reflection of the current enthusiasm for the blocky Scotch Shorthorns than of a sudden preference for roan and white as such. If preference for these colors played a role, it was probably on the part of Cruickshank and the Scottish breeders.

FIG. 16.3. The Bates Shorthorn bull, "Earl of Oxford 3rd" (51186). From Wallace (1923, fig. 12b).

FIG. 16.4. The Cruickshank bull, "Field Marshall" (47870). From Wallace (1923, fig. 17a).

Other Breeds of Beef Cattle

The courses of inbreeding in the American representatives of two other beef breeds are compared in table 16.10. The American Herefords, studied by Willham (1937), trace to imports in the 1880s, and later, from the white-faced red British Herefords. The latter are considered to have been the first improved breed of cattle in Great Britain. Their recognized history began with Richard Tomkins (died 1723), whose herd was continued by his son Benjamin (1714–89) and the sons of the latter, contemporaries of the Colling brothers. Related animals bred by William Hewer, another contemporary of the Collings, were the most influential in the later history of the breed. No single bull was more than half as important as Favourite was for the Short-horns as far as is known, but the first Hereford herdbook was not published until 1846, 24 years later than the *Coates's Herd Book* for the younger breed. The American herdbook began to be published in 1881.

Willham traced his pedigrees back to about 1860, so that the inbreeding recorded in table 16.10 is in relation to this date, more than a century and a half after the foundation of the improved breed and thus not at all comparable with the coefficients recorded for the Shorthorns. Table 16.10 shows a fairly steady increase in F_{ST} from 0 to 4.8% in the course of 70 years. There was very little consanguine mating (F_{IS}) up to 1920, average 1.3%, but a considerable amount (3.5%) in 1930.

The genes of the early core herds had undoubtedly become so generally diffused by 1860 that they could have had nothing to do with the rise from 1900 to 1930. Its source is indicated by the rapid increase in the correlation of the bull Beau Brummell (51817), born in 1890, with the American breed as a whole. This started from about the same low values as between random animals in 1880–1900, but it had reached 24.6% by 1930, the equivalent of that of an individual to his grandchildren in a randomly breeding population, and the highest in the breed. The breed was clearly being made over by the herd of which this bull was the leading representative in much the same way that the Cruickshank Shorthorns, represented by Champion of England, had made over the Shorthorn breed in Great Britain, and later, America. Beau Brummel was the grandson of the bulls Anxiety 4th (9904) and North-pole (8946), imported by Gudgell and Simpson in 1881. They inbred the descendants of these bulls rather closely until the dispersion of the herd in 1916 and the so-called straight-bred line, derived wholly from it, was continued by others. This core cluster was the most important direct and indirect source of bulls throughout the breed during the period considered.

The inbreeding history of the Aberdeen-Angus breed, studied by Stonaker (1943), goes back closer to the foundation period than in the case of the

TABLE 16.10. Inbreeding coefficients for Hereford and Aberdeen-Angus cattle registered in the American herdbooks. The percentage relationship to the whole breed is shown and also that to the bull Beau Brummel (BB) (born 1850) for Herefords, and that to Black Prince of Tillyfour (BP of T) (born 1860) for the Aberdeen-Angus breed.

| | HEREFORD | | | | | ABERDEEN-ANGUS | | | | |
| | Inbreeding (%) | | | Relationship (%) | | Inbreeding (%) | | | Relationship (%) | |
	F_{IT}	F_{ST}	F_{IS}	Breed	BB	F_{IT}	F_{ST}	F_{IS}	Breed	BP of T
1850	0	0	0			0	0	0		
1860						(4.9)				
1870	1.2	0.7		1.4						
1880	2.9 ± 0.5	1.1 ± 0.3	+1.8	2.1 ± 0.5	1.2					
1890	3.4 ± 0.6	1.3 ± 0.4	+2.1	2.6 ± 0.7	2.6					
1900	2.7 ± 0.5	2.6 ± 0.4	+0.1	5.2 ± 0.7	5.0	8.9 ± 0.9	5.1	+4.0	9.4	24.1
1910	4.9 ± 0.7	2.8 ± 0.5	+2.2	5.3 ± 1.0	10.4	12.7 ± 1.1	9.5	+3.5	16.8	29.1
1920	4.6 ± 0.7	3.7 ± 0.6	+0.9	7.1 ± 1.1	17.0	10.8 ± 1.0	6.8	+4.3	12.2	25.9
1930	8.1 ± 0.8	4.8 ± 0.6	+3.5	8.8 ± 1.2	24.6	14.2 ± 1.2	9.2	+5.5	16.1	28.8
1939						11.3 ± 1.1	7.2	+4.4	13.3	24.1

SOURCE: Data from Herefords from Willham 1937 and for Aberdeen-Angus from Stonaker 1943.

Herefords. Improvement of the black-polled cattle of northern Scotland began more than a century later than that of the Hereford cattle. It centered in the herd of Hugh Watson of Keillor from 1818 to 1860 and that of William McCombie of Tillyfour from 1830 to 1880. The herdbook began to be published in 1862. Stonaker traced the American breed to an average birth date of 1850, although the number registered in America was too small for study before 1900. Imports had begun in 1873.

There was no one animal in the foundation stock comparable to Favourite in the Shorthorns. The coefficient F_{IT} of the Scotish breed must, however, have been in the phase of rapid increase between 1850 and 1860, since Stonaker states that that for 1930 would have been only 9.8% instead of 14.2% if the pedigree lines had been traced only to 1860. This implies a value of F_{IT} of 4.9% in 1860, relative to the 1850 population.

The observed cumulative inbreeding (F_{ST}) fluctuated between 1900 and 1939 with only a slow upward trend. At first, it would seem that F_{ST} should never show any real decline, but this does not allow for the heterogeneity of the breed due to high F_{ST} in the core herds and relatively low values in the peripheral ones, in the process of grading up by bulls from the former. The correlation of the American breed with McCombie's Black Prince of Tillyfour (77), always the highest in the American breed, showed fluctuations that paralleled those of F_{ST}. Stonaker showed that the years in which the correlations were low (1900, 1920, 1939) all followed several years of rapidly expanding sales, while sales were static from 1900 to 1910 and fell sharply before becoming static between 1920 and 1930. Thus the changes in the coefficient can be explained as due to expansion of the peripheral herds in boom periods, and contraction under depressed conditions, relative to core herds.

The current inbreeding (F_{IS}) in the American Aberdeen-Angus breed was consistently high from 1900 to 1939, with average 4.3%, much higher than in the Herefords (1.6%). This suggests that the former included multiple relatively independent core herds all tending to be inbred to a considerable extent within themselves, to contrast with the situation in the American Herefords with a single cluster.

Dairy Breeds

The inbreeding histories of two dairy breeds in America, the Holstein-Friesians (Lush, Holbert, and Willham 1936) and the Brown Swiss (Yoder and Lush 1937) are compared in table 16.11. Both are ancient European breeds with no definite periods of modern improvement. The herdbook of the Friesians in Holland was first published in 1873, and that of the American

TABLE 16.11, Inbreeding coefficients (as percentages) for Holstein-Friesians and Brown Swiss registered in American herdbooks. The percentage relationships to the whole breed are shown in both breeds. Those to the cow DeKol 2nd, born 1884, are shown for the Holstein-Friesians.

| | Holstein-Friesian | | | | | Brown Swiss | | | |
| | Inbreeding (%) | | | Relationship (%) | | Inbreeding (%) | | | Relationship (%) |
	F_{IT}	F_{ST}	F_{IS}	Breed	DeKol 2nd	F_{IT}	F_{ST}	F_{IS}	Breed
1881–83	0	0	0		0.1	0	0	0	
1889	2.4 ± 0.4			0.7 ± 0.3	3.6				
1899	2.1 ± 0.3	0.4 ± 0.2	+1.7	2.6 ± 0.5	9.0				
1909	4.7 ± 0.5	1.4 ± 0.3	+3.3	2.4 ± 0.3	8.2	5.0 ± 0.7	1.9 ± 0.5	+3.1	3.6 ± 0.9
1919	3.4 ± 0.4	1.3 ± 0.2	+2.1	3.4 ± 0.6	12.2	4.0 ± 0.6	1.6 ± 0.4	+2.4	3.9 ± 0.8
1928	4.7 ± 0.5	1.8 ± 0.3	+2.9						
1929	...	...	...	...		3.8 ± 0.6	2.3 ± 0.5	+1.5	4.3 ± 1.0
1931	4.0 ± 0.5	1.7 ± 0.2	+2.3	3.4 ± 0.4	8.1				

SOURCE: Data for Holstein-Friesians from Lush, Holbert, and Willham 1936 and for Brown Swiss from Yoder and Lush 1937.

Holstein-Friesians based on imports from Holland, in 1871. Imports reached their maximum in 1883–85. The average birth date to which the American pedigrees traced was 1881.

Table 16.11 shows a moderate amount of current inbreeding (average $F_{IS} = 2.6\%$) but little cumulative inbreeding (average $F_{ST} = 1.75\%$ in the last four years). The breed shows much less correlation between random animals in the last year (3.4%), relative to the imported animals, than in either the Herefords (8.8%) or the Aberdeen-Angus (13.3%). The animal that showed the closest relationship to the breed was in this case a cow, DeKol 2nd (234), imported in 1885. This coefficient rose from 0.1% in 1889 to 10.1% in the last four years. She established world records for milk production, and four of her daughters had records that were highest or close to it for the time. The 11 next highest coefficients were shown by her descendants. The breed, however, seems to have been too numerous, with some 110,000 registrations per year, 1919–29, for an influence on F_{ST} comparable to that of the Anxiety 4th-Beau Brummel line in the American Herefords, although the number of the latter reached comparable values by 1920.

A study of Brown Swiss cattle in Switzerland by Sciuchetti (1935) showed very little total inbreeding ($F_{IT} = 1.0\%$). The inter se relationship was, however, enough to indicate a slightly higher cumulative inbreeding, $F_{ST} = 1.7\%$. These imply avoidance of even average consanguinity in making matings ($F_{IS} = 0.7\%$).

The pedigrees of American Brown Swiss cattle traced to importations from 1869 to 1906 and formed widely separated herds with only occasional crossbreeding. The lines traced to animals with average birth dates of 1881 for the 1909 sample; 1883–84 for the two later samples.

The amounts of current inbreeding in 1929 (average $F_{IS} = 1.5\%$) and of cumulative inbreeding (average 2.3%) were low after 45 years. The highest relationship coefficients were with the foundation bull William Tell (1), born 1868, 9.2% in 1909, declining later, and with the bull College Boy (4316, born 1913, rising to 9.1% by 1929). The rise in the cumulative inbreeding cannot be attributed to any one core herd, but rather to influence spreading from several.

Buchanan Smith (1928) studied the total inbreeding coefficient, F_{IT}, of British Jersey cattle by ten-year periods from 1876–85 to 1916–25. The values for successive periods were 2.6%, 2.3%, 2.9%, 3.2%, and 3.9%, with standard errors of about 0.4%. These are somewhat less than in the American Holstein-Friesian and Brown Swiss, which themselves were low. A higher figure (5.3%) was found by Fowler (1932) for British Ayrshires.

Robertson and Mason (1954) have studied the inbreeding in the Red

Danish breed (two-thirds of the cattle of Denmark). The breed originated in the middle of the 19th century from crosses between local island cattle and bulls from Slesvig. From about the beginning of this century, extensive use began to be made of progeny-tested bulls.

The authors studied two groups. One consisted of 50 out of 61 bulls whose daughters were tested in special stations in 1949–50 and 1950–51. A sample of 50 cows was taken at random from the cow herdbook of 1951.

Taking the breed in 1911 as the starting point, the bull samples showed an inbreeding coefficient of 11.2%, while the cows showed only 4.6%. The bulls showed the highest relationships (26.5%) to a bull Eske Braugstrup (2456) born in 1927 and 18.0% to the bull Højager (2168) born in 1923. The latter bull and some of his descendants were blacklisted because of being heterozygous for a gene causing paralysis of the hind quarters, which was reaching seriously high frequencies.

The Red Danish is an example of a breed in which rapid improvement is being brought about by grading up of the cow population by use of bulls of a rather closely knit nucleus of proven superiority; a process that is not without danger, as illustrated above.

Breeds of Sheep

The inbreeding histories of two breeds of sheep in America are compared in table 16.12. The Rambouillets were studied by Dickson and Lush (1933) and the Hampshires by Carter (1940).

TABLE 16.12. Inbreeding coefficients and relationship inter se (percentages) (R) in Rambouillet and Hampshire sheep registered in American flock books.

	RAMBOUILLET					HAMPSHIRE			
	Inbreeding (%)			(R%)		Inbreeding (%)			(R%)
	F_{IT}	F_{ST}	F_{IS}			F_{IT}	F_{ST}	F_{IS}	
1890					1911	0			
1896	2.2 ± 0.4	2.7	−0.5	5.2 ± 0.8	1925	1.4 ± 0.4	0.0	1.4	0.0
1906	3.8 ± 0.3	0.8	+3.0	1.5 ± 0.6	1935	2.9 ± 0.6	0.3	2.6	0.5
1916	3.7 ± 0.3	1.4	+2.3	2.7 ± 0.4					
1926	5.5 ± 0.4	1.4	+4.1	2.6 ± 0.4					

SOURCE: Data for Rambouillet from Dickson and Lush 1933 and for Hampshire from Carter 1940.

The Rambouillets trace to Merinos imported to France and Germany from Spain in the 18th century. They were imported into America as early as 1840, but the principal imports were in the 1880s and 1890s. The American herdbook was established in 1891. Any base date is unsatisfactory because of the wide range of dates to which pedigrees were traced, but most were before 1894. About 45% of the ancestral lines trace to sheep bred by Von Homeyer in Germany. The relationship among these could not be taken into account.

There was an appreciable amount of consanguine mating (average F_{IS}, 1906–26, 3.1%) but little cumulative inbreeding ($F_{ST} = 1.4\%$ in 1926). What there was must have traced to multiple flocks since no animal showed a relationship to the breed greater than 8.0%.

The Hampshire breed originated in Hants and adjacent counties in southern England but was not improved until the latter half of the 19th century. The first English herdbook was issued in 1889; the first American herdbook in 1890 was based on sheep tracing in all lines to recognized English flocks. The number grew slowly, but after 1927 about 20,000 were registered annually.

Analysis indicates little inbreeding of any sort (average $F_{IS} = 2.0\%$, average $F_{ST} = 0.2\%$, 1925–35). Only two animals had a relationship coefficient as great as 2.4% and 3.0% in 1935.

Swine

The Poland China breed of swine arose from a mixture of types in southwestern Ohio about 1850. The first breed association was formed in 1878. The breed included one-third of all swine registered in America when analyzed by Lush and Anderson (1939) (table 16.13).

TABLE 16.13. Inbreeding coefficients and relationship inter se (percentages) in Poland China swine (American) and in the Danish Landrace.

	POLAND CHINA SWINE					DANISH LANDRACE			
	Inbreeding (%)			R (%)		Inbreeding (%)			R (%)
	F_{IT}	F_{ST}	F_{IS}	Breed		F_{IT}	F_{ST}	F_{IS}	Breed
1885	0	0	0	0	1897	0	0	0	0
1900	2.0 ± 0.5	1.6 ± 0.5	0.4	3.2 ± 0.9					
1910	5.6 ± 0.6	3.4 ± 0.6	2.3	6.4 ± 1.2					
1920	6.6 ± 0.9	4.7 ± 0.7	2.0	8.9 ± 1.4					
1929	9.8 ± 1.0	7.8 ± 0.9	2.2	14.2 ± 1.7					
					1930	6.9 ± 0.7	8.6 ± 0.9	−1.9	16 ± 1.7

SOURCE: Data for Poland China swine from Lush and Anderson 1939 and for Danish Landrace from Rottensten 1937.

It was developed as an extreme lard type in accordance with market demands up to 1900. In spite of very large numbers, particular boars (King Tecumseh [11793], born 1885), and four generations of descendants, played a major role in the period before 1900 in building up a cumulative inbreeding coefficient of 1.6%. From this year on, the market demand was shifting toward a leaner type, met especially by Peter Mouw's "big type." According to Lush and Anderson, Mouw's herd "played much the same role as the Cruickshank herd in British Shorthorn history." This herd, however, traced to the same core lines as before. The most influential boar, Chief Price (61861), born 1898, was of this line. The cumulative inbreeding thus continued to rise in spite of the change in type and reached 7.8% by 1929, which is surprisingly high, considering the number of registered Poland Chinas. The average amount of current inbreeding averaged 2.2% for 1910 to 1929.

The amount of inbreeding in the Danish Landrace was studied by Rottensten (1937) for 1930, relative to the breed of 1897. The amount of cumulative inbreeding (8.6%) was somewhat greater than in the Poland Chinas, although the number of generations was less. There was avoidance of current inbreeding (-1.9%) in 1930 so that the total ($F_{IT} = 6.9\%$) was less than in the Poland Chinas.

Horses

Calder (1927) studied the total inbreeding, F_{IT}, from complete pedigrees of 676 stallions of the relatively small Clydesdale breed of draft horses, drawn at random from the studbooks at five-year intervals. The first studbook was issued in 1878 but gave pedigrees back to about 1865. F_{ST} was not calculated but probably differed little from F_{IT} because mating of close relatives was generally avoided.

The foundation stock was highly heterogeneous. There was little increase in F_{IT} in the first 25 years (table 16.14), but from 1890 there was a roughly linear rise with no consistent sex difference to 6.25% in 1920–25. Up to 1905–10, this increase was almost wholly due to common ancestry of Prince of Wales (born 1866) and, somewhat more, to Darnley (born 1872). After this, increments were due largely to male descendants of Darnley, especially his great-grandson, Baron's Pride (born 1895), and in the last period a son of the latter, Baron of Buchlyvie (born 1900). In this breed, the high degree of homogeneity obtained in the course of some 60 years seems to have been due to individual selection, associated with strong line breeding to a single stallion, Darnley.

Steele (1944) calculated both total inbreeding (F_{IT}) and that expected under random mating (F_{ST}) from five-generation pedigrees of three breeds of light horses, 1931–35. His samples of Thoroughbreds were not random ones but consisted about equally of the leading stake winners and the horses

TABLE 16.14. Inbreeding coefficients (percentages) in Clydesdale horses (Scotland) and three breeds of light horses (America).

CLYDESDALES (SCOTLAND)			LIGHT HORSES, FIVE GENERATIONS TO 1931–35 (UNITED STATES)			
				Inbreeding (%)		
Period	F_{IT} (%)	Period	F_{IT} (%)		F_{IT}	F_{ST}	F_{IS}
1865–70	0.1	1895–1900	2.3	Thoroughbred, winners	8.3 ± 0.8	8.8 ± 0.8	−0.5
1870–75	0.2	1900–05	2.6	last	8.3 ± 0.9	7.6 ± 1.1	+0.8
1875–80	0.3	1905–10	3.4				
1880–85	0.3	1910–15	3.8	Standardbred (2.05% or better)	4.0 ± 0.6	4.3 ± 0.6	−0.3
1885–90	1.1	1915–20	4.7				
1890–95	1.2	1920–	6.2	Saddle horse	3.2 + 0.4	3.1 + 0.4	+0.1

SOURCE: Data for Clydesdale horses from Calder 1927 and for light horses from Steele 1944.

that ran last. Total inbreeding was essentially the same, 8.3% in both, and the cumulative coefficient differed little, 8.8% and 7.6%, respectively.

Among Standardbreds, he took those with records of 2:05 for the mile, or better. There was about half as much inbreeding as in the Thoroughbreds, $F_{IT} = 4.0\%$, $F_{ST} = 4.3\%$.

A random sample of Saddle horses, a breed with frequent outcrosses to other breeds, gave somewhat lower figures, 3.2% and 3.1%, respectively.

Comparison of Breeds with Respect to Inbreeding

Lush (1946) made estimates of the effective numbers of males per generation on the basis of the estimates of total inbreeding (F_{IT}) in most of the cases discussed here. These are revised in table 16.15 on the basis of cumulative inbreeding, F_{ST}. The differences from Lush's results are unimportant in most cases. The figures given for length of generation were derived from their data by most of the authors. The Shorthorns are here assumed to have the same generation length as other beef cattle, and the Clydesdales are assigned the numbers doubtfully attributed to them by Lush. The long history of the Shorthorns is divided into two periods, and the Aberdeens are also so divided. Because of the considerable standard errors of F_{ST} in many cases, the estimates of N_e and N_m $(= 0.25 N_e)$ (effective number of males) are to be taken as merely indicative of the order of magnitude. Those for the Hampshire sheep merely indicate that these were very large.

Summary

The frequencies of the colors red, roan, and white in Shorthorn cattle have been studied primarily in order to distinguish between alternative hypotheses on the mode of inheritance by supplementing the conclusions from pedigree analysis by the population properties. These indicate unequivocally that only one major pair of alleles is involved, with roan the heterozygote of the other two. The difficulty in interpretation arose from phenotypic overlapping due to minor factor. Hardy-Weinberg frequencies are exhibited approximately, in spite of strong selection of sires. Polymorphism is maintained by some preference of the majority of breeders for roan over red and stronger preference over white, supplemented by a slight natural selection against white.

The results are of interest from the standpoint of breed history in showing that a gene frequency may change enormously in a single generation ($\Delta q = 0.10$ in America in 1921) as a by-product of the wholesale use of sires from a particular strain.

The chief interest in the inbreeding histories of various breeds of livestock is less in the inbreeding as such that in the light thrown on the process of improvement of domestic animals.

We can only speculate on the process by which the wild ancestors, domesticated by Neolithic man, evolved up to the point at which they provided material for the development of the modern breeds in the last two or three centuries. No doubt there was a selection for docility and later a selection of individuals as parents on the basis of the usefulness of their characters. Many localities came to be characterized by types of horses, cattle, sheep, and swine recognized by color, conformation, or other characteristics. There was probably always spreading of influence of strains of recognized superiority. The importation of superior cattle from the Netherlands to northeast England in the 16th and 17th centuries is an example that probably contributed to the formation of the Shorthorn breed of cattle.

This breed is especially favorable for study because of the publication of a herdbook while records of its foundation period were still available. No doubt there were many herds in the region in which it originated, differentiated somewhat by the varying ideals of the breeders, but near the end of the 18th century one small nucleus, the herds of the Colling brothers, was to have overwhelmingly the most influence in the formation of this great breed. Charles Colling's bull, Favourite, born in 1793, shows a random correlation of 0.55 with random Shorthorns of 1920. There were many breeders in the first half of the 19th century who carried out the initial expansion, but a few herds (Bates, the Booths) dominated the breeding to such an extent that the effective size of the whole breed was equivalent to that of a population of about 50 or better of a large cow population headed by only about 13 bulls per generation. Again in the latter part of the century and in America in the early 20th century, the breed (in which over 170,000 had been registered in Britain, over a million in the United States, and over 200,000 in Canada by 1921) was radically transformed in type by sires tracing to a single herd—that of Cruickshank, whose leading bull, Champion of England, born in 1859, showed a relationship to the breed as a whole of 0.26 for 1850 but 0.46 by 1920.

The improvement of the breed was brought about mainly by the development of high excellence in a few herds at each period and the subsequent diffusion of excellence, principally by the use of bulls from the core herds, rather than by mass selection throughout the breed as a whole, although the latter, no doubt, played a role (Wright 1922a, 1923a).

The study of the inbreeding history of other breeds of livestock confirms this mode of improvement, although the records of none of the others go back so close to the foundation period. The effective number of sires per

TABLE 16.15. Estimates of effective size of population, N_e, and of effective number of males, N_m ($= N_e/4$). The contemporary inbreeding F_{IS} is also shown.

	Period	Length of Generation (years)	No. of Generations	F_{ST} (%)	F_{ST} per Generation (%)	N_e	N_m	F_{IS} (%)
Cattle								
Shorthorn U.K.	1780–1850	5.4	13.0	20.0	1.54	32	8	+3.1
Shorthorn U.K.	1850–1920	5.4	13.0	4.9	0.38	132	33	+0.4
Hereford U.S.	1860–1930	5.4	13.0	4.8	0.37	135	34	+1.6
Aberdeen Angus U.S.	1850–1900	5.4	9.3	5.1	0.55	91	23	...
Aberdeen-Angus U.S.	1900–39	5.4	7.2	2.1	0.29	172	43	+4.3
Holstein-Friesian U.S.	1881–1931	4.5	11.1	1.7	0.15	330	83	+2.5
Brown Swiss U.S.	1883–1929	5.4	8.5	2.3	0.27	185	46	+2.3
Sheep								
Rambouillet U.S.	1890–1926	4.2	8.6	1.4	0.16	310	78	+3.1
Hampshire U.S.	1911–35	3.7	6.5	0.3	0.05	(1,000)	(250)	+2.0
Swine								
Poland China U.S.	1885–1929	2.5	17.6	7.8	0.44	114	29	+1.7
Landrace Denmark	1897–1930	2.2	15.0	8.6	0.57	88	22	−1.9
Horses								
Clydesdale Scotland	1865–1922	11	5.2	6.2	1.19	42	11	...
Thoroughbred U.S.	1931–35	12.1	5	8.2	1.64	30	8	+0.2
Standardbred U.S.	1931–35	12.4	5	4.3	0.86	58	15	−0.3
Saddle horse U.S.	1931–35	11.3	5	3.1	0.62	81	20	+0.1

Source: Contemporary inbreeding F_{IS} is revised from Lush 1946 in order to base on F_{ST} instead of F_{IT}.

generation (American herdbooks if not otherwise stated) is found to be about 20–40 in other breeds of beef cattle (Hereford, Aberdeen Angus), 50–100 in breeds of dairy cattle (Holstein-Friesian, Brown Swiss, British Ayrshires, and Jerseys); 10–20 in breeds of horses (Thoroughbreds, Standardbreds, Saddle horse, British Clydesdales); 20–30 in breeds of swine (Poland China, Danish Landrace); about 80 in Rambouillet sheep (and much larger only in Hampshire sheep and Brown Swiss cattle in Switzerland). In most of these breeds, the numbers registered are enormously greater than these figures.

The shifting balance theory discussed in chapter 13 was an extension to evolution in nature of essentially the same process that seemed to be characteristic of the improvement of domestic animals.

CHAPTER 17

General Conclusions

Experimental studies of the effects of self-fertilization have been carried out with various plants (especially maize); sib mating and other forms of close inbreeding with laboratory rodents, domestic animals, and insects have also been studied. These have involved a great diversity of characters: growth, morphology, viability, fecundity, color, reactions to transplantation, resistance to disease, temperament, and intelligence.

The results all agree with the hypothesis: these characters primarily reflect the tendency toward random fixation of one or another allele at each heterallelic locus. This obviously explains the profound differentiation of all characters among inbred lines, all maintained under the same conditions.

The depression of all aspects of vigor, which is the traditional consequence of close inbreeding, is not an obvious consequence of random fixation. It has been shown, however, that the depression is far from uniform among inbred lines or among different characters in the same line. Most lines tend, indeed, to show some depression in all respects, but each acquires its own peculiar combination of negligible to severe depression of different aspects of vigor, and occasionally lines are superior to the random bred foundation stock in some respects.

This tendency toward depression in each respect on the average seems to be due primarily to a prevailing correlation between favorableness of effect and tendency toward dominance. This correlation has both physiological and evolutionary explanations, which were discussed in chapter 15 after a review of the evidence on the effects of both inbreeding and selection and of their evolutionary implications. With respect to immediate physiology, gene products tend to be involved in interactions that generally imply dosage-reponse curves expected to exhibit diminishing returns for each positive effect; they thus tend to bring about an approach to dominance of at least the most essential processes. From the standpoint of population dynamics, moreover, unfavorable more or less dominant alleles are kept at much lower frequencies than equally unfavorable recessives at the same rate of recurrence by mutation. Also from the evolutionary standpoint, genes with favorable modifying effects on the heterozygotes of a gene of major importance, which thereby tend to shift the phenotype of the latter toward the favorable homozygote, may be expected to be favored by selection as long as these heterozygotes are sufficiently abundant for this process to outweigh selection of their effects in the homozygotes. The efficacy of selection for dominance of favorable mutations has never been questioned and has often been demonstrated experimentally. There has, however, been much controversy on the reason for prevailing recessiveness of the large class of mutations with such deleterious effects that they are kept at very low frequencies. In this case the selection pressure on modifiers of the rare heterozygotes is of the order of the mutation pressure or less, and thus is generally of lower order than the probable effects on the abundant type homozygotes. The most common sorts of major deleterious mutation, however, probably consist of inactivations of essential genes and thus may be expected to approach recessiveness from the nature of the dosage-response curves. Discussion of this matter has been much confused by the failure to distinguish these cases.

The tendency toward dominance of favorable effects leads to over-dominance where each of two alleles has both favorable and unfavorable pleiotropic effects. Here the fixation of either homozygote brings about a reduction in vigor.

The effect of inbreeding on the amount of heterozygosis, y, is measured by the inbreeding coefficient, F. This equals $(y - y_0)/y_0$, in which y_0 is the amount under random mating at the same gene frequencies. The amount of differentiation among inbred lines in the absence of differential selection is measured by the same coefficient, $F = \sigma_q^2/q(1 - q)$, in which σ_q^2 is the variance of the frequencies of one of the alleles and $q(1 - q)$ is the limiting variance under complete fixation. With multiple alleles, $F = \sum \sigma_q^2/\sum q(1 - q)$, in which the summations are with respect to the values for all alleles, each opposed by its array of alleles collectively. The coefficient F is also the double-entry correlation between uniting gametes on assigning arbitrary values to all alleles, a quantiy easily deduced for any mating system by path analysis.

Low values of F, such as for progeny of first-cousin matings ($F = 0.0625$), typically give rise to appreciable depression of mean fitness but increased variability, both mainly by extension of the lower tail of the phenotypic distribution from increased segregation of deleterious recessives or near-recessives. Some extension of the upper tail, however, is possible by increased segregation of favorable homozygotes.

The data from relatively neutral characters indicate that the actual progress toward fixation under different systems of inbreeding agrees well with that expected. It requires, for example, 3.27 generations to make the same progress under sib mating as under self-fertilization. In general, the rate of decrease of heterozygosis per generation in a population of given size, N, is $(y - y^1)/y^1 = (F - F^1)/(1 - F^1) = 1/(2N)$ for neutral loci.

Selection favoring heterozygotes tends, of course, to delay fixation, or even to prevent it, if sufficiently strong. Experiments with rats have shown that fixation may be delayed at least fourfold under sib mating by strong selection for vigor within progenies and especially among the multiple branching lines.

The immediate restoration of at least average vigor in the progeny of crosses between closely inbred lines can be interpreted as merely the converse of the inbreeding depression, the bringing about of 100% heterozygosis at all loci in which the crossed lines had become homozygous for different alleles. The loss of half of the excess F_1 vigor in F_2 in diploids, but not in tetraploids, confirms this interpretation. So does the absence of appreciable inbreeding decline in self-fertilizing species or of heterosis on crossing different lines of such species, except such as may be accounted for by the gradual accumulation of different mutations.

The question as to the importance of overdominant loci in both inbreeding decline and heterosis is still controversial. The experimental demonstration that most of the extraordinary heterosis manifested by crosses between inbred lines of maize depends on repulsion linkage of unfavorable genes in foundation lines that were still far from equilibrium, rather than an overdominance of single loci, at least with respect to alleles with major differential effects, has tended to shift the emphasis away from overdominance.

Long-continued inbreeding tends to bring about uniformity within strains in characters of which variability is wholly genetic, but otherwise merely reduces variability to the proportion that was initially nongenetic. There are many abnormalities that appear only when certain factors, genetic and nongenetic, fall below a threshold. Fixation of genotypes leads to a situation in which inbred lines are likely to become characterized by the production of particular abnormalities at particular percentages under given environmental conditions.

The various aspects of vigor are also likely to vary continuously only above a certain threshold or breaking point, below which there is failure. In such respects, inbred lines are likely to exhibit even more variability than a random bred stock with the same frequency array of genes.

An important question is the degree of persistence of the fixation brought about by close inbreeding. There is enough persistence to have made closely inbred lines very valuable in experimental studies in which genetic uniformity is important. Nevertheless, experiments have shown that branches of such lines, maintained by close inbreeding, drift apart at least in mammals in the course of generations at a rate that implies either mutation rates greater by two or more orders of magnitude than found for genes with conspicuous effects (usually some 10^{-5} to 10^{-6} per generation in higher organisms) or that the character under consideration is affected by a correspondingly enormous number of loci, or a combination. Inbred lines of *Drosophila* also drift apart but apparently more slowly. The same is true of parthenogenetic or vegetatively propagated lines of lower animals and plants. The difference in number of loci may be important here.

Selection experiments have also been conducted with organisms of the most diverse sorts: mammals, birds, insects, especially *Drosophila* and *Tribolium*, parthenogenetic lines of Cladocera, hydra, and various protozoa, many higher plants, bacteria, and even RNA in the test tube. Characters of the most diverse sorts have been studied: size and form, color, viability, fecundity, sex ratio, resistance to neomorphic and infective diseases, leukocyte counts, reactions to light and gravity, intelligence, and so on.

The most important conclusion is the practically invariable success of directional selection in bringing about drastic changes even from the narrowest foundation stocks, such as the descendants of single pairs of wild *Drosophila*. There is evidently an enormous amount of heterallelism in wild species.

The theoretical courses of directional selection have been worked out for a great variety of conditions. The course in experiments is not, however, at all closely predictable because of the large number of parameters required for characterization of multiple sets of alleles, which may vary widely in effects, degrees of dominance, pleiotropy, linkage relations, and initial gene frequencies. These are, in general, impossible to determine in advance. In the special case of a population derived from F_1 of a cross between isogenic lines, there is, however, the simplification that the gene frequency array for all heterallelic loci is 0.50:0.50.

The regression of offspring on parents in the first generation of selection usually gives a moderately adequate empirical basis for prediction for a few generations, but realized heritability is usually soon found to be much less than that estimated initially. Unexpected pleiotropic effects usually appear and often bring an experiment to an end because of low viability or fecundity, or both. If not, a plateau is usually reached after a moderate number of generations, not because genetic variability has been exhausted, but because natural selection prevents its utilization. The reality of such selection has often been demonstrated by the loss of part of the selective gains on relaxation of the artificial selection.

In some cases, the deleterious effects responsible for a plateau have been clearly demonstrated to be pleiotropic. Thus progress may be blocked by lethality or sterility of the homozygotes of genes with desirable heterozygous effects. Similarly, negative physiological correlations between desirable characters have been the source of a discouraging "slippage" in many attempts at improving livestock.

In other cases, there has been a secondary rise sometime after a plateau has been reached. This has been interpreted, and in some cases clearly demonstrated, to be due to the occurrence of a rare crossover that releases a favorable gene from linkage with an unfavorable one. This has been especially the case in experiments with *Drosophila* species in which, as already noted, linkage plays a much more important role than usual.

There have been many experiments on natural selection in the laboratory in which the course of change in the frequencies of known genes have been observed under specified conditions of competition. The results are usually of the general sort expected although usually not quantitatively predictable because of the difficulty of determining the genetic parameters with respect to viability and fecundity in side experiments under conditions fully comparable to those in the selection experiment, and especially because of linkage relations. Experiments with *Drosophila* have again illustrated the great importance of linkage and occasional crossing-over. The results of experiments give no basis, however, for questioning the mathematical theory.

Cage experiments started with known frequencies of different chromosome arrangements have been very instructive with respect to the nature of natural selection. There is evidence for both heterozygous advantage and rarity advantage, the latter especially with respect to selective mating.

There have been extensive experiments with systems of mating, designed to maximize the advantages from overdominance by selecting strains for complementarity either with a known tester stock or with each other. The results have been rather disappointing, probably because overdominance of major loci has proved to be less important than had been supposed.

The one system that has an unquestionable advantage over mass selection in a single broadly based population has been selection among a number of such strains, which takes advantage of their inevitable, more or less random, differentiation. The history of the improved breeds of livestock indicates the value of selection both within and among strains.

Experiments with known alleles in small populations of constant size, subject only to natural selection and sampling drift, have shown that the theory of stochastic distributions applies reasonably well in practice, on allowing for a somewhat reduced effective population number.

An important question to which the answer is not yet wholly clear is the order of magnitude of the number of loci in higher organisms. In those organisms that have been studied most extensively, such as *Drosophila* species and maize, the continual recurrence of mutations at the same loci indicates that the number of loci with easily recognizable mutations is limited to a few thousand. On the other hand, the amount of DNA in the genomes indicates room for some 200,000 in *Drosophila* and numbers of this order in many invertebrates, including protozoa; a few million in some other invertebrates and most vertebrates; and even tens of millions in some amphibia. There is a similar range in eukaryotic plants. The numbers are several orders of magnitude less in prokaryotes. All of these figures are based on the assumption that genes include some thousand nucleotides.

Direct studies in vitro (shearing and reassociation) indicate that 30% or more of the DNA consists of enormous numbers of replications of a few, presumably heterochromatic genes. There is, moreover, the possibility that the loci capable of easily recognizable mutation constitute only a small fraction of all loci. Another possibility is, however, that most recognized loci are complexes of more or less differentiated simple loci, which is certainly true of many. Finally there is the possibility that there may be enormous amounts of "nonsense" DNA that does not code for any protein molecules.

Since all loci are undoubtedly susceptible to mutation, all must be at least weakly heterallelic in a species with hundreds of thousands of individuals.

The experimental conclusions on mutation, selection, and inbreeding provide firm support for the interpretation of evolution as determined by natural selection of variability, provided ultimately by mutation of Mendelian genes. Two major modes of selection within sexually reproducing populations have been discussed. One is the selection among individuals (or families) within essentially panmictic populations; the other is that among differentiated local populations by differential population growth and diffusion (the shifting balance process).

Since gene frequencies tend to approach equilibrium under long-persisting conditions, continuing evolution by individual selection depends either on continually changing conditions or on the occurrence of novel favorable mutations. Conditions are, indeed, continually changing in terms of geologic time, but the ensuing evolution is somewhat like a treadmill, old adaptations continually being undone in favor of new ones. A more serious limitation is the restriction of selection to the net effects of genes (except where two or more are strongly linked), because of the rapid breaking up of gene combinations in meiosis.

Novel favorable mutations no doubt occur from time to time, but so rarely that evolution on this basis must be enormously slower than from changing conditions. Even under changing conditions, however, the rate of gene substitution is necessarily slow because the rate of selective change of frequency at any locus is limited by the amount of reproductive excess. That used in one gene replacement is unavailable for others. This is the limitation imposed by the so-called cost of evolution.

The shifting balance process is important because the selection relates to the adaptiveness of genetic systems as wholes and because limitation by "the cost of evolution" is relatively small. There is no cost associated with the random drift that leads to the local crossing of a saddle leading to a superior selective peak. There is indeed limitation of the rate of further establishment of the superior peaks in each of the more successful local populations, but little in the diffusion process that establishes superior systems throughout the species.

Selection among uniparentally reproducing clones in a species in which variability is maintained by occasional crosses, resembles an extreme form of the shifting balance process in that it is according to genotypes as wholes. It is a process that permits immediate adaptation to new situations insofar as such adaptations are formed by segregation from the occasional crosses. Favorable combinations are, however, broken up by further crossing and segregation. This process is less effective in the long run than the shifting balance process under biparental reproduction.

The summaries of chapters 12 and 13 provide more detailed theoretical conclusions. The store of variability in nature and the relative importance of the different kinds of selection depend on the prevalence of favorable population structures and cannot be profitably discussed before consideration of the variability within and among local populations in actual species, which will be the subject of volume 4.

BIBLIOGRAPHY

Agar, W. E. 1914. Experiments on inheritance in parthenogenesis. *Phil. Trans. Roy. Soc. (London) Ser. B* 205:421–89.

Anderson, E. 1949. *Introgressive hybridization.* New York: John Wiley & Sons, Inc.

Anderson, E., and Brown, W. L. 1952. The origin of cornbelt maize and its genetic significance. In *Heterosis,* ed. J. W. Gowen. Ames, Iowa: Iowa State College Press. Pp. 134–48.

Anderson, W. W. 1966. Genetic divergence in M. Vetukhiv's experimental populations of *Drosophila pseudoobscura. Genet. Res. Camb.* 7:255–66.

———. 1969. Polymorphism arising from the mating advantage of rare male genotypes. *Proc. Nat. Acad. Sci. U.S.* 64:190–97.

Ayala, F. J. 1966. Evolution of fitness: I. Improvement in the productivity and size of irradiated populations of *Drosophila serrata and Drosophila birchii. Genetics* 53:883–95.

———. 1967. Evolution of fitness: III. Improvement of fitness in irradiated populations of *Drosophila serrata. Proc. Nat. Acad. Sci. U.S.* 58:1919–23.

———. 1969. Evolution of fitness: V. Ratio of evolution of irradiated populations of *Drosophila. Proc. Nat. Acad. Sci. U.S.* 63:790–93.

Banta, A. M. 1921. *Selection in Cladocera on the basis of a physiological character.* Publication 305. Carnegie Institution of Washington.

———. 1939. *Studies on the physiological genetics and evolution of some Cladocera.* Publication 513. Carnegie Institution of Washington.

Barker, J. S. F. 1962. Studies of selective mating using the yellow mutant of *Drosophila melanogaster. Genetics* 47:623–40.

Barnett, S. A., and Coleman, E. A. 1960. Heterosis in F_1 mice in a cold environment. *Genet Res. Camb.* 1:25–38.

Barrington, A., and Pearson, K. 1906. On the inheritance of coat colour in cattle. *Biometrica* 4:427–64.

Bateman, A. J. 1959. The viability of near-normal irradiated chromosomes. *Intern. J. Radiation Biol.* 1:170–80.

Bateson, W. 1909. *Mendel's principles of heredity.* Cambridge: Cambridge University Press.

Beadle, G. W., and Tatum, E. L. 1941. Genetic control of biochemical reactions in *Neurospora*. *Proc. Nat. Acad. Sci. U.S.* 27:499–506.

Bell, A. E. 1968. Selection studies in *Tribolium*. *Proc. XII Intern. Congr. Genet.* 2:202–3.

Bell, A. E., and McNary, H. W. 1963. Genetic correlations and asymmetry of the correlated response from selection for increased body weight of *Tribolium* in two environments. *Proc. XI Intern. Congr. Genetic* 1:256.

Bell, A. E., and Moore, C. H. 1958. Further comparisons of reciprocal recurrent selection with conventional methods of selection for the improvement of quantitative characteristics. *Proc. X Intern. Congr. Genet.* 2:20–21.

Bell, A. E.; Moore, C. H.; and Warren, D. C. 1955. The evolution of new methods for the improvement of quantitative characters. *Cold Spring Harbor Symp. Quant. Biol.* 20:197–211.

Benedict, R. C. 1915. Some modern variations of the Boston fern and their source. *J. N. Y. Botan. Gardens* 16:189.

Bennett, J. 1956. Inexpensive population cages. *Drosophila Inform. Serv.* 30:159–60.

Bhalla, S. C., and Sokal, R. R. 1964. Competition among genotypes in the house-fly at varying densities and proportions (the green strain). *Evolution* 18:312–30.

Bittner, J. J. 1931. A genetic study of the transplantation of tumors arising in hybrid mice. *Am. J. Cancer* 15:2202–47.

Bliss, C. J. 1935. The calculation of the dosage mortality curve. *Ann. Appl. Biol.* 22:134–67.

Bodmer, W. F. 1963. Natural selection for modifiers of heterozygote fitness. *J. Theoret. Biol.* 4:86–97.

Bodmer, W. F., and Parsons, P. A. 1962. Linkage and recombination in evolution. *Advan. Genet.* 11:1–100.

Boivin, A.; Vandrely, R.; and Vandrely, C. 1948. L'acide désoxyribonucléique du noyau cellulaire dépositaire des caractères héréditaires: Arguments d'ordre analytique. *Compt. Rend. Acad. Sci.* 226:1061–63.

Bonnier, G., and Jonsson, U. B. 1957. Studies on X-ray-induced detrimentals in the second chromosome of *Drosophila melanogaster*. *Hereditas* 43:441–61.

Boorman, S. A. 1974. Island models for takeover by a social trait facing a frequency-dependent selection barrier, in a Mendelian population. *Proc. Nat. Acad. Sci. U.S.* 71:2103–7.

Bowman, J. C., and Falconer, D. S. 1960. Inbreeding depression and heterosis of litter size in mice. *Genet. Res. Camb.* 1:262–74.

Bray, D. F.; Bell, A. E.; and King, S. C. 1962. The importance of genotype by environment interaction with reference to control populations. *Genet. Res. Camb.* 3:282–302.

Bridges, C. B. 1913. Nondisjunction of the sex chromosomes of *Drosophila*. *J. Exp. Zool.* 15:587–606.

———. 1935. Salivary chromosome maps with a key to the banding of the chromosomes. *J. Heredity* 26:60–64.

Brink, R. A. 1960. Paramutation and chromosome organization. *Quart. Rev. Biol.* 35:120–37.

Brink, R. A., and Nilan, R. A. 1952. The relation between light variegated and medium variegated pericarp in maize. *Genetics* 37:519–44.

Britten, R. J.; and Kohne, D. E. 1968. Repeated sequences in DNA. *Science* 161:529–40.

Bruce, A. B. 1910. The Mendelian theory of heredity and the augmentation of vigor. *Science.* 32:627–28.

Brues, A. 1964. The cost of evolution vs. the cost of not evolving. *Evolution* 18:379–83.

Buri, P. 1956. Gene frequency in small populations of mutant *Drosophila*. *Evolution* 10:367–402.

Buzzati-Traverso, A. 1952. Heterosis in population genetics. In *Heterosis*, ed. J. W. Gowen. Ames, Iowa: Iowa State College Press, Pp. 149–60.

Cain, A. J., and Currey, J. D. 1963. Area effects in *Cepaea*. *Phil. Trans. Roy. Soc.* (*London*) *Ser. B* 246:1–81.

Calder, A. 1927. The role of inbreeding in the development of the Clydesdale breed of horses. *Proc. Roy. Soc. Edinburgh* 47:118–40.

Carpenter, J. R.; Grüneberg, H.; and Russell, E. S. 1957. Genetical differentiation involving morphological characters in an inbred strain of mice: II. American branches at the C57BL and C57BR strains. *J. Morphol.* 100:377–88.

Carson, H. L. 1958. Responses to selection under different conditions of recombination in *Drosophila*. *Cold Spring Harbor Symp. Quant. Biol.* 23:291–306.

———. 1967. Selection for parthenogenesis in *Drosophila mercatorum*. *Genetics* 55:157–71.

Carter, R. C. 1940. A genetic history of Hampshire sheep. *J. Heredity* 31:89–93.

Castle, W. E. 1905. *Heredity of coat characters in guinea pigs and rabbits*. Publication 23. Carnegie Institution of Washington.

———. 1906. *The origin of a polydactylous race of guinea pigs*. Publication 49. Carnegie Institution of Washington.

———. 1907. Color varieties of the rabbit and of other rodents: Their origin and inheritance. *Science* 26:287–91.

———. 1912a. On the origin of a pink-eyed guinea pig with a colored coat. *Science* 35:508–10.

———. 1912b. Some biological principles of animal breeding. *Am. Breeders Mag.* 3:279–82.

———. 1919. Piebald rats and selection: A correction. *Am. Naturalist* 53:370–76.

———. 1941. Influence of certain color mutations on body size in mice, rats and rabbits. *Genetics* 26:177–91.

———. 1951. Variations in the hooded pattern of rats and a new allele of hooded. *Genetics* 36:254–66.

Castle, W. E.; Carpenter, F. W.; Clark, A. H.; Mast, S. O.; and Barrows, W. M. 1906. The effects of inbreeding, crossbreeding and selection upon the fertility and variability of *Drosophila*. *Proc. Am. Acad. Arts Sci.* 41:731–86.

Castle, W. E., and Phillips, J. C. 1914. *Piebald rats and selection*. Publication 195. Carnegie Institution of Washington.

Castle, W. E., and Wright, S. 1916. *Studies of inheritance in guinea pigs and rats*. Publication 26. Carnegie Institution of Washington.

Cavalli-Sforza, L. L., and Lederberg, J. 1956. Isolation of preadaptive mutants in bacteria by sib selection. *Genetics* 41:317–81.

Chabora, A. J. C. 1968. Disruptive selection for sternopleural chaeta number in various strains of *Drosophila melanogaster*. *Am. Naturalist* 102:525–32.

Chai, C. K. 1957. Developmental homeostasis of body growth in mice. *Am. Naturalist* 91:49–55.

———. 1966. Selection for leucocyte counts in mice. *Genet. Res. Camb.* 8:125–42.

———. 1969. Effects of inbreeding in rabbits: Inbred lines, discrete characters, breeding performance and morbidity. *J. Heredity* 60:64–70.

———. 1970. Genetic basis of leucocyte production in mice. *J. Heredity* 61:67–71.

Chetverikov, I. 1926. On certain aspects of the evolutionary process from the standpoint of modern genetics. *Zh. Exp. Biol.* 1:3–54 (in Russian). *Proc. Am. Phil. Soc.* 105:167–95 (English translation by M. Barker).

Claringbold, B. J., and Biggers, J. D. 1955. The response of inbred mice to oestrogens. *J. Endocrinol.* 12:9–14.

Clayton, G. A.; Knight, G. R.; Morris, J. A.; and Robertson, A. 1957. An experimental check on quantitative genetical theory: III. Correlated responses. *J. Genet.* 55:171–80.

Clayton, G. A., Morris, J. A.; and Robertson, A. 1953. Selection for abdominal chaetae in a large population of *Drosophila melanogaster*. In *Symposium on genetics of population structure*. International Union of the Biological Sciences. Pavia, Italy: Premiata Tipografia Successori Filli Fusi. Pp. 7–15.

———. 1957. An experimental check on quantitative genetical theory: I. Short-term responses to selection. *J. Genet.* 55:131–51.

Clayton, G. A., and Robertson, A. 1955. Mutation and cumulative variation. *Am. Naturalist* 89:151–55.

———. 1957. An experimental check on quantitative genetical theory: II. Long-term effects of selection. *J. Genet.* 55:152–70.

Cloudman, A. M. 1932. A comparative study of transplantability of eight mammary gland tumors arising in inbred mice. *Am. J. Cancer* 16:568–630.

Collins, G. N. 1921. Dominance and the vigor of first generation hybrids. *Am. Naturalist* 55:116–33.

Comstock, R. F., and Robinson, H. F. 1948. The components of genetic variance in populations of biparental progenies and their use in estimating the average degree of dominance. *Biometrics* 4:254–66.

———. 1952. Estimation of average dominance of genes. In *Heterosis*, ed. J. W. Gowen. Ames, Iowa: Iowa State College Press. Pp. 484–516.

Comstock, R. F.; Robinson, H. F.; and Cockerham, C. C. 1957. Quantitative genetics project report. Mimeo Series 167. Raleigh: North Carolina State College Institute of Statistics.

Comstock, R. F.; Robinson, H. F.; and Harvey, P. H. 1949. A breeding procedure designed to make maximum use of both general and specific combining ability. *Agron. J.* 41:360–67.

Constantino, R. F.; Bell, A. E.; and Rogler, J. C. 1967. Genetic analysis of a population of *Tribolium*: I. Corn oil sensitivity and selection response. *Heredity* 22:529–39.

Crampe, H. 1883. Zuchtversuche mit zahmen Wanderratten. *Landwirtsch. Jahrb.* 12:389–458.

Crosby, J. L. 1963. The evolution and nature of dominance. *J. Theoret. Biol.* 5:35–51.

Crow, J. F. 1948. Alternative hypotheses of hybrid vigor. *Genetics* 33:477–87.

———. 1952. Dominance and overdominance. In *Heterosis*, ed. J. W. Gowen. Ames, Iowa: Iowa State College Press. Pp. 282–97.

———. 1954. Analysis of DDT-resistant strains of *Drosophila*. *J. Econ. Entomol.* 47:393–98.

———. 1957. Genetics of insect resistance to chemicals. *Ann. Rev. Entomd.* 2:227–46.

———. 1970. Genetic loads and the cost of natural selection. In *Mathematical topics in population genetics*, ed. K. Kojima. Berlin: Springer-Verlag. Pp. 128–77.

Crow, J. F., and Kimura, M. 1956. Some genetic problems in natural populations. *Proc. Third Berkeley Symp. Math. Stat. Prob.* 4:1–22.

———. 1965*a*. Evolution in sexual and asexual populations. *Am. Naturalist* 99:439–50.

———. 1965*b*. The theory of genetic loads. *Proc. XI Intern. Congr. Genet.* 3:495–505.

———. 1969. Evolution in sexual and asexual populations: A reply. *Am. Naturalist* 103:89–91.

———. 1970. *An introduction to population genetics theory.* New York: Harper & Row Publishers, Inc.

Crow, J. F., and Morton, N. E. 1955. Measurement of gene frequency drift in small populations. *Evolution* 9:202–14.

Crow, J. F., and Temin, R. G. 1964. Evidence for the partial dominance of recessive lethal genes in natural populations of *Drosophila*. *Am. Naturalist* 98:21–33.

Cruden, D. 1949. The computation of inbreeding coefficients in closed populations. *J. Heredity* 40:248–51.

Cuénot, L. 1904. L'hérédité de la pigmentation chez les souris: 3me note: Les formules héréditaire. *Arch. Zool. Expér. Gén.*, ser. 4, tome 2:45–56.

Danforth, C. 1923. The frequency of mutation and the incidence of hereditary traits in man: Eugenics, genetics and the family. *Proc. Second Intern. Congr. Eugenics* 1:120–28.

Darwin, C. 1859. *The origin of species by means of natural selection.* London: John Murray. 6th ed. London: D. Appleton, 1910.

———. 1868. *The variation of animals and plants under domestication.* London: John Murray. 2d ed. London: D. Appleton, 1883.

————. 1876. *The effects of cross- and self-fertilization in the vegetable kingdom.* London: John Murray Publishers, Ltd.

Davenport, C. B. 1908. Degeneration, albinism and inbreeding. *Science* 28:454–55.

Demerec, M. 1940. Genetic behavior of euchromatic segments inserted in heterochromatin. *Genetics* 25:618–27.

Deol, M. S.; Grüneberg, H.; Searle, A. G.; and Truslove, G. M. 1957. Genetical differentiation involving morphological characteristics in an inbred strain of mice: I. A British branch of the C57BL strain. *J. Morphol.* 100:345–76.

————. 1960. How pure are our inbred strains of mice? *Genet. Res. Camb.* 1:50–58.

Detlefsen, J. A., and Roberts, E. 1921. Studies on crossing-over: I. The effect of selection on crossover values. *J. Exp. Zool.* 32:333–54.

Dickerson, G. E. 1955. Genetic slippage in response to selection for multiple objectives. *Cold Spring Harbor Symp. Quant. Biol.* 20:213–24.

Dickerson, G. E.; Blunn, C. T.; Chapman, A. B.; Kottman, R. M.; Krider, J. L.; Warwick, E. J.; and Whatley Jr., J. A.; in collaboration with Baker, M. L.; Lush, J. L.; and Winters, L. M. 1954. Evaluation of selection in developing inbred lines of swine. *Univ. Missouri Coll. Agr. Res. Bull.* 551.

Dickson, R. C. 1941. Inheritance of resistance to hydrocyanic acid fumigation in the California red scale. *Hilgardia* 13:515–21.

Dickson, W. F., and Lush, J. L. 1933. Inbreeding and the genetic history of Rambouillet sheep in America. *J. Heredity* 24:19–33.

Diederich, G. W. 1941. Nonrandom mating between yellow-white and wild type in *Drosophila melanogaster*. *Genetics* 26:148.

Dobzhansky, T. 1937, 1941, 1951. *Genetics and the origin of species.* 1st, 2d, 3d ed. New York: Columbia University Press.

————. 1946. Genetics of natural populations: XIII. Recombination and variability in populations of *Drosophila pseudoobscura*. *Genetics* 31:269–90.

————. 1947. Genetics of natural populations: XIV. A response of certain gene arrangements in the third chromosome of *Drosophila pseudoobscura* to natural selection. *Genetics* 32:142–60.

————. 1952. Nature and origin of heterosis. In *Heterosis*, ed. J. W. Gowen. Ames, Iowa: Iowa State College Press. Pp. 218–23.

————. 1955. A review of some fundamental concepts and problems of population genetics. *Cold Spring Harbor Symp. Quant. Biol.* 20:1–15.

————. 1970. *Genetics of the evolutionary process.* New York: Columbia University Press.

Dobzhansky, T., and Levine, M. 1948. Genetics of natural populations: XVII. Proof of operation of natural selection in wild populations of *Drosophila pseudoobscura*. *Genetics* 33:537–47.

Dobzhansky, T., and Pavlovsky, O. 1957. An experimental study of interaction between genetic drift and natural selection. *Evolution* 11:311–19.

Dobzhansky, T., and Spassky, B. 1947. Evolutionary changes in laboratory cultures of *Drosophila pseudoobscura*. *Evolution* 1:191–216.

――――. 1960. Release of genetic variability through recombination: V. Breakup of synthetic lethals by crossing-over in *Drosophila pseudoobscura*. *Zool. Jahrb. Abt. Syst.* 88:57–66.

――――. 1962. Selection for geotaxis in monomorphic and polymorphic populations of *Drosophila pseudoobscura*. *Proc. Nat. Acad. Sci. U.S.* 48:1704–12.

――――. 1967a. An experiment on migration and simultaneous selection for several traits in *Drosophila pseudoobscura*. *Genetics* 55:723–34.

――――. 1967b. Effects of selection and migration on geotactic and phototactic behaviour of *Drosophila*: I. *Proc. Roy. Soc. (London) Ser. B* 168:22–47.

――――. 1969. Artificial and natural selection for two behavioral traits in *Drosophila pseudoobscura*. *Proc. Nat. Acad. Sci. U.S.* 62:75–80.

Dobzhansky, T.; Spassky, B.; and Sved, J. 1969. Effects of selection and migration on geotactic and phototactic behavior of *Drosophila*: II. *Proc. Roy. Soc. (London) Ser. B* 173:191–207.

Dobzhansky, T., and Spassky, N. P. 1962. Genetic drift and natural selection in experimental populations of *Drosophila pseudoobscura*. *Proc. Nat. Acad. Sci. U.S.* 48:148–56.

Dodson, E. O. 1962. Note on the cost of natural selection. *Am. Naturalist* 96:123–26.

Druger, M. 1962. Selection and body size in *Drosophila pseudoobscura* at different temperatures. *Genetics* 47:209–22.

Dudley, J. W., and Lambert, R. J. 1969. Genetic variability after sixty-five generations of selection in Illinois high oil, low oil, high protein, and low protein strains of *Zea mays* L. *Crop Sci.* 9:179–81.

Dun, R. B., and Fraser, A. S. 1959. Selection for an invariant character, vibrissa number, in the house mouse. *Australian J. Biol. Sci.* 21:506–23.

Dunn, L. C. 1956. Analysis of a complex gene in the house mouse. *Cold Spring Harbour Symp. Quant. Biol.* 21:187–95.

Durrant, A., and Mather, K. 1954. Heritable variation in a long inbred line of *Drosophila*. *Genetica* 27:97–119.

East, E. M. 1908. Inbreeding in corn. *(New Haven) Conn. Agr. Exp. Sta. Rept.*, 1907. Pp. 419–28.

――――. 1909. The distinction between development and heredity in inbreeding. *Am. Naturalist* 43:173–81.

――――. 1910. The transmission of variations in the potato in asexual reproduction. Public document no. 24. *Rept. (Hartford) Conn. Agr. Exp. Sta.*, 1909, 1910. Pp. 119–60.

――――. 1935. Genetic reactions in *Nicotiana*: III. Dominance. *Genetics* 20:443–51.

――――. 1936. Heterosis. *Genetics* 21:375–97.

East, E. M., and Jones, D. F. 1919. *Inbreeding and outbreeding*. Philadelphia: J. B. Lippincott Co.

Eaton, O. N. 1932. *Correlations of hereditary and other factors affecting growth in guinea pigs*. Technical bulletin 279. Washington, D.C.: U.S. Department of Agriculture.

————. 1953. Heterosis in the performance of mice. *Genetics* 38:609–29.

Ehrman, L. 1964. Genetic divergence in M. Vetukhiv's experimental populations of *Drosophila pseuboobscura*. *Genet. Res. Camb.* 5:150–57.

————. 1967. Further studies on genotype frequency and mating success in *Drosophila*. *Am. Naturalist* 101:415–24.

————. 1969. The sensory basis of mate selection in *Drosophila*. *Evolution* 23:59–64.

Emerson, R. A. 1917. Genetical analysis of variegated pericarp in maize. *Genetics* 2:1–35.

Emerson, R. A.; Beadle, G. W.; and Fraser, A. C. 1935. *A summary of linkage studies in maize*. Publication 180. Ithaca, N.Y.: Cornell University Agricultural Experimental Station.

Emerson, R. A., and Smith, H. H. 1950. *Inheritance of number of kernel rows in maize*. Memoir 298. Ithaca, N.Y.: Cornell University Agricultural Experimental Station.

Enfield, F. D.; Comstock, R. E.; and Braskerud, O. 1966. Selection for pupa weight in *Tribolium castaneum*: I. Parameters in base populations. *Genetics* 54:523–33.

Eshel, I. 1972. On the neighbor effect and the evolution of altruistic traits. *Theoret. Pop. Biol.* 1:258–77.

Ewing, H. E. 1916. Eighty-seven generations in a parthenogenetic pure line of *Aphis avenae* (Fab). *Biol. Bull.* 31:53–112.

Falconer, D. S. 1953. Selection for large and small size in mice. *J. Genet.* 5:470–501.

————. 1954. Selection for sex ratio in mice and *Drosophila*. *Am. Naturalist* 87:385–97.

————. 1955. Patterns of response in selection experiments with mice. *Cold Spring Harbor Symp. Quant. Biol.* 20:178–96.

————. 1957. Selection for phenotypic intermediates in *Drosophila*. *J. Genet.* 55:551–61.

————. 1960a. The genetics of litter size in mice. *J. Cellular Comp. Physiol.* 56 (suppl. 1): 153–67.

————. 1960b. Selection of mice for growth on high and low planes of nutrition. *Genet. Res. Camb.* 1:91–113.

————. 1960c. *Introduction to quantitative genetics*. New York: Ronald Press Co.

————. 1964. Maternal effects and selective response. *Proc. XI Intern. Congr. Genet.* Pp. 763–74.

Falconer, D. S., and Bloom, J. L. 1962. A genetic study of induced lung tumours in mice. *Brit. J. Cancer* 16:665–85.

————. 1964. Changes in susceptibility to urethane-induced lung tumours produced by selective breeding in mice. *Brit. J. Cancer* 18:322–32.

Falconer, D. S., and King, J. W. B. 1953. A study of selection limits in the mouse. *J. Genet.* 51:561–81.

Falconer, D. S., and Latyszewski, M. 1952. The environment in relation to selection for size in mice. *J. Genet.* 51:67–80.

Falconer, D. S., and Roberts, R. C. 1960. Effect of inbreeding on evolutionary rate and foetal mortality in mice. *Genet. Res. Camb.* 1:422–30.

Falk, R. 1955. Studies on X-ray-induced viability mutations in the third chromosome of *Drosophila melanogaster*. *Hereditas* 11:259–77.

Feller, W. 1967. On fitness and the cost of natural selection. *Genet. Res. Camb.* 9:1–15.

Felsenstein, J. 1971. On the biological significance of the cost of gene substitution. *Am. Naturalist* 105:1–12.

Fisher, R. A. 1918. The correlations between relatives on the supposition of Mendelian inheritance. *Trans. Roy. Soc. Edinburgh* 52:399–433.

———. 1925. *Statistical methods for research workers*. Edinburgh: Oliver and Boyd.

———. 1928a. The possible modification of the response of wild type to recurrent mutations. *Am. Naturalist* 62:115–26.

———. 1928b. Two further notes on the origin of dominance. *Am. Naturalist* 62:571–74.

———. 1929. The evolution of dominance: A reply to Professor Sewall Wright. *Am. Naturalist* 63:553–56.

———. 1930. *The genetical theory of natural selection*. Oxford: Clarendon Press. 2d rev. ed. 1958 Dover Publ.

———. 1931. The evolution of dominance. *Biol. Rev.* 6:345–68.

———. 1934. Professor Wright on the theory of dominance. *Am. Naturalist* 68:370–74.

———. 1941. Average excess and average effect of a gene substitution. *Ann. Eugen.* 11:53–63.

———. 1949. *The theory of inbreeding*. Edinburgh: Oliver and Boyd. 2d ed. 1965.

Fisher, R. A., and Ford, E. B. 1947. The spread of a gene in natural conditions in a colony of the moth, *Panaxia dominula* L. *Heredity* 1:143–74.

———. 1950. The Sewall Wright effect. *Heredity* 4:47–49.

Fisher, R. A., and Holt, S. B. 1944. The experimental modification of dominance in Danforth's short-tailed mice. *Ann. Eugen.* 12:102–20.

Ford, E. B. 1930. The theory of dominance. *Am. Naturalist* 64:560–66.

———. 1940. Genetic research in the Lepidoptera. *Ann. Eugen.* 10:227–52.

———. 1964. *Ecological genetics*. London: Methuen & Co., Ltd.

Fowler, A. B. 1932. The Ayrshire breed: A genetic study. *J. Dairy Res.* 4:11–27.

Fox, A. S. 1949. Immunogenetic studies of *Drosophila melanogaster*: II. Interactions between the *Rb* and *V* loci in production of antigens. *Genetics* 34:647–64.

Frahm, R. R., and Kojima, K. 1966. Comparison of selection responses on body weight under divergent larval density conditions in *Drosophila pseudoobscura*. *Genetics* 54:625–37.

Frydenberg, O. 1962. The modification of polymorphism in some artificial populations of *Drosophila melanogaster*. *Hereditas* 48:423–41.

———. 1963. Population studies of a lethal mutant in *Drosophila melanogaster*: I. Behaviour in populations with discrete generations. *Hereditas* 50:89–116.

————. 1964a. Population studies of a lethal mutant in *Drosophila melanogaster*: II. Behaviour in populations with overlapping generations. *Hereditas* 51:31–73.

————. 1964b. Long-term instability of an ebony polymorphism in artificial populations of *Drosophila melanogaster*. *Hereditas* 51:198–206.

Fuller, J. L., and Thompson, W. R. 1960. *Behavior genetics*. New York: John Wiley & Sons, Inc.

Gaertner, C. F. von 1849. *Versuche und Beobachtungen über die Bastarderzeugung im Pflanzenreich*. Stuttgart: n.p.

Gardner, C. O. 1963. Estimates of genetic parameters in cross-fertilizing plants and their implications in plant breeding. In *Statistical genetics and plant breeding*, ed. W. D. Hanson and H. F. Robinson. Publication no. 982. Washington, D.C.: National Academy of Science–National Research Council. Pp. 225–52.

Gardner, C. O.; Harvey, D. H.; Comstock, R. E.; and Robinson, H. F. 1953. Dominance of genes controlling quantitative characters in maize. *Agron. J.* 45:186–91.

Gardner, C. O., and Lonnquist, J. M. 1959. Linkage and the degree of dominance of genes controlling quantitative characters of maize. *Agron. J.* 51:524–28.

Garrod, A. E. 1902. The incidence of alkaptonuria, a study of chemical individuality. *Lancet*, 13 December 1902.

————. 1908. Inborn errors of metabolism. *Lancet*, 4, 11, 18, 25 January 1908.

Gershenson, S. 1928. A new sex ratio abnormality in *Drosophila obscura*. *Genetics* 13:488–507.

Gibson, J. B., and Thoday, J. M. 1962. Effects of disruptive selection: VI. A second chromosome polymorphism. *Heredity* 17:1–26.

Goldschmidt, R. 1911. *Einführung in die Vererbungswissenschaft*. Leipzig: Verlag von Wilhelm Engelmann.

————. 1938. *Physiological genetics*. New York: McGraw-Hill Book Co.

————. 1940. *The material basis of evolution*. New Haven: Yale University Press.

Goodale, H. D. 1937. Can artificial selection produce unlimited change? *Am. Naturalist* 7:433–59.

————. 1938. A study of the inheritance of weight in the albino mouse by selection. *J. Heredity* 28:101–12.

————. 1941. Progress report on possibilities in progeny-test breeding. *Science* 94:442–43.

————. 1942. Further progress with artificial selection. *Am. Naturalist* 76:515–19.

Gorer, P. A. 1937. The genetic and antigenic basis of tumour transplantation. *J. Pathol. Biol.* 44:691–97.

Gowen, J. W. 1952. Hybrid vigor in *Drosophila*. In *Heterosis*, ed. J. W. Gowen. Ames, Iowa: Iowa State College Press. Pp. 474–93.

Gravatt, P. 1914. A raddish-cabbage hybrid. *J. Heredity* 5:265–72.

Green, E. L. 1941. Genetic and nongenetic factors which influence the type of the skeleton in an inbred strain of mice. *Genetics* 26:192–222.

————. 1953. A skeletal difference between sublines of the C3H strain of mice. *Science* 117:81–82.

Greenberg, R., and Crow, J. F. 1960. Comparison of the effects of lethal and detrimental chromosomes from *Drosophila* populations. *Genetics* 45:1153–68.

Gregory, P. W. 1932. The potential and actual fecundity of some breeds of rabbits. *J. Exp. Zool.* 62:271–85.

Grewal, M. S. 1962. The rate of genetic divergence of sublines in the C57BL strain of mice. *Genet. Res. Camb.* 3:226–37.

Griffing, B., and Langridge, J. 1963. Phenotypic stability of growth in the self-fertilized species, *Arabidopsis thaliana*. In *Statistical genetics and plant breeding*, ed. W. D. Hanson and H. F. Robinson. Publication no. 982. Washington, D.C.: National Academy of Science–National Research Council. Pp. 368–94.

Grüneberg, H. 1951. The genetics of a tooth defect in the mouse. *Proc. Roy. Soc. (London) Ser. B* 138:437–51.

————. 1952. Genetical studies on the skeleton of the mouse. *J. Genet.* 51:95–114.

————. 1954. Variation within inbred strains of mice. *Nature* 173:674–76.

Grüneberg, H.; Bains, G. S.; Berry, R. J.; Riles, L.; Smith, C. A. B.; and Weiss, R. A. 1966. *A search for genetic effects of high natural radioactivity in South India*. Special Report Series 307. London: Medical Research Council.

Guaita, G. von 1898. Versuche mit Kreuzungen von verschieden Rassen der Hausmaus. *Ber. Naturforsch. Ges. Freiburg* 10:317–32.

————. 1900. Zweite Mittheilung über Versuche mit Kreuzungen von verschieden Hausmaus Rassen. *Ber. Naturforsch. Ges. Freiburg* 11:131–43.

Gulick, J. T. 1872. On diversity of evolution under one set of external conditions. *Linnean Soc. J. Zool.* 11:496–505.

————. 1905. *Evolution, racial and habitudinal*. Publication 25. Carnegie Institution of Washington.

Gustafsson, Å. 1946. The effect of heterozygosity on variability and vigour. *Hereditas* 32:263–86.

————. 1947a. The advantageous effect of deleterious mutations. *Hereditas* 33:573–75.

————. 1947b. Mutations in agricultural plants. *Hereditas* 33:1–106.

Hadler, N. M. 1964. Genetic influence on phototaxis in *Drosophila melanogaster*. *Biol. Bull.* 126:264–73.

Haldane, J. B. S. 1924. A mathematical theory of natural and artificial selection: Part 1. *Cambridge Phil. Soc. Trans.* 23:19–41.

————. 1927. A mathematical theory of natural and artificial selection: V. Selection and mutation. *Proc. Cambridge Phil. Soc.* 23:838–44.

————. 1930. A note on Fisher's theory of the origin of dominance. *Am. Naturalist* 64:87–90.

————. 1932. *The causes of evolution*. London: Harper and Brothers.

————. 1933. The part played by recurrent mutation in evolution. *Am. Naturalist* 67:5–19.

————. 1937. The effect of variation on fitness. *Am. Naturalist* 71:337–49.

————. 1939. The theory of the evolution of dominance. *J. Genet.* 37:365–74.

————. 1956. The theory of selection for melanism in Lepidoptera. *Proc. Roy. Soc. (London) Ser. B* 145:303–8.

————. 1957. The cost of natural selection. *J. Genet.* 55:511–24.

————. 1960. More precise expressions for the cost of natural selection. *J. Genet.* 57:351–60.

————. 1964. A defense of beanbag genetics. *Perspectives Biol. Med.* 7:343–59.

Hamilton, W. D. 1963. The evolution of altruistic behavior. *Am. Naturalist* 97:354–56.

————. 1964. The genetical evolution of social behavior. *J. Theoret. Biol.* 7:1–16.

Hammond, J. 1947. Animal breeding in relation to nutrition and environmental conditions. *Biol. Rev.* 22:195–213.

Hanel, E. 1907. Vererbung bei ungeschlechticher Fortpflanzung von *Hydra grisea. Jenaische Zeit. Naturwiss.* 43:321–72.

Hardin, R. T., and Bell, A. E. 1967. Two-way selection for body weight in *Tribolium* on two levels of nutrition. *Genet. Res. Camb.* 9:309–30.

Harlan, H. V., and Martini, M. L. 1938. The effect of natural selection in a mixture of barley varieties. *J. Agr. Res.* 57:189–99.

Hase, A. 1909. Ueber die deutscher Süswasser Polypen, *Hydra fusca* L. *Hydra grisea* L. and *Hydra viridis* L. *Arch. Rass. Ges. Biol.* 6:721–53.

Hazel, L. N., and Lush, J. L. 1942. The efficiency of three methods of selection. *J. Heredity* 33:393–95.

Hazel, L. N., and Terrell, C. E. 1946. Effects of some environmental factors on fleece characteristics of range Rambouillet yearling ewes. *J. Agr. Sci.* 5:382–88.

Hegner, R. W. 1919. Heredity, variation and the appearance of diversities during the vegetative reproduction of *Arcella dentata. Genetics* 4:95–150.

Helman, B. 1949. Étude de la vitalité relative du génotype sauvage Oregon et de génotype comportant le gène Stubble chez *Drosophila melanogaster. Compt. Rend. Acad. Sci.* 228:2057–58.

Hetzer, H. O., and Harvey, W. R. 1967. Selection for high and low fatness in swine. *J. Animal Sci.* 26:1244–51.

Hill, W. G., and Robertson, A. 1966. The effect of linkage on limits to artificial selection. *Genet. Res. Camb.* 8:269–94.

Hiraizumi, Y., and Crow, J. F. 1960. Heterozygous effects on viability, fertility, rate of development and longevity of *Drosophila* chromosomes that are lethal when homozygous. *Genetics* 45:1071–83.

Hirsch, J. 1962. Individual differences in behavior and their genetic bases. In *Roots of behavior*, vol. 3., ed. E. L. Bliss. New York: Harper and Brothers. P. 23.

Hull, F. H. 1945. Recurrent selection for specific combining ability in corn. *J. Am. Soc. Agron.* 37:134–45.

————. 1952. Recurrent selection and overdominance. In *Heterosis*, ed. J. W. Gowen. Ames, Iowa: Iowa State College Press. Pp. 451–73.

Hunt, H. R., and Goodman, H. O. 1962. The inheritance of resistance and susceptibility to dental caries. *Intern. Dental J.* 12:306–21.

Hunt, H. R.; Hoppert, C. A.; and Erwin, W. G. 1944. Inheritance of susceptibility to caries in albino rats (*Mus norvegicus*). *J. Dental Res.* 23:385–401.

Hunt, H. R.; Hoppert, C. A.; and Rosen, S. 1955. Genetic factors in experimental
rat caries. In *Advances in experimental caries research*, ed. R. F. Sognnaes.
Washington, D.C.: American Association for the Advancement of Science.
Pp. 66–81.

Hyde, R. H. 1914. Fertility and sterility in *Drosophila ampelophila*. *J. Exp.
Zool.* 17:141–71, 173–212.

Jennings, H. S. 1908. Heredity, variation and evolution in protozoa: II. Heredity
and variation in size and form in *Paramecium* with studies of growth,
environmental action and selection. *Proc. Am. Phil. Soc.* 47:393–546.

———. 1916. Heredity, variation and the results of selection in the uniparental
reproduction of *Difflugia corona*. *Genetics* 1:467–534.

Johannsen, W. 1903. *Ueber Erblichkeit in Population und in reinen Linien*. Jena,
Germany: Gustav Fischer.

———. 1909. *Elemente der exakten Erblichkeitslehre*. Jena, Germany: Gustav
Fischer.

Jollos, V. 1934. Dauermodifikationen und Mutationen bei Protozoen. *Arch.
Protistenk.* 83:197–219.

Jones, D. F. 1917. Dominance of linked factors as a means of accounting for
heterosis. *Genetics* 2:466–79.

———. 1918. *The effects of inbreeding and crossbreeding upon development*.
Bulletin 207. New Haven: Connecticut Agricultural Experimental Station.

———. 1945. Heterosis resulting from degenerative changes. *Genetics* 30:527–42.

———. 1957. Gene action in heterosis. *Genetics* 42:93–103.

Judd, B. H., and Young, M. W. 1973. An examination of the one cistron-one
chromomere concept. *Cold Spring Harbor Symp. Quant. Biol.* 38:573–79.

Käfer, E. 1952. Vitalitätsmutationen ausgelöst durch Röntgenstrahlen bei *Drosophila melanogaster*. *Z. ind. Abst. Vererb.* 84:508–35.

Karper, R. E. 1930. The effect of a single gene upon development in the heterozygote in *Sorghum*. *J. Heredity* 2:187–92.

Keeble, F., and Pellew, C. 1910. The mode of inheritance of stature and of time
of flowering in peas (*Pisum sativum*). *J. Genet.* 1:47–56.

Kelley, T. L. 1923. *Statistical method*. New York: Macmillan Co.

Kempthorne, O. 1957. *An introduction to genetic statistics*. New York: John Wiley
& Sons, Inc.

Kerkis, J. 1938. The frequency of mutations affecting viability. *Bull. Acad. Sci.
U.S.S.R.* (*Biol.*) 1938:75–96.

Kerr, W. E., and Wright, S. 1954a. Experimental studies of the distribution of
gene frequencies in very small populations of *Drosophila melanogaster*: I.
Forked. *Evolution* 8:172–77.

———. 1954b. Experimental studies of the distribution of gene frequencies in
very small populations of *Drosophila melanogaster*: III. Aristapedia and
spineless. *Evolution* 8:293–301.

Kimura, M. 1956. A model of a genetic system which leads to closer linkage by natural selection. *Evolution* 10:278–87.

———. 1957. Some problems of stochastic processes in genetics. *Ann. Math. Statist.* 28:882–901.

———. 1958. On the change of population fitness by natural selection. *Heredity* 12:145–67.

———. 1961. Some calculations on mutational load. *Japan. J. Genet.* 36 (suppl. 1): 179–90.

———. 1962. On the probability of fixation of mutant genes in a population. *Genetics* 47:713–19.

———. 1965. Attainment of quasi-linkage equilibrium when gene frequencies are changing by natural selection. *Genetics* 52:875–90.

———. 1968. Evolutionary rate at the molecular level. *Nature* 217:624–26.

Kimura, M., and Crow, J. F. 1963. The measurement of effective population number. *Evolution* 17:279–88.

Kimura, M.; Maruyama, T.; and Crow, J. F. 1963. The mutation load in small populations. *Genetics* 48:1303–12.

Kimura, M., and Ohta, T. 1971. *Theoretical aspects of population genetics*. Princeton, N.J.: Princeton University Press.

Kindred, B. 1963. Selection for an invariant character, vibrissa number, in the house mouse: V. Selection on nontabby segregants from tabby selection lines. *Genetics* 55:365–73.

———. 1967. Selection for canalization in mice. *Genetics* 55:635–44.

King, H. D. 1918*a*. Studies on inbreeding: I. The effects of inbreeding on the growth and variability in body weight of the albino rat. *J. Exp. Zool.* 26:1–54.

———. 1918*b*. Studies on inbreeding: II. The effects of inbreeding on the fertility and on the constitutional vigor of the albino rat. *J. Exp. Zool.* 26:55–98.

———. 1918*c*. Studies on inbreeding: III. The effects of inbreeding with selection on the sex ratio of the albino rat. *J. Exp. Zool.* 27:1–35.

———. 1919. Studies on inbreeding: IV. A further study of the effects of inbreeding on the growth and variability in body weight of the albino rat. *J. Exp. Zool.* 29:134–75.

King, J. C. 1955*a*. Evidence of the integration of the gene pool from studies of DDT resistance in *Drosophila*. *Cold Spring Harbor Symp. Quant. Biol.* 20:311–17.

———. 1955*b*. Integration of the gene pool as demonstrated by resistance to DDT. *Am. Naturalist* 89:39–46.

King, J. C., and Sømne, L. 1958. Chromosomal analysis of the genetic factors for resistance to DDT in two resistant lines of *Drosophila melanogaster*. *Genetics* 43:577–93.

King, J. L. 1967. Continuously distributed factors affecting fitness. *Genetics* 55:483–92.

King, J. L., and Jukes, T. H. 1969. Non-Darwinian evolution. *Science* 164:788–98.

Kinman, M. L., and Sprague, G. F. 1945. Relation between number of parental lines and the theoretical performance of synthetic varieties of corn. *J. Am Soc. Agron.* 37:341–51.

Koelreuter, J. G. 1766. *Vorläufigen Nachricht von einigen das Geschlecht der Pflanzen betreffenden Versuchen und Beobachtungen.* Leipzig: n.p.

Kojima, K. 1969. Genetic variability and selection responses in quantitative traits. *Japan. J. Genet.* 44:294–98.

Kojima, K., and Kelleher, T. M. 1963. A comparison of purebred and crossbred selection schemes with two populations of *Drosophila pseudoobscura.* *Genetics* 48:57–72.

Krimbas, C. B. 1960. Synthetic sterility in *Drosophila willistoni.* *Proc. Nat. Acad. Sci. U.S.* 46:832–33.

Kyle, W. H., and Goodale, H. D. 1963. Selection progress toward an absolute limit for amount of white hair on mice. *Proc. XI Intern. Congr. Genet.* 1:154–55.

Lancefield, D. E. 1918. An autosomal bristle modifier affecting a sex-linked character. *Am. Naturalist* 52:462–64.

Lashley, K. S. 1915. Inheritance in the asexual reproduction of *Hydra.* *J. Exp. Zool.* 19:157–210.

Latter, B. D. H., and Novitski, C. E. 1969. Selection in finite populations with multiple alleles: I. Limits to directional selection. *Genetics* 62:859–76.

Latter, B. D. H., and Robertson, A. 1962. The effect of inbreeding and artificial selection on reproductive fitness. *Genet. Res. Camb.* 3:110–38.

Lawrence, W. J. C., and Scott-Moncrieff, R. 1935. The genetics and chemistry of flower colour in *Dahlia*: A new theory of specific pigmentation. *J. Genet.* 30:155–226.

Lederberg, J., and Lederberg, E. M. 1952. Replica plating and indirect selection of bacterial mutants. *J. Bacteriol.* 63:399–406.

Leng, E. R. 1962. Results of long-term selection for chemical composition in maize and their significance in evaluating breeding systems. *Z. Pflanzenz.* 47:67–91.

Lerner, I. M. 1954. *Genetic homeostasis.* New York: John Wiley & Sons, Inc.

———. 1958. *The genetic basis of selection.* New York: John Wiley & Sons, Inc.

Levine, H.; Pavlovsky, O.; and Dobzhansky, T. 1954. Interaction of adaptive values in polymorphic experimental populations of *Drosophila pseudoobscura.* *Evolution* 8:335–49.

———. 1958. Dependence of the adaptive values of certain genotypes of *Drosophila melanogaster* on the composition of the gene pool. *Evolution* 12:10–23.

Lewontin, R. C. 1955. The effect of population density on viability in *Drosophila melanogaster.* *Evolution* 9:27–41.

———. 1963. Interaction of genotypes determining viability in *Drosophila buskii.* *Proc. Nat. Acad. Sci. U.S.* 48:270–78.

L'Héritier, P. L.; Neefs, Y.; and Teissier, G., 1937. Aptérisme des insectes et sélection naturelle. *Compt. Rend. Acad. Sci.* 204:907–9.

L'Héritier, P. L., and Teissier, G. 1933. Etude d'une population de *Drosophiles* en equilibre. *Compt. Rend. Acad. Sci.* 197:1765.

———. 1934. Une expérience de sélection naturelle: Courbe d'élimination du gène "Bar" dans une population de *Drosophiles* en equilibre. *Compt. Rend. Soc. Biol.* 117:1049–51.

———. 1937. Elimination des formes mutantes dans les populations de *Drosophiles. Compt. Rend. Sci. Soc. Biol.* 124:880–84.

Li, C. C. 1955a. *Population genetics.* Chicago: University of Chicago Press.

———. 1955b. The stability of an equilibrium and the excess fitness of a population. *Am. Naturalist* 89:281–95.

Liljedahl. L. E. 1968. Studies of inbreeding and ageing in female mice. *Lantbrukshögskolans Ann.* 34:225–335.

Lindsey, M. F.; Lonnquist, J. H.; and Gardner, C. O. 1961. Estimates of genetic variance in open pollinated varieties of cornbelt corn. *Crop. Sci.* 2:105–8.

Lindsley, D. L., and Grell, E. H. 1968. *Genetic variations of* Drosophila melanogaster. Publication 627. Carnegie Institution of Washington.

Lindstrom, E. W. 1941. Analysis of modern maize breeding principles and methods. *Proc. VII Intern. Congr. Genet.* Pp. 151–56.

Little, C. C., and Johnson, B. W. 1922. The inheritance of susceptibility to implants of splenic tissue in mice. *Proc. Soc. Exp. Biol. Med.* 19:163–67.

Little, C. C., and Strong, L. C. 1924. Genetic studies on the transplantation of two adenocarcinomata. *J. Exp. Zool.* 41:93–114.

Little, C. C., and Tyzzer, E. E. 1916. Further experimental studies on the inheritance of susceptibility to a transplantable tumor carcinoma (J.w.A.) of the Japanese waltzing mouse. *J. Med. Res.* 33:393–453.

Livesay, E. A. 1930. An experimental study of hybrid vigor or heterosis in rats. *Genetics* 15:17–54.

Loeb, L. 1901. On transplantation of tumors. *J. Med. Res.* 1:28–38.

Loeb, L., and King, H. D. 1927. Transplantation and individuality differentials in strains of inbred rats. *Am. J. Pathol.* 3:143–67.

———. 1931. Individuality differentials in strains of inbred rats. *Arch. Pathol.* 12:203–21.

Loeb, L.; King, H. D.; and Blumenthal, H. T. 1943. Transplantation and individuality differentials in inbred strains of rats. *Biol. Bull.* 84:1–12.

Loeb, L., and Wright, S. 1927. Transplantation and individuality differentials in inbred strains of guinea pigs. *Am. J. Pathol.* 3:251–83.

Lotsy, J. D. 1908. *Vorlesungen über Descendenztheorien mit besonderer Berücksichtigung der Botanischen seite der Frage.* Jena, Germany: Gustav Fischer.

Ludwin, I. 1951. Natural selection in *Drosophila melanogaster* under laboratory conditions. *Evolution* 5:231–42.

Lundqvist, A. 1966. Heterosis and inbreeding depression in autotetraploids. *Hereditas* 56:317–66.

———. 1969. Some effects of continued inbreeding in an autotetroploid high bred strain of rye. *Hereditas* 61:361–69.

Luria, S. E., and Delbrück, M. 1943. Mutations of bacteria from virus sensitivity to virus resistance. *Genetics* 28:491–511.

Lush, J. L. 1946. Chance as a cause of changes in gene frequencies within pure breeds of livestock. *Am. Naturalist* 80:318–42.

———. 1947. Family merit and individual merit as bases for selection. *Am. Naturalist* 81:241–61, 362–79.

Lush, J. L., and Anderson, A. L. 1939. A genetic history of Poland China swine. *J. Heredity* 30:149–56, 219–24.

Lush, J. L.; Holbert, J. C.; and Willham, D. S. 1936. Genetic history of the Holstein-Friesian cattle in the United States. *J. Heredity* 27:61–72.

Lynch, C. J.; Pierce-Chase, C.; and Dubos, R. 1965. A genetic study of susceptibility to experimental tuberculosis in mice infected with mammalian tubercle bacilli. *J. Exp. Med.* 121:1051–70.

Lyon, M. F. 1961. Gene action in the X chromosome of the mouse (*Mus musculus* L). *Nature* 190:372–73.

MacArthur, J. W. 1944a. Genetics of body size and related characters: I. Selecting small and large races of the laboratory mouse. *Am. Naturalist* 78:142–57.

———. 1944b. Genetics of body size and related characters: II. Satellite characters associated with body size in mice. *Am. Naturalist* 78:224–37.

———. 1949. Selection for small and large body size in the house mouse. *Genetics* 34:194–209.

MacCurdy, H., and Castle, W. E. 1907. *Selection and crossbreeding in relation to the inheritance of coat pigments and coat patterns in rats and guinea pigs.* Publication 70. Carnegie Institution of Washington.

MacDowell, E. C. 1915. Bristle inheritance in *Drosophila*: I. Extra bristles. *J. Exp. Zool.* 19:61–98.

———. 1917. Bristle inheritance in *Drosophila*: II. Selection. *J. Exp. Zool.* 23:109–46.

———. 1920. Bristle inheritance in *Drosophila*: III. Correlation. *J. Exp. Zool.* 30:419–60.

Magalhães, L. E.; da Cunha, A. B.; de Toledo, J. S.; de Toledo, S. P.; Souza, H. L. de; Targa, H. J.; Setzer, V.; and Pavan, C. 1964. On lethals and their suppressors in experimental populations of *Drosophila willistoni*. *Mutation Res.* 2:45–54.

Magalhães, L. E.; de Toledo, J. S.; and da Cunha, A. B. 1965. The nature of lethals of *Drosophila willistoni*. *Genetics* 52:600–608.

Malin, D. F. 1923. *The evolution of breeds*. Des Moines, Iowa: Wallace Publ. Co.

Mather, K. 1936. Segregation and linkage in autotetraploids. *J. Genet.* 32:287–314.

———. 1941. Variation and selection of polygenic characters. *J. Genet.* 41:159–93.

———. 1949. *Biometric genetics: The study of continuous variations.* New York: Dover Publications, Inc.

———. 1955. Response to selection. *Cold Spring Harbor Symp. Quant. Biol.* 20:158–65.

Mather, K., and Harrison, B. J. 1949. The manifold effects of selection. *Heredity* 3:1–52, 131–62.

Mather, K., and Wigan, L. G. 1942. The selection of invisible mutations. *Proc. Roy. Soc. (London) Ser. B* 131:50–64.

Matzinger, D. F. 1963. Experimental estimates of genetic parameters and their applications in self-fertilizing plants. In *Statistical genetics and plant breeding*, ed. W. D. Hanson and H. F. Robinson. Publication no. 982. Washington, D.C.: National Academy of Science—National Research Council. Pp. 253–79.

Mayo, O. 1966. The evolution of dominance. *Heredity* 21:469–571.

Mayr, E. 1942. *Systematics and the origin of species.* New York: Columbia University Press.

———. 1959. Where are we? *Cold Spring Harbor Symp. Quant. Biol.* 24:1–14.

———. 1963. *Animal species and evolution.* Cambridge, Mass.: Harvard University Press.

McAlpin, S.; Mukai, T.; and Burdick, A. B. 1960. Heterozygote viability at a second chromosome recessive lethal in *Drosophila melanogaster. Genetics* 45:315–29.

McClintock, B. 1952. Chromosome organization and genic expression. *Cold Spring Harbor Symp. Quant. Biol.* 16:13–47.

———. 1956. Controlling elements and the gene. *Cold Spring Harbor Symp. Quant. Biol.* 21:197–216.

McLaren, A., and Michie, D. 1954. Are inbred strains suitable for bioassay? *Nature* 173:686–88.

———. 1956. Variability of response in experimental animals: A comparison of the reactions of inbred, F_1 hybrid and random bred mice to a narcotic drug. *J. Genetic.* 54:440–55.

McPhee, H. C., and Eaton, O. N. 1931. *Genetic growth differentiation in guinea pigs.* Technical bulletin 232. Washington, D.C.: U.S. Department of Agriculture.

McPhee, H. C.; Russel, E. Z.; and Zeller, J. 1931. An inbreeding experiment with Poland China swine. *J. Heredity* 22:383–403.

McPhee, H. C., and Wright, S. 1925. Mendelian analysis of the pure breeds of livestock: III. The Shorthorns. *J. Heredity* 16:205–15.

———. 1926. Mendelian analysis of the pure breeds of livestock: IV. The British Dairy Shorthorns. *J. Heredity* 17:396–401.

Mendel, G. 1866. Versuche über Pflanzen-Hybriden. *Verhandl. Naturforsch. Ver. Brünn* 4 (1965): 3–47.

Merrell, D. J. 1949. Selective mating in *Drosophila melanogaster. Genetics* 34:370–89.

———. 1953. Selective mating as a cause of gene frequency changes in laboratory populations of *Drosophila melanogaster. Evolution* 7:287–96.

———. 1960. Heterosis in DDT-resistant and susceptible populations of *Drosophila melanogaster. Genetics* 46:573–81.

———. 1965. Competition involving dominant mutants in experimental populations of *Drosophila melanogaster. Genetics* 52:165–89.

Merrell, D. J., and Underhill, J. C. 1956. Competition between mutants in experimental populations of *Drosophila melanogaster. Genetics* 41:469–85.

Metz, C. W. 1938. Chromosome behavior, inheritance and sex determination in *Sciara. Am. Naturalist* 72:485–520.

———. 1947. Duplication of chromosome parts as a factor in evolution. *Am. Naturalist* 81:81–103.

Middleton, A. R. 1915. Heritable variation and the results of selection in the fission rate of *Stylonychia pustulata. J. Exp. Zool.* 19:451–503.

Milkman, R. D. 1967. Heterosis as a major cause of heterozygosity in nature. *Genetics* 55:493–95.

Miller, C. P., and Bohnhoff, M. 1947. The development of streptomycin-resistant variants of Meningococcus. *Science* 105:620–21.

Millicent, E., and Thoday, J. M. 1961. Effects of disruptive selection: IV. Gene flow and divergence. *Heredity* 16:199–217.

Mills, D. R.; Peterson, R. L.; and Spiegelman, S. 1967. An extra-cellular Darwinian experiment with a self-duplicating nucleic acid molecule. *Proc. Nat. Acad. Sci. U.S.* 58:217–24.

Mirsky, A. E., and Ris. H. 1949. Variable and constant components of chromosomes. *Nature* 163:666–67.

———. 1951. The composition and structure of isolated chromosomes. *J. Gen. Physiol.* 34:475–97.

Moenkhaus, W. J. 1911. The effects of inbreeding and selection on fertility, vigor and sex ratio of *Drosophila ampelophila. J. Morphol.* 22:123–54.

Moll, R. H.; Lonnquist, J. H.; Fortuno, J. V.; and Johnson, E. C. 1965. The relationship of heterosis and genetic divergence. *Genetics* 52:139–44.

Montgomery, E. G. 1912. *Competition in cereals.* Bulletin 127. Lincoln: Nebraska Agricultural Experimental Station. Pp. 3–22.

Moran, P. A. P. 1962. *The statistical processes of evolutionary theory.* Oxford: Clarendon Press.

———. 1964. On the nonexistence of adaptive topographies. *Ann. Human Genet.* 27:383–93.

———. 1967. Haldane's dilemma on the rate of evolution. *Ann. Human Genet.* 33:245–49.

Morgan, T. H. 1919. *The physical basis of heredity.* Philadelphia: J. B. Lippincott Co.

———. 1932. *The scientific basis of evolution.* New York: W. W. Norton & Co, Inc.

Morgan, T. H.; Bridges, C. B.; and Sturtevant, A. H. 1925. *The genetics of* Drosophila. The Hague: Martinus Nijhoff.

Morley, F. H. W. 1954. Selection for economic characters in Australian Merino sheep. *Australian J. Agr. Res.* 5:305–16.

Morton, N. E. 1956. Empirical risks in consanguine marriage: Birth weight, gestation time and measurement of infants. *Am. J. Human Genet.* 10:344–49.

Morton, N. E.; Crow, J. F.; and Muller, H. J. 1956. An estimate of the mutational damage in man from data on consanguineous marriages. *Proc. Nat. Acad. Sci. U.S.* 42:855–63.

Mourad, A. F. 1965. Genetic divergence in M. Vetukhiv's experimental populations of *Drosophila pseudoobscura*. *Genet. Res. Camb.* 6:139–46.

Mukai, T. 1964. The genetic structure of natural populations of *Drosophila melanogaster*: I. Spontaneous mutation rate of polygenes controlling viability. *Genetics* 50:1–19.

———. 1968. Experimental studies of the mechanism involved in the maintenance of genetic variability in *Drosophila* populations. *Japan. J. Genet.* 43:399–413.

———. 1969a. Maintenance of polygenic and isoallelic variation in populations. *Proc. XII Intern. Congr. Genet.* 3:293–308.

———. 1969b. The genetic structure of natural populations of *Drosophila melanogaster*: VI. Further studies on the optimum heterozygosity hypothesis. *Genetics* 61:479–95.

Mukai, T., and Burdick, A. B. 1959. Single gene heterosis associated with a second chromosome recessive lethal in *Drosophila melanogaster*. *Genetics* 44:211–32.

Mukai, T.; Chigusa, S.; Mettler, L. E.; and Crow, J. F. 1972. Mutation rate and dominance of genes affecting viability in *Drosophila melanogaster*. *Genetics* 72:335–55.

Mukai, T.; Chigusa, S.; and Yoshikawa, I. 1964. The genetic structure of natural populations of *Drosophila melanogaster*: II. Overdominance of spontaneous mutant polygenes controlling viability in homozygous background. *Genetics* 50:711–15.

———. 1965. The genetic structure of natural populations of *Drosophila melanogaster*: III. Dominance effect of spontaneous mutant polygenes controlling viability in heterozygous genetic background. *Genetics* 52:493–501.

Mukai, T., and Yamazaki, T. 1967. The genetic structure of natural populations of *Drosophila melanogaster*: V. Coupling-repulsion effect of spontaneous mutant polygenes controlling viability. *Genetics* 59:513–35.

Mukai, T.; Yoshikawa, I.; and Sano, K. 1966. The genetic structure of natural populations of *Drosophila melanogaster*: IV. Heterozygous effects of radiation-induced mutations on viability in various genetic backgrounds. *Genetics* 53:513–39.

Muller, H. J. 1918. Genetic variability in twin hybrids and constant hybrids in a case of balanced lethal factors. *Genetics* 3:422–99.

———. 1927. Artificial transmutation of the gene. *Science* 66:84–87.

———. 1929. The method of evolution. *Sci. Monthly* 29:481–505.

———. 1932a. Further studies on the nature and causes of gene mutation. *Proc. VI Intern. Congr. Genet.* 1:213–55.

———. 1932b. Some genetic aspects of sex. *Am. Naturalist* 68:118–38.

———. 1950. Our load of mutations. *Am. J. Human Genet.* 2:111–76.

Muller, H. J.; League, B. B.; and Offermann, C. A. 1931. Effects of dosage change of sex-linked genes and the compensatory effect of other gene differences between male and female. *Anat. Record* 51 (suppl.):110.

Neal, N. P. 1935. The decrease in yielding capacity in advanced generations of hybrid corn. *J. Am. Soc. Agron.* 27:666–70.

Nei, M. 1963. Effect of selection on the components of genetic variance. In *Statistical genetics and plant breeding*, ed. W. D. Hanson and H. F. Robinson. Publication no. 982. Washington, D.C.: National Academy of Science–National Research Council. Pp. 501–11.

———. 1967. Modification of linkage intensity by natural selection. *Genetics* 57:625–41.

Norton, H. T. J. 1915. In *Mimicry in butterflies*, R. C. Punnett. Cambridge: Cambridge University Press.

Novick, A., and Szilard, L. 1950a. Experiments with the chemostat on spontaneous mutation of bacteria. *Proc. Nat. Acad. Sci. U.S.* 36:708–19.

———. 1950b. Description of the chemostat. *Science* 112:715–16.

———. 1951. Genetic mechanisms and bacterial viruses: I. Experiments on spontaneous and chemically induced mutations of bacteria growing in the chemostat. *Cold Spring Harbor Symp. Quant. Biol.* 16:337–47.

———. 1952. Anti-mutagens. *Nature* 170:926.

Nozawa, K. 1958. Competition between the brown gene and its wild type allele in *Drosophila melanogaster*: I. Effect of population density. *Japan. J. Genet.* 58:262–71.

O'Donald, P. 1967a. The evolution of selective advantage in a deleterious mutation. *Genetics* 56:399–404.

———. 1967b. On the evolution of dominance, over-dominance and balanced polymorphism. *Proc. Roy. Soc. (London) Ser. B* 268:216–28.

———. 1968. The evolution of dominance by selection for an optimum. *Genetics* 58:451–60.

———. 1969. Haldane's dilemma and the rate of natural selection. *Nature* 221:815–16.

Oshima, C. 1958. The resistance of strains of *Drosophila melanogaster* to DDT and dieldrin. *Am. Naturalist* 92:171–82.

Park, T.; Mertz, D. B.; and Petruswicz, K. 1961. Genetic strains of *Tribolium*: Their primary characteristics. *Physiol. Zool.* 34:62–80.

Park, Y. I.; Hansen, C. T.; Chung, C. S.; and Chapman, A. B. 1966. Influence of feeding regime on the effects of selection for postweaning gain in the rat. *Genetics* 54:1315–27.

Parsons, P. A., and Bodmer, W. F. 1961. The evolution of overdominance: Natural selection and heterozygote advantage. *Nature* 190:7–12.

Pavlovsky, O., and Dobzhansky, T. 1966. Genetics of natural populations: XXXVII. The coadapted system of chromosomal variants in a population of *Drosophila pseudoobscura*. *Genetics* 53:843–54.

Payne, F. 1918a. The effect of artificial selection on bristle number in *Drosophila ampelophila* and its interpretation. *Proc. Nat. Acad. Sci. U.S.* 4:55–58.

———. 1918b. *An experiment to test the nature of variation on which selection acts*. Indiana University Studies no. 36, vol. 5, pp. 3–45. Bloomington: Indiana University.

———. 1920. Selection for high and low bristle number in the mutant strain reduced. *Genetics* 3:501–42.

Pearson, K. 1904. On a generalized theory of alternative inheritance with special reference to Mendel's laws. *Phil. Trans. Roy. Soc. (London) Ser. A* 203:53–86.

———. 1909. On the ancestral genetic correlations of a Mendelian population mating at random. *Proc. Roy. Soc. (London) Ser. B* 81:225–29.

———. 1910. Darwinism, biometry and some recent biology. *Biometrika* 7:368–85.

———. 1913. On the probable error of a coefficient of correlation as found from a fourfold table. *Biometrika* 9:22–27.

Petit, C. 1951. La role de l'isolement sexuel dans l'evolution des populations de *Drosophila melanogaster*. *Bull. Biol. France Belg.* 85:352–418.

———. 1958. Le determinisme génétique et psychophysiologique de la compétition sexuelle chez *Drosophila melanogaster*. *Bull. Biol. France Belg.* 92:248–329.

Plunkett, C. R. 1932a. A contribution to the theory of dominance. *Am. Naturalist* 67:84–85.

———. 1932b. Temperature as a tool of research in phenogenetics: Methods and results. *Proc. VI Intern. Congr. Genet.* 2:158–60.

Pollak, E.; Robinson, H. F.; and Comstock, R. E. 1957. Interpopulation hybrids in open-pollinated varieties of maize. *Am. Naturalist* 91:387–91.

Powers, L. 1942. The nature of the series of environmental variances and the estimation of the genetic variances and the geometric means in crosses involving species of *Lycopersicon*. *Genetics* 27:561–75.

Prout, T. 1962. The effects of stabilizing selection on the time of development in *Drosophila melanogaster*. *Genet. Res. Camb.* 3:364–82.

Punnett, R. C. 1915. *Mimicry in butterflies*. Cambridge: Cambridge University Press.

Quayle, H. J. 1938. The development of resistance to hydrocyanic gas in certain scale insects. *Hilgardia* 11:183–209.

Rahnefeld, G. W.; Boylan, W. J.; Comstock, R. E.; and Singh, M. 1963. Mass selection for post-weaning growth in mice. *Genetics* 48:1567–83.

Randolph, L. F. 1928. *Chromosome numbers in* Zea mays *L.* Memoir 117. Ithaca, N.Y.: Cornell University Agricultural Experimental Station.

———. 1941. An evaluation of induced polyploidy as a method of breeding crop plants. *Am. Naturalist* 75:347–63.

———. 1942. The influence of heterozygosity on fertility and vigor in autotetraploid maize. *Genetics* 27:163.

Rasmussen, J. E. 1958. Persistance of mutants in laboratory populations of *Drosophila melanogaster*. *Proc. X Intern. Congr. Genet.* 2:227–28.

Rasmussen, J. M. 1933. A contribution to the theory of quantitative character inheritance. *Hereditas* 18:245–61.

Rasmussen, M. 1955. Selection for bristle numbers in some unrelated strains of *Drosophila melanogaster*. *Acta Zool.* 36:1–49.

Reed, S. C., and Reed, E. W. 1948. Natural selection in laboratory populations of *Drosophila*. *Evolution* 2:176–86.

———. 1950. Natural selection in laboratory populations of *Drosophila:* II.

Competition between a white eye gene and its wild type allele. *Evolution* 4:34–42.

Reeve, E. C. R., and Robertson, F. W. 1953. Studies in quantitative inheritance: II. Analysis of a strain of *Drosophila melanogaster* selected for long wings. *J. Genet.* 51:276–316.

———. 1954. Studies in quantitative inheritance: VI. Sternite chaeta number in *Drosophila:* A metameric quantitative character. *Z. ind. Abst. Vererb.* 86:269–88.

Rendel, J. M. 1952. White heifer disease in a herd of Dairy Shorthorns. *J. Genet.* 51:89–94.

———. 1959. Variation and dominance at the scute locus in *Drosophila melanogaster. Australian J. Biol. Soc.* 12:524–33.

Rendel, J. M.; Sheldon, B. L.; and Finlay, D. E. 1965. Canalization of development of scutellar bristles of *Drosophila* by control of the scute locus. *Genetics* 52:1137–51.

Richardson, R. H., and Kojima, K. 1965. The kinds of genetic variability in relation to selection responses in *Drosophila* fecundity. *Genetics* 52:583–98.

Richey, F. D., and Sprague, G. F. 1931. *Experiments on hybrid vigor and convergent improvement in corn.* Technical bulletin 267. Washington, D.C.: U.S. Department of Agriculture. Pp. 1–22..

Ris, H., and Kubai, D. F. 1970. Chromosome structure. *Ann. Rev. Genet.* 4:263–94.

Ritzema-Bos, J. 1894. Untersuchungen über die Folgen der Zucht in engster Blutverwandtschaft. *Biol. Cent.* 14:75–81.

Roberts, R. C. 1960. The effects on litter size of crossing lines of mice inbred without selection. *Genet. Res. Camb.* 1:239–52.

Robertson, A. 1960. A theory of limits in artificial selection. *Proc. Roy. Soc. (London) Ser. B* 153:234–49.

———. 1964. The effect of nonrandom mating within inbred lines on the rate of inbreeding. *Genet. Res. Camb.* 5:164–67.

Robertson, A., and Mason, I. L. 1954. A genetic analysis of the Red Danish breed of cattle. *Acta Agr. Scand.* 4:257–65.

Robertson, F. W. 1954. Studies in quantitative genetics: V. Chromosome analyses of crosses between selected and unselected lines of different body size in *Drosophila melanogaster. J. Genet.* 52:494–520.

———. 1955. Selection reponse and the properties of genetic variation. *Cold Spring Harbor Symp. Quant. Biol.* 20:166–77.

———. 1960a. The ecological genetics of growth of *Drosophila:* I. Body size and developmental time on different diets. *Gen. Res. Camb.* 1:288–304.

———. 1960b. The ecological genetics of growth of *Drosophila:* II. Selection for large body size on different diets. *Gen. Res. Camb.* 1:305–18.

Robertson, F. W., and Reeve, E. C. R. 1952a. Studies in quantitative inheritance: I. The effects of selection of wing and thorax length in *Drosophila melanogaster. J. Genet.* 50:414–48.

———. 1952b. Heterozygosity, environmental variation and heterosis. *Nature* 170:296–98.

———. 1953. Studies in quantitative inheritance: IV. The effects of substituting chromosomes from selected strains in different genetic backgrounds in *Drosophila melanogaster. J. Genet.* 51:586–610.

———. 1955a. Studies in quantitative inheritance: VII. Crosses between strains of different body size in *Drosophila melanogaster. Z. ind. Abst. Vererb.* 86:424–38.

———. 1955b. Studies in quantitative inheritance: VIII. Further analysis of heterosis in crosses between inbred lines of *Drosophila melanogaster. Z. ind. Abst. Vererb.* 86:439–58.

Robinson, H. F.; Cockerham, C. C.; and Moll, R. H. 1960. Studies on estimation of dominance variance and effects of linkage bias. In *Biometrical genetics.* New York: Pergamon Press, Inc. Pp. 171–77.

Robinson, H. F.; Comstock, R. E.; and Harvey, P. H. 1949. Estimation of heritability and the degree of dominance in corn. *Agron. J.* 41:353–59.

———. 1955. Genetic variance of open-pollinated varieties of corn. *Genetics* 40:45–60.

Robinson, H. F.; Mann, T. J.; and Comstock, R. E. 1954. An analysis of quantitative variability in *Nicotiana tabacum. Heredity* 8:365–76.

Roderick, T. H. 1963. Selection for radiation resistance in mice. *Genetics* 48:205–16.

Root, F. M. 1918. Inheritance in the asexual reproduction of *Centropyxis aculenta. Genetics* 3:173–206.

Rottensten, K. 1937. Inbreeding in Danish Landrace swine (translated title). *Nord. Jordbrugsforskning* 3–4:94–114.

Russell, L. B. 1961. Genetics of mammalian sex chromosomes. *Science* 133:1795–1803.

———. 1964. Another look at the single-active X hypothesis. *Trans. N.Y. Acad. Sci. Ser. II* 26:726–36.

Russell, W. L. 1941. Inbred and hybrid animals and their value in research. In *Biology of the laboratory mouse,* ed. G. D. Snell. Philadelphia: Blakiston Co. Pp. 325–48.

———. 1963. The effect of radiation dose rate and fractionation on mutation in mice. In *Repair from genetic radiation,* ed. F. H. Sobels. New York: Pergamon Press, Inc.

Sakai, K. 1955. Competition in plants and its relation to selection. *Cold Spring Harbor Symp. Quant. Biol.* 20:137–57.

———. 1961. Competitive ability in plants: Its inheritance and some related problems. *Symp. Soc. Exp. Biol.* 15:247–63.

Sandler, L.; Hiraizumi, Y.; and Sandler, I. 1959. Meiotic drive in natural populations of *Drosophila melanogaster*: The cytologic basis of segregation-distortion. *Genetics* 44:233–50.

Sandler, L., and Novitski, E. 1957. Meiotic drive as an evolutionary force. *Am. Naturalist* 91:105–10.

Sankaranarayanan, K. 1964. Genetic loads in irradiated experimental populations of *Drosophila melanogaster*. *Genetics* 50:131–50.

Scharloo, W. 1964. The effects of disruptive and stabilizing selection on the expression of a cubitus interruptus mutant in *Drosophila*. *Genetics* 50:553–62.

Scharloo, W.; Boer, M.; and Hoogmoed, M. S. 1967. Disruptive selection on sternopleural chaeta number in *Drosophila melanogaster*. *Genet. Res. Camb.* 9:115–18.

Scharloo, W.; Hoogmoed, M. S.; and Kuile, A. ter 1967. Stabilizing and disruptive selection on a mutant character in *Drosophila:* I. The phenotypic variance and its components. *Genetics* 56:709–36.

Schindewolf, O. H. 1950. *Grundfragen der Paläontologie*. Stuttgart: Schwarzerbart.

Schoffner, R. N. 1948. The reaction of the fowl to inbreeding. *Poultry Sci.* 27:448–52.

Schull, W. J., and Neel, J. V. 1965. *The effects of inbreeding on Japanese children*. New York: Harper & Row Publishers, Inc.

Sciuchetti, A. 1935. Beitrag zur genetischen Analyse der Schweizischen Braunrasse. *Arch. Julius Klaus-Stiftung Vererbungsforsch. Sozialanthropol. Rassenhyg.* 10:85–99.

Scossirolli, R. E. 1954. Effectiveness of artificial selection under irradiation of plateaued populations of *Drosophila melanogaster*. *I.U.B.S. Symp. Genet. Pop. Structure B* 15:42–66.

Scott, J. P., and Fuller, J. L. 1965. *Genetics and the social behavior of the dog*. Chicago: University of Chicago Press.

Sentz, J. C.; Robinson, H. F.; and Comstock, R. F. 1954. Relation between heterozygosis and performance in maize. *Agron. J.* 46:514–20.

Sheldon, B. L.; Rendel, J. M.; and Finlay, D. E. 1964. The effect of homozygosity on developmental stability. *Genetics* 49:471–84.

Sheppard, P. M., and Ford, E. B. 1966. Natural selection and the evolution of dominance. *Heredity* 21:139–47.

Shull, A. F. 1912. The influence of inbreeding on vigor in *Hydatina senta*. *Biol. Bull.* 24:1–13.

Shull, G. H. 1908. The composition of a field of maize. *Am. Breeding Assoc. Rept.* 4:296–301.

———. 1910. Hybridization methods in corn breeding. *Am. Breeders Mag.* 1:98–107.

———. 1911. The genotypes of maize. *Am. Naturalist* 45:234–53.

———. 1914. Duplicate genes for capsule form in *Bursa bursa-pastoris*. *Z. ind. Abst. Vererb.* 12:97–149.

———. 1952. Beginnings of the heterosis concept. In *Heterosis*, ed. J. W. Gowen. Ames, Iowa: Iowa State College Press. Pp. 14–48.

Simmons, A. S. 1966. Experiments on random genetic drift and natural selection in *Drosophila pseudoobscura*. *Evolution* 20:100–103.

Simpson, G. G. 1944. *Tempo and mode in evolution*. New York: Columbia University Press.

———. 1953. *The major features of evolution.* New York: Columbia University Press.

———. 1961. *Principles of animal taxonomy.* New York: Columbia University Press.

Sismanidis, A. 1942. Selection for an almost invariable character in *Drosophila*. *J. Genet.* 44:204–15.

Sittman, K.; Abplanalp, H.; and Fraser, R. A. 1966. Inbreeding depression in Japanese quail. *Genetics* 54:371–79.

Smathers, K. M. 1961. The contribution of heterozygosity of certain gene loci to fitness of laboratory populations of *Drosophila melanogaster*. *Am. Naturalist* 95:27–38.

Smith, A. D. B. 1928. Inbreeding in Jersey cattle. *Brit. Assoc. Advan. Sci. Rept.* 22:649–55.

Smith, J. Maynard. 1968a. Haldane's dilemma and the rate of evolution. *Nature* 219:1114–19.

———. 1968b. Evolution in sexual and asexual populations. *Am. Naturalist* 102:469–73.

Snell, G. D., ed. 1941. *Biology of the laboratory mouse.* Philadelphia: Blakiston Co.

Snell, G. D., and Higgins, G. F. 1951. Alleles at the histocompatibility locus in the mouse as determined by tumor transplantation. *Genetics* 36:306–10.

Sokal, R. R., and Huber, I. 1963. Competition among genotypes in *Tribolium castaneum* at varying densities and gene frequencies (the sooty locus). *Am. Naturalist* 97:169–84.

Sokal, R. R., and Hunter, P. E. 1954. Reciprocal selection for correlated quantitative characters in *Drosophila*. *Science* 119:649–51.

Sokal, R. R., and Karten, I. 1964. Competition among genotypes in *Tribolium castaneum* at varying densities and gene frequencies (the black locus). *Genetics* 49:195–211.

Sokal, R. R., and Sullivan, R. L. 1963. Competition between mutant and wild type housefly strains of varying densities. *Ecology* 43:314–22.

Sonneborn, T. W. 1947. Recent advances in the genetics of *Paramecium* and *Euplotes*. *Advan. Genet.* 1:264–358.

———. 1957. *Breeding systems, reproductive methods and species problems in Protozoa: The species problem.* Washington, D.C.: American Association for the Advancement of Science. Pp. 153–324.

Souza, H. L. de; da Cunha, A. B.; and dos Santos, E. P. 1968. Adaptive polymorphism of behavior developed in laboratory populations of *Drosophila willistoni*. *Am. Naturalist* 102:583–86.

Spiess, E. B. 1957. Relation between frequencies and adaptive values of chromosomal arrangements in *Drosophila persimilis*. *Evolution* 11:84–93.

———. 1968. Low frequency advantage in mating of *Drosophila pseudoobscura* karyotypes. *Am. Naturalist* 102:363–79.

Spofford, J. B. 1956. The relation between expressivity and selection against eyeless in *Drosophila melanogaster*. *Genetics* 41:938–59.

Sprague, G. F. 1963. Orientation and objectives. In *Statistical genetics and plant breeding*, ed. W. D. Hanson and H. F. Robinson. Publication no. 982. Washington, D.C.: National Academy of Science–National Research Council. Pp. IX–XV.

Sprague, G. F., and Brimhall, B. 1949. Quantitative inheritance of oil in the corn kernel. *Agron. J.* 41:30–33.

Sprague, G. F.; Russell, W. A.; and Penny, L. H. 1960. Mutations affecting quantitative traits in the selfed progeny of doubled monoploid maize stocks. *Genetics* 45:855–66.

Stalker, H. D. 1954. Parthenogenesis in *Drosophila. Genetics* 39:434.

———. 1956. On the evolution of parthenogenesis in *Lonchoptera* (Diptera). *Evolution* 10:345–59.

Stebbins, G. L. 1950. *Variation and evolution in plants.* New York: Columbia University Press.

Steele, D. 1944. A genetic analysis of recent Thoroughbreds, Standardbreds and American Saddle horses. *Kentucky Agr. Exp. Sta. Bull.* 462.

Stern, C. 1929. Ueber die additive Wirkung multiple Allelen. *Biol. Zentr.* 49:261–90.

Stern, C.; Carson, B.; Kinst, M.; Novitski, E.; and Uphoff, D. 1952. The viability of heterozygotes for lethals. *Genetics* 37:413–49.

Stern, C., and Sherwood, E. R. 1966. *The origin of genetics.* San Francisco: W. H. Freeman & Co.

Stonaker, H. H. 1943. The breeding situation of the Aberdeen-Angus breed. *J. Heredity* 24:323–28.

Stout, A. B. 1915. *The establishment of varieties in* Coleus *by the selection of somatic variations.* Publication 218. Carnegie Institution of Washington.

Strandskov, H. H. 1939. Inheritance of internal organ differences in guinea pigs. *Genetics* 24:722–27.

———. 1942. Skeletal variations in guinea pigs and their inheritance. *J. Mammalogy* 23:65–75.

Streams, F. A., and Pimentel, D. 1961. Effects of immigration on the evolution of populations. *Am. Naturalist* 95:201–10.

Stringfield, G. H. 1950. Heterozygosis and hybrid vigor in maize. *Agron. J.* 42:45–112.

Strong, L. C. 1929. Transplantation studies on tumors arising spontaneously in heterozygous individuals. *J. Cancer Res.* 13:103–15.

Sturtevant, A. H. 1918. *An analysis of the effects of selection.* Publication 264. Carnegie Institution of Washington.

Sturtevant, A. H., and Dobzhansky, T. 1936. Geographical distribution and cytology of "sex ratio" in *Drosophila pseudoobscura* and related species. *Genetics* 21:473–90.

Sullivan, R. L., and Sokal, R. R. 1963. The effects of larval density on several strains of the house fly. *Ecology* 44:120–30.

Suneson, C. A., and Wiebe, G. A. 1942. Survival of barley and wheat varieties in mixtures. *J. Am. Soc. Agron.* 34:1052–56.

Susman, M., and Carson, H. L. 1958. Development of balanced polymorphism in laboratory populations of *Drosophila melanogaster*. *Am. Naturalist* 93:359–64.

Sved, J. A. 1968. Possible rates of gene substitution in evolution. *Am. Naturalist* 102:283–93.

Sved, J. A., and Mayo, O. 1970. The evolution of dominance. In *Mathematical topics in population genetics*, ed. K. Kojima, Berlin: Springer-Verlag. Pp. 289–316.

Sved, J. A.; Reed, T. E.; and Bodmer, W. F. 1967. The number of balanced polymorphisms that can be maintained in a natural population. *Genetics* 55:469–81.

Swift, H. H. 1950. The constancy of desoxyribose nucleic acid in plant nuclei. *Proc. Nat. Acad. Sci. U.S.* 36:643–54.

Tantawy, A. D., and Reeve, E. C. R. 1956. Studies in quantitative inheritance: IX. The effects of inbreeding at different rates in *Drosophila melanogaster*. *Z. ind. Abst. Vererb.* 87:648–67.

Teissier, G. 1942. Persistence d'un gène léthal dans une population de *Drosophiles*. *Compt. Rend. Acad. Sci.* 214:327–30.

———. 1947. Variation de la frequence de gène ebony dans une population Stationaire de *Drosophiles*. *Compt. Rend. Acad. Sci.* 224:676–77, 1788–89.

Temin, R. G. 1966. Homozygous viability and fertility loads in *Drosophila melanogaster*. *Genetics* 53:27–46.

Temin, R. G.; Meyer, H. D.; Dawson, P. S.; and Crow, J. F. 1969. The influence of epistasis on homozygous viability depression in *Drosophila melanogaster*. *Genetics* 61:497–515.

Thoday, J. M. 1958a. Homeostasis in selection experiments. *Heredity* 12:401–15.

———. 1958b. Effects of disruptive selection. *Proc. X Intern. Congr. Genet.* 2:294.

———. 1958c. Effects of disruptive selection: The experimental production of a polymorphic population. *Nature* 181:1124–25.

———. 1959. Effects of disruptive selection: I. Genetic flexibility. *Heredity* 13:187–203.

———. 1960. Effects of disruptive selection: III. Coupling and repulsion. *Heredity* 14:35–49.

———. 1965. Effects of selection for genetic diversity. *Proc. XI Intern. Congr. Genet.* 3:533–540.

Thoday, J. M., and Boam, T. 1958. Effects of disruptive selection: II. Polymorphism and divergence without isolation. *Heredity* 14:215–18.

———. 1961. Regular responses to selection: I. Description of responses. *Genet. Res. Camb.* 2:161–76.

Thoday, J. M., and Gibson, J. B. 1962. Isolation by disruptive selection. *Nature* 193:1164–66.

———. 1970. The probability of isolation by disruptive selection. *Am Naturalist* 104:219–30.

Thompson, W. R. 1954. The inheritance and development of intelligence. *Assoc. Nerv. Ment. Dis. Res.* 33:209–331.

Thomson, J. A. 1961. Interallelic selection in experimental populations of *Drosophila melanogaster*: White and satsuma. *Genetics* 40:1435–42.

Timofeeff-Ressovsky, N. W. 1927. Studies on the phenotypic manifestation of hereditary factors: I. On the phenotypic manifestation of the genovariation, radius incompletus, in *Drosophila funebris*. *Genetics* 12:128–98.

———. 1934. Ueber den Einfluss des genotypischen Milieus und der Ausenbedingungen auf die Realisation des Genotyps. *Nachr. Ges. Wiss. Göttingen Biol. NF* 1:53–106.

———. 1935. Auslösung von vitalitätsmutationen durch Röntgen bestrahlung bei *Drosophila melanogaster. Nachr. Ges. Wiss. Göttingen Biol. NF* 1:163–80.

Turner, J. R. G. 1967. Why does the genotype not congeal? *Evolution* 21:645–56.

———. 1970. Changes in mean fitness under natural selection. In *Mathematical topics in population genetics*, ed. K. Kojima. Berlin: Springer-Verlag. Pp. 32–78.

———. 1971. Wright's adaptive surface and some general rules for equilibria in complex polymorphisms. *Am Naturalist* 105:267–78.

Tyler, W. J.; Chapman, A. B.; and Dickerson, G. E. 1949. Growth and production of inbred and outbred Holstein-Friesian cattle. *J. Dairy Sci.* 32:247–56.

Vetukhiv, M. 1954. Integration of the genotype in local populations of three species of *Drosophila. Evolution* 8:241–51.

———. 1956. Fecundity of hybrids between geographical populations of *Drosophila pseudoobscura. Evolution* 10:139–46.

Vries, H. de. 1901–3. *Die Mutationstheorie.* 2 vols. Leipzig: Veit u. Co.

———. 1905. *Species and varieties: Their origin by mutation.* Chicago: Open Court Publishing Co.

Waddington, C. H. 1942. The canalization of development and the inheritance of acquired characters. *Nature* 150:563.

———. 1955. On a case of quantitative variation on either side of the wild type. *Z. ind. Abst. Vererb.* 87:208–28.

Wallace, B. 1948. Studies on "sex-ratio" in *Drosophila pseudoobscura:* Selection and "sex-ratio." *Evolution* 2:189–217.

———. 1950. Autosomal lethals in experimental populations of *Drosophila melanogaster. Evolution* 4:172–74.

———. 1956. Studies on irradiated populations of *Drosophila melanogaster. J. Genet.* 54:280–95.

———. 1958. The average effect of radiation-induced mutations on viability in *Drosophila melanogaster. Evolution* 12:532–52.

———. 1959. The role of heterozygosity in *Drosophila* populations. *Proc. X Intern. Congr. Genet.* 1:408–19.

———. 1963. Further data on the overdominance of induced mutations. *Genetics* 96:633–51.

————. 1965. The viability effects of spontaneous mutations in *Drosophila melanogaster*. *Am. Naturalist* 99:335–48.

Wallace, B., and Madden, C. 1965. Studies on inbred strains of *Drosophila melanogaster*. *Am. Naturalist* 99:495–509.

Wallace, Robert. 1923. *Farm livestock of Great Britain*. 5th ed. Edinburgh: Oliver and Boyd.

Watanabe, T.; Anderson, W. W.; Dobzhansky, T.; and Pavlovsky, O. 1970. Selection in experimental populations of *Drosophila pseudoobscura* with different initial frequencies of chromosomal variants. *Genet. Res. Camb.* 15:123–29.

Weinberg, W. 1909. Ueber Vererbungsgestze beim Menschen. *Zeit. ind. Abst. Vererb.* 1:277–330, 377–92, 440–60; 2:276–330.

————. 1910. Weitere Beiträge zur Theorie der Vererbung. *Arch. Rass. Ges. Biol.* 7:35–49, 169–73.

Weir, J. A. 1953. Association of blood pH with sex ratio in mice. *J. Heredity* 44:133–38.

————. 1960. A sex ratio factor in the house mouse that is transmitted by the male. *Genetics* 45:1539–52.

Weismann, A. 1904. *The evolution theory*, trans. J. Arthur Thomson and M. R. Thomson. London: Edward Arnold.

Wentworth, E. N. 1913a. The segregation of fecundity factors in *Drosophila*. *J. Genet.* 3:113–20.

————. 1913b. Color in Shorthorn cattle. *Am. Breeders Mag.* 4:202–8.

White, M. J. D. 1954. *Animal cytology and evolution*. Cambridge: Cambridge University Press. 6th ed. 1973.

Whitney, D. D. 1912. Reinvigoration produced by cross-fertilization in *Hydatina senta*. *J. Exp. Zool.* 12:337–62.

Wiedemann, G. 1936. Modellversuche zum Selektionswirking von Faktormutationen bei *Drosophila melanogaster*. *Genetica* 18:277–90.

Wigan, L. G. 1941. Polygenic variation in wild *Drosophila melanogaster*. *Nature* 148:373.

Willham, D. S. 1937. A genetic history of Hereford cattle in the United States. *J. Heredity* 28:283–94.

Williams, W. 1959. The isolation of pure lines from F_1 hybrids of the tomato: The problem of heterosis in inbreeding crop species. *J. Agron Sci.* 53:347–53.

————. 1960. Relative variability of inbred lines and F_1 hybrids in *Lycopersicum esculentum*. *Genetics* 45:1457–65.

Williams, W., and Gilbert, N. 1960. Heterosis and the inheritance of yield in the tomato. *Heredity* 14:133–49.

Willis, J. C. 1922. *Age and area*. Cambridge: Cambridge University Press.

————. 1940. *The course of evolution*. Cambridge: Cambridge University Press.

Wilson, J. 1908. Mendelian characters among Shorthorn cattle. *Sci. Proc. Roy. Dublin Soc.* 11:317–24.

Wilson, S. P.; Blair, P. V.; Kyle, W. H.; and Bell, A. E. 1968. The influence of selection and mating systems on larval weight in *Tribolium. J. Heredity* 59:313–17.

Wilson, S. P.; Goodale, H. D.; Kyle, W. H.; and Godfrey, E. F. 1971. Long-term selection for body weight in mice. *J. Heredity* 62:228–34.

Wilson, S. P.; Kyle, W. H.; and Bell, A. E. 1965. The effects of mating system on pupa weight in *Tribolium. Genet. Res. Camb.* 6:341–51.

Wriedt, C. 1925. Formalism in breeding livestock in relation to genetics. *J. Heredity* 16:19–24.

————. 1930. *Heredity in livestock*. London: Macmillan & Co.

Wright, S. 1915. The albino series of allelomorphis in guinea pigs. *Am. Naturalist* 49:140–48.

————. 1916. *An intensive study of the inheritance of color and of other coat characters in guinea pigs with especial reference to graded variation*. Publication 241. Carnegie Institution of Washington. Pp. 59–160.

————. 1917a. Color inheritance in mammals: III. The rat. *J. Heredity* 8:426–30.

————. 1917b. Color inheritance in mammals: VI. Cattle. *J. Heredity* 8:521–27.

————. 1920. The relative importance of heredity and environment in determining the piebald pattern of guinea pigs. *Proc. Nat. Acad. Sci. U.S.* 6:320–32.

————. 1922a. Coefficients of inbreeding and relationship. *Am. Naturalist* 56:330–38.

————. 1922b. *The effects of inbreeding and crossbreeding on guinea pigs: I. Decline in vigor*. Bulletin 1090. Washington, D.C.: U.S. Department of Agriculture. Pp. 1–36.

————. 1922c. *The effects of inbreeding and crossbreeding on guinea pigs: II. Differentiation among inbred families*. Bulletin 1090. Washington, D.C.: U.S. Department of Agriculture. Pp. 37–63.

————. 1922d. *The effects of inbreeding and crossbreeding on guinea pigs: III. Crosses between highly inbred families*. Bulletin 1121. Washington, D.C.: U.S. Department of Agriculture.

————. 1923a. Mendelian analysis of the pure breeds of livestock: I. The measurement of inbreeding and relationship. *J. Heredity* 14:339–48.

————. 1923b. Mendelian analysis of the pure breeds of livestock: II. The Duchess family of Shorthorns as bred by Thomas Bates. *J. Heredity* 14:405–22.

————. 1923c. The relation between piebald and tortoiseshell color patterns in guinea pigs. *Anat. Record* 23:393.

————. 1925. The factors of the albino series of guinea pigs and their effects on black and yellow pigmentation. *Genetics* 10:223–60.

————. 1926. Effects of age of parents on characteristics of the guinea pig. *Am. Naturalist* 60:552–59.

————. 1927. The effects in combination of the major color factors of the guinea pig. *Genetics* 12:530–69.

————. 1928. An eight-factor cross in the guinea pig. *Genetics* 13:508–31.

————. 1929a. Fisher's theory of dominance. *Am. Naturalist* 63:274–79.

————. 1929b. The evolution of dominance. *Am. Naturalist* 63:556–61.

———. 1929c. Evolution in a Mendelian population. *Anat. Record* 44:287.

———. 1931. Evolution in Mendelian populations. *Genetics* 16:97–159.

———. 1932a. On the evaluation of dairy sires. *Proc. Am. Soc. Animal Prod.* 1931:71–78.

———. 1932b. The roles of mutation, inbreeding, crossbreeding and selection in evolution. *Proc. VI Intern. Congr. Genet.* 1:356–66.

———. 1934a. Physiological and evolutionary theories of dominance. *Am. Naturalist* 68:25–53.

———. 1934b. On the genetics of subnormal development of the head (otocephaly) in the guinea pig. *Genetics* 19:471–505.

———. 1934c. An analysis of variability in number of digits in an inbred strain of guinea pigs. *Genetics* 19:506–36.

———. 1934d. The results of crosses between inbred strains of guinea pigs differing in number of digits. *Genetics* 19:537–51.

———. 1934e. Professor Fisher on the theory of dominance. *Am. Naturalist* 68:562–65.

———. 1935a. A mutation of the guinea pig, tending to restore the pentadactyl foot when heterozygous, producing a monstrosity when homozygous. *Genetics* 20:84–107.

———. 1935b. Evolution in populations in approximate equilibrium. *J. Genet.* 30:257–66.

———. 1938. Size of population and breeding structure in relation to evolution. *Science* 87:430–31.

———. 1939. Statistical genetics in relation to evolution. In *Actualités scientifiques et industrielles*, no. 802. Paris: Hermann et Cie. Pp. 5–64.

———. 1940. Breeding structure of populations in relation to speciation. *Am. Naturalist* 74:232–48.

———. 1941a. The "age and area" concept extended. (A review of a book by J. C. Willis.) *Ecology* 22:345–47.

———. 1941b. *The material basis of evolution* (by R. Goldschmidt): A review. *Sci. Monthly* 53:165–70.

———. 1941c. On the probability of fixation of reciprocal translocations. *Am. Naturalist* 75:513–22.

———. 1942. Statistical genetics and evolution. *Bull. Am. Math. Soc.* 48:223–46.

———. 1945. Tempo and mode in evolution: A critical review. *Ecology* 26:415–19.

———. 1948. On the roles of directed and random changes in gene frequency in the genetics of populations. *Evolution* 2:279–94.

———. 1949a. Adaptation and selection. In *Genetics, paleontology, and evolution*, ed. G. L. Jepson, G. G. Simpson, and E. Mayr. Princeton, N.J.: Princeton University Press. Pp. 365–89.

———. 1949b. Population structure in evolution. *Proc. Am. Phil. Soc.* 93:471–78.

———. 1949c. On the genetics of hair direction in the guinea pig: I. Variability in the patterns found in combinations of *R* and *M* loci. *J. Exp. Zool.* 112:303–24.

———. 1949*d*. Differentiation of strains of guinea pigs under inbreeding. *Proc. First Nat. Cancer Conf.* Pp. 13–27.

———. 1950*a*. On the genetics of hair direction in the guinea pig: III. Interaction between the processes due to the loci *R* and *St*. *J. Exp. Zool.* 113:33–64.

———. 1950*b*. Discussion on population genetics and radiation. *J. Cellular Comp. Physiol.* 35:187–210.

———. 1951*a*. The genetical structure of populations. *Ann. Eugenics* 15:323–54.

———. 1951*b*. Fisher and Ford on the Sewall Wright effect. *Am. Scientist* 39:452–58.

———. 1954. Summary of patterns of mammalian gene action. *J. Nat. Cancer. Inst.* 15:837–51.

———. 1955. *Discussion on responses of populations to radiation: Conference on genetics, Argonne National Laboratory, Nov. 19–20, 1954.* Washington, D.C.: U.S. Atomic Energy Commission, Division of Biology and Medicine. Pp. 59–62.

———. 1956. Classification of the factors of evolution. *Cold Spring Harbor Symp. Quant. Biol.* (1955) 20:16–24D.

———. 1959*a*. Silvering (*si*) and diminution (*dm*) of coat color of the guinea pig, and male sterility of the white or near white combination of these. *Genetics* 44:563–90.

———. 1959*b*. Genetics, the gene, and the hierarchy of biological sciences. *Proc. X Intern. Congr. Genet.* 1:475–89. (Also, *Science* 30:959–65.)

———. 1960*a*. On the appraisal of genetic effects of radiation in man. In *The biological effects of atomic radiation.* Summary reports (1960). Washington, D.C.: National Academy of Science-National Research Council. Pp. 18–24.

———. 1960*b*. The genetics of vital characters of the guinea pig. *J. Cellular Comp. Physiol.* 56 (suppl. 1):123–51.

———. 1960*c*. Genetics and twentieth-century Darwinism: A review and discussion. *Am. J. Human Genet.* 12:365–72.

———. Genic interaction. In *Methodology in mammalian genetics*, ed. W. J. Burdette. San Francisco: Holden-Day, Inc. Pp. 151–92.

———. 1963*b*. Plant and animal improvement in the presence of multiple selective peaks. In *Statistical genetics and plant breeding*, ed. W. D. Hanson and H. F. Robinson. Publication no. 982. Washington, D.C.: National Academy of Science-National Research Council. Pp. 116–22.

———. 1964*a*. Pleiotropy in the evolution of structural reduction and dominance. *Am. Naturalist* 98:65–69.

———. 1964*b*. Stochastic processes in evolution. In *Stochastic models in medicine and biology*, ed. J. Gurland. Madison: University of Wisconsin Press. Pp. 199–244.

———. 1965*a*. Factor interaction and linkage in evolution. *Proc. Roy. Soc. (London) Ser B* 162:80–104.

———. 1965*b*. The interpretation of population structure by F-statistics with special regard to systems of mating. *Evolution* 19:355–420.

————. 1966. Mendel's ratios. In *The origin of genetics*, ed. C. Stern and E. R. Sherwood. San Francisco: W. H. Freeman & Co.

————. 1967*a*. The foundations of population genetics. In *Heritage from Mendel*, ed. R. Alexander Brink. Madison: University of Wisconsin Press.

————. 1967*b*. *Comments on the preliminary working papers of Eden and Waddington: Mathematical challenges to the Neo-Darwinian interpretation of evolution.* Wistar Symposium Monograph, no. 5. Philadelphia: Wistar Institute.

————. 1968. *Evolution and the genetics of populations.* Vol. 1, *Genetic and biometric foundations.* Chicago: University of Chicago Press.

————. 1969*a*. *Evolution and the genetics of populations.* Vol. 2, *The theory of gene frequencies.* Chicago: University of Chicago Press.

————. 1969*b*. The theoretical course of directional selection. *Am. Naturalist* 103:561–74.

————. 1970. Random drift and the shifting balance theory of evolution. In *Mathematical topics in population genetics*, ed. K. Kojima. Berlin: Springer-Verlag.

————. 1973. *The origin of the F-statistics for describing the genetic aspects of population structure.* Population Genetics Monographs, vol. 3, *Genetic structure of populations*, ed. N. E. Morton. Honolulu: University Press of Hawaii.

Wright, S., and Chase, H. B. 1936. On the genetics of the spotted pattern of the guinea pig. *Genetics* 21:758–87.

Wright, S., and Dobzhansky, T. 1946. Genetics of natural populations: XII. Experimental production of some of the changes caused by natural selection in certain populations of *Drosophila pseudoobscura*. *Genetics* 31:125–56.

Wright, S., and Eaton, O. N. 1924. Factors which determine otocephaly in guinea pigs. *J. Agr. Res.* 26:161–82.

————. 1929. *The persistence of differentiation among inbred families of guinea pigs.* Technical Bulletin 103. Washington, D.C.: U.S. Department of Agriculture.

Wright, S., and Kerr, W. E. 1954. Experimental studies of the distribution of gene frequencies in very small populations of *Drosophila melanogaster*: II. Bar. *Evolution* 8:225–40.

Wright, S., and Lewis, P. A. 1921. Factors in the resistance of guinea pigs to tuberculosis with especial regard to inbreeding and heredity. *Am. Naturalist* 55:20–50.

————. 1922. Abstract. *Anat. Record* 23:93–94.

Wright, S., and McPhee, H. C. 1925. An approximate method of calculating coefficients of inbreeding and relationship from livestock pedigrees. *J. Agr. Res.* 31:377–83.

Wynne-Edwards, V. C. 1962. *Animal dispersion in relation to social behavior.* Edinburgh: Oliver and Boyd.

————. 1963. Intergroup selection in the evolution of social systems. *Nature* 200:623–26.

Yamada, Y., and Bell, A. E. 1969. Selection for larval growth in *Tribolium* under two levels of nutrition. *Genet. Res. Camb.* 13:175–95.

Yoder, D. M., and Lush, J. L. 1937. A genetic history of the Brown Swiss cattle in the United States. *J. Heredity* 28:154–60.

Yule, G. U. 1902. Mendel's laws and their probable relation to intraracial heredity. *New Phytol.* 1:192–207, 222–38.

———. 1906. On the theory of inheritance of quantitative compound characters and the basis of Mendel's law: A preliminary note. *Proc. III Intern. Congr. Genet.* Pp. 140–42.

———. 1924. A mathematical theory of evolution based on the conclusions of Dr. J. C. Willis, F. R. S. *Phil. Trans. Roy. Soc. (London) Ser. B* 215:21–87.

Zeleny, C. 1922. The effect of selection for eye facet numbers in the white-bar eye race of *Drosophila melanogaster. Genetics* 7:1–115.

Zirkle, C. 1952. Early ideas on inbreeding and crossbreeding. In *Heterosis*, ed. J. W. Gowen, Ames, Iowa: Iowa State College Press. Pp. 1–13.

CORRECTIONS FOR VOLUMES 1 AND 2

Volume 1

p. 37, line 11 up:	replace closing parenthesis after "alleles" by comma
p. 47, line 30:	*Nepeta* (not *Napeta*)
p. 73, first table:	$EPB\ c^d c^a$: grade at birth, 40 (not 46)
p. 81, line 14:	colon at end of line after "factors"
p. 112, line 4 up:	6.4A (not 6.4B)
p. 115, line 6 up:	strains (not strain)
p. 134, figure 6.9B:	scutellar (not dorsocentral)
p. 135, line 1:	scutellar (not dorsocentral)
p. 143, line 14 up:	percentiles or quantiles (quartiles correct elsewhere)
p. 190, line 24:	"by Park" after "Baklemischew"
p. 193, bottom line:	1921 (not 1920)
p. 205, table 9.11:	after 13_N No. = 1,688 (not 1,680)
p. 224, table 10.4:	note that r means point product moment correlation (not tetrachoric)
p. 235, line 4 below figure 10.5:	log x (not log X)
p. 243, table 10.17 after 1916, below 35:	2.062 (not 2.066)
p. 243, table 10.17 after 1918, below 32:	2.075 (not 2.074)
p. 243, table 10.17 after 1924, below 32:	2.156 (not 2.168)
p. 262, line 3 up:	scutellar (not dorsocentral)
p. 263, figure 11.8A:	scutellar (not dorsocentral)
p. 270, line 10 below heading:	from (not for)
p. 271, line 2 up:	scutellar (not dorsocentral)
p. 281, equation 12.20:	$\sum^N$ not $\sum^n$
p. 287, line 22:	(1900b) (not 1901)
p. 287, line 25:	insert r^5 after last term

p. 297, (12.53) Rows	insert C before $\overset{R}{\sum} (M_{i.} - \overline{M})^2$, replace $C\sigma^2_{M_{ij}}$ by $C\sigma^2_{M_i}$.
p. 297 Columns:	insert R before $\overset{C}{\sum} (M_{ij} - \overline{M})^2$, replace $R\sigma^2_{M_{ij}}$ by $R\sigma^2_{M._j}$
p. 309, (13.23):	$p_{ti} = c_{ti}\sigma_i/\sigma_t$ (not $c_{ti}\sigma_i/\sigma_s$)
p. 309, (13.24):	$r_{st} = \sum p_{si}p_{ti} + \sum p_{si}r_{ij}p_{tj}, \; j \neq i$
p. 311, figure 13.11:	There should be independent factors δA_1 for A_1, δA_m for A_m, δB_1 for B_1, δB_n for B_n (not common factors δA and δB, respectively)
p. 318, line 16:	hyperplane (not hypoplane)
p. 322, table 13.2, last figure after $\sum p^2_{xi}$:	5.9998 (not 0.59998)
p. 341, line 4 up:	0.556, 0.570, 0.576 and 0.623 (average 0.581) instead of 0.428, 0.512, 0.619 and 0.740 (average 0.575).
p. 342, et seq:	captions of figures 14.1–14.3, tables 14.14, 14.15, 14.17, 14.21: Wright 1960e (not Wright 1960a)
p. 347, table 14.16:	Wright (1921) (not Wright (1934f))
p. 381, line 13 up:	(1921) not (1922)
p. 407, line 15 up:	3.56 (not 3.2)
p. 407 line 16 up:	54.40 (not 54.04)
p. 421, line 3 up:	complex (not complexed)
p. 427, line 4:	insert "integral" after "probability"
p. 463, digit number:	add references to page 94–96 (genetics), 128 (maternal age), 224 (in litters)
p. 467:	pleiotropy (not pleitropy)
p. 469, weight:	add guinea pig, 111, 203–4, 240–43, 250–52, 288–89, 341–60
p. 469, white spotting:	path analysis 337–340

Volume 2

p. 51, line 6 up:	maximum (not minimum)
p. 81, figure 4.4:	add broken line from AB/CD to $ABCD$
p. 120, lines 4–6 under General Formulas:	substitute "such that $\overline{W} = N/N'$" for "measured $\dots W_i = N_i/N'_i$" (no use is made of the incorrect statement)
p. 186, line 10 up:	P's (not p's)
pp. 191, 192	exchange figures 7.12 and 7.13, but leave captions on the pages where they are

p. 276, figure 11.2:   connect E_1 and H_1, E_2 and H_2 with two-headed arrows with values $-e_1h_1$ and $-e_2h_2$, respectively, to avoid spurious correlations

p. 278, figure 11.4: L'_{A_1} (not L'_A), L'_{B_1} (not L'_B)

p. 280, (11.28): σ_p^2 (not σ_p)

p. 286, line 4: σ_H^2 (not σ_H^3)

p. 292, (12.4): $[1 - q_D + m(q_D - q_T)]$ not $[(1 - q_D + m)(q_D - q_T)]$

p. 294, (12.14): $2(1 - F_{IS})[q_T(1 - q_T) - \sigma_{q(ST)}^2]$ (not $2[(1 - F_{IS})q_T(1 - q_T) - \sigma_{q(ST)}^2])$

p. 296, (12.20, in last term): $\dfrac{1}{N_S - 1} \prod\limits_{X=1}^{S-2}$ (not $\dfrac{1}{N_S} \prod\limits_{X=1}^{S-1}$)

p. 297, line 3: $t_S - 1$ (not t_s)

p. 298, (12.25): $1 < X < S$ (not $1 < X \leqslant S$)

p. 299, caption for figure 12.2: $10^{7/2}$ (not 10^7)

p. 300, (12.34): $\sum\limits_{X-1}^{S-1} \log$ (not $\log \sum\limits_{X=1}^{S-1}$)

p. 304, (12.48): $\Gamma(2a + 1)$ not $\Gamma(3a + 1)$

p. 304, line 18: -0.81 (not 0.25)

p. 309, last line: $(0.3/m)N_1$ (not $(3/m)N_1$)

pp. 317, 318:   exchange figures 12.13 and 12.14, but leave captions on the pages where they are

p. 320, line 5 up: sib mating (not self-fertilization)

p. 322, figure 12.15a: delete n's in designations of abscissa s (as in 12.15b), insert k as designation of ordinates, $M = kN$ (not $M = kn$)

p. 365:   omit a and b on figures on second row

p. 399, (14.14):   insert brace at end

p. 403, table 14.3   after $N = 10^3$, $u = 10^{-4}$, $s = 0.1$: 8.2 under n_e (not 11.2)

p. 473, lines 5 and 4 up: delete sentence, "It is the average ratio. . . ."

p. 475, line 25 in two places: v not r

p. 481, line 14 (equation): $\sigma_{\Delta q}^2$ (not $\sigma_{\nabla q}^2$)

p. 485, line 15 up (equation): $\left[\dfrac{n}{\sum} \dfrac{1}{N}\right]$ not $\left[\dfrac{n}{\sum} \dfrac{1}{n}\right]$

p. 486, line 6 up $F = \sigma_q^2/Q(1 - Q)$, not $\sigma_q^2 Q(1 - Q)$

AUTHOR INDEX

604

SUBJECT INDEX